The Complete Art of War

The Art of War By Sun Tzu translated by Lionel Giles
On War by Carl von Clausewitz translated by Colonel J.J. Graham
The Art of Warby Niccolò Machiavelli
The Art of War by Baron De Jomini translated by G.H. Mendell, and
W.P. Craighill

The Complete
Art of War

Wilder Publications, LLC.
PO Box 3005
Radford VA 24143-3005

ISBN 10: 1-60459-360-1
ISBN 13: 978-1-5154-3628-7

First Edition

10 9 8 7 6 5 4 3 2 1

The Art of War

By Sun Tzu
Translated and commented on by Lionel Giles

Table of Contents

Laying Plans

Sun Tzu said: The art of war is of vital importance to the State.

It is a matter of life and death, a road either to safety or to ruin. Hence it is a subject of inquiry which can on no account be neglected.

The art of war, then, is governed by five constant factors, to be taken into account in one's deliberations, when seeking to determine the conditions obtaining in the field.

These are: (1) The Moral Law; (2) Heaven; (3) Earth; (4) The Commander; (5) Method and discipline. [It appears from what follows that Sun Tzu means by "Moral Law" a principle of harmony, not unlike the Tao of Lao Tzu in its moral aspect. One might be tempted to render it by "morale," were it not considered as an attribute of the ruler.]

The *moral law* causes the people to be in complete accord with their ruler, so that they will follow him regardless of their lives, undismayed by any danger. [Tu Yu quotes Wang Tzu as saying: "Without constant practice, the officers will be nervous and undecided when mustering for battle; without constant practice, the general will be wavering and irresolute when the crisis is at hand."]

Heaven signifies night and day, cold and heat, times and seasons. [The commentators, I think, make an unnecessary mystery of two words here. Meng Shih refers to "the hard and the soft, waxing and waning" of Heaven. Wang Hsi, however, may be right in saying that what is meant is "the general economy of Heaven," including the five elements, the four seasons, wind and clouds, and other phenomena.]

Earth comprises distances, great and small; danger and security; open ground and narrow passes; the chances of life and death.

The *Commander* stands for the virtues of wisdom, sincerely, benevolence, courage and strictness. [The five cardinal virtues of the

Chinese are (1) humanity or benevolence; (2) uprightness of mind; (3) self-respect, self-control, or "proper feeling;" (4) wisdom; (5) sincerity or good faith. Here "wisdom" and "sincerity" are put before "humanity or benevolence," and the two military virtues of "courage" and "strictness" substituted for "uprightness of mind" and "self-respect, self-control, or 'proper feeling.'"]

By *Method and Discipline* are to be understood the marshaling of the army in its proper subdivisions, the graduations of rank among the officers, the maintenance of roads by which supplies may reach the army, and the control of military expenditure.

These five heads should be familiar to every general: he who knows them will be victorious; he who knows them not will fail.

Therefore, in your deliberations, when seeking to determine the military conditions, let them be made the basis of a comparison, in this wise:

(a) Which of the two sovereigns is imbued with the Moral law? [I.e., "is in harmony with his subjects..]

(b) Which of the two generals has most ability?

(c) With whom lie the advantages derived from Heaven and Earth?

(d) On which side is discipline most rigorously enforced? [Tu Mu alludes to the remarkable story of Ts`ao Ts`ao (A.D. 155-220), who was such a strict disciplinarian that once, in accordance with his own severe regulations against injury to standing crops, he condemned himself to death for having allowed him horse to shy into a field of corn! However, in lieu of losing his head, he was persuaded to satisfy his sense of justice by cutting off his hair. Ts`ao Ts`ao's own comment on the present passage is characteristically curt: "when you lay down a law, s ee that it is not disobeyed; if it is disobeyed the offender must be put to death."]

(e) Which army is stronger? [Morally as well as physically. As Mei Yao-ch`en puts it, freely rendered, "*Espirit De Corps* and 'big battalions.'"]

(f) On which side are officers and men more highly trained? [Tu Yu quotes Wang Tzu as saying: "Without constant practice, the officers will be nervous and undecided when mustering for battle; without constant practice, the general will be wavering and irresolute when the crisis is at hand."]

(g) In which army is there the greater constancy both in reward and punishment? [On which side is there the most absolute certainty that merit will be properly rewarded and misdeeds summarily punished?]

By means of these seven considerations I can forecast victory or defeat.

The general that hearkens to my counsel and acts upon it, will conquer: —let such a one be retained in command! The general that hearkens not to my counsel nor acts upon it, will suffer defeat: —let such a one be dismissed! [The form of this paragraph reminds us that Sun Tzu's treatise was composed expressly for the benefit of his patron Ho Lu, king of the Wu State.]

While heeding the profit of my counsel, avail yourself also of any helpful circumstances over and beyond the ordinary rules.

According as circumstances are favorable, one should modify one's plans. [Sun Tzu, as a practical soldier, will have none of the "bookish theoric." He cautions us here not to pin our faith to abstract principles; "for," as Chang Yu puts it, "while the main laws of strategy can be stated clearly enough for the benefit of all and sundry, you must be guided by the actions of the enemy in attempting to secure a favorable position in actual warfare." On the eve of the battle of Waterloo, Lord Uxbridge, commanding the cavalry, went to the Duke of Wellington in order to learn what his plans and calculations were for the morrow, because, as he explained, he might suddenly find himself Commander-in-chief and would be unable to frame new plans in a critical moment. The Duke listened quietly and then said: "Who will attack the fir st tomorrow — I or Bonaparte?" "Bonaparte," replied Lord Uxbridge. "Well," continued the Duke, "Bonaparte has not given me any idea of his projects; and as my plans will depend upon his, how can you expect me to tell you what mine are?"]

All warfare is based on deception. [The truth of this pithy and profound saying will be admitted by every soldier. Col. Henderson tells us that Wellington, great in so many military qualities, was especially distinguished by "the extraordinary skill with which he concealed his movements and deceived both friend and foe."]

Hence, when able to attack, we must seem unable; when using our forces, we must seem inactive; when we are near, we must make the enemy believe we are far away; when far away, we must make him believe we are near.

Hold out baits to entice the enemy. Feign disorder, and crush him. [All commentators, except Chang Yu, say, "When he is in disorder, crush him." It is more natural to suppose that Sun Tzu is still illustrating the uses of deception in war.]

If he is secure at all points, be prepared for him. If he is in superior strength, evade him.

If your opponent is of choleric temper, seek to irritate him. Pretend to be weak, that he may grow arrogant. [Wang Tzu, quoted by Tu Yu, says that the good tactician plays with his adversary as a cat plays with a mouse, first feigning weakness and immobility, and then suddenly pouncing upon him.]

If he is taking his ease, give him no rest. [This is probably the meaning though Mei Yao-ch`en has the note: "while we are taking our ease, wait for the enemy to tire himself out." The *Yu Lan* has "Lure him on and tire him out."]

If his forces are united, separate them. [Less plausible is the interpretation favored by most of the commentators: "If sovereign and subject are in accord, put division between them."]

Attack him where he is unprepared, appear where you are not expected.

These military devices, leading to victory, must not be divulged beforehand.

Now the general who wins a battle makes many calculations in his temple ere the battle is fought. [Chang Yu tells us that in ancient times it was customary for a temple to be set apart for the use of a general who was about to take the field, in order that he might there elaborate his plan of campaign.]

The general who loses a battle makes but few calculations beforehand. Thus do many calculations lead to victory, and few calculations to defeat: how much more no calculation at all! It is by attention to this point that I can foresee who is likely to w in or lose.

Waging War

Sun Tzu said: In the operations of war, where there are in the field a thousand swift chariots, as many heavy chariots, and a hundred thousand mail-clad soldiers, with provisions enough to carry them a thousand *li*, the expenditure at home and at the front, including entertainment of guests, small items such as glue and paint, and sums spent on chariots and armor, will reach the total of a thousand ounces of silver per day. Such is the cost of raising an army of 100 ,000 men. [The "swift chariots" were lightly built and, according to Chang Yu, used for the attack; the "heavy chariots" were heavier, and designed for purposes of defense. Li Ch`uan, it is true, says that the latter were light, but this seems hardly probable. I t is interesting to note the analogies between early Chinese warfare and that of the Homeric Greeks. In each case, the war-chariot was the important factor, forming as it did the nucleus round which was grouped a certain number of foot-soldiers. With regard to the numbers given here, we are informed that each swift chariot was accompanied by 75 footmen, and each heavy chariot by 25 footmen, so that the whole army would be divided up into a thousand battalions, each consisting of two chariots and a hundred men.] [2.78 modern *li* go to a mile. The length may have varied slightly since Sun Tzu's time.]

When you engage in actual fighting, if victory is long in coming, then men's weapons will grow dull and their ardor will be damped. If you lay siege to a town, you will exhaust your strength.

Again, if the campaign is protracted, the resources of the State will not be equal to the strain.

Now, when your weapons are dulled, your ardor damped, your strength exhausted and your treasure spent, other chieftains will spring up to take advantage of your extremity. Then no man, however wise, will be able to avert the consequences that must ensue.

Thus, though we have heard of stupid haste in war, cleverness has never been seen associated with long delays. [This concise and difficult sentence is not well explained by any of the commentators. Ts`ao Kung, Li Ch`uan, Meng Shih, Tu Yu, Tu Mu and Mei Yao-ch`en have notes to the effect that a general, though naturally stupid, may nevertheless conquer through sheer force of rapidity. Ho Shih says: "Haste may be stupid, but at any rate it saves expenditure of energy and treasure; protracted operations may be very clever, but they bring calamity in their train." Wang Hsi evades the difficulty by remarking: "Lengthy operations mean an army growing old, wealth being expended, an empty exchequer and distress among the people; true cleverness insures against the occurrence of such calamities." Chang Yu says: "So long as victory can be attained, stupid haste is preferable to clever dilatoriness."] [Now Sun Tzu says nothing whatever, except possibly by implication, about ill-considered haste being better than ingenious but lengthy operations. What he does say is something much more guarded, namely that, while speed may sometimes be injudicious, tardiness can never be anything but foolish — if only because it means impoverishment to the nation. In considering the point raised here by Sun Tzu, the classic example of Fabius Cunctator will inevitably occur to the mind. That general deliberately measured the endurance of Rome against that of Hannibals's isolated army, because it seemed to him that the latter was more likely to suffer from a long campaign in a strange country. But it is quite a moot question whether his tactics would have proved successful in the long run. Their reversal it is true, led to Cannae; but this only establishes a negative presumption in their favor.]

There is no instance of a country having benefitted from prolonged warfare.

It is only one who is thoroughly acquainted with the evils of war that can thoroughly understand the profitable way of carrying it on. [That is, with rapidity. Only one who knows the disastrous effects of a long war can realize the supreme importance of rapidity in bringing it to a close. Only two commentators seem to favor this interpretation, but it fits well into the logic of the context, whereas the rendering, "He who does not know the evils of war cannot appreciate its benefits," is distinctly pointless.]

The skillful soldier does not raise a second levy, neither are his supply-wagons loaded more than twice. [Once war is declared, he will not waste precious time in waiting for reinforcements, nor will he return his army back for fresh supplies, but crosses the enemy's frontier without delay. This may seem an audacious policy to recommend, but with all great strategists, from Julius Caesar to Napoleon Bonaparte, the value of time — that is, being a little ahead of your opponent —has counted for more than either numerical superiority or the nicest calculations with regard to commissariat.]

Bring war material with you from home, but forage on the enemy. Thus the army will have food enough for its needs. [The Chinese word translated here as "war material" literally means "things to be used", and is meant in the widest sense. It includes all the impedimenta of an army, apart from provisions.]

Poverty of the State exchequer causes an army to be maintained by contributions from a distance. Contributing to maintain an army at a distance causes the people to be impoverished. [The beginning of this sentence does not balance properly with the next, though obviously intended to do so. The arrangement, moreover, is so awkward that I cannot help suspecting some corruption in the text. It never seems to occur to Chinese commentators that an emendation may be necessary for the sense, and we get no help from them there. The Chinese words Sun Tzu used to indicate the cause of the people's impoverishment clearly have reference to some system by which the husbandmen sent their contributions of corn to the army direct. But why should it fall on them to maintain an army in this way, except because the State or Government is too poor to do so?]

On the other hand, the proximity of an army causes prices to go up; and high prices cause the people's substance to be drained away. [Wang Hsi says high prices occur before the army has left its own territory. Ts`ao Kung understands it of an army that has already crossed the frontier.]

When their substance is drained away, the peasantry will be afflicted by heavy exactions.

With this loss of substance and exhaustion of strength, the homes of the people will be stripped bare, and three-tenths of their income will be dissipated; [Tu Mu and Wang Hsi agree that the people are not mulcted not of 3/10, but of 7/10, of their income. But this is hardly to be extracted from our text. Ho Shih has a characteristic tag: "The *people* being regarded as the essential part of the State, and *food* as the people's heaven, is it not right that those in authority should value and be careful of both?"] while government expenses for broken chariots, worn-out horses, breast-plates and helmets, bows and arrows, spears and shields, protective mantles, drought-oxen and heavy wagons, will amount to four-tenths of its total revenue.

Hence a wise general makes a point of foraging on the enemy. One cartload of the enemy's provisions is equivalent to twenty of one's own, and likewise a single *picul* of his provender is equivalent to twenty from one's own store. [Because twenty cartloads will be consumed in the process of transporting one cartload to the front. A *picul* is a unit of measure equal to 133.3 pounds (65.5 kilograms).]

Now in order to kill the enemy, our men must be roused to anger; that there may be advantage from defeating the enemy, they must have their rewards. [Tu Mu says: "Rewards are necessary in order to make the soldiers see the advantage of beating the enemy; thus, when you capture spoils from the enemy, they must be used as rewards, so that all your men may have a keen desire to fight, each on his own account."]

Therefore in chariot fighting, when ten or more chariots have been taken, those should be rewarded who took the first. Our own flags should be substituted for those of the enemy, and the chariots mingled and used in conjunction with ours. The captured soldiers should be kindly treated and kept.

This is called, using the conquered foe to augment one's own strength.

In war, then, let your great object be victory, not lengthy campaigns. [As Ho Shih remarks: "War is not a thing to be trifled with." Sun Tzu here reiterates the main lesson which this chapter is intended to enforce."]

Attack by Stratagem

Sun Tzu said: In the practical art of war, the best thing of all is to take the enemy's country whole and intact; to shatter and destroy it is not so good. So, too, it is better to recapture an army entire than to destroy it, to capture a regimen t, a detachment or a company entire than to destroy them. [The equivalent to an army corps, according to Ssu-ma Fa, consisted nominally of 12500 men; according to Ts`ao Kung, the equivalent of a regiment contained 500 men, the equivalent to a detachment consists from any number between 100 and 500, and the equivalent of a company contains from 5 to 100 men. For the last two, however, Chang Yu gives the exact figures of 100 and 5 respectively.]

Hence to fight and conquer in all your battles is not supreme excellence; supreme excellence consists in breaking the enemy's resistance without fighting. [Here again, no modern strategist but will approve the words of the old Chinese general. Moltke's greatest triumph, the capitulation of the huge French army at Sedan, was won practically without bloodshed.]

Thus the highest form of generalship is to balk the enemy's plans; [Perhaps the word "balk" falls short of expressing the full force of the Chinese word, which implies not an attitude of defense, whereby one might be content to foil the enemy's stratagems one after another, but an active policy of counter-attack. Ho Sh ih puts this very clearly in his note: "When the enemy has made a plan of attack against us, we must anticipate him by delivering our own attack first."] the next best is to prevent the junction of the enemy's forces; [Isolating him from his allies. We must not forget that Sun Tzu, in speaking of hostilities, always has in mind the numerous states or principalities into which the China of his day was split up.] the next in order is to attack the enemy's army in the field; [When he is already at full strength.] and the worst policy of all is to besiege walled cities.

The rule is, not to besiege walled cities if it can possibly be avoided. [Another sound piece of military theory. Had the Boers acted upon it in 1899, and refrained from dissipating their strength before Kimberley, Mafeking, or even Ladysmith, it is more than probable that they would have been masters of the situation before the British were ready seriously to oppose them.]

The preparation of mantlets, movable shelters, and various implements of war, will take up three whole months; [It is not quite clear what the Chinese word, here translated as "mantlets", described. Ts`ao Kung simply defines them as "large shields," but we get a better idea of them from Li Ch`uan, who says they were to protect the heads of those who were assaulting the city walls at close quarters. This seems to suggest a sort of Roman *Testudo*, ready made. Tu Mu says they were wheeled vehicles used in repelling attacks, but this is denied by Ch`en Hao. See supra II. 14. The name is also applied to turrets on city walls. Of the "movable shelters" we get a fairly clear description from several commentators. They were wooden missile-proof structures on four wheels, propelled from within, covered over with raw hides, and used in sieges to convey parties of me n to and from the walls, for the purpose of filling up the encircling moat with earth. Tu Mu adds that they are now called "wooden donkeys."] and the piling up of mounds over against the walls will take three months more. [These were great mounds or ramparts of earth heaped up to the level of the enemy's walls in order to discover the weak points in the defense, and also to destroy the fortified turrets mentioned in the preceding note.]

The general, unable to control his irritation, will launch his men to the assault like swarming ants, [This vivid simile of Ts`ao Kung is taken from the spectacle of an army of ants climbing a wall. The meaning is that the general, losing patience at the long delay, may make a premature attempt to storm the place before his engines of war are ready.] with the result that one-third of his men are slain, while the town still remains untaken. Such are the disastrous effects of a siege. [We are reminded of the terrible losses of the Japanese before Port Arthur, in the most recent siege which history has to record.]

Therefore the skillful leader subdues the enemy's troops without any fighting; he captures their cities without laying siege to them; he overthrows their kingdom without lengthy operations in the field. [Chia Lin notes that he only overthrows the Government, but does no harm to individuals. The classical instance is Wu Wang, who after having put an end to the Yin dynasty was acclaimed "Father and mother of the people."]

With his forces intact he will dispute the mastery of the Empire, and thus, without losing a man, his triumph will be complete. [Owing to the double meanings in the Chinese text, the latter part of the sentence is susceptible of quite a different meaning: "And thus, the weapon not being blunted by use, its keenness remains perfect."]

This is the method of attacking by stratagem.

It is the rule in war:

a) If our forces are ten to the enemy's one, to surround him;

b) If five to one, to attack him; [Straightway, without waiting for any further advantage.]

c) If twice as numerous, to divide our army into two. [Tu Mu takes exception to the saying; and at first sight, indeed, it appears to violate a fundamental principle of war. Ts'ao Kung, however, gives a clue to Sun Tzu's meaning: "Being two to the enemy's one, we may use one part of our army in the regular way, and the other for some special diversion." Chang Yu thus further elucidates the point: "If our force is twice as numerous as that of the enemy, it should be split up into two divisions, one to meet the enemy in front, and one to fall upon his r ear; if he replies to the frontal attack, he may be crushed from behind; if to the rearward attack, he may be crushed in front." This is what is meant by saying that 'one part may be used in the regular way, and the other for some special diversion.' Tu Mu does not understand that dividing one's army is simply an irregular, just as concentrating it is the regular, strategical method, and he is too hasty in calling this a mistake."]

d) If equally matched, we can offer battle; [Li Ch`uan, followed by Ho Shih, gives the following paraphrase: "If attackers and attacked are equally matched in strength, only the able general will fight."]

e) If slightly inferior in numbers, we can avoid the enemy; [The meaning, "we can *watch* the enemy," is certainly a great improvement on the above; but unfortunately there appears to be no very good authority for the variant. Chang Yu reminds us that the saying only applies if the other factors are equal; a small difference in numbers is often more than counterbalanced by superior energy and discipline.]

f) If quite unequal in every way, we can flee from him.

Hence, though an obstinate fight may be made by a small force, in the end it must be captured by the larger force.

Now the general is the bulwark of the State; if the bulwark is complete at all points; the State will be strong; if the bulwark is defective, the State will be weak. [As Li Ch`uan tersely puts it: "Gap indicates deficiency; if the general's ability is not perfect (i.e. if he is not thoroughly versed in his profession), his army will lack strength."]

There are three ways in which a ruler can bring misfortune upon his army:—

a) By commanding the army to advance or to retreat, being ignorant of the fact that it cannot obey. This is called hobbling the army. [Li Ch`uan adds the comment: "It is like tying together the legs of a thoroughbred, so that it is unable to gallop." One would naturally think of "the ruler" in this passage as being at home, and trying to direct the movements of his army from a distance. But the commentators understand just the reverse, and quote the saying of T`ai Kung: "A kingdom should not be governed from without, and army should not be directed from within." Of course it is true that, during an engagement, or when in close touch with the enemy, the general should not be in the thick of his own troops, but a little distance apart. Otherwise, he will be liable to misjudge the position as a whole, and give wrong orders.]

(b) By attempting to govern an army in the same way as he administers a kingdom, being ignorant of the conditions which obtain in an army. This causes restlessness in the soldier's minds. [Ts`ao Kung's note is, freely translated: "The military sphere and the civil sphere are wholly distinct; you can't handle an army in kid gloves." And Chang Yu says: "Humanity and justice are the principles on which to govern a state, but not an army; opportunism and flexibility, on the other hand, are military rather than civil virtues to assimilate the governing of an army"—to that of a State, understood.]

c) By employing the officers of his army without discrimination, [That is, he is not careful to use the right man in the right place.] through ignorance of the military principle of adaptation to circumstances. This shakes the confidence of the soldiers. [I follow Mei Yao-ch`en here. The other commentators refer not to the ruler, but to the officers he employs. Thus Tu Yu says: "If a general is ignorant of the principle of adaptability, he must not be entrusted with a position of authority." Tu Mu quotes: "The skillful employer of men will employ the wise man, the brave man, the covetous man, and the stupid man. For the wise man delights in establishing his merit, the brave man likes to show his courage in action, the covetous man is quick at seizing advantages, and the stupid man has no fear of death."]

But when the army is restless and distrustful, trouble is sure to come from the other feudal princes. This is simply bringing anarchy into the army, and flinging victory away.

Thus we may know that there are five essentials for victory:

a) He will win who knows when to fight and when not to fight. [Chang Yu says: If he can fight, he advances and takes the offensive; if he cannot fight, he retreats and remains on the defensive. He will invariably conquer who knows whether it is right to take the offensive or the defensive.]

b) He will win who knows how to handle both superior and inferior forces. [This is not merely the general's ability to estimate numbers correctly, as Li Ch`uan and others make out. Chang Yu expounds the saying more satisfactorily: "By applying the art of war, it is possible with a lesser force to defeat a greater, and vice versa. The secret lies in an eye for locality, and in not letting the right moment slip. Thus Wu Tzu says: 'With a superior force, make for easy ground; with an inferior one, make for difficult ground.'"]

c) He will win whose army is animated by the same spirit throughout all its ranks.

d) He will win who, prepared himself, waits to take the enemy unprepared.

e) He will win who has military capacity and is not interfered with by the sovereign. [Tu Yu quotes Wang Tzu as saying: "It is the sovereign's function to give broad instructions, but to decide on battle it is the function of the general." It is needless to dilate on the military disasters which have been caused by undue interference wit h operations in the field on the part of the home government. Napoleon undoubtedly owed much of his extraordinary success to the fact that he was not hampered by central authority.]

Hence the saying: If you know the enemy and know yourself, you need not fear the result of a hundred battles. If you know yourself but not the enemy, for every victory gained you will also suffer a defeat. [Li Ch`uan cites the case of Fu Chien, prince of Ch`in, who in 383 A.D. marched with a vast army against the Chin Emperor. When warned not to despise an enemy who could command the services of such men as Hsieh An and Huan Ch`ung, he boastfully replied: "I have the population of eight provinces at my back, infantry and horsemen to the number of one million; why, they could dam up the Yangtsze River itself by merely throwing their whips into the stream. What danger have I to fear?" Neverthe less, his forces were soon after disastrously routed at the Fei River, and he was obliged to beat a hasty retreat.]

If you know neither the enemy nor yourself, you will succumb in every battle. [Chang Yu said: "Knowing the enemy enables you to take the offensive, knowing yourself enables you to stand on the defensive." He adds: "Attack is the secret of defense; defense is the planning of an attack." It would be hard to find a better epitome of the root-principle of war.]

Thus it may be known that the leader of armies is the arbiter of the people's fate, the man on whom it depends whether the nation shall be in peace or in peril.

Tactical Dispositions

Sun Tzu said: The good fighters of old first put themselves beyond the possibility of defeat, and then waited for an opportunity of defeating the enemy.

To secure ourselves against defeat lies in our own hands, but the opportunity of defeating the enemy is provided by the enemy himself. [That is, of course, by a mistake on the enemy's part.]

Thus the good fighter is able to secure himself against defeat, [Chang Yu says this is done, "By concealing the disposition of his troops, covering up his tracks, and taking unremitting precautions."] but cannot make certain of defeating the enemy.

Hence the saying: One may *Know* how to conquer without being able to *do* it.

Security against defeat implies defensive tactics; ability to defeat the enemy means taking the offensive. [I retain the sense found in a similar passage in in spite of the fact that the commentators are all against me. The meaning they give, "He who cannot conquer takes the defensive," is plausible enough.]

Standing on the defensive indicates insufficient strength; attacking, a superabundance of strength.

The general who is skilled in defense hides in the most secret recesses of the earth; [Literally, "hides under the ninth earth," which is a metaphor indicating the utmost secrecy and concealment, so that the enemy may not know his whereabouts."] he who is skilled in attack flashes forth from the topmost heights of heaven. [Another metaphor, implying that he falls on his adversary like a thunderbolt, against which there is no time to prepare. This is the opinion of most of the commentators.]

Thus on the one hand we have ability to protect ourselves; on the other, a victory that is complete.

To see victory only when it is within the ken of the common herd is not the acme of excellence. [As Ts`ao Kung remarks, "the thing is to see the plant before it has germinated," to foresee the event before the action has begun. Li Ch`uan alludes to the story of Han Hsin who, when about to attack the vastly superior army of Chao, which was strong ly entrenched in the city of Ch`eng-an, said to his officers: "Gentlemen, we are going to annihilate the enemy, and shall meet again at dinner." The officers hardly took his words seriously, and gave a very dubious assent. But Han Hsin had already worked out in his mind the details of a clever stratagem, whereby, as he foresaw, he was able to capture the city and inflict a crushing defeat on his adversary."]

Neither is it the acme of excellence if you fight and conquer and the whole Empire says, "Well done!" [True excellence being, as Tu Mu says: "To plan secretly, to move surreptitiously, to foil the enemy's intentions and balk his schemes, so that at last the day may be won without shedding a drop of blood." Sun Tzu reserves his approbation for things that "the world's coarse thumb And finger fail to plumb."]

To lift an autumn hair is no sign of great strength; ["Autumn" hair" is explained as the fur of a hare, which is finest in autumn, when it begins to grow afresh. The phrase is a very common one in Chinese writers.] to see the sun and moon is no sign of sharp sight; to hear the noise of thunder is no sign of a quick ear. [Ho Shih gives as real instances of strength, sharp sight and quick hearing: Wu Huo, who could lift a tripod weighing 250 stone; Li Chu, who at a distance of a hundred paces could see objects no bigger than a mustard seed; and Shih K`uang, a blind musician who could hear the footsteps of a mosquito.]

What the ancients called a clever fighter is one who not only wins, but excels in winning with ease. [The last half is literally "one who, conquering, excels in easy conquering." Mei Yao-ch`en says: "He who only sees the obvious, wins his battles with difficulty; he who looks below the surface of things, wins with ease."]

Hence his victories bring him neither reputation for wisdom nor credit for courage. [Tu Mu explains this very well: "Inasmuch as his victories are gained over circumstances that have not come to light, the world as large knows nothing of them, and he wins no reputation for wisdom; inasmuch as the hostile state submits before there has been any bloodshed, he receives no credit for courage."]

He wins his battles by making no mistakes. [Ch`en Hao says: "He plans no superfluous marches, he devises no futile attacks." The connection of ideas is thus explained by Chang Yu: "One who seeks to conquer by sheer strength, clever though he may be at winning pitched battles, is also liable on occasion to be vanquished; whereas he who can look into the future and discern conditions that are not yet manifest, will never make a blunder and therefore invariably win."]

Making no mistakes is what establishes the certainty of victory, for it means conquering an enemy that is already defeated.

Hence the skillful fighter puts himself into a position which makes defeat impossible, and does not miss the moment for defeating the enemy. [A "counsel of perfection" as Tu Mu truly observes. "Position" need not be confined to the actual ground occupied by the troops. It includes all the arrangements and preparations which a wise general will make to increase the safety of his army.]

Thus it is that in war the victorious strategist only seeks battle after the victory has been won, whereas he who is destined to defeat first fights and afterwards looks for victory. [Ho Shih thus expounds the paradox: "In warfare, first lay plans which will ensure victory, and then lead your army to battle; if you will not begin with stratagem but rely on brute strength alone, victory will no longer be assured."]

The consummate leader cultivates the moral law, and strictly adheres to method and discipline; thus it is in his power to control success.

In respect of military method, we have, firstly, Measurement; secondly, Estimation of quantity; thirdly, Calculation; fourthly, Balancing of chances; fifthly, Victory.

Measurement owes its existence to Earth; Estimation of quantity to Measurement; Calculation to Estimation of quantity; Balancing of chances to Calculation; and Victory to Balancing of chances. [It is not easy to distinguish the four terms very clearly in the Chinese. The first seems to be surveying and measurement of the ground, which enable us to form an estimate of the enemy's strength, and to make calculations based on the data thus obtain ed; we are thus led to a general weighing-up, or comparison of the enemy's chances with our own; if the latter turn the scale, then victory ensues. The chief difficulty lies in third term, which in the Chinese some commentators take as a calculation of *numbers*, thereby making it nearly synonymous with the second term. Perhaps the second term should be thought of as a consideration of the enemy's general position or condition, while the third term is the estimate of his numerical strength. On the other hand, Tu Mu says: "The question of relative strength having been settled, we can bring the varied resources of cunning into play." Ho Shih seconds this interpretation, but weakens it. However, it points to the third term as being a calculation of numbers.]

A victorious army opposed to a routed one, is as a pound's weight placed in the scale against a single grain. [Literally, "a victorious army is like an *i* (20 oz.) weighed against a *shu* (1/24 oz.); a routed army is a *shu* weighed against an I." The point is simply the enormous advantage which a disciplined force, flushed with victory, has over one demoralized by defeat." Legge, in his note on Mencius, I. 2. ix. 2, makes the I to be 24 Chinese ounces, and corrects Chu Hsi's statement that it equaled 20 oz. only. But Li Ch`uan of the T`ang dynasty here gives the same figure as Chu Hsi.]

The onrush of a conquering force is like the bursting of pent-up waters into a chasm a thousand fathoms deep.

Energy

Sun Tzu said: The control of a large force is the same principle as the control of a few men: it is merely a question of dividing up their numbers. [That is, cutting up the army into regiments, companies, etc., with subordinate officers in command of each. Tu Mu reminds us of Han Hsin's famous reply to the first Han Emperor, who once said to him: "How large an army do you think I could lead?" "Not more than 100,000 men, your Majesty." "And you?" asked the Emperor. "Oh!" he answered, "the more the better."]

Fighting with a large army under your command is nowise different from fighting with a small one: it is merely a question of instituting signs and signals.

To ensure that your whole host may withstand the brunt of the enemy's attack and remain unshaken — this is effected by maneuvers direct and indirect. [We now come to one of the most interesting parts of Sun Tzu's treatise, the discussion of the *Cheng* and the *Ch`i*." As it is by no means easy to grasp the full significance of these two terms, or to render them consistently by good English equivalents; it may be as well to tabulate some of the commentators' remarks on the subject before proceeding further. Li Ch`uan: "Facing the enemy is *Cheng*, making lateral diversion is *Ch`i*. Chia Lin: "In presence of the enemy, your troops should be arrayed in normal fashion, but in order to secure victory abnormal maneuvers must be employed." Mei Yao-ch`en: "*Ch`i* is active, *Cheng* is passive; passivity means waiting for an opportunity, activity beings the victory itself." Ho Shih: "We must cause the enemy to regard our straightforward attack as one that is secretly designed, and vice versa; thus *Cheng* may also be *Ch`i*, and *Ch`i* may also be *Cheng*."]

[He instances the famous exploit of Han Hsin, who when marching ostensibly against Lin-chin (now Chao-I in Shensi), suddenly threw a large force across the Yellow River in wooden tubs, utterly disconcerting his opponent. [Ch`ien Han Shu, ch. 3.] Here, w e are told, the march on Lin-chin was *Cheng*, and the surprise maneuver was *Ch`i*." Chang Yu gives the following summary of opinions on the words: "Military writers do not agree with regard to the meaning of *Ch`i* and *Cheng*. Wei Liao Tzu [4th cent. B.C.] says: 'Direct warfare favors frontal attacks, indirect warfare attacks from the rear.' Ts`ao Kung says: 'Going straight out to join battle is a direct operation; appearing on the enemy's rear is an indirect maneuver.' Li Wei-kung [6th and 7th cent . A.D.] says: 'In war, to march straight ahead is *Cheng*; turning movements, on the other hand, are *Ch`i*.' These writers simply regard *Cheng* as *Cheng*, and *Ch`i* as *Ch`i*; they do not note that the two are mutually interchangeable and run into each other like the two sides of a circle]

[A comment on the T`ang Emperor T`ai Tsung goes to the root of the matter: 'A *Ch`i* maneuver may be *Cheng*, if we make the enemy look upon it as *Cheng*; then our real attack will be *Ch`i*, and vice versa. The whole secret lies in confusing the enemy, so that he cannot fathom our real intent.'" To put it perhaps a little more clearly: any attack or other operation is *Cheng*, on which the enemy has had his attention fixed; whereas that is *Ch`i*," which takes him by surprise or comes from an unexpected quarter. If the enemy perceives a movement which is meant to be *Ch`i*," it immediately becomes *Cheng*."]

That the impact of your army may be like a grindstone dashed against an egg — this is effected by the science of weak points and strong.

In all fighting, the direct method may be used for joining battle, but indirect methods will be needed in order to secure victory. [Chang Yu says: "Steadily develop indirect tactics, either by pounding the enemy's flanks or falling on his rear." A brilliant example of "indirect tactics" which decided the fortunes of a campaign was Lord Roberts' night march round the Peiwar Ko tal in the second Afghan war

Indirect tactics, efficiently applied, are inexhaustible as Heaven and Earth, unending as the flow of rivers and streams; like the sun and moon, they end but to begin anew; like the four seasons, they pass away to return once more. [Tu Yu and Chang Yu understand this of the permutations of Ch`i and Cheng." But at present Sun Tzu is not speaking of Cheng at all, unless, indeed, we suppose with Cheng Yu-hsien that a clause relating to it has fallen out of the text. Of course, as has already been pointed out, the two are so inextricably interwoven in all military operations, that they cannot really be considered apart. Here we simply have an expression, in figurative language, of the almost infinite resource of a great leader.]

There are not more than five musical notes, yet the combinations of these five give rise to more melodies than can ever be heard.

There are not more than five primary colors (blue, yellow, red, white, and black), yet in combination they produce more hues than can ever been seen.

There are not more than five cardinal tastes (sour, acrid, salt, sweet, bitter), yet combinations of them yield more flavors than can ever be tasted.

In battle, there are not more than two methods of attack — the direct and the indirect; yet these two in combination give rise to an endless series of maneuvers.

The direct and the indirect lead on to each other in turn. It is like moving in a circle — you never come to an end. Who can exhaust the possibilities of their combination?

The onset of troops is like the rush of a torrent which will even roll stones along in its course.

The quality of decision is like the well-timed swoop of a falcon which enables it to strike and destroy its victim. [The Chinese here is tricky and a certain key word in the context it is used defies the best efforts of the translator. Tu Mu defines this word as "the measurement or estimation of distance." But this meaning does not quite fit the illustrative simile. Applying this definition to the falcon, it seems to me to denote that instinct of *self restraint* which keeps the bird from swooping on its quarry until the right moment, together with the power of judging when the right moment has arrived. Th e analogous quality in soldiers is the highly important one of being able to reserve their fire until the very instant at which it will be most effective. When the

"Victory" went into action at Trafalgar at hardly more than drifting pace, she was for several minutes exposed to a storm of shot and shell before replying with a single gun. Nelson coolly waited until he was within close range, when the broadside he brought to bear worked fearful havoc on the enemy's nearest ships.]

Therefore the good fighter will be terrible in his onset, and prompt in his decision. [The word "decision" would have reference to the measurement of distance mentioned above, letting the enemy get near before striking. But I cannot help thinking that Sun Tzu meant to use the word in a figurative sense comparable to our own idiom "short and sharp." Cf. Wang Hsi's note, which after describing the falcon's mode of attack, proceeds: "This is just how the 'psychological moment' should be seized in war."]

Energy may be likened to the bending of a crossbow; decision, to the releasing of a trigger. [None of the commentators seem to grasp the real point of the simile of energy and the force stored up in the bent cross-bow until released by the finger on the trigger.]

Amid the turmoil and tumult of battle, there may be seeming disorder and yet no real disorder at all; amid confusion and chaos, your array may be without head or tail, yet it will be proof against defeat. [Mei Yao-ch`en says: "The subdivisions of the army having been previously fixed, and the various signals agreed upon, the separating and joining, the dispersing and collecting which will take place in the course of a battle, may give the appearance of disorder when no real disorder is possible. Your formation may be without head or tail, your dispositions all topsy-turvy, and yet a rout of your forces quite out of the question."]

Simulated disorder postulates perfect discipline, simulated fear postulates courage; simulated weakness postulates strength. [In order to make the translation intelligible, it is necessary to tone down the sharply paradoxical form of the original. Ts`ao Kung throws out a hint of the meaning in his brief note: "These things all serve to destroy formation and conceal one's condition." But Tu Mu is the first to put it quite plainly: "If you wish to feign confusion in order to lure the enemy on, you must first have perfect discipline; if you wish to display timidity in order to entrap the enemy, you must have extreme courag e; if you wish to parade your weakness in order to make the enemy over-confident, you must have exceeding strength."]

Hiding order beneath the cloak of disorder is simply a question of subdivision; concealing courage under a show of timidity presupposes a fund of latent energy; [The commentators strongly understand a certain Chinese word here differently than anywhere else in this chapter. Thus Tu Mu says: "seeing that we are favorably circumstanced and yet make no move, the enemy will believe that we are really afraid."] masking strength with weakness is to be effected by tactical dispositions. [Chang Yu relates the following anecdote of Kao Tsu, the first Han Emperor: "Wishing to crush the Hsiung-nu, he sent out spies to report on their condition. But the Hsiung-nu, forewarned, carefully concealed all their able-bodied men and well-fed horses, and only allowed infirm soldiers and emaciated cattle to be seen. The result was that spies one and all recommended the Emperor to deliver his attack. Lou Ching alone opposed them, saying: "When two countries go to war, they are naturally inclined to make an ostentatious display of their strength. Yet our spies have seen nothing but old age and infirmity. This is surely some ruse on the part of the enemy, and it would be unwise for us to attack." The Emperor, however, disregarding this advice, fell into the trap and found himself surrounded at Po-teng."]

Thus one who is skillful at keeping the enemy on the move maintains deceitful appearances, according to which the enemy will act. [Ts`ao Kung's note is "Make a display of weakness and want." Tu Mu says: "If our force happens to be superior to the enemy's, weakness may be simulated in order to lure him on; but if inferior, he must be led to believe that we are strong, in order that he may keep off. In fact, all the enemy's movements should be determined by the signs that we choose to give him." Note the following anecdote of Sun Pin, a descendent of Sun Wu: In 341 B.C., the Ch`I State being at war with Wei, sent T`ien Chi and Sun Pin against the general P`ang Chuan, who happened to be a deadly personal enemy of the later. Sun Pin said: "The Ch`I State has a reputation for cowardice, and therefore our adversary despises us. Let us turn this circumstance to account."]

[Accordingly, when the army had crossed the border into Wei territory, he gave orders to show 100,000 fires on the first night, 50,000 on the next, and the night after only 20,000. P`ang Chuan pursued them hotly, saying to himself: "I knew these men of Ch`I were cowards: their numbers have already fallen away by more than half." In his retreat, Sun Pin came to a narrow defile, with he calculated that his pursuers would reach after dark.]

[Here he had a tree stripped of its bark, and inscribed upon it the words: "Under this tree shall P`ang Chuan die." Then, as night began to fall, he placed a strong body of archers in ambush near by, with orders to shoot directly they saw a light. Later on, P`ang Chuan arrived at the spot, and noticing the tree, struck a light in order to read what was written on it. His body was immediately riddled by a volley of arrows, and his whole army thrown into confusion. [The above is Tu Mu's version of the story; the *Shih Chi*, less dramatically but probably with more historical truth, makes P`ang Chuan cut his own throat with an exclamation of despair, after the rout of his army.]

He sacrifices something, that the enemy may snatch at it.

By holding out baits, he keeps him on the march; then with a body of picked men he lies in wait for him. [With an emendation suggested by Li Ching, this then reads, "He lies in wait with the main body of his troops."]

The clever combatant looks to the effect of combined energy, and does not require too much from individuals. [Tu Mu says: "He first of all considers the power of his army in the bulk; afterwards he takes individual talent into account, and uses each men according to his capabilities. He does not demand perfection from the untalented."]

Hence his ability to pick out the right men and utilize combined energy.

When he utilizes combined energy, his fighting men become as it were like unto rolling logs or stones. For it is the nature of a log or stone to remain motionless on level ground, and to move when on a slope; if four-cornered, to come to a standstill, but if round-shaped, to go rolling down. [Ts`au Kung calls this "the use of natural or inherent power."]

Thus the energy developed by good fighting men is as the momentum of a round stone rolled down a mountain thousands of feet in height. So much on the subject of energy. [The chief lesson of this chapter, in Tu Mu's opinion, is the paramount importance in war of rapid evolutions and sudden rushes. "Great results," he adds, "can thus be achieved with small forces."]

Weak Points and Strong

Sun Tzu said: Whoever is first in the field and awaits the coming of the enemy, will be fresh for the fight; whoever is second in the field and has to hasten to battle will arrive exhausted.

Therefore the clever combatant imposes his will on the enemy, but does not allow the enemy's will to be imposed on him. [One mark of a great soldier is that he fight on his own terms or fights not at all.]

By holding out advantages to him, he can cause the enemy to approach of his own accord; or, by inflicting damage, he can make it impossible for the enemy to draw near. [In the first case, he will entice him with a bait; in the second, he will strike at some important point which the enemy will have to defend.]

If the enemy is taking his ease, he can harass him; [This passage may be cited as evidence against Mei Yao-Ch`en's interpretation of I.] if well supplied with food, he can starve him out; if quietly encamped, he can force him to move.

Appear at points which the enemy must hasten to defend; march swiftly to places where you are not expected.

An army may march great distances without distress, if it marches through country where the enemy is not. [Ts`ao Kung sums up very well: "Emerge from the void [q.d. like "a bolt from the blue"], strike at vulnerable points, shun places that are defended, attack in unexpected quarters."]

You can be sure of succeeding in your attacks if you only attack places which are undefended. [Wang Hsi explains "undefended places" as "weak points; that is to say, where the general is lacking in capacity, or the soldiers in spirit; where the walls are not strong enough, or the precautions not strict enough; where relief comes too late, or provisions are too scanty, or the defenders are variance amongst themselves."]

You can ensure the safety of your defense if you only hold positions that cannot be attacked. [i.e., where there are none of the weak points mentioned above. There is rather a nice point involved in the interpretation of this later clause. Tu Mu, Ch`en Hao, and Mei Yao-ch`en assume the meaning to be: "In order to make your defense quite safe, you must defend *even* those places that are not likely to be attacked;" and Tu Mu adds: "How much more, then, those that will be attacked." Taken thus, however, the clause balances less well with the preceding—always a consideration I n the highly antithetical style which is natural to the Chinese. Chang Yu, therefore, seems to come nearer the mark in saying: "He who is skilled in attack flashes forth from the topmost heights of heaven, making it impossible for the enemy to guard against him. This being so, the places that I shall attack are precisely those that the enemy cannot defend . . . He who is skilled in defense hides in the most secret recesses of the earth, making it impossible for the enemy to estimate h is whereabouts. This being so, the places that I shall hold are precisely those that the enemy cannot attack."]

Hence that general is skillful in attack whose opponent does not know what to defend; and he is skillful in defense whose opponent does not know what to attack. [An aphorism which puts the whole art of war in a nutshell.]

O divine art of subtlety and secrecy! Through you we learn to be invisible, through you inaudible; [Literally, "without form or sound," but it is said of course with reference to the enemy.] and hence we can hold the enemy's fate in our hands.

You may advance and be absolutely irresistible, if you make for the enemy's weak points; you may retire and be safe from pursuit if your movements are more rapid than those of the enemy.

If we wish to fight, the enemy can be forced to an engagement even though he be sheltered behind a high rampart and a deep ditch. All we need do is attack some other place that he will be obliged to relieve. [Tu Mu says: "If the enemy is the invading party, we can cut his line of communications and occupy the roads by which he will have to return; if we are the invaders, we may direct our attack against the sovereign himself." It is clear that Sun Tzu, unlike certain generals in the late Boer war, was no believer in frontal attacks.]

If we do not wish to fight, we can prevent the enemy from engaging us even though the lines of our encampment be merely traced out on the ground. All we need do is to throw something odd and unaccountable in his way. [This extremely concise expression is intelligibly paraphrased by Chia Lin: "even though we have constructed neither wall nor ditch." Li Ch`uan says: "we puzzle him by strange and unusual dispositions;" and Tu Mu finally clinches the meaning by three illustrative anecdotes—one of Chu-ko Liang, who when occupying Yang-p`ing and about to be attacked by Ssu-ma I, suddenly struck his colors, stopped the beating of the drums, and flung open the city gates, showing only a few men engaged in sweeping and sprinkling the ground. This unexpected proceeding had the intended effect; for Ssu-ma I, suspecting an ambush, actually drew off his army and retreated. What Sun Tzu is advocating here, therefore, is nothing more nor less than the timely use of "bluff."]

By discovering the enemy's dispositions and remaining invisible ourselves, we can keep our forces concentrated, while the enemy's must be divided. [The conclusion is perhaps not very obvious, but Chang Yu (after Mei Yao-ch`en) rightly explains it thus: "If the enemy's dispositions are visible, we can make for him in one body; whereas, our own dispositions being kept secret, the enemy will be obliged to divide his forces in order to guard against attack from every quarter."]

We can form a single united body, while the enemy must split up into fractions. Hence there will be a whole pitted against separate parts of a whole, which means that we shall be many to the enemy's few.

And if we are able thus to attack an inferior force with a superior one, our opponents will be in dire straits.

The spot where we intend to fight must not be made known; for then the enemy will have to prepare against a possible attack at several different points; [Sheridan once explained the reason of General Grant's victories by saying that "while his opponents were kept fully employed wondering what he was going to do, *he* was thinking most of what he was going to do himself."] and his forces being thus distributed in many directions, the numbers we shall have to face at any given point will be proportionately few.

For should the enemy strengthen his van, he will weaken his rear; should he strengthen his rear, he will weaken his van; should he strengthen his left, he will weaken his right; should he strengthen his right, he will weaken his left. If he sends reinforcements everywhere, he will everywhere be weak. [In Frederick the Great's *instructions to his generals* we read: "A defensive war is apt to betray us into too frequent detachment. Those generals who have had but little experience attempt to protect every point, while those who are better acquainted with their profession, having only the capital object in view, guard against a decisive blow, and acquiesce in small misfortunes to avoid greater."]

Numerical weakness comes from having to prepare against possible attacks; numerical strength, from compelling our adversary to make these preparations against us. [The highest generalship, in Col. Henderson's words, is "to compel the enemy to disperse his army, and then to concentrate superior force against each fraction in turn."]

Knowing the place and the time of the coming battle, we may concentrate from the greatest distances in order to fight. [What Sun Tzu evidently has in mind is that nice calculation of distances and that masterly employment of strategy which enable a general to divide his army for the purpose of a long and rapid march, and afterwards to effect a junction at precisely the right spot and the right hour in order to confront the enemy in overwhelming strength. Among many such successful junctions which military history records, one of the most dramatic and decisive was the appearance of Blucher just at the critical moment on t he field of Waterloo.]

But if neither time nor place be known, then the left wing will be impotent to succor the right, the right equally impotent to succor the left, the van unable to relieve the rear, or the rear to support the van. How much more so if the furthest port ions of the army are anything under a hundred *li* apart, and even the nearest are separated by several *li*! [The Chinese of this last sentence is a little lacking in precision, but the mental picture we are required to draw is probably that of an army advancing towards a given rendezvous in separate columns, each of which has orders to be there on a fixed date . If the general allows the various detachments to proceed at haphazard, without precise instructions as to the time and place of meeting, the enemy will be able to annihilate the army in detail. Chang Yu's note may be worth quoting here: "If we do no t know the place where our opponents mean to concentrate or the day on which they will join battle, our unity will be forfeited through our preparations for defense, and the positions we hold will be insecure. Suddenly happening upon a powerful foe, we shall be brought to battle in a flurried condition, and no mutual support will be possible between wings, vanguard or rear, especially if there is any great distance between the foremost and hindmost divisions of the army."]

Though according to my estimate the soldiers of Yueh exceed our own in number, that shall advantage them nothing in the matter of victory. I say then that victory can be achieved. [Alas for these brave words! The long feud between the two states ended in 473 B.C. with the total defeat of Wu by Kou Chien and its incorporation in Yueh. This was doubtless long after Sun Tzu's death. With his present assertion Chang Yu is the only one to point out the seeming discrepancy, which he thus goes on to explain: "In the chapter on Tactical Dispositions it is said, 'One may *know* how to conquer without being able to *do* it,' whereas here we have the statement that 'victory' can be achieved.' The explanation is, that in the former chapter, where the offensive and defensive are under discussion, it is said that if the enemy is fully prepared, one cannot make certain of beating him. But the present passage refers particularly to the soldiers of Yueh who, according to Sun Tzu's calculations, will be kept in ignorance of the time and place of the impending struggle. That is why he says here that victory can be achieved."]

Though the enemy be stronger in numbers, we may prevent him from fighting. Scheme so as to discover his plans and the likelihood of their success. [An alternative reading offered by Chia Lin is: "Know beforehand all plans conducive to our success and to the enemy's failure."]

Rouse him, and learn the principle of his activity or inactivity. [Chang Yu tells us that by noting the joy or anger shown by the enemy on being thus disturbed, we shall be able to conclude whether his policy is to lie low or the reverse. He instances the action of Cho-ku Liang, who sent the scornful present of a woman 's head-dress to Ssu-ma I, in order to goad him out of his Fabian tactics.]

Force him to reveal himself, so as to find out his vulnerable spots.

Carefully compare the opposing army with your own, so that you may know where strength is superabundant and where it is deficient.

In making tactical dispositions, the highest pitch you can attain is to conceal them; conceal your dispositions, and you will be safe from the prying of the subtlest spies, from the machinations of the wisest brains. [The piquancy of the paradox evaporates in translation. Concealment is perhaps not so much actual invisibility as "showing no sign" of what you mean to do, of the plans that are formed in your brain. Tu Mu explains: "Though the enemy may have clever and capable officers, they will not be able to lay any plans against us."]

How victory may be produced for them out of the enemy's own tactics—that is what the multitude cannot comprehend.

All men can see the tactics whereby I conquer, but what none can see is the strategy out of which victory is evolved. [i.e., everybody can see superficially how a battle is won; what they cannot see is the long series of plans and combinations which has preceded the battle. Do not repeat the tactics which have gained you one victory, but let your methods be regulated by the infinite variety of circumstances]

[As Wang Hsi sagely remarks: "There is but one root-principle underlying victory, but the tactics which lead up to it are infinite in number." With this compare Col. Henderson: "The rules of strategy are few and simple. They may be learned in a week. They may be taught by familiar illustrations or a dozen diagrams. But such knowledge will no more teach a man to lead an army like Napoleon than a knowledge of grammar will teach him to write like Gibbon."]

Military tactics are like unto water; for water in its natural course runs away from high places and hastens downwards.

So in war, the way is to avoid what is strong and to strike at what is weak. [Like water, taking the line of least resistance.]

Water shapes its course according to the nature of the ground over which it flows; the soldier works out his victory in relation to the foe whom he is facing.

Therefore, just as water retains no constant shape, so in warfare there are no constant conditions.

He who can modify his tactics in relation to his opponent and thereby succeed in winning, may be called a heaven-born captain.

The five elements (water, fire, wood, metal, earth) are not always equally predominant; [That is, as Wang Hsi says: "they predominate alternately."] the four seasons make way for each other in turn. [Literally, "have no invariable seat."]

There are short days and long; the moon has its periods of waning and waxing. [The purport of the passage is simply to illustrate the want of fixity in war by the changes constantly taking place in Nature. The comparison is not very happy, however, because the regularity of the phenomena which Sun Tzu mentions is by no means paralleled in war.]

Maneuvering

Sun Tzu said: In war, the general receives his commands from the sovereign.

Having collected an army and concentrated his forces, he must blend and harmonize the different elements thereof before pitching his camp. ["Chang Yu says: "the establishment of harmony and confidence between the higher and lower ranks before venturing into the field;" and he quotes a saying of Wu Tzu (chap. 1 adinit.): "Without harmony in the State, no military expedition can be undertaken; without harmony in the army, no battle array can be formed." In an historical romance Sun Tzu is represented as saying to Wu Yuan: "As a general rule, those who are waging war should get rid of all the domestic troubles before proceeding to attack the external foe."]

After that, comes tactical maneuvering, than which there is nothing more difficult. [I have departed slightly from the traditional interpretation of Ts`ao Kung, who says: "From the time of receiving the sovereign's instructions until our encampment over against the enemy, the tactics to be pursued are most difficult." It seems to me t hat the tactics or maneuvers can hardly be said to begin until the army has sallied forth and encamped, and Ch`ien Hao's note gives color to this view: "For levying, concentrating, harmonizing and entrenching an army, there are plenty of old rules which will serve. The real difficulty comes when we engage in tactical operations." Tu Yu also observes that "the great difficulty is to be beforehand with the enemy in seizing favorable position."]

The difficulty of tactical maneuvering consists in turning the devious into the direct, and misfortune into gain. [This sentence contains one of those highly condensed and somewhat enigmatical expressions of which Sun Tzu is so fond. This is how it is explained by Ts`ao Kung:

"Make it appear that you are a long way off, then cover the distance rapidly and arrive o n the scene before your opponent." Tu Mu says: "Hoodwink the enemy, so that he may be remiss and leisurely while you are dashing along with utmost speed." Ho Shih gives a slightly different turn: "Although you may have difficult ground to traverse and natural obstacles to encounter this is a drawback which can be turned into actual advantage by celerity of movement." Signal examples of this saying are afforded by the two famous passages across the Alps—that of Hannibal, which laid Italy at his mercy, and that of Napoleon two thousand years later, which resulted in the great victory of Marengo.]

Thus, to take a long and circuitous route, after enticing the enemy out of the way, and though starting after him, to contrive to reach the goal before him, shows knowledge of the artifice of *deviation*. [Tu Mu cites the famous march of Chao She in 270 B.C. to relieve the town of O-yu, which was closely invested by a Ch`in army. The King of Chao first consulted Lien P`o on the advisability of attempting a relief, but the latter thought the distance too great, and the intervening country too rugged and difficult. His Majesty then turned to Chao She, who fully admitted the hazardous nature of the march, but finally said: "We shall be like two rats fighting in a whole—and the pluckier one will win!" So he left the capital with his army, but had only gone a distance of 30 *li* when he stopped and began throwing up entrenchments. For 28 days he continued strengthening his fortifications, and took care that spies should carry the intelligence to the enemy. The Ch`in general was overjoyed, and attributed his adversary's tardiness to the fact that the beleaguered city was in the Han State, and thus not actually part of Chao territory. But the spies had no sooner departed than Chao She began a forced march lasting for two days and one night, and arrive on the scene of action with such astonishing rapidity that he was able to occupy a commanding position on the "North hill" before the enemy had got wind of his movements. A crushing defeat followed for the Ch`in forces, who were obliged to raise the siege of O-yu in all haste and retreat across the border.]

Maneuvering with an army is advantageous; with an undisciplined multitude, most dangerous. [I adopt the reading of the *T`uang Tien*, Cheng Yu-hsien and the *T`u Shu*, since they appear to apply the exact nuance required in order to make sense. The commentators using the standard text take this line to mean that maneuvers may be profitable, or they may be dangerous: it all depends on the ability of the general.]

If you set a fully equipped army in march in order to snatch an advantage, the chances are that you will be too late. On the other hand, to detach a flying column for the purpose involves the sacrifice of its baggage and stores. [Some of the Chinese text is unintelligible to the Chinese commentators, who paraphrase the sentence. I submit my own rendering without much enthusiasm, being convinced that there is some deep-seated corruption in the text. On the whole, it is clear t hat Sun Tzu does not approve of a lengthy march being undertaken without supplies.]

Thus, if you order your men to roll up their buff-coats, and make forced marches without halting day or night, covering double the usual distance at a stretch, [The ordinary day's march, according to Tu Mu, was 30 *li*; but on one occasion, when pursuing Liu Pei, Ts`ao Ts`ao is said to have covered the incredible distance of 300 *li* within twenty-four hours.] doing a hundred *li* in order to wrest an advantage, the leaders of all your three divisions will fall into the hands of the enemy.

The stronger men will be in front, the jaded ones will fall behind, and on this plan only one-tenth of your army will reach its destination. [The moral is, as Ts`ao Kung and others point out: Don't march a hundred *li* to gain a tactical advantage, either with or without impedimenta. Maneuvers of this description should be confined to short distances. Stonewall Jackson said: "The hardships of forced marches are often more painful than the dangers of battle." He did not often call upon his troops for extraordinary exertions. It was only when he intended a surprise, or when a rapid retreat was imperative, that he sacrificed everything for speed.]

If you march fifty *li* in order to outmaneuver the enemy, you will lose the leader of your first division, and only half your force will reach the goal. [Literally, "the leader of the first division will be *torn away*."]

If you march thirty *li* with the same object, two-thirds of your army will arrive. [In the *T`uang Tien* is added: "From this we may know the difficulty of maneuvering."]

We may take it then that an army without its baggage-train is lost; without provisions it is lost; without bases of supply it is lost. [I think Sun Tzu meant "stores accumulated in depots." But Tu Yu says "fodder and the like," Chang Yu says "Goods in general," and Wang Hsi says "fuel, salt, foodstuffs, etc."]

We cannot enter into alliances until we are acquainted with the designs of our neighbors.

We are not fit to lead an army on the march unless we are familiar with the face of the country—its mountains and forests, its pitfalls and precipices, its marshes and swamps.

We shall be unable to turn natural advantage to account unless we make use of local guides.

In war, practice dissimulation, and you will succeed. [In the tactics of Turenne, deception of the enemy, especially as to the numerical strength of his troops, took a very prominent position.]

Whether to concentrate or to divide your troops, must be decided by circumstances.

Let your rapidity be that of the wind, [The simile is doubly appropriate, because the wind is not only swift but, as Mei Yao-ch`en points out, "invisible and leaves no tracks."] your compactness that of the forest. [Meng Shih comes nearer to the mark in his note: "When slowly marching, order and ranks must be preserved"—so as to guard against surprise attacks. But natural forest do not grow in rows, whereas they do generally possess the quality of density or compactness.]

In raiding and plundering be like fire, is immovability like a mountain. [That is, when holding a position from which the enemy is trying to dislodge you, or perhaps, as Tu Yu says, when he is trying to entice you into a trap.]

Let your plans be dark and impenetrable as night, and when you move, fall like a thunderbolt. [Tu Yu quotes a saying of T`ai Kung which has passed into a proverb: "You cannot shut your ears to the thunder or your eyes to the lighting—so rapid are they." Likewise, an attack should be made so quickly that it cannot be parried.]

When you plunder a countryside, let the spoil be divided amongst your men; [Sun Tzu wishes to lessen the abuses of indiscriminate plundering by insisting that all booty shall be thrown into a common stock, which may afterwards be fairly divided amongst all.] when you capture new territory, cut it up into allotments for the benefit of the soldiery. [Ch`en Hao says "quarter your soldiers on the land, and let them sow and plant it." It is by acting on this principle, and harvesting the lands they invaded, that the Chinese have succeeded in carrying out some of their most memorable and triumphant expeditions, such as that of Pan Ch`ao who penetrated to the Caspian, and in more recent years, those of Fu-k`ang-an and Tso Tsung-t`ang.]

Ponder and deliberate before you make a move. [Chang Yu quotes Wei Liao Tzu as saying that we must not break camp until we have gained the resisting power of the enemy and the cleverness of the opposing general.]

He will conquer who has learnt the artifice of deviation.

Such is the art of maneuvering. [With these words, the chapter would naturally come to an end. But there now follows a long appendix in the shape of an extract from an earlier book on War, now lost, but apparently extant at the time when Sun Tzu wrote. The style of this fragment is not noticeable different from that of Sun Tzu himself, but no commentator raises a doubt as to its genuineness.]

The Book of Army Management says: [It is perhaps significant that none of the earlier commentators give us any information about this work. Mei Yao-Ch`en calls it "an ancient military classic," and Wang Hsi, "an old book on war." Considering the enormous amount of fighting that had gone on for centuries before Sun Tzu's time between the various kingdoms and principalities of China, it is not in itself improbable that a collection of military maxims should have been made and written down at some earlier period.]

On the field of battle, [Implied, though not actually in the Chinese.] the spoken word does not carry far enough: hence the institution of gongs and drums. Nor can ordinary objects be seen clearly enough: hence the institution of banners and flags.

Gongs and drums, banners and flags, are means whereby the ears and eyes of the host may be focused on one particular point. [Chang Yu says: "If sight and hearing converge simultaneously on the same object, the evolutions of as many as a million soldiers will be like those of a single man."!]

The host thus forming a single united body, is it impossible either for the brave to advance alone, or for the cowardly to retreat alone. [Chuang Yu quotes a saying: "Equally guilty are those who advance against orders and those who retreat against orders." Tu Mu tells a story in this connection of Wu Ch`I, when he was fighting against the Ch`in State. Before the battle had begun, one of his soldiers, a man of matchless daring, sallied forth by himself, captured two heads from the enemy, and returned to camp. Wu Ch`I had the man instantly executed, whereupon an officer ventured to remonstrate, saying: "This man was a good soldier, and ought not to have been beheaded." Wu Ch`I replied: "I fully believe he was a good soldier, but I had him beheaded because he acted without orders."]

This is the art of handling large masses of men.

In night-fighting, then, make much use of signal-fires and drums, and in fighting by day, of flags and banners, as a means of influencing the ears and eyes of your army. [Ch`en Hao alludes to Li Kuang-pi's night ride to Ho-yang at the head of 500 mounted men; they made such an imposing display with torches, that though the rebel leader Shih Ssu-ming had a large army, he did not dare to dispute their passage.]

A whole army may be robbed of its spirit; ["In war," says Chang Yu, "if a spirit of anger can be made to pervade all ranks of an army at one and the same time, its onset will be irresistible. Now the spirit of the enemy's soldiers will be keenest when they have newly arrived on the scene, and it is therefore our cue not to fight at once, but to wait until their ardor and enthusiasm have worn off, and then strike. It is in this way that they may be robbed of their keen spirit." Li Ch`uan and others tell an anecdote (to be found in the *Tso C Huan*, year 10.) of Ts`ao Kuei, a protégé of Duke Chuang of Lu. The latter State was attacked by Ch`I, and the duke was about to join battle at Ch`ang-cho, after the first roll of the enemy's drums, when Ts`ao said: "Not just yet."] [Only after their drums had beaten for the third time, did he give the word for attack. Then they fought, and the men of Ch`I were utterly defeated. Questioned afterwards by the Duke as to the meaning of his delay, Ts`ao Kuei replied: "In battle, a courageous spirit is everything. Now the first roll of the drum tends to create this spirit, but with the second it is already on the wane, and after the third it is gone altogether. I attacked when their spirit was gone and ours was at its height. Hence our victory." Wu Tzu (chap. 4) puts "spirit" first among the "four important influences" in war, and continues: "The value of a whole army—a mighty host of a million men—is dependent on one man alone: such is the influence of spirit!"] a commander-in-chief may be robbed of his presence of mind. [Chang Yu says: "Presence of mind is the general's most important asset. It is the quality which enables him to discipline disorder and to inspire courage into the panic-stricken." The great general Li Ching (A.D. 571-649) has a saying: "Attacking does not merely consist in assaulting walled cities or striking at an army in battle array; it must include the art of assailing the enemy's mental equilibrium."]

Now a solider's spirit is keenest in the morning; [Always provided, I suppose, that he has had breakfast. At the battle of the Trebia, the Romans were foolishly allowed to fight fasting, whereas Hannibal's men had breakfasted at their leisure.] by noonday it has begun to flag; and in the evening, his mind is bent only on returning to camp.

A clever general, therefore, avoids an army when its spirit is keen, but attacks it when it is sluggish and inclined to return. This is the art of studying moods.

Disciplined and calm, to await the appearance of disorder and hubbub amongst the enemy:—this is the art of retaining self-possession.

To be near the goal while the enemy is still far from it, to wait at ease while the enemy is toiling and struggling, to be well-fed while the enemy is famished:—this is the art of husbanding one's strength.

To refrain from intercepting an enemy whose banners are in perfect order, to refrain from attacking an army drawn up in calm and confident array:—this is the art of studying circumstances.

It is a military axiom not to advance uphill against the enemy, nor to oppose him when he comes downhill.

Do not pursue an enemy who simulates flight; do not attack soldiers whose temper is keen.

Do not swallow bait offered by the enemy. [Li Ch`uan and Tu Mu, with extraordinary inability to see a metaphor, take these words quite literally of food and drink that have been poisoned by the enemy.

Ch`en Hao and Chang Yu carefully point out that the saying has a wider application.]

Do not interfere with an army that is returning home. [The commentators explain this rather singular piece of advice by saying that a man whose heart is set on returning home will fight to the death against any attempt to bar his way, and is therefore too dangerous an opponent to be tackled. Chang Yu quote s the words of Han Hsin: "Invincible is the soldier who hath his desire and returneth homewards." A marvelous tale is told of Ts`ao Ts`ao's courage and resource in ch. 1 of the *San Kuo Chi*: In 198 A.D., he was besieging Chang Hsiu in Jang, when Liu Pi ao sent reinforcements with a view to cutting off Ts`ao's retreat. The latter was obliged to draw off his troops, only to find himself hemmed in between two enemies, who were guarding each outlet of a narrow pass in which he had engaged himself. In this desperate plight Ts`ao waited until nightfall, when he bored a tunnel into the mountain side and laid an ambush in it. As soon as the whole army had passed by, the hidden troops fell on his rear, while Ts`ao himself turned and met his pursuers in front, so that they were thrown into confusion and annihilated. Ts`ao Ts`ao said afterwards: "The brigands tried to check my army in its retreat and brought me to battle in a desperate position: hence I knew how to overcome them."]

When you surround an army, leave an outlet free. [This does not mean that the enemy is to be allowed to escape. The object, as Tu Mu puts it, is "to make him believe that there is a road to safety, and thus prevent his fighting with the courage of despair." Tu Mu adds pleasantly: "After that, you ma y crush him."]

Do not press a desperate foe too hard. [Ch`en Hao quotes the saying: "Birds and beasts when brought to bay will use their claws and teeth." Chang Yu says: "If your adversary has burned his boats and destroyed his cooking-pots, and is ready to stake all on the issue of a battle, he must not be pushed to extremities." Ho Shih illustrates the meaning by a story taken from the life of Yen-ch`ing. That general, together with his colleague Tu Chung-wei was surrounded by a vastly superior army of Khitans in the year 945 A.D. The country was bare and desert-like, and the little Chinese force was soon in dire straits for want of water. The wells they bored ran dry, and the men were reduced to squeezing lumps of mud and sucking out the moisture. Their ranks thinned rapidly, until at last Fu Yen -ch`ing exclaimed: "We are desperate men. Far better to die for our country than to go with fettered hands into captivity!" A strong gale happened to be blowing from the northeast and darkening the air with dense clouds of sandy dust. To Chung-wei was for waiting until this had abated before deciding on a final attack; but luckily another officer, Li Shou-cheng by name, was quicker to see an opportunity, and said: "They are many and we are few, but in the midst of this sandstorm our numbers will not be discernible; victory will go to the strenuous fighter, and the wind will be our best ally." Accordingly, Fu Yen-ch`ing made a sudden and wholly unexpected onslaught with his cavalry, routed the barbarians and succeeded in breaking through to safety.]

Such is the art of warfare.

Variation in Tactics

Sun Tzu said: In war, the general receives his commands from the sovereign, collects his army and concentrates his forces. [It may have been interpolated here merely in order to supply a beginning to the chapter.]

When in difficult country, do not encamp. In country where high roads intersect, join hands with your allies. Do not linger in dangerously isolated positions. [The last situation is not one of the Nine Situations as given in the beginning of chap. XI, but occurs later on. Chang Yu defines this situation as being situated across the frontier, in hostile territory. Li Ch`uan says it is "country in which there are no springs or wells, flocks or herds, vegetables or firewood;" Chia Lin, "one of gorges, chasms and precipices, without a road by which to advance."]

In hemmed-in situations, you must resort to stratagem. In desperate position, you must fight.

There are roads which must not be followed, ["Especially those leading through narrow defiles," says Li Ch`uan, "where an ambush is to be feared."] armies which must be not attacked, [More correctly, perhaps, "there are times when an army must not be attacked." Ch`en Hao says: "When you see your way to obtain a rival advantage, but are powerless to inflict a real defeat, refrain from attacking, for fear of overtaxing your men's strength."] towns which must not be besieged, [Ts`ao Kung gives an interesting illustration from his own experience. When invading the territory of Hsu-chou, he ignored the city of Hua-pi, which lay directly in his path, and pressed on into the heart of the country. This excellent strategy was rewarded by the subsequent capture of no fewer than fourteen important district cities. Chang Yu says: "No town should be attacked which, if taken, cannot be held, or if left alone, will not cause any trouble." Hsun Ying, when urged to attack Pi-yang, replied: "The city is small and well-fortified; even if I succeed in taking it, it will be no great feat of arms; whereas if I fail, I shall make myself a laughing-stock." In the seventeenth century, sieges still formed a large proportion of war. It was Turenne who directed attention to the importance of marches, countermarches and maneuvers. He said: "It is a great mistake to waste men in taking a town when the same expenditure of soldiers will gain a province."] positions which must not be contested, commands of the sovereign which must not be obeyed. [This is a hard saying for the Chinese, with their reverence for authority, and Wei Liao Tzu (quoted by Tu Mu) is moved to exclaim: "Weapons are baleful instruments, strife is antagonistic to virtue, a military commander is the negation of civil order!" The unpalatable fact remains, however, that even Imperial wishes must be subordinated to military necessity.]

The general who thoroughly understands the advantages that accompany variation of tactics knows how to handle his troops.

The general who does not understand these, may be well acquainted with the configuration of the country, yet he will not be able to turn his knowledge to practical account. [Literally, "get the advantage of the ground," which means not only securing good positions, but availing oneself of natural advantages in every possible way. Chang Yu says: "Every kind of ground is characterized by certain natural features, and also gives scope for a certain variability of plan. How it is possible to turn these natural features to account unless topographical knowledge is supplemented by versatility of mind?"]

So, the student of war who is unversed in the art of war of varying his plans, even though he be acquainted with the Five Advantages, will fail to make the best use of his men. [Chia Lin tells us that these imply five obvious and generally advantageous lines of action, namely: "if a certain road is short, it must be followed; if an army is isolated, it must be attacked; if a town is in a parlous condition, it must be besieged; if a position can be stormed, it must be attempted; and if consistent with military operations, the ruler's commands must be obeyed." But there are circumstances which sometimes forbid a general to use these advantages. For instance, "a certain road ma y be the shortest way for him, but if he knows that it abounds in natural obstacles, or that the enemy has laid an ambush on it, he will not follow that road. A hostile force may be open to attack, but if he knows that it is hard-pressed and likely to fight with desperation, he will refrain from striking," and so on.]

Hence in the wise leader's plans, considerations of advantage and of disadvantage will be blended together. ["Whether in an advantageous position or a disadvantageous one," says Ts`ao Kung, "the opposite state should be always present to your mind. "]

If our expectation of advantage be tempered in this way, we may succeed in accomplishing the essential part of our schemes. [Tu Mu says: "If we wish to wrest an advantage from the enemy, we must not fix our minds on that alone, but allow for the possibility of the enemy also doing some harm to us, and let this enter as a factor into our calculations."]

If, on the other hand, in the midst of difficulties we are always ready to seize an advantage, we may extricate ourselves from misfortune. [Tu Mu says: "If I wish to extricate myself from a dangerous position, I must consider not only the enemy's ability to injure me, but also my own ability to gain an advantage over the enemy. If in my counsels these two considerations are properly blend ed, I shall succeed in liberating myself.... For instance; if I am surrounded by the enemy and only think of effecting an escape, the nervelessness of my policy will incite my adversary to pursue and crush me; it would be far better to encourage my men to deliver a bold counter-attack, and use the advantage thus gained to free myself from the enemy's toils."]

Reduce the hostile chiefs by inflicting damage on them; [Chia Lin enumerates several ways of inflicting this injury, some of which would only occur to the Oriental mind:—"Entice away the enemy's best and wisest men, so that he may be left without counselors. Introduce traitors into his country, that the government policy may be rendered futile. Foment intrigue and deceit, and thus sow dissension between the ruler and his ministers. By means of every artful contrivance, cause deterioration amongst his men and waste of his treasure. Corrupt his moral s by insidious gifts leading him into excess. Disturb and unsettle his mind by presenting him with lovely women." Chang Yu (after Wang Hsi) makes a different interpretation of Sun Tzu here: "Get the enemy into a position where he must suffer injury, an d he will submit of his own accord."] and make trouble for them, [Tu Mu, in this phrase, in his interpretation indicates that trouble should be make for the enemy affecting their "possessions," or, as we might say, "assets," which he considers to be "a large army, a rich exchequer, harmony amongst the soldiers, punctual fulfillment of commands." These give us a whip-hand over the enemy.] and keep them constantly engaged; [Literally, "make servants of them." Tu Yu says "prevent the from having any rest."] hold out specious allurements, and make them rush to any given point. [Meng Shih's note contains an excellent example of the idiomatic use of: "cause them to forget *pien* (the reasons for acting otherwise than on their first impulse), and hasten in our direction."]

The art of war teaches us to rely not on the likelihood of the enemy's not coming, but on our own readiness to receive him; not on the chance of his not attacking, but rather on the fact that we have made our position unassailable.

There are five dangerous faults which may affect a general:

(a) Recklessness, which leads to destruction; ["Bravery without forethought," as Ts`ao Kung analyzes it, which causes a man to fight blindly and desperately like a mad bull. Such an opponent, says Chang Yu, "must not be encountered with brute force, but may be lured into an ambush and slain. In estimating the character of a general, men are wont to pay exclusive attention to his courage, forgetting that courage is only one out of many qualities which a general should possess. The merely brave man is prone to fight recklessly; and he who fights recklessly, without any perception of what is expedient, must be condemned." Ssu-ma Fa, too, make the incisive remark: "Simply going to one's death does not bring about victory."]

(b) Cowardice, which leads to capture; [Ts`ao Kung defines the Chinese word translated here as "cowardice" as being of the man "whom timidity prevents from advancing to seize an advantage," and Wang Hsi adds "who is quick to flee at the sight of danger." Meng Shih gives the closer paraphrase "he who is bent on returning alive," this is, the man who will never take a risk. But, as Sun Tzu knew, nothing is to be achieved in war unless you are willing to take risks. T`ai Kung said: "He who lets an advantage slip will subsequently bring upon himself real disaster." In 404 A.D., Liu Yu pursued the rebel Huan Hsuan up the Yangtsze and fought a naval battle with him at the island of Ch`eng-hung. The loyal troops numbered only a few thousands, while their opponents were in great force. But Hu an Hsuan, fearing the fate which was in store for him should be overcome, had a light boat made fast to the side of his war-junk, so that he might escape, if necessary, at a moment's notice. The natural result was that the fighting spirit of his soldiers was utterly quenched, and when the loyalists made an attack from windward with fireships, all striving with the utmost ardor to be first in the fray, Huan Hsuan's forces were routed, had to burn all their baggage and fled for two days and nights without stopping. Chang Yu tells a somewhat similar story of Chao Ying-ch`I, a general of the Chin State who during a battle with the army of Ch`u in 597 B.C. had a boat kept in readiness for him on the river, wishing in case of defeat to be the first t o get across.]

(c) A hasty temper, which can be provoked by insults; [Tu Mu tells us that Yao Hsing, when opposed in 357 A.D. by Huang Mei, Teng Ch`iang and others shut himself up behind his walls and refused to fight. Teng Ch`iang said: "Our adversary is of a choleric temper and easily provoked; let us make constant s allies and break down his walls, then he will grow angry and come out. Once we can bring his force to battle, it is doomed to be our prey." This plan was acted upon, Yao Hsiang came out to fight, was lured as far as San-yuan by the enemy's pretended flight, and finally attacked and slain.]

d) A delicacy of honor which is sensitive to shame; [This need not be taken to mean that a sense of honor is really a defect in a general. What Sun Tzu condemns is rather an exaggerated sensitiveness to slanderous reports, the thin-skinned man who is stung by opprobrium, however undeserved. Mei Yao-ch`en truly observes, though somewhat paradoxically: "The seek after glory should be careless of public opinion."]

(e) Over-solicitude for his men, which exposes him to worry and trouble. [Here again, Sun Tzu does not mean that the general is to be careless of the welfare of his troops. All he wishes to emphasize is the danger of sacrificing any important military advantage to the immediate comfort of his men. This is a shortsighted poli cy, because in the long run the troops will suffer more from the defeat, or, at best, the prolongation of the war, which will be the consequence. A mistaken feeling of pity will often induce a general to relieve a beleaguered city, or to reinforce a hard-pressed detachment, contrary to his military instincts. It is now generally admitted that our repeated efforts to relieve Ladysmith in the South African War were so many strategical blunders which defeated their own purpose. And in the end, relief came through the very man who started out with the distinct resolve no longer to subordinate the interests of the whole to sentiment in favor of a part. An old soldier of one of our generals who failed most conspicuously in this war, tried once, I remember, to defend him to me on the ground that he was always "so good to his men." By this plea, had he but known it, he was only condemning him out of Sun Tzu's mouth.]

These are the five besetting sins of a general, ruinous to the conduct of war.

When an army is overthrown and its leader slain, the cause will surely be found among these five dangerous faults. Let them be a subject of meditation.

The Army on the March

Sun Tzu said: We come now to the question of encamping the army, and observing signs of the enemy. Pass quickly over mountains, and keep in the neighborhood of valleys. [The idea is, not to linger among barren uplands, but to keep close to supplies of water and grass. "Abide not in natural ovens," i.e. "the openings of valleys." Chang Yu tells the following anecdote: Wu-tu Ch`iang was a robber captain in the time of the Later Han, and Ma Yuan was sent to exterminate his gang. Ch`iang having found a refuge in the hills, Ma Yuan made no attempt to force a battle, but seized all the favorable positions commanding supplies of water and forage. Ch` iang was soon in such a desperate plight for want of provisions that he was forced to make a total surrender. He did not know the advantage of keeping in the neighborhood of valleys."]

Camp in high places, [Not on high hills, but on knolls or hillocks elevated above the surrounding country.] facing the sun. [Tu Mu takes this to mean "facing south," and Ch`en Hao "facing east."]

Do not climb heights in order to fight. So much for mountain warfare.

After crossing a river, you should get far away from it. ["In order to tempt the enemy to cross after you," according to Ts`ao Kung, and also, says Chang Yu, "in order not to be impeded in your evolutions." The T`uang Tien reads, "If the Enemy crosses a river," etc. But in view of the next sentence, this is almost certainly an interpolation.]

When an invading force crosses a river in its onward march, do not advance to meet it in mid-stream. It will be best to let half the army get across, and then deliver your attack. [Li Ch`uan alludes to the great victory won by Han Hsin over Lung Chu at the Wei River. Turning to the ch`ien han shu, ch. 34, fol. 6 verso, we find the battle described as follows: "The two armies were drawn up on opposite sides of the river. In the night, Han Hsin ordered his men to take some ten thousand sacks filled with sand and construct a dam higher up. Then, leading half his army across, he attacked Lung Chu; but after a time, pretending to have failed in his attempt, he hastily withdrew t o the other bank. Lung Chu was much elated by this unlooked-for success, and exclaiming: "I felt sure that Han Hsin was really a coward!" he pursued him and began crossing the river in his turn. Han Hsin now sent a party to cut open the sandbags, thus releasing a great volume of water, which swept down and prevented the greater portion of Lung Chu's army from getting across. He then turned upon the force which had been cut off, and annihilated it, Lung Chu himself being amongst the slain. The rest of the army, on the further bank, also scattered and fled in all directions.]

If you are anxious to fight, you should not go to meet the invader near a river which he has to cross. [For fear of preventing his crossing.]

Moor your craft higher up than the enemy, and facing the sun. [The repetition of these words in connection with water is very awkward. Chang Yu has the note: "Said either of troops marshaled on the river-bank, or of boats anchored in the stream itself; in either case it is essential to be high er than the enemy and facing the sun." The other commentators are not at all explicit.]

Do not move up-stream to meet the enemy. [Tu Mu says: "As water flows downwards, we must not pitch our camp on the lower reaches of a river, for fear the enemy should open the sluices and sweep us away in a flood. Chu-ko Wu-hou has remarked that 'in river warfare we must not advance against th e stream,' which is as much as to say that our fleet must not be anchored below that of the enemy, for then they would be able to take advantage of the current and make short work of us." There is also the danger, noted by other commentators, that the enemy may throw poison on the water to be carried down to us.]

So much for river warfare.

In crossing salt-marshes, your sole concern should be to get over them quickly, without any delay. [Because of the lack of fresh water, the poor quality of the herbage, and last but not least, because they are low, flat, and exposed to attack.]

If forced to fight in a salt-marsh, you should have water and grass near you, and get your back to a clump of trees. [Li Ch`uan remarks that the ground is less likely to be treacherous where there are trees, while Tu Mu says that they will serve to protect the rear.] So much for operations in salt-marches.

In dry, level country, take up an easily accessible position with rising ground to your right and on your rear, [Tu Mu quotes T`ai Kung as saying: "An army should have a stream or a marsh on its left, and a hill or tumulus on its right."] so that the danger may be in front, and safety lie behind. So much for campaigning in flat country.

These are the four useful branches of military knowledge [Those, namely, concerned with (1) mountains, (2) rivers, (3) marshes, and (4) plains. Compare Napoleon's "Military Maxims," no. 1.] which enabled the Yellow Emperor to vanquish four several sovereigns. [Regarding the "Yellow Emperor": Mei Yao-ch`en asks, with some plausibility, whether there is an error in the text as nothing is known of Huang Ti having conquered four other Emperors. The *Shih Chi* (ch. 1 ad init.) speaks only of his victories over Y en Ti and Ch`ih Yu. In the *li*U T`AO it is mentioned that he "fought seventy battles and pacified the Empire." Ts`ao Kung's explanation is, that the Yellow Emperor was the first to institute the feudal system of vassals princes, each of whom (to the number of four) originally bore the title of Emperor. Li Ch`uan tells us that the art of war originated under Huang Ti, who received it from his Minister Feng Hou.]

All armies prefer high ground to low. ["High Ground," says Mei Yao-ch`en, "is not only more agreement and salubrious, but more convenient from a military point of view; low ground is not only damp and unhealthy, but also disadvantageous for fight ing."] and sunny places to dark.

If you are careful of your men, [Ts`ao Kung says: "Make for fresh water and pasture, where you can turn out your animals to graze."] and camp on hard ground, the army will be free from disease of every kind, [Chang Yu says: "The dryness of the climate will prevent the outbreak of illness."] and this will spell victory.

When you come to a hill or a bank, occupy the sunny side, with the slope on your right rear. Thus you will at once act for the benefit of your soldiers and utilize the natural advantages of the ground.

When, in consequence of heavy rains up-country, a river which you wish to ford is swollen and flecked with foam, you must wait until it subsides.

Country in which there are precipitous cliffs with torrents running between, deep natural hollows, [The latter defined as "places enclosed on every side by steep banks, with pools of water at the bottom.] confined places, [Defined as "natural pens or prisons" or "places surrounded by precipices on three sides—easy to get into, but hard to get out of."] tangled thickets, [Defined as "places covered with such dense undergrowth that spears cannot be used."] quagmires [Defined as "low-lying places, so heavy with mud as to be impassable for chariots and horsemen."] and crevasses, [Defined by Mei Yao-ch`en as "a narrow difficult way between beetling cliffs." Tu Mu's note is "ground covered with trees and rocks, and intersected by numerous ravines and pitfalls." This is very vague, but Chia Lin explains it clearly enough as a defile or narrow pass, and Chang Yu takes much the same view. On the whole, the weight of the commentators certainly inclines to the rendering "defile." But the ordinary meaning of the Chinese in one place is "a crack or fissure" and the fact that the meaning of the Chinese elsewhere in the sentence indicates something in the nature of a defile, make me think that Sun Tzu is here speaking of crevasses.] should be left with all possible speed and not approached.

While we keep away from such places, we should get the enemy to approach them; while we face them, we should let the enemy have them on his rear.

If in the neighborhood of your camp there should be any hilly country, ponds surrounded by aquatic grass, hollow basins filled with reeds, or woods with thick undergrowth, they must be carefully routed out and searched; for these are places where men in ambush or insidious spies are likely to be lurking. [Chang Yu has the note: "We must also be on our guard against traitors who may lie in close covert, secretly spying out our weaknesses and overhearing our instructions."]

When the enemy is close at hand and remains quiet, he is relying on the natural strength of his position. [Here begin Sun Tzu's remarks on the reading of signs, much of which is so good that it could almost be included in a modern manual like Gen. Baden-Powell's "Aids to Scouting."]

When he keeps aloof and tries to provoke a battle, he is anxious for the other side to advance. [Probably because we are in a strong position from which he wishes to dislodge us. "If he came close up to us, says Tu Mu, "and tried to force a battle, he would seem to despise us, and there would be less probability of our responding to the challenge. "]

If his place of encampment is easy of access, he is tendering a bait.

Movement amongst the trees of a forest shows that the enemy is advancing. [Ts`ao Kung explains this as "felling trees to clear a passage," and Chang Yu says: "Every man sends out scouts to climb high places and observe the enemy. If a scout sees that the trees of a forest are moving and shaking, he may know that they are being cut down to clear a passage for the enemy's march."]

The appearance of a number of screens in the midst of thick grass means that the enemy wants to make us suspicious. [Tu Yu's explanation, borrowed from Ts`ao Kung's, is as follows: "The presence of a number of screens or sheds in the midst of thick vegetation is a sure sign that the enemy has fled and, fearing pursuit, has constructed these hiding-places in order t o make us suspect an ambush." It appears that these "screens" were hastily knotted together out of any long grass which the retreating enemy happened to come across.]

The rising of birds in their flight is the sign of an ambuscade. [Chang Yu's explanation is doubtless right: "When birds that are flying along in a straight line suddenly shoot upwards, it means that soldiers are in ambush at the spot beneath."] Startled beasts indicate that a sudden attack is coming.

When there is dust rising in a high column, it is the sign of chariots advancing; when the dust is low, but spread over a wide area, it betokens the approach of infantry. ["High and sharp," or rising to a peak, is of course somewhat exaggerated as applied to dust. The commentators explain the phenomenon by saying that horses and chariots, being heavier than men, raise more dust, and also follow one another in the same wheel-track, whereas foot-soldiers would be marching in ranks, many abreast. According to Chang Yu, "every army on the march must have scouts some way in advance, who on sighting dust raised by the enemy, will gallop back and report it to the commander- in-chief." Cf. Gen. Baden-Powell: "As you move along, say, in a hostile country, your eyes should be looking afar for the enemy or any signs of him: figures, dust rising, birds getting up, glitter of arms, etc."]

When it branches out in different directions, it shows that parties have been sent to collect firewood. A few clouds of dust moving to and fro signify that the army is encamping. [Chang Yu says: "In apportioning the defenses for a cantonment, light horse will be sent out to survey the position and ascertain the weak and strong points all along its circumference. Hence the small quantity of dust and its motion."]

Humble words and increased preparations are signs that the enemy is about to advance. ["As though they stood in great fear of us," says Tu Mu. "Their object is to make us contemptuous and careless, after which they will attack us." Chang Yu alludes to the story of T`ien Tan of the Ch`i-mo against the Yen forces, led by Ch`i Chieh. In ch. 82 of the *Shih Chi* we read: "T`ien Tan openly said: 'My only fear is that the Yen army may cut off the noses of their Ch`i prisoners and place them in the front rank to fight against us; that would be the undoing of our city.' The other side being informed of this speech, at once acted on the suggestion; but those within the city were enraged at seeing their fellow-countrymen thus mutilated, and fearing only lest they should fall into the enemy's hands, were nerved to defend themselves more obstinately than ever. Once again T`ien Tan sent back converted spies who reported these words to the enemy: "What I dread most is that the men of Yen may dig up the ancestral tombs outside the town, and by inflicting this indignity on our forefathers cause us to become faint-hearted.']

[Forthwith the besiegers dug up all the graves and burned the corpses lying in them. And the inhabitants of Chi-mo, witnessing the outrage from the city-walls, wept passionately and were all impatient to go out and fight, their fury being increased tenfold. T`ien Tan knew then that his soldiers were ready for any enterprise. But instead of a sword, he himself too a mattock in his hands, and ordered others to be distributed amongst his best warriors, while the ranks were filled up with their wives an d concubines. He then served out all the remaining rations and bade his men eat their fill. The regular soldiers were told to keep out of sight, and the walls were manned with the old and weaker men and with women. This done, envoys were dispatched to the enemy's camp to arrange terms of surrender, whereupon the Yen army began shouting for joy.]

[T`ien Tan also collected 20,000 ounces of silver from the people, and got the wealthy citizens of Chi-mo to send it to the Yen general with the prayer that, when the town capitulated, he would allow their homes to be plundered or their women to be maltreated. Ch`i Chieh, in high good humor, granted their prayer; but his army now became increasingly slack and careless. Meanwhile, T`ien Tan got together a thousand oxen, decked them with pieces of red silk, painted their bodies, dragon-like, with colored stripes, and fastened sharp blades on their horns and well-greased rushes on their tails. When night came on, he lighted the ends of the rushes, and drove the oxen through a number of holes which he had pierced in the walls, backing them up with a force of 5000 picked warriors. The animals, maddened with pain, dashed furiously into the enemy's camp where they caused the utmost confusion and dismay; for their tails acted as torches, showing up the hideous pattern on their bodies, and the weapons on t heir horns killed or wounded any with whom they came into contact.]

[In the meantime, the band of 5000 had crept up with gags in their mouths, and now threw themselves on the enemy. At the same moment a frightful din arose in the city itself, all those that remained behind making as much noise as possible by banging drum s and hammering on bronze vessels, until heaven and earth were convulsed by the uproar. Terror-stricken, the Yen army fled in disorder, hotly pursued by the men of Ch`i, who succeeded in slaying their general Ch`i Chien . . . The result of the battle was the ultimate recovery of some seventy cities which had belonged to the Ch`i State."]

Violent language and driving forward as if to the attack are signs that he will retreat.

When the light chariots come out first and take up a position on the wings, it is a sign that the enemy is forming for battle.

Peace proposals unaccompanied by a sworn covenant indicate a plot. [The reading here is uncertain. Li Ch`uan indicates "a treaty confirmed by oaths and hostages." Wang Hsi and Chang Yu, on the other hand, simply say "without reason," "on a frivolous pretext."]

When there is much running about [Every man hastening to his proper place under his own regimental banner.] and the soldiers fall into rank, it means that the critical moment has come.

When some are seen advancing and some retreating, it is a lure.

When the soldiers stand leaning on their spears, they are faint from want of food.

If those who are sent to draw water begin by drinking themselves, the army is suffering from thirst. [As Tu Mu remarks: "One may know the condition of a whole army from the behavior of a single man."]

If the enemy sees an advantage to be gained and makes no effort to secure it, the soldiers are exhausted.

If birds gather on any spot, it is unoccupied. [A useful fact to bear in mind when, for instance, as Ch`en Hao says, the enemy has secretly abandoned his camp.] Clamor by night betokens nervousness.

If there is disturbance in the camp, the general's authority is weak. If the banners and flags are shifted about, sedition is afoot. If the officers are angry, it means that the men are weary. [Tu Mu understands the sentence differently: "If all the officers of an army are angry with their general, it means that they are broken with fatigue" owing to the exertions which he has demanded from them.]

When an army feeds its horses with grain and kills its cattle for food, [In the ordinary course of things, the men would be fed on grain and the horses chiefly on grass.] and when the men do not hang their cooking-pots over the camp-fires, showing that they will not return to their tents, you may know that they are determined to fight to the death. [I may quote here the illustrative passage from the *Hou Han Shu*, ch. 71, given in abbreviated form by the *P`ei Wen Yun fu*: "The rebel Wang Kuo of Liang was besieging the town of Ch`en-ts`ang, and Huang-fu Sung, who was in supreme command, and Tung Cho were sent out against him. The latter pressed for hasty measures, but Sung turned a deaf ear to his counsel. At last the rebels were utterly worn out, and began to throw down their weapons of their own accord. Sung was not advancing to the attack, but Cho said: 'It is a principle of war not to pursue desperate men and not to press a retreating host.' Sung answered: 'That does not apply here. What I am about to attack is a jaded army, not a retreating host; with disciplined troops I am falling on a disorganized multitude, not a band of desperate men.' Thereupon he advances to the attack unsupported by his colleague, and routed the enemy, Wang Kuo being slain."]

The sight of men whispering together in small knots or speaking in subdued tones points to disaffection amongst the rank and file.

Too frequent rewards signify that the enemy is at the end of his resources; [Because, when an army is hard pressed, as Tu Mu says, there is always a fear of mutiny, and lavish rewards are given to keep the men in good temper.] too many punishments betray a condition of dire distress. [Because in such case discipline becomes relaxed, and unwonted severity is necessary to keep the men to their duty.]

To begin by bluster, but afterwards to take fright at the enemy's numbers, shows a supreme lack of intelligence. [I follow the interpretation of Ts`ao Kung, also adopted by Li Ch`uan, Tu Mu, and Chang Yu. Another possible meaning set forth by Tu Yu, Chia Lin, Mei Tao-ch`en and Wang Hsi, is: "The general who is first tyrannical towards his men, and then in terror lest they should mutiny, etc." This would connect the sentence with what went before about rewards and punishments.]

When envoys are sent with compliments in their mouths, it is a sign that the enemy wishes for a truce. [Tu Mu says: "If the enemy open friendly relations be sending hostages, it is a sign that they are anxious for an armistice, either because their strength is exhausted or for some other reason." But it hardly needs a Sun Tzu to draw such an obvious inference.]

If the enemy's troops march up angrily and remain facing ours for a long time without either joining battle or taking themselves off again, the situation is one that demands great vigilance and circumspection. [Ts`ao Kung says a maneuver of this sort may be only a ruse to gain time for an unexpected flank attack or the laying of an ambush.]

If our troops are no more in number than the enemy, that is amply sufficient; it only means that no direct attack can be made. [Literally, "no martial advance." That is to say, *Cheng* tactics and frontal attacks must be eschewed, and stratagem resorted to instead.]

What we can do is simply to concentrate all our available strength, keep a close watch on the enemy, and obtain reinforcements. [This is an obscure sentence, and none of the commentators succeed in squeezing very good sense out of it. I follow Li Ch`uan, who appears to offer the simplest explanation: "Only the side that gets more men will win." Fortunately we have Chang Yu to expound its meaning to us in language which is lucidity itself: "When the numbers are even, and no favorable opening presents itself, although we may not be strong enough to deliver a sustained attack, we can find additional recruits amongst our sutlers and camp-followers, and then, concentrating our forces and keeping a close watch on the enemy, contrive to snatch the victory. But we must avoid borrowing foreign soldiers to help us." He then quotes from Wei Liao Tzu, ch. 3: "The nominal strength o f mercenary troops may be 100,000, but their real value will be not more than half that figure."]

He who exercises no forethought but makes light of his opponents is sure to be captured by them. [Ch`en Hao, quoting from the *Tso Chuan*, says: "If bees and scorpions carry poison, how much more will a hostile state! Even a puny opponent, then, should not be treated with contempt."]

If soldiers are punished before they have grown attached to you, they will not prove submissive; and, unless submissive, then will be practically useless. If, when the soldiers have become attached to you, punishments are not enforced, they will still be unless.

Therefore soldiers must be treated in the first instance with humanity, but kept under control by means of iron discipline. [Yen Tzu [B.C. 493] said of Ssu-ma Jang-chu: "His civil virtues endeared him to the people; his martial prowess kept his enemies in awe." Cf. Wu Tzu, ch. 4 init.: "The ideal commander unites culture with a warlike temper; the profession of arms requires a combination of hardness and tenderness."]

This is a certain road to victory.

If in training soldiers commands are habitually enforced, the army will be well-disciplined; if not, its discipline will be bad.

If a general shows confidence in his men but always insists on his orders being obeyed, [Tu Mu says: "A general ought in time of peace to show kindly confidence in his men and also make his authority respected, so that when they come to face the enemy, orders may be executed and discipline maintained, because they all trust and look up to him." What Sun Tzu has said, however, would lead one rather to expect something like this: "If a general is always confident that his orders will be carried out," etc.] the gain will be mutual. [Chang Yu says: "The general has confidence in the men under his command, and the men are docile, having confidence in him. Thus the gain is mutual" He quotes a pregnant sentence from Wei Liao Tzu, ch. 4: "The art of giving orders is not to try to rectify minor blunders and not to be swayed by petty doubts." Vacillation and fussiness are the surest means of sapping the confidence of an army.]

Terrain

Sun Tzu said: We may distinguish six kinds of terrain, to wit:

(a) Accessible ground; [Mei Yao-ch`en says: "plentifully provided with roads and means of communications."]

(b) Entangling ground; [The same commentator says: "Net-like country, venturing into which you become entangled."]

(c) Temporizing ground; [Ground which allows you to "stave off" or "delay."]

(d) Narrow passes;

(e) Precipitous heights;

(f) Positions at a great distance from the enemy. [It is hardly necessary to point out the faultiness of this classification. A strange lack of logical perception is shown in the Chinese unquestioning acceptance of glaring cross-divisions such as the above.]

Ground which can be freely traversed by both sides is called *accessible*.

With regard to ground of this nature, be before the enemy in occupying the raised and sunny spots, and carefully guard your line of supplies. [The general meaning of the last phrase is doubtlessly, as Tu Yu says, "not to allow the enemy to cut your communications." In view of Napoleon's dictum, "the secret of war lies in the communications," we could wish that Sun Tzu had done more than skirt the edge of this important subject here and in I. Col. Henderson says: "The line of supply may be said to be as vital to the existence of an army as the heart to the life of a human being. Just as the duelist who finds his adversary's point menacing him with certain death, and his own guard astray, is compelled to conform to his adversary's movements, and to content himself with warding off his thrusts, so the commander whose communications are suddenly threatened finds himself in a false position, and he will be fortunate if he has not to change all his plans, to split up his force into more or less isolated detachments, and to fight with inferior numbers on ground which he has not had time to prepare, and where defeat will not be an ordinary failure, but will entail the ruin or surrender of his whole army."]

Then you will be able to fight with advantage.

Ground which can be abandoned but is hard to re-occupy is called *entangling*.

From a position of this sort, if the enemy is unprepared, you may sally forth and defeat him. But if the enemy is prepared for your coming, and you fail to defeat him, then, return being impossible, disaster will ensue.

When the position is such that neither side will gain by making the first move, it is called *temporizing* ground. [Tu Mu says: "Each side finds it inconvenient to move, and the situation remains at a deadlock."]

In a position of this sort, even though the enemy should offer us an attractive bait, [Tu Yu says, "turning their backs on us and pretending to flee." But this is only one of the lures which might induce us to quit our position.] it will be advisable not to stir forth, but rather to retreat, thus enticing the enemy in his turn; then, when part of his army has come out, we may deliver our attack with advantage.

With regard to *narrow passes*, if you can occupy them first, let them be strongly garrisoned and await the advent of the enemy. [Because then, as Tu Yu observes, "the initiative will lie with us, and by making sudden and unexpected attacks we shall have the enemy at our mercy."]

Should the army forestall you in occupying a pass, do not go after him if the pass is fully garrisoned, but only if it is weakly garrisoned.

With regard to *precipitous heights*, if you are beforehand with your adversary, you should occupy the raised and sunny spots, and there wait for him to come up. [Ts`ao Kung says: "The particular advantage of securing heights and defiles is that your actions cannot then be dictated by the enemy." Chang Yu tells the following anecdote of P`ei Hsing-chien (A.D. 619-682), who was sent on a punitive expedition against the Turkic tribes. "At night he pitched his camp as usual, and it had already been completely fortified by wall and ditch, when suddenly he gave orders that the army should shift its quarters to a hill near by. This was highly displeasing to his officers, who protested loudly against the extra fatigue which it would entail on the men. P`ei Hsing-chien, however, paid no heed to their remonstrances and had the camp moved as quickly as possible. The same night, a terrific storm came on, which flooded their former place of encampment to the depth of over twelve feet. The recalcitrant officers were amazed at the sight, and owned that they had been in the wrong. 'How did you know what was going to happen?' they asked. P`ei Hsing-chien replied: 'From this time forward be content to obey orders without asking unnecessary questions.' From this it may be seen," Chang Yu continues, "that high and sunny places are advantageous not only for fighting, but also because they are immune from disastrous floods."]

If the enemy has occupied them before you, do not follow him, but retreat and try to entice him away. [The turning point of Li Shih-min's campaign in 621 A.D. against the two rebels, Tou Chien-te, King of Hsia, and Wang Shih-ch`uang, Prince of Cheng, was his seizure of the heights of Wu-lao, in spike of which Tou Chien-te persisted in his attempt to relieve his ally in Lo-yang, was defeated and taken prisoner.]

If you are situated at a great distance from the enemy, and the strength of the two armies is equal, it is not easy to provoke a battle, [The point is that we must not think of undertaking a long and wearisome march, at the end of which, as Tu Yu says, "we should be exhausted and our adversary fresh and keen."] and fighting will be to your disadvantage.

These six are the principles connected with Earth. The general who has attained a responsible post must be careful to study them.

Now an army is exposed to six several calamities, not arising from natural causes, but from faults for which the general is responsible. These are: a) Flight; b) Insubordination; c) Collapse; d) Ruin; e) Disorganization; e) Rout.

Other conditions being equal, if one force is hurled against another ten times its size, the result will be the *flight* of the former.

When the common soldiers are too strong and their officers too weak, the result is *insubordination*. [Tu Mu cites the unhappy case of T`ien Pu, who was sent to Wei in 821 A.D. with orders to lead an army against Wang T`ing-ts`ou. But the whole time he was in command, his soldiers treated him with the utmost contempt, and open ly flouted his authority by riding about the camp on donkeys, several thousands at a time. T`ien Pu was powerless to put a stop to this conduct, and when, after some months had passed, he made an attempt to engage the enemy, his troops turned tail and dispersed in every direction. After that, the unfortunate man committed suicide by cutting his throat.]

When the officers are too strong and the common soldiers too weak, the result is *collapse*. [Ts`ao Kung says: "The officers are energetic and want to press on, the common soldiers are feeble and suddenly collapse."]

When the higher officers are angry and insubordinate, and on meeting the enemy give battle on their own account from a feeling of resentment, before the commander-in-chief can tell whether or no he is in a position to fight, the result is *ruin*. [Wang Hsi`s note is: "This means, the general is angry without cause, and at the same time does not appreciate the ability of his subordinate officers; thus he arouses fierce resentment and brings an avalanche of ruin upon his head."]

When the general is weak and without authority; when his orders are not clear and distinct; [Wei Liao Tzu says: "If the commander gives his orders with decision, the soldiers will not wait to hear them twice; if his moves are made without vacillation, the soldiers will not be in two minds about doing their duty." General Baden-Powell says, italicizing the words: "The secret of getting successful work out of your trained men lies in one nutshell—in the clearness of the instructions they receive." "the most fatal defect in a military leader is difference; the worst calamities that befall an army arise from hesitation."] when there are no fixes duties assigned to officers and men, [Tu Mu says: "Neither officers nor men have any regular routine."] and the ranks are formed in a slovenly haphazard manner, the result is utter *disorganization*.

When a general, unable to estimate the enemy's strength, allows an inferior force to engage a larger one, or hurls a weak detachment against a powerful one, and neglects to place picked soldiers in the front rank, the result must be *rout*. [Chang Yu paraphrases the latter part of the sentence and continues: "Whenever there is fighting to be done, the keenest spirits should be appointed to serve in the front ranks, both in order to strengthen the resolution of our own men and to demoralize the enemy."]

These are six ways of courting defeat, which must be carefully noted by the general who has attained a responsible post.

The natural formation of the country is the soldier's best ally; [Ch`en Hao says: "The advantages of weather and season are not equal to those connected with ground."] but a power of estimating the adversary, of controlling the forces of victory, and of shrewdly calculating difficulties, dangers and distances, constitutes the test of a great general.

He who knows these things, and in fighting puts his knowledge into practice, will win his battles. He who knows them not, nor practices them, will surely be defeated.

If fighting is sure to result in victory, then you must fight, even though the ruler forbid it; if fighting will not result in victory, then you must not fight even at the ruler's bidding. [Huang Shih-kung of the Ch`in dynasty, who is said to have been the patron of Chang Liang and to have written the *San Lueh*, has these words attributed to him: "The responsibility of setting an army in motion must devolve on the general alone; if advance and retreat are controlled from the Palace, brilliant results will hardly be achieved. Hence the god-like ruler and the enlightened monarch are content to play a humble part in furthering their country's cause [lit., kneel down t o push the chariot wheel]." This means that "in matters lying outside the zenana, the decision of the military commander must be absolute." Chang Yu also quote the saying: "Decrees from the Son of Heaven do not penetrate the walls of a camp."]

The general who advances without coveting fame and retreats without fearing disgrace, [It was Wellington, I think, who said that the hardest thing of all for a soldier is to retreat.] whose only thought is to protect his country and do good service for his sovereign, is the jewel of the kingdom. [A noble presentiment, in few words, of the Chinese "happy warrior." Such a man, says Ho Shih, "even if he had to suffer punishment, would not regret his conduct."]

Regard your soldiers as your children, and they will follow you into the deepest valleys; look upon them as your own beloved sons, and they will stand by you even unto death. [In this connection, Tu Mu draws for us an engaging picture of the famous general Wu Ch`i, from whose treatise on war I have frequently had occasion to quote: "He wore the same clothes and ate the same food as the meanest of his soldier s, refused to have either a horse to ride or a mat to sleep on, carried his own surplus rations wrapped in a parcel, and shared every hardship with his men. One of his soldiers was suffering from an abscess, and Wu Ch`i himself sucked out the virus. Th e soldier's mother, hearing this, began wailing and lamenting. Somebody asked her, saying: 'Why do you cry? Your son is only a common soldier, and yet the commander-in-chief himself has sucked the poison from his sore.' The woman replied, 'Many years ago, Lord Wu performed a similar service for my husband, who never left him afterwards, and finally met his death at the hands of the enemy. And now that he has done the same for my son, he too will fall fighting I know not where.'" Li Ch`uan mentions the Viscount of Ch`u, who invaded the small state of Hsiao during the winter. The Duke of Shen said to him: "Many of the soldiers are suffering severely from the cold." So he made a round of the whole army, comforting and encouraging the men; and straightway they felt as if they were clothed in garments lined with floss silk.]

If, however, you are indulgent, but unable to make your authority felt; kind-hearted, but unable to enforce your commands; and incapable, moreover, of quelling disorder: then your soldiers must be likened to spoilt children; they are useless for any practical purpose. [Li Ching once said that if you could make your soldiers afraid of you, they would not be afraid of the enemy. Tu Mu recalls an instance of stern military discipline which occurred in 219 A.D., when Lu Meng was occupying the town of Chiang-ling. He had given stringent orders to his army not to molest the inhabitants nor take anything from them by force. Nevertheless, a certain officer serving under his banner, who happened to be a fellow-townsman, ventured to appropriate a bamboo hat belonging to one of the people, in order to wear it over his regulation helmet as a protection against the rain. Lu Meng considered that the fact of his being also a native of Ju-nan should not be allowed to palliate a clear breach of discipline, and accordingly he order ed his summary execution, the tears rolling down his face, however, as he did so. This act of severity filled the army with wholesome awe, and from that time forth even articles dropped in the highway were not picked up.]

If we know that our own men are in a condition to attack, but are unaware that the enemy is not open to attack, we have gone only halfway towards victory. [That is, Ts`ao Kung says, "the issue in this case is uncertain."]

If we know that the enemy is open to attack, and also know that our men are in a condition to attack, but are unaware that the nature of the ground makes fighting impracticable, we have still gone only halfway towards victory.

Hence the experienced soldier, once in motion, is never bewildered; once he has broken camp, he is never at a loss. [The reason being, according to Tu Mu, that he has taken his measures so thoroughly as to ensure victory beforehand. "He does not move recklessly," says Chang Yu, "so that when he does move, he makes no mistakes."]

Hence the saying: If you know the enemy and know yourself, your victory will not stand in doubt; if you know Heaven and know Earth, you may make your victory complete. [Li Ch`uan sums up as follows: "Given a knowledge of three things—the affairs of men, the seasons of heaven and the natural advantages of earth—, victory will invariably crown your battles."]

The Nine Situations

Sun Tzu said: The art of war recognizes nine varieties of ground:

(1) Dispersive ground — When a chieftain is fighting in his own territory, it is dispersive ground; (2) Facile ground — When he has penetrated into hostile territory, but to no great distance, it is facile ground; (3) Contentious ground — Ground the possession of which imports great advantage to either side, is contentious ground; (4) Open ground — Ground on which each side has liberty of movement is open ground; (5) Ground of intersecting highways — Ground which forms the key to three contiguous states, so that he who occupies it first has most of the Empire at his command, is a ground of intersecting highways; (6) Serious ground — When an army has penetrated into the heart of a hostile country, leaving a number of fortified cities in its rear, it is serious ground; (7) Difficult ground — Mountain forests, rugged steeps, marshes and fens—all country that is hard to traverse: this is difficult ground; (8) Hemmed-in ground — Ground which is reached through narrow gorges, and from which we can only retire by tortuous paths, so that a small number of the enemy would suffice to crush a large body of our men: this is hemmed in ground; (9) Desperate ground — Ground on which we can only be saved from destruction by fighting without delay, is desperate ground.

On dispersive ground, therefore, fight not. On facile ground, halt not. On contentious ground, attack not. On open ground, do not try to block the enemy's way. On the ground of intersecting highways, join hands with your allies. On serious ground, gather in plunder. In difficult ground, keep steadily on the march. On hemmed-in ground, resort to stratagem. On desperate ground, fight.

Those who were called skillful leaders of old knew how to drive a wedge between the enemy's front and rear; to prevent co-operation between his large and small divisions; to hinder the good troops from rescuing the bad, the officers from rallying their me n. When the enemy's men were united, they managed to keep them in disorder. When it was to their advantage, they made a forward move; when otherwise, they stopped still.

If asked how to cope with a great host of the enemy in orderly array and on the point of marching to the attack, I should say: "Begin by seizing something which your opponent holds dear; then he will be amenable to your will." Rapidity is the essence of war: take advantage of the enemy's unreadiness, make your way by unexpected routes, and attack unguarded spots.

The following are the principles to be observed by an invading force: The further you penetrate into a country, the greater will be the solidarity of your troops, and thus the defenders will not prevail against you.

(1) Make forays in fertile country in order to supply your army with food; (2) Carefully study the well-being of your men, and do not overtax them. Concentrate your energy and hoard your strength. Keep your army continually on the move, and devise unfathomable plans; (3) Throw your soldiers into positions whence there is no escape, and they will prefer death to flight. If they will face death, there is nothing they may not achieve. Officers and men alike will put forth their uttermost strength; (4) Soldiers when in desperate straits lose the sense of fear. If there is no place of refuge, they will stand firm. If they are in hostile country, they will show a stubborn front. If there is no help for it, they will fight hard.

Thus, without waiting to be marshaled, the soldiers will be constantly on the qui vive; without waiting to be asked, they will do your will; without restrictions, they will be faithful; without giving orders, they can be trusted. Prohibit the taking of omens, and do away with superstitious doubts. Then, until death itself comes, no calamity need be feared.

If our soldiers are not overburdened with money, it is not because they have a distaste for riches; if their lives are not unduly long, it is not because they are disinclined to longevity.

On the day they are ordered out to battle, your soldiers may weep, those sitting up bedewing their garments, and those lying down letting the tears run down their cheeks. But let them once be brought to bay, and they will display the courage of a Chu or a Kuei.

The skillful tactician may be likened to the shuai-jan. Now the shuai-jan is a snake that is found in the ChUng mountains. Strike at its head, and you will be attacked by its tail; strike at its tail, and you will be attacked by its head; strike at its middle, and you will be attacked by head and tail both. Asked if an army can be made to imitate the shuai-jan, I should answer, Yes. For the men of Wu and the men of Yueh are enemies; yet if they are crossing a river in the same boat and are caught by a storm, they will come to each other's assistance just as the left hand helps the right.

Hence it is not enough to put one's trust in the tethering of horses, and the burying of chariot wheels in the ground.

The principle on which to manage an army is to set up one standard of courage which all must reach. How to make the best of both strong and weak—that is a question involving the proper use of ground. Thus the skillful general conducts his army just as though he were leading a single man, willy-nilly, by the hand.

It is the business of a general to be quiet and thus ensure secrecy; upright and just, and thus maintain order. He must be able to mystify his officers and men by false reports and appearances, and thus keep them in total ignorance.

By altering his arrangements and changing his plans, he keeps the enemy without definite knowledge. By shifting his camp and taking circuitous routes, he prevents the enemy from anticipating his purpose.

At the critical moment, the leader of an army acts like one who has climbed up a height and then kicks away the ladder behind him. He carries his men deep into hostile territory before he shows his hand.

He burns his boats and breaks his cooking-pots; like a shepherd driving a flock of sheep, he drives his men this way and that, and nothing knows whither he is going.

To muster his host and bring it into danger:—this may be termed the business of the general.

The different measures suited to the nine varieties of ground; the expediency of aggressive or defensive tactics; and the fundamental laws of human nature: these are things that must most certainly be studied.

When invading hostile territory, the general principle is, that penetrating deeply brings cohesion; penetrating but a short way means dispersion.

When you leave your own country behind, and take your army across neighborhood territory, you find yourself on critical ground. When there are means of communication on all four sides, the ground is one of intersecting highways.

When you penetrate deeply into a country, it is serious ground. When you penetrate but a little way, it is facile ground.

When you have the enemy's strongholds on your rear, and narrow passes in front, it is hemmed-in ground. When there is no place of refuge at all, it is desperate ground.

Therefore, on dispersive ground, I would inspire my men with unity of purpose. On facile ground, I would see that there is close connection between all parts of my army.

On contentious ground, I would hurry up my rear.

On open ground, I would keep a vigilant eye on my defenses. On ground of intersecting highways, I would consolidate my alliances.

On serious ground, I would try to ensure a continuous stream of supplies. On difficult ground, I would keep pushing on along the road.

On hemmed-in ground, I would block any way of retreat. On desperate ground, I would proclaim to my soldiers the hopelessness of saving their lives.

For it is the soldier's disposition to offer an obstinate resistance when surrounded, to fight hard when he cannot help himself, and to obey promptly when he has fallen into danger.

We cannot enter into alliance with neighboring princes until we are acquainted with their designs. We are not fit to lead an army on the march unless we are familiar with the face of the country—its mountains and forests, its pitfalls and precipices, its marshes and swamps. We shall be unable to turn natural advantages to account unless we make use of local guides.

To be ignored of any one of the following four or five principles does not befit a warlike prince.

When a warlike prince attacks a powerful state, his generalship shows itself in preventing the concentration of the enemy's forces. He overawes his opponents, and their allies are prevented from joining against him.

Hence he does not strive to ally himself with all and sundry, nor does he foster the power of other states. He carries out his own secret designs, keeping his antagonists in awe. Thus he is able to capture their cities and overthrow their kingdoms.

Bestow rewards without regard to rule, issue orders without regard to previous arrangements; and you will be able to handle a whole army as though you had to do with but a single man.

Confront your soldiers with the deed itself; never let them know your design. When the outlook is bright, bring it before their eyes; but tell them nothing when the situation is gloomy.

Place your army in deadly peril, and it will survive; plunge it into desperate straits, and it will come off in safety.

For it is precisely when a force has fallen into harm's way that is capable of striking a blow for victory.

Success in warfare is gained by carefully accommodating ourselves to the enemy's purpose.

By persistently hanging on the enemy's flank, we shall succeed in the long run in killing the commander-in-chief.

This is called ability to accomplish a thing by sheer cunning.

On the day that you take up your command, block the frontier passes, destroy the official tallies, and stop the passage of all emissaries.

Be stern in the council-chamber, so that you may control the situation.

If the enemy leaves a door open, you must rush in.

Forestall your opponent by seizing what he holds dear, and subtly contrive to time his arrival on the ground.

Walk in the path defined by rule, and accommodate yourself to the enemy until you can fight a decisive battle.

At first, then, exhibit the coyness of a maiden, until the enemy gives you an opening; afterwards emulate the rapidity of a running hare, and it will be too late for the enemy to oppose you.

The Attack by Fire

Sun Tzu said: There are five ways of attacking with fire. The first is to burn soldiers in their camp; the second is to burn stores; the third is to burn baggage trains; the fourth is to burn arsenals and magazines; the fifth is to hurl dropping fi re amongst the opponent.

In order to carry out an attack, we must have means available. The material for raising fire should always be kept in readiness. There is a proper season for making attacks with fire, and special days for starting a conflagration. The proper season is when the weather is very dry; the special days are those when the moon is in the constellations of the Sieve, the Wall, the Wing or the Cross-bar; for these four are all days of rising wind.

In attacking with fire, one should be prepared to meet five possible developments: (1) When fire breaks out inside to opponent's camp, respond at once with an attack from without; (2) If there is an outbreak of fire, but the opponent's soldiers remain qui et, bide your time and do not attack; (3) When the force of the flames has reached its height, follow it up with an attack, if that is practicable; if not, stay where you are; (4) If it is possible to make an assault with fire from without, do not wait fo r it to break out within, but deliver your attack at a favorable moment; (5) When you start a fire, be to windward of it. Do not attack from the leeward.

A wind that rises in the daytime lasts long, but a night breeze soon falls. In every army, the five developments connected with fire must be known, the movements of the stars calculated, and a watch kept for the proper days. Hence those who use fire as an aid to the attack show intelligence; those who use water as an aid to the attack gain an accession of strength. By means of water, an opponent may be intercepted, but not robbed of all his belongings.

Unhappy is the fate of one who tries to win his battles and succeed in his attacks without cultivating the spirit of enterprise; for the result is waste of time and general stagnation. Hence the saying: The enlightened ruler lays his plans well ahead; the good general cultivates his resources.

Move not unless you see an advantage; use not your troops unless there is something to be gained; fight not unless the position is critical. If it is to your advantage, make a forward move; if not, stay where you are. Anger may in time change to gladness; vexation may be succeeded by content.

No leader should put troops into the field merely to gratify his own spleen; no leader should fight a battle simply out of pique. But a kingdom that has once been destroyed can never come again into being; nor can the dead ever be brought back to life. Hence the enlightened leader is heedful, and the good leader full of caution.

The Use of Spies

Sun Tzu said: Raising a host of a hundred thousand men and engaging them in war entails heavy loss on the people and a drain on the resources. The daily expenditure will amount to a thousand ounces of silver. There will be commotion at home and abroad, and men will drop out exhausted.

Opposing forces may face each other for years, striving for the victory which may be decided in a single day. This being so, to remain in ignorance of the enemy's condition simply because one grudges the outlay of a hundred ounces of silver is the height of stupidity.

One who acts thus is no leader of men, no present help to his cause, no master of victory. Thus, what enables the wise commander to strike and conquer, and achieve things beyond the reach of ordinary men, is foreknowledge. Now this foreknowledge cannot be elicited from spirits; it cannot be obtained inductively from experience, nor by any deductive calculation. Knowledge of the enemy's dispositions can only be obtained from other men.

Hence the use of spies, of whom there are five classes: (1) Local spies — Having local spies means employing the services of the inhabitants of an enemy territory; (2) Moles — Having moles means making use of officials of the enemy; (3) Double agents — Having double agents means getting hold of the enemy's spies and using them for our own purposes; (4) Doomed spies — Having doomed spies means doing certain things openly for purposes of deception, and allowing our spies to know of them and report them to t he enemy; (5) Surviving spies — Surviving spies means are those who bring back news from the enemy's camp.

When these five kinds of spy are all at work, none can discover the secret system. This is called "divine manipulation of the threads." It is the commander's most precious faculty. Hence it is that which none in the whole army are more intimate relations to be maintained than with spies. None should be more liberally rewarded. In no other fields should greater secrecy be preserved.

(1) Spies cannot be usefully employed without a certain intuitive sagacity; (2) They cannot be properly managed without benevolence and straight forwardness; (3) Without subtle ingenuity of mind, one cannot make certain of the truth of their reports; (4) Be subtle! be subtle! and use your spies for every kind of warfare; (5) If a secret piece of news is divulged by a spy before the time is ripe, he must be put to death together with the man to whom the secret was told.

Whether the object be to crush an enemy, to storm a territory, or to kill an enemy general, it is always necessary to begin by finding out the names of the attendants, the aides-de-camp, and door-keepers and sentries of the general in command. Our spies must be commissioned to ascertain these.

The enemy's spies who have come to spy on us must be sought out, tempted with bribes, led away and comfortably housed. Thus they will become double agents and available for our service. It is through the information brought by the double agent that we are able to acquire and employ local and inward spies. It is owing to his information, again, that we can cause the doomed spy to carry false tidings to the enemy.

Lastly, it is by his information that the surviving spy can be used on appointed occasions. The end and aim of spying in all its five varieties is knowledge of the enemy; and this knowledge can only be derived, in the first instance, from the double agent . Hence it is essential that the double agent be treated with the utmost liberality.

Hence it is only the enlightened and wise general who will use the highest intelligence of the army for purposes of spying and thereby they achieve great results. Spies are the most important asset, because on them depends an army's ability to march.

On War

by General Carl von Clausewitz

Translated by Colonel J.J. Graham
Introduction and Notes by Colonel F.N. Maude

Table of Contents

Book IV: The Combat

Introduction

The Germans interpret their new national colors— black, red, and white— by the saying, "Durch Nacht und Blut zur licht." ("Through night and blood to light"), and no work yet written conveys to the thinker a clearer conception of all that the red streak in their flag stands for than this deep and philosophical analysis of "War" by Clausewitz.

It reveals "War," stripped of all accessories, as the exercise of force for the attainment of a political object, unrestrained by any law save that of expediency, and thus gives the key to the interpretation of German political aims, past, present, and future, which is unconditionally necessary for every student of the modern conditions of Europe. Step by step, every event since Waterloo follows with logical consistency from the teachings of Napoleon, formulated for the first time, some twenty years afterwards, by this remarkable thinker.

What Darwin accomplished for Biology generally Clausewitz did for the Life-History of Nations nearly half a century before him, for both have proved the existence of the same law in each case, viz., "The survival of the fittest"—the "fittest," as Huxley long since pointed out, not being necessarily synonymous with the ethically "best." Neither of these thinkers was concerned with the ethics of the struggle which each studied so exhaustively, but to both men the phase or condition presented itself neither as moral nor immoral, any more than are famine, disease, or other natural phenomena, but as emanating from a force inherent in all living organisms which can only be mastered by understanding its nature. It is in that spirit that, one after the other, all the Nations of the Continent, taught by such drastic lessons as Koniggrätz and Sedan, have accepted the lesson, with the result that to-day Europe is an armed camp, and peace is maintained by the equilibrium of forces, and will continue just as long as this equilibrium exists, and no longer.

Whether this state of equilibrium is in itself a good or desirable thing may be open to argument. I have discussed it at length in my "War and the World's Life"; but I venture to suggest that to no one would a renewal of the era of warfare be a change for the better, as far as existing humanity is concerned. Meanwhile, however, with every year that elapses the forces at present in equilibrium are changing in magnitude—the pressure of populations which have to be fed is rising, and an explosion along the line of least resistance is, sooner or later, inevitable.

As I read the teaching of the recent Hague Conference, no responsible Government on the Continent is anxious to form in themselves that line of least resistance; they know only too well what War would mean; and we alone, absolutely unconscious of the trend of the dominant thought of Europe, are pulling down the dam which may at any moment let in on us the flood of invasion.

Now no responsible man in Europe, perhaps least of all in Germany, thanks us for this voluntary destruction of our defences, for all who are of any importance would very much rather end their days in peace than incur the burden of responsibility which War would entail. But they realise that the gradual dissemination of the principles taught by Clausewitz has created a condition of molecular tension in the minds of the Nations they govern analogous to the "critical temperature of water heated above boiling-point under pressure," which may at any moment bring about an explosion which they will be powerless to control.

The case is identical with that of an ordinary steam boiler, delivering so and so many pounds of steam to its engines as long as the envelope can contain the pressure; but let a breach in its continuity arise—relieving the boiling water of all restraint—and in a moment the whole mass flashes into vapour, developing a power no work of man can oppose.

The ultimate consequences of defeat no man can foretell. The only way to avert them is to ensure victory; and, again following out the principles of Clausewitz, victory can only be ensured by the creation in peace of an organisation which will bring every available man, horse, and gun (or ship and gun, if the war be on the sea) in the shortest possible time, and with the utmost possible momentum, upon the decisive field of action—which in turn leads to the final doctrine formulated by Von der Goltz in excuse for the action of the late President Kruger in 1899:

"The Statesman who, knowing his instrument to be ready, and seeing War inevitable, hesitates to strike first is guilty of a crime against his country."

It is because this sequence of cause and effect is absolutely unknown to our Members of Parliament, elected by popular representation, that all our efforts to ensure a lasting peace by securing efficiency with economy in our National Defences have been rendered nugatory.

This estimate of the influence of Clausewitz's sentiments on contemporary thought in Continental Europe may appear exaggerated to those who have not familiarised themselves with M. Gustav de Bon's exposition of the laws governing the formation and conduct of crowds I do not wish for one minute to be understood as asserting that Clausewitz has been conscientiously studied and understood in any Army, not even in the Prussian, but his work has been the ultimate foundation on which every drill regulation in Europe, except our own, has been reared. It is this ceaseless repetition of his fundamental ideas to which one-half of the male population of every Continental Nation has been subjected for two to three years of their lives, which has tuned their minds to vibrate in harmony with his precepts, and those who know and appreciate this fact at its true value have only to strike the necessary chords in order to evoke a response sufficient to overpower any other ethical conception which those who have not organised their forces beforehand can appeal to.

The recent set-back experienced by the Socialists in Germany is an illustration of my position. The Socialist leaders of that country are far behind the responsible Governors in their knowledge of the management of crowds. The latter had long before (in 1893, in fact) made their arrangements to prevent the spread of Socialistic propaganda beyond certain useful limits. As long as the Socialists only threatened capital they were not seriously interfered with, for the Government knew quite well that the undisputed sway of the employer was not for the ultimate good of the State. The standard of comfort must not be pitched too low if men are to be ready to die for their country. But the moment the Socialists began to interfere seriously with the discipline of the Army the word went round, and the Socialists lost heavily at the polls.

If this power of predetermined reaction to acquired ideas can be evoked successfully in a matter of internal interest only, in which the "obvious interest" of the vast majority of the population is so clearly on the side of the Socialist, it must be evident how enormously greater it will prove when set in motion against an external enemy, where the "obvious interest" of the people is, from the very nature of things, as manifestly on the side of the Government; and the Statesman who failed to take into account the force of the "resultant thought wave" of a crowd of some seven million men, all trained to respond to their ruler's call, would be guilty of treachery as grave as one who failed to strike when he knew the Army to be ready for immediate action.

As already pointed out, it is to the spread of Clausewitz's ideas that the present state of more or less immediate readiness for war of all European Armies is due, and since the organisation of these forces is uniform this "more or less" of readiness exists in precise proportion to the sense of duty which animates the several Armies. Where the spirit of duty and self-sacrifice is low the troops are unready and inefficient; where, as in Prussia, these qualities, by the training of a whole century, have become instinctive, troops really are ready to the last button, and might be poured down upon any one of her neighbours with such rapidity that the very first collision must suffice to ensure ultimate success—a success by no means certain if the enemy, whoever he may be, is allowed breathing-time in which to set his house in order.

An example will make this clearer. In 1887 Germany was on the very verge of War with France and Russia. At that moment her superior efficiency, the consequence of this inborn sense of duty—surely one of the highest qualities of humanity—was so great that it is more than probable that less than six weeks would have sufficed to bring the French to their knees. Indeed, after the first fortnight it would have been possible to begin transferring troops from the Rhine to the Niemen; and the same case may arise again. But if France and Russia had been allowed even ten days' warning the German plan would have been completely defeated. France alone might then have claimed all the efforts that Germany could have put forth to defeat her.

Yet there are politicians in England so grossly ignorant of the German reading of the Napoleonic lessons that they expect that Nation to sacrifice the enormous advantage they have prepared by a whole century of self-sacrifice and practical patriotism by an appeal to a Court of Arbitration, and the further delays which must arise by going through the medieaeval formalities of recalling Ambassadors and exchanging ultimatums.

Most of our present-day politicians have made their money in business—a "form of human competition greatly resembling War," to paraphrase Clausewitz. Did they, when in the throes of such competition, send formal notice to their rivals of their plans to get the better of them in commerce? Did Mr. Carnegie, the arch-priest of Peace at any price, when he built up the Steel Trust, notify his competitors when and how he proposed to strike the blows which successively made him master of millions? Surely the Directors of a Great Nation may consider the interests of their shareholders—i.e., the people they govern—as sufficiently serious not to be endangered by the deliberate sacrifice of the preponderant position of readiness which generations of self-devotion, patriotism and wise forethought have won for them?

As regards the strictly military side of this work, though the recent researches of the French General Staff into the records and documents of the Napoleonic period have shown conclusively that Clausewitz had never grasped the essential point of the Great Emperor's strategic method, yet it is admitted that he has completely fathomed the spirit which gave life to the form; and notwithstandingthe variations in application which have resulted from the progress of invention in every field of national activity (not in the technical improvements in armament alone), this spirit still remains the essential factor in the whole matter. Indeed, if anything, modern appliances have intensified its importance, for though, with equal armaments on both sides, the form of battles must always remain the same, the facility and certainty of combination which better methods of communicating orders and intelligence have conferred upon the Commanders has rendered the control of great masses immeasurably more certain than it was in the past.

Men kill each other at greater distances, it is true—but killing is a constant factor in all battles. The difference between "now and then" lies in this, that, thanks to the enormous increase in range (the essential feature in modern armaments), it is possible to concentrate by surprise, on any chosen spot, a man-killing power fully twentyfold greater than was conceivable in the days of Waterloo; and whereas in Napoleon's time this concentration of man-killing power (which in his hands took the form of the great case-shot attack) depended almost entirely on the shape and condition of the ground, which might or might not be favourable, nowadays such concentration of fire-power is almost independent of the country altogether.

Thus, at Waterloo, Napoleon was compelled to wait till the ground became firm enough for his guns to gallop over; nowadays every gun at his disposal, and five times that number had he possessed them, might have opened on any point in the British position he had selected, as soon as it became light enough to see.

Or, to take a more modern instance, viz., the battle of St. Privat-Gravelotte, August 18, 1870, where the Germans were able to concentrate on both wings batteries of two hundred guns and upwards, it would have been practically impossible, owing to the section of the slopes of the French position, to carry out the old-fashioned case-shot attack at all. Nowadays there would be no difficulty in turning on the fire of two thousand guns on any point of the position, and switching this fire up and down the line like water from a fire-engine hose, if the occasion demanded such concentration.

But these alterations in method make no difference in the truth of the picture of War which Clausewitz presents, with which every soldier, and above all every Leader, should be saturated.

Death, wounds, suffering, and privation remain the same, whatever the weapons employed, and their reaction on the ultimate nature of man is the same now as in the struggle a century ago. It is this reaction that the Great Commander has to understand and prepare himself to control; and the task becomes ever greater as, fortunately for humanity, the opportunities for gathering experience become more rare.

In the end, and with every improvement in science, the result depends more and more on the character of the Leader and his power of resisting "the sensuous impressions of the battlefield." Finally, for those who would fit themselves in advance for such responsibility, I know of no more inspiring advice than that given by Krishna to Arjuna ages ago, when the latter trembled before the awful responsibility of launching his Army against the hosts of the Pandav's:

This Life within all living things, my Prince, Hides beyond harm. Scorn thou to suffer, then, For that which cannot suffer. Do thy part! Be mindful of thy name, and tremble not. Nought better can betide a martial soul Than lawful war. Happy the warrior To whom comes joy of battle.... . . . . But if thou shunn'st This honourable field—a Kshittriya— If, knowing thy duty and thy task, thou bidd'st Duty and task go by—that shall be sin! And those to come shall speak thee infamy From age to age. But infamy is worse For men of noble blood to bear than death! . . . . . . Therefore arise, thou Son of Kunti! Brace Thine arm for conflict; nerve thy heart to meet, As things alike to thee, pleasure or pain, Profit or ruin, victory or defeat. So minded, gird thee to the fight, for so Thou shalt not sin!

Col. F. N. Maude, C.B., late R.E.

Preface to the First Edition

It will naturally excite surprise that a preface by a female hand should accompany a work on such a subject as the present. For my friends no explanation of the circumstance is required; but I hope by a simple relation of the cause to clear myself of the appearance of presumption in the eyes also of those to whom I am not known.

The work to which these lines serve as a preface occupied almost entirely the last twelve years of the life of my inexpressibly beloved husband, who has unfortunately been torn too soon from myself and his country. To complete it was his most earnest desire; but it was not his intention that it should be published during his life; and if I tried to persuade him to alter that intention, he often answered, half in jest, but also, perhaps, half in a foreboding of early death: "Thou shalt publish it." These words (which in those happy days often drew tears from me, little as I was inclined to attach a serious meaning to them) make it now, in the opinion of my friends, a duty incumbent on me to introduce the posthumous works of my beloved husband, with a few prefatory lines from myself; and although here may be a difference of opinion on this point, still I am sure there will be no mistake as to the feeling which has prompted me to overcome the timidity which makes any such appearance, even in a subordinate part, so difficult for a woman.

It will be understood, as a matter of course, that I cannot have the most remote intention of considering myself as the real editress of a work which is far above the scope of my capacity: I only stand at its side as an affectionate companion on its entrance into the world. This position I may well claim, as a similar one was allowed me during its formation and progress. Those who are acquainted with our happy married life, and know how we shared everything with each other—not only joy and sorrow, but also every occupation, every interest of daily life—will understand that my beloved husband could not be occupied on a work of this kind without its being known to me. Therefore, no one can like me bear testimony to the zeal, to the love with which he laboured on it, to the hopes which he bound up with it, as well as the manner and time of its elaboration. His richly gifted mind had from his early youth longed for light and truth, and, varied as were his talents, still he had chiefly directed his reflections to the science of war, to which the duties of his profession called him, and which are of such importance for the benefit of States. Scharnhorst was the first to lead him into the right road, and his subsequent appointment in 1810 as Instructor at the General War School, as well as the honour conferred on him at the same time of giving military instruction to H.R.H. the Crown Prince, tended further to give his investigations and studies that direction, and to lead him to put down in writing whatever conclusions he arrived at. A paper with which he finished the instruction of H.R.H. the Crown Prince contains the germ of his subsequent works. But it was in the year 1816, at Coblentz, that he first devoted himself again to scientific labours, and to collecting the fruits which his rich experience in those four eventful years had brought to maturity. He wrote down his views, in the first place, in short essays, only loosely connected with each other. The following, without date, which has been found amongst his papers, seems to belong to those early days.

"In the principles here committed to paper, in my opinion, the chief things which compose Strategy, as it is called, are touched upon. I looked upon them only as materials, and had just got to such a length towards the moulding them into a whole.

"These materials have been amassed without any regularly preconceived plan. My view was at first, without regard to system and strict connection, to put down the results of my reflections upon the most important points in quite brief, precise, compact propositions. The manner in which Montesquieu has treated his subject floated before me in idea. I thought that concise, sententious chapters, which I proposed at first to call grains, would attract the attention of the intelligent just as much by that which was to be developed from them, as by that which they contained in themselves. I had, therefore, before me in idea, intelligent readers already acquainted with the subject. But my nature, which always impels me to development and systematising, at last worked its way out also in this instance. For some time I was able to confine myself to extracting only the most important results from the essays, which, to attain clearness and conviction in my own mind, I wrote upon different subjects, to concentrating in that manner their spirit in a small compass; but afterwards my peculiarity gained ascendency completely—I have developed what I could, and thus naturally have supposed a reader not yet acquainted with the subject.

"The more I advanced with the work, and the more I yielded to the spirit of investigation, so much the more I was also led to system; and thus, then, chapter after chapter has been inserted.

"My ultimate view has now been to go through the whole once more, to establish by further explanation much of the earlier treatises, and perhaps to condense into results many analyses on the later ones, and thus to make a moderate whole out of it, forming a small octavo volume. But it was my wish also in this to avoid everything common, everything that is plain of itself, that has been said a hundred times, and is generally accepted; for my ambition was to write a book that would not be forgotten in two or three years, and which any one interested in the subject would at all events take up more than once."

In Coblentz, where he was much occupied with duty, he could only give occasional hours to his private studies. It was not until 1818, after his appointment as Director of the General Academy of War at Berlin, that he had the leisure to expand his work, and enrich it from the history of modern wars. This leisure also reconciled him to his new avocation, which, in other respects, was not satisfactory to him, as, according to the existing organisation of the Academy, the scientific part of the course is not under the Director, but conducted by a Board of Studies. Free as he was from all petty vanity, from every feeling of restless, egotistical ambition, still he felt a desire to be really useful, and not to leave inactive the abilities with which God had endowed him. In active life he was not in a position in which this longing could be satisfied, and he had little hope of attaining to any such position: his whole energies were therefore directed upon the domain of science, and the benefit which he hoped to lay the foundation of by his work was the object of his life. That, notwithstanding this, the resolution not to let the work appear until after his death became more confirmed is the best proof that no vain, paltry longing for praise and distinction, no particle of egotistical views, was mixed up with this noble aspiration for great and lasting usefulness.

Thus he worked diligently on, until, in the spring of 1830, he was appointed to the artillery, and his energies were called into activity in such a different sphere, and to such a high degree, that he was obliged, for the moment at least, to give up all literary work. He then put his papers in order, sealed up the separate packets, labelled them, and took sorrowful leave of this employment which he loved so much. He was sent to Breslau in August of the same year, as Chief of the Second Artillery District, but in December recalled to Berlin, and appointed Chief of the Staff to Field-Marshal Count Gneisenau (for the term of his command). In March 1831, he accompanied his revered Commander to Posen. When he returned from there to Breslau in November after the melancholy event which had taken place, he hoped to resume his work and perhaps complete it in the course of the winter. The Almighty has willed it should be otherwise. On the 7th November he returned to Breslau; on the 16th he was no more; and the packets sealed by himself were not opened until after his death.

The papers thus left are those now made public in the following volumes, exactly in the condition in which they were found, without a word being added or erased. Still, however, there was much to do before publication, in the way of putting them in order and consulting about them; and I am deeply indebted to several sincere friends for the assistance they have afforded me, particularly Major O'Etzel, who kindly undertook the correction of the Press, as well as the preparation of the maps to accompany the historical parts of the work. I must also mention my much-loved brother, who was my support in the hour of my misfortune, and who has also done much for me in respect of these papers; amongst other things, by carefully examining and putting them in order, he found the commencement of the revision which my dear husband wrote in the year 1827, and mentions in the Notice hereafter annexed as a work he had in view. This revision has been inserted in the place intended for it in the first book (for it does not go any further).

There are still many other friends to whom I might offer my thanks for their advice, for the sympathy and friendship which they have shown me; but if I do not name them all, they will, I am sure, not have any doubts of my sincere gratitude. It is all the greater, from my firm conviction that all they have done was not only on my own account, but for the friend whom God has thus called away from them so soon.

If I have been highly blessed as the wife of such a man during one and twenty years, so am I still, notwithstanding my irreparable loss, by the treasure of my recollections and of my hopes, by the rich legacy of sympathy and friendship which I owe the beloved departed, by the elevating feeling which I experience at seeing his rare worth so generally and honourably acknowledged.

The trust confided to me by a Royal Couple is a fresh benefit for which I have to thank the Almighty, as it opens to me an honourable occupation, to which I devote myself. May this occupation be blessed, and may the dear little Prince who is now entrusted to my care, some day read this book, and be animated by it to deeds like those of his glorious ancestors.

Written at the Marble Palace, Potsdam, 30th June, 1832.

Marie von Clausewitz,

Born Countess Bruhl, Oberhofmeisterinn to H.R.H. the Princess William.

Notice

I look upon the first six books, of which a fair copy has now been made, as only a mass which is still in a manner without form, and which has yet to be again revised. In this revision the two kinds of War will be everywhere kept more distinctly in view, by which all ideas will acquire a clearer meaning, a more precise direction, and a closer application. The two kinds of War are, first, those in which the object is the *overthrow of the enemy*, whether it be that we aim at his destruction, politically, or merely at disarming him and forcing him to conclude peace on our terms; and next, those in which our object is *Merely to make some conquests on the frontiers of his country*, either for the purpose of retaining them permanently, or of turning them to account as matter of exchange in the settlement of a peace. Transition from one kind to the other must certainly continue to exist, but the completely different nature of the tendencies of the two must everywhere appear, and must separate from each other things which are incompatible.

Besides establishing this real difference in Wars, another practically necessary point of view must at the same time be established, which is, that *war is only a continuation of state policy by other means*. This point of view being adhered to everywhere, will introduce much more unity into the consideration of the subject, and things will be more easily disentangled from each other. Although the chief application of this point of view does not commence until we get to the eighth book, still it must be completely developed in the first book, and also lend assistance throughout the revision of the first six books. Through such a revision the first six books will get rid of a good deal of dross, many rents and chasms will be closed up, and much that is of a general nature will be transformed into distinct conceptions and forms.

The seventh book—on attack—for the different chapters of which sketches are already made, is to be considered as a reflection of the sixth, and must be completed at once, according to the above-mentioned more distinct points of view, so that it will require no fresh revision, but rather may serve as a model in the revision of the first six books.

For the eighth book—on the Plan of a War, that is, of the organisation of a whole War in general—several chapters are designed, but they are not at all to be regarded as real materials, they are merely a track, roughly cleared, as it were, through the mass, in order by that means to ascertain the points of most importance. They have answered this object, and I propose, on finishing the seventh book, to proceed at once to the working out of the eighth, where the two points of view above mentioned will be chiefly affirmed, by which everything will be simplified, and at the same time have a spirit breathed into it. I hope in this book to iron out many creases in the heads of strategists and statesmen, and at least to show the object of action, and the real point to be considered in War.

Now, when I have brought my ideas clearly out by finishing this eighth book, and have properly established the leading features of War, it will be easier for me to carry the spirit of these ideas in to the first six books, and to make these same features show themselves everywhere. Therefore I shall defer till then the revision of the first six books.

Should the work be interrupted by my death, then what is found can only be called a mass of conceptions not brought into form; but as these are open to endless misconceptions, they will doubtless give rise to a number of crude criticisms: for in these things, every one thinks, when he takes up his pen, that whatever comes into his head is worth saying and printing, and quite as incontrovertible as that twice two make four. If such a one would take the pains, as I have done, to think over the subject, for years, and to compare his ideas with military history, he would certainly be a little more guarded in his criticism.

Still, notwithstanding this imperfect form, I believe that an impartial reader thirsting for truth and conviction will rightly appreciate in the first six books the fruits of several years' reflection and a diligent study of War, and that, perhaps, he will find in them some leading ideas which may bring about a revolution in the theory of War.

Berlin, 10th July, 1827.

Besides this notice, amongst the papers left the following unfinished memorandum was found, which appears of very recent date:

The manuscript on the conduct of the Grande Guerre, which will be found after my death, in its present state can only be regarded as a collection of materials from which it is intended to construct a theory of War. With the greater part I am not yet satisfied; and the sixth book is to be looked at as a mere essay: I should have completely remodelled it, and have tried a different line.

But the ruling principles which pervade these materials I hold to be the right ones: they are the result of a very varied reflection, keeping always in view the reality, and always bearing in mind what I have learnt by experience and by my intercourse with distinguished soldiers.

The seventh book is to contain the attack, the subjects of which are thrown together in a hasty manner: the eighth, the plan for a War, in which I would have examined War more especially in its political and human aspects.

The first chapter of the first book is the only one which I consider as completed; it will at least serve to show the manner in which I proposed to treat the subject throughout.

The theory of the Grande Guerre, or Strategy, as it is called, is beset with extraordinary difficulties, and we may affirm that very few men have clear conceptions of the separate subjects, that is, conceptions carried up to their full logical conclusions. In real action most men are guided merely by the tact of judgment which hits the object more or less accurately, according as they possess more or less genius.

This is the way in which all great Generals have acted, and therein partly lay their greatness and their genius, that they always hit upon what was right by this tact. Thus also it will always be in action, and so far this tact is amply sufficient. But when it is a question, not of acting oneself, but of convincing others in a consultation, then all depends on clear conceptions and demonstration of the inherent relations, and so little progress has been made in this respect that most deliberations are merely a contention of words, resting on no firm basis, and ending either in every one retaining his own opinion, or in a compromise from mutual considerations of respect, a middle course really without any value.[Herr Clausewitz evidently had before his mind the endless consultations at the Headquarters of the Bohemian Army in the Leipsic Campaign 1813.]

Clear ideas on these matters are therefore not wholly useless; besides, the human mind has a general tendency to clearness, and always wants to be consistent with the necessary order of things.

Owing to the great difficulties attending a philosophical construction of the Art of War, and the many attempts at it that have failed, most people have come to the conclusion that such a theory is impossible, because it concerns things which no standing law can embrace. We should also join in this opinion and give up any attempt at a theory, were it not that a great number of propositions make themselves evident without any difficulty, as, for instance, that the defensive form, with a negative object, is the stronger form, the attack, with the positive object, the weaker—that great results carry the little ones with them—that, therefore, strategic effects may be referred to certain centres of gravity—that a demonstration is a weaker application of force than a real attack, that, therefore, there must be some special reason for resorting to the former—that victory consists not merely in the conquest on the field of battle, but in the destruction of armed forces, physically and morally, which can in general only be effected by a pursuit after the battle is gained—that successes are always greatest at the point where the victory has been gained, that, therefore, the change from one line and object to another can only be regarded as a necessary evil—that a turning movement is only justified by a superiority of numbers generally or by the advantage of our lines of communication and retreat over those of the enemy—that flank positions are only justifiable on similar grounds—that every attack becomes weaker as it progresses.

The Introduction of the Author

That the conception of the scientific does not consist alone, or chiefly, in system, and its finished theoretical constructions, requires nowadays no exposition. System in this treatise is not to be found on the surface, and instead of a finished building of theory, there are only materials.

The scientific form lies here in the endeavour to explore the nature of military phenomena to show their affinity with the nature of the things of which they are composed. Nowhere has the philosophical argument been evaded, but where it runs out into too thin a thread the Author has preferred to cut it short, and fall back upon the corresponding results of experience; for in the same way as many plants only bear fruit when they do not shoot too high, so in the practical arts the theoretical leaves and flowers must not be made to sprout too far, but kept near to experience, which is their proper soil.

Unquestionably it would be a mistake to try to discover from the chemical ingredients of a grain of corn the form of the ear of corn which it bears, as we have only to go to the field to see the ears ripe. Investigation and observation, philosophy and experience, must neither despise nor exclude one another; they mutually afford each other the rights of citizenship. Consequently, the propositions of this book, with their arch of inherent necessity, are supported either by experience or by the conception of War itself as external points, so that they are not without abutments. [That this is not the case in the works of many military writers especially of those who have aimed at treating of War itself in a scientific manner, is shown in many instances, in which by their reasoning, the pro and contra swallow each other up so effectually that there is no vestige of the tails even which were left in the case of the two lions.]

It is, perhaps, not impossible to write a systematic theory of War full of spirit and substance, but ours hitherto, have been very much the reverse. To say nothing of their unscientific spirit, in their striving after coherence and completeness of system, they overflow with commonplaces, truisms, and twaddle of every kind. If we want a striking picture of them we have only to read Lichtenberg's extract from a code of regulations in case of fire.

If a house takes fire, we must seek, above all things, to protect the right side of the house standing on the left, and, on the other hand, the left side of the house on the right; for if we, for example, should protect the left side of the house on the left, then the right side of the house lies to the right of the left, and consequently as the fire lies to the right of this side, and of the right side (for we have assumed that the house is situated to the left of the fire), therefore the right side is situated nearer to the fire than the left, and the right side of the house might catch fire if it was not protected before it came to the left, which is protected. Consequently, something might be burnt that is not protected, and that sooner than something else would be burnt, even if it was not protected; consequently we must let alone the latter and protect the former. In order to impress the thing on one's mind, we have only to note if the house is situated to the right of the fire, then it is the left side, and if the house is to the left it is the right side.

In order not to frighten the intelligent reader by such commonplaces, and to make the little good that there is distasteful by pouring water upon it, the Author has preferred to give in small ingots of fine metal his impressions and convictions, the result of many years' reflection on War, of his intercourse with men of ability, and of much personal experience. Thus the seemingly weakly bound-together chapters of this book have arisen, but it is hoped they will not be found wanting in logical connection. Perhaps soon a greater head may appear, and instead of these single grains, give the whole in a casting of pure metal without dross.

Brief Memoir of General Clausewitz

The Author of the work here translated, General Carl Von Clausewitz, was born at Burg, near Magdeburg, in 1780, and entered the Prussian Army as Fahnenjunker (i.e., ensign) in 1792. He served in the campaigns of 1793-94 on the Rhine, after which he seems to have devoted some time to the study of the scientific branches of his profession. In 1801 he entered the Military School at Berlin, and remained there till 1803. During his residence there he attracted the notice of General Scharnhorst, then at the head of the establishment; and the patronage of this distinguished officer had immense influence on his future career, and we may gather from his writings that he ever afterwards continued to entertain a high esteem for Scharnhorst. In the campaign of 1806 he served as Aide-de-camp to Prince Augustus of Prussia; and being wounded and taken prisoner, he was sent into France until the close of that war. On his return, he was placed on General Scharnhorst's Staff, and employed in the work then going on for the reorganisation of the Army. He was also at this time selected as military instructor to the late King of Prussia, then Crown Prince. In 1812 Clausewitz, with several other Prussian officers, having entered the Russian service, his first appointment was as Aide-de-camp to General Phul. Afterwards, while serving with Wittgenstein's army, he assisted in negotiating the famous convention of Tauroggen with York. Of the part he took in that affair he has left an interesting account in his work on the "Russian Campaign." It is there stated that, in order to bring the correspondence which had been carried on with York to a termination in one way or another, the Author was despatched to York's headquarters with two letters, one was from General d'Auvray, the Chief of the Staff of Wittgenstein's army, to General Diebitsch, showing the arrangements made to cut off York's corps from Macdonald (this was necessary in order to give York a plausible excuse for seceding from the French); the other was an intercepted letter from Macdonald to the Duke of Bassano. With regard to the former of these, the Author says, "it would not have had weight with a man like York, but for a military justification, if the Prussian Court should require one as against the French, it was important."

The second letter was calculated at the least to call up in General York's mind all the feelings of bitterness which perhaps for some days past had been diminished by the consciousness of his own behaviour towards the writer.

As the Author entered General York's chamber, the latter called out to him, "Keep off from me; I will have nothing more to do with you; your d——d Cossacks have let a letter of Macdonald's pass through them, which brings me an order to march on Piktrepohnen, in order there to effect our junction. All doubt is now at an end; your troops do not come up; you are too weak; march I must, and I must excuse myself from further negotiation, which may cost me my head." The Author said that be would make no opposition to all this, but begged for a candle, as he had letters to show the General, and, as the latter seemed still to hesitate, the Author added, "Your Excellency will not surely place me in the embarrassment of departing without having executed my commission." The General ordered candles, and called in Colonel von Roeder, the chief of his staff, from the ante-chamber. The letters were read. After a pause of an instant, the General said, "Clausewitz, you are a Prussian, do you believe that the letter of General d'Auvray is sincere, and that Wittgenstein's troops will really be at the points he mentioned on the 31st?" The Author replied, "I pledge myself for the sincerity of this letter upon the knowledge I have of General d'Auvray and the other men of Wittgenstein's headquarters; whether the dispositions he announces can be accomplished as he lays down I certainly cannot pledge myself; for your Excellency knows that in war we must often fall short of the line we have drawn for ourselves." The General was silent for a few minutes of earnest reflection; then he held out his hand to the Author, and said, "You have me. Tell General Diebitsch that we must confer early to-morrow at the mill of Poschenen, and that I am now firmly determined to separate myself from the French and their cause." The hour was fixed for 8 A.M. After this was settled, the General added, "But I will not do the thing by halves, I will get you Massenbach also." He called in an officer who was of Massenbach's cavalry, and who had just left them. Much like Schiller's Wallenstein, he asked, walking up and down the room the while, "What say your regiments?" The officer broke out with enthusiasm at the idea of a riddance from the French alliance, and said that every man of the troops in question felt the same.

"You young ones may talk; but my older head is shaking on my shoulders," replied the General. ["Campaign in Russia in 1812"; translated from the German of General Von Clausewitz (by Lord Ellesmere).]

After the close of the Russian campaign Clausewitz remained in the service of that country, but was attached as a Russian staff officer to Blucher's headquarters till the Armistice in 1813.

In 1814, he became Chief of the Staff of General Walmoden's Russo-German Corps, which formed part of the Army of the North under Bernadotte. His name is frequently mentioned with distinction in that campaign, particularly in connection with the affair of Goehrde.

Clausewitz re-entered the Prussian service in 1815, and served as Chief of the Staff to Thielman's corps, which was engaged with Grouchy at Wavre, on the 18th of June.

After the Peace, he was employed in a command on the Rhine. In 1818, he became Major-General, and Director of the Military School at which he had been previously educated.

In 1830, he was appointed Inspector of Artillery at Breslau, but soon after nominated Chief of the Staff to the Army of Observation, under Marshal Gneisenau on the Polish frontier.

The latest notices of his life and services are probably to be found in the memoirs of General Brandt, who, from being on the staff of Gneisenau's army, was brought into daily intercourse with Clausewitz in matters of duty, and also frequently met him at the table of Marshal Gneisenau, at Posen.

Amongst other anecdotes, General Brandt relates that, upon one occasion, the conversation at the Marshal's table turned upon a sermon preached by a priest, in which some great absurdities were introduced, and a discussion arose as to whether the Bishop should not be made responsible for what the priest had said. This led to the topic of theology in general, when General Brandt, speaking of himself, says, "I expressed an opinion that theology is only to be regarded as an historical process, as a *moment* in the gradual development of the human race. This brought upon me an attack from all quarters, but more especially from Clausewitz, who ought to have been on my side, he having been an adherent and pupil of Kiesewetter's, who had indoctrinated him in the philosophy of Kant, certainly diluted—I might even say in homoeopathic doses." This anecdote is only interesting as the mention of Kiesewetter points to a circumstance in the life of Clausewitz that may have had an influence in forming those habits of thought which distinguish his writings.

"The way," says General Brandt, "in which General Clausewitz judged of things, drew conclusions from movements and marches, calculated the times of the marches, and the points where decisions would take place, was extremely interesting. Fate has unfortunately denied him an opportunity of showing his talents in high command, but I have a firm persuasion that as a strategist he would have greatly distinguished himself. As a leader on the field of battle, on the other hand, he would not have been so much in his right place, from a *manque d'habitude du commandement*, he wanted the art *d'enlever les troupes*."

After the Prussian Army of Observation was dissolved, Clausewitz returned to Breslau, and a few days after his arrival was seized with cholera, the seeds of which he must have brought with him from the army on the Polish frontier. His death took place in November 1831.

His writings are contained in nine volumes, published after his death, but his fame rests most upon the three volumes forming his treatise on "War." In the present attempt to render into English this portion of the works of Clausewitz, the translator is sensible of many deficiencies, but he hopes at all events to succeed in making this celebrated treatise better known in England, believing, as he does, that so far as the work concerns the interests of this country, it has lost none of the importance it possessed at the time of its first publication.

J. J. Graham (col.)

What Is War?

1. Introduction

We propose to consider first the single elements of our subject, then each branch or part, and, last of all, the whole, in all its relations—therefore to advance from the simple to the complex. But it is necessary for us to commence with a glance at the nature of the whole, because it is particularly necessary that in the consideration of any of the parts their relation to the whole should be kept constantly in view.

2. Definition

We shall not enter into any of the abstruse definitions of War used by publicists. We shall keep to the element of the thing itself, to a duel. War is nothing but a duel on an extensive scale. If we would conceive as a unit the countless number of duels which make up a War, we shall do so best by supposing to ourselves two wrestlers. Each strives by physical force to compel the other to submit to his will: each endeavours to throw his adversary, and thus render him incapable of further resistance.

War therefore is an act of violence intended to compel our opponent to fulfil our will.

Violence arms itself with the inventions of Art and Science in order to contend against violence. Self-imposed restrictions, almost imperceptible and hardly worth mentioning, termed usages of International Law, accompany it without essentially impairing its power. Violence, that is to say, physical force (for there is no moral force without the conception of States and Law), is therefore the MEANS; the compulsory submission of the enemy to our will is the ultimate object. In order to attain this object fully, the enemy must be disarmed, and disarmament becomes therefore the immediate OBJECT of hostilities in theory. It takes the place of the final object, and puts it aside as something we can eliminate from our calculations.

3. Utmost Use of Force

Now, philanthropists may easily imagine there is a skilful method of disarming and overcoming an enemy without great bloodshed, and that this is the proper tendency of the Art of War. However plausible this may appear, still it is an error which must be extirpated; for in such dangerous things as War, the errors which proceed from a spirit of benevolence are the worst. As the use of physical power to the utmost extent by no means excludes the co-operation of the intelligence, it follows that he who uses force unsparingly, without reference to the bloodshed involved, must obtain a superiority if his adversary uses less vigour in its application. The former then dictates the law to the latter, and both proceed to extremities to which the only limitations are those imposed by the amount of counter-acting force on each side.

This is the way in which the matter must be viewed and it is to no purpose, it is even against one's own interest, to turn away from the consideration of the real nature of the affair because the horror of its elements excites repugnance.

If the Wars of civilised people are less cruel and destructive than those of savages, the difference arises from the social condition both of States in themselves and in their relations to each other. Out of this social condition and its relations War arises, and by it War is subjected to conditions, is controlled and modified. But these things do not belong to War itself; they are only given conditions; and to introduce into the philosophy of War itself a principle of moderation would be an absurdity.

Two motives lead men to War: instinctive hostility and hostile intention. In our definition of War, we have chosen as its characteristic the latter of these elements, because it is the most general. It is impossible to conceive the passion of hatred of the wildest description, bordering on mere instinct, without combining with it the idea of a hostile intention. On the other hand, hostile intentions may often exist without being accompanied by any, or at all events by any extreme, hostility of feeling. Amongst savages views emanating from the feelings, amongst civilised nations those emanating from the understanding, have the predominance; but this difference arises from attendant circumstances, existing institutions, etc., and, therefore, is not to be found necessarily in all cases, although it prevails in the majority. In short, even the most civilised nations may burn with passionate hatred of each other.

We may see from this what a fallacy it would be to refer the War of a civilised nation entirely to an intelligent act on the part of the Government, and to imagine it as continually freeing itself more and more from all feeling of passion in such a way that at last the physical masses of combatants would no longer be required; in reality, their mere relations would suffice—a kind of algebraic action.

Theory was beginning to drift in this direction until the facts of the last War [Clausewitz alludes here to the "Wars of Liberation," 1813,14,15.] taught it better. If War is an *act* of force, it belongs necessarily also to the feelings. If it does not originate in the feelings, it *reacts*, more or less, upon them, and the extent of this reaction depends not on the degree of civilisation, but upon the importance and duration of the interests involved.

Therefore, if we find civilised nations do not put their prisoners to death, do not devastate towns and countries, this is because their intelligence exercises greater influence on their mode of carrying on War, and has taught them more effectual means of applying force than these rude acts of mere instinct. The invention of gunpowder, the constant progress of improvements in the construction of firearms, are sufficient proofs that the tendency to destroy the adversary which lies at the bottom of the conception of War is in no way changed or modified through the progress of civilisation.

We therefore repeat our proposition, that War is an act of violence pushed to its utmost bounds; as one side dictates the law to the other, there arises a sort of reciprocal action, which logically must lead to an extreme. This is the first reciprocal action, and the first extreme with which we meet (*first reciprocal action*).

4. The Aim Is to Disarm the Enemy

We have already said that the aim of all action in War is to disarm the enemy, and we shall now show that this, theoretically at least, is indispensable.

If our opponent is to be made to comply with our will, we must place him in a situation which is more oppressive to him than the sacrifice which we demand; but the disadvantages of this position must naturally not be of a transitory nature, at least in appearance, otherwise the enemy, instead of yielding, will hold out, in the prospect of a change for the better. Every change in this position which is produced by a continuation of the War should therefore be a change for the worse. The worst condition in which a belligerent can be placed is that of being completely disarmed. If, therefore, the enemy is to be reduced to submission by an act of War, he must either be positively disarmed or placed in such a position that he is threatened with it. From this it follows that the disarming or overthrow of the enemy, whichever we call it, must always be the aim of Warfare. Now War is always the shock of two hostile bodies in collision, not the action of a living power upon an inanimate mass, because an absolute state of endurance would not be making War; therefore, what we have just said as to the aim of action in War applies to both parties. Here, then, is another case of reciprocal action. As long as the enemy is not defeated, he may defeat me; then I shall be no longer my own master; he will dictate the law to me as I did to him. This is the second reciprocal action, and leads to a second extreme (*second reciprocal action*).

5. Utmost Exertion of Powers

If we desire to defeat the enemy, we must proportion our efforts to his powers of resistance. This is expressed by the product of two factors which cannot be separated, namely, the sum of available means and the strength of the Will. The sum of the available means may be estimated in a measure, as it depends (although not entirely) upon numbers; but the strength of volition is more difficult to determine, and can only be estimated to a certain extent by the strength of the motives. Granted we have obtained in this way an approximation to the strength of the power to be contended with, we can then take of our own means, and either increase them so as to obtain a preponderance, or, in case we have not the resources to effect this, then do our best by increasing our means as far as possible. But the adversary does the same; therefore, there is a new mutual enhancement, which, in pure conception, must create a fresh effort towards an extreme. This is the third case of reciprocal action, and a third extreme with which we meet (*third reciprocal action*).

6. Modification in the Reality

Thus reasoning in the abstract, the mind cannot stop short of an extreme, because it has to deal with an extreme, with a conflict of forces left to themselves, and obeying no other but their own inner laws. If we should seek to deduce from the pure conception of War an absolute point for the aim which we shall propose and for the means which we shall apply, this constant reciprocal action would involve us in extremes, which would be nothing but a play of ideas produced by an almost invisible train of logical subtleties. If, adhering closely to the absolute, we try to avoid all difficulties by a stroke of the pen, and insist with logical strictness that in every case the extreme must be the object, and the utmost effort must be exerted in that direction, such a stroke of the pen would be a mere paper law, not by any means adapted to the real world.

Even supposing this extreme tension of forces was an absolute which could easily be ascertained, still we must admit that the human mind would hardly submit itself to this kind of logical chimera. There would be in many cases an unnecessary waste of power, which would be in opposition to other principles of statecraft; an effort of Will would be required disproportioned to the proposed object, which therefore it would be impossible to realise, for the human will does not derive its impulse from logical subtleties.

But everything takes a different shape when we pass from abstractions to reality. In the former, everything must be subject to optimism, and we must imagine the one side as well as the other striving after perfection and even attaining it. Will this ever take place in reality? It will if,

(1) War becomes a completely isolated act, which arises suddenly, and is in no way connected with the previous history of the combatant States.

(2) If it is limited to a single solution, or to several simultaneous solutions.

(3) If it contains within itself the solution perfect and complete, free from any reaction upon it, through a calculation beforehand of the political situation which will follow from it.

7. War Is Never an Isolated Act

With regard to the first point, neither of the two opponents is an abstract person to the other, not even as regards that factor in the sum of resistance which does not depend on objective things, viz., the Will. This Will is not an entirely unknown quantity; it indicates what it will be to-morrow by what it is to-day. War does not spring up quite suddenly, it does not spread to the full in a moment; each of the two opponents can, therefore, form an opinion of the other, in a great measure, from what he is and what he does, instead of judging of him according to what he, strictly speaking, should be or should do. But, now, man with his incomplete organisation is always below the line of absolute perfection, and thus these deficiencies, having an influence on both sides, become a modifying principle.

8. War Does Not Consist of a Single Instantaneous Blow

The second point gives rise to the following considerations:—

If War ended in a single solution, or a number of simultaneous ones, then naturally all the preparations for the same would have a tendency to the extreme, for an omission could not in any way be repaired; the utmost, then, that the world of reality could furnish as a guide for us would be the preparations of the enemy, as far as they are known to us; all the rest would fall into the domain of the abstract. But if the result is made up from several successive acts, then naturally that which precedes with all its phases may be taken as a measure for that which will follow, and in this manner the world of reality again takes the place of the abstract, and thus modifies the effort towards the extreme.

Yet every War would necessarily resolve itself into a single solution, or a sum of simultaneous results, if all the means required for the struggle were raised at once, or could be at once raised; for as one adverse result necessarily diminishes the means, then if all the means have been applied in the first, a second cannot properly be supposed. All hostile acts which might follow would belong essentially to the first, and form, in reality only its duration.

But we have already seen that even in the preparation for War the real world steps into the place of mere abstract conception—a material standard into the place of the hypotheses of an extreme: that therefore in that way both parties, by the influence of the mutual reaction, remain below the line of extreme effort, and therefore all forces are not at once brought forward.

It lies also in the nature of these forces and their application that they cannot all be brought into activity at the same time. These forces are *the armies actually on foot, the country,* with its superficial extent and its population, *and the allies.*

In point of fact, the country, with its superficial area and the population, besides being the source of all military force, constitutes in itself an integral part of the efficient quantities in War, providing either the theatre of war or exercising a considerable influence on the same.

Now, it is possible to bring all the movable military forces of a country into operation at once, but not all fortresses, rivers, mountains, people, etc.—in short, not the whole country, unless it is so small that it may be completely embraced by the first act of the War. Further, the co-operation of allies does not depend on the Will of the belligerents; and from the nature of the political relations of states to each other, this co-operation is frequently not afforded until after the War has commenced, or it may be increased to restore the balance of power.

That this part of the means of resistance, which cannot at once be brought into activity, in many cases, is a much greater part of the whole than might at first be supposed, and that it often restores the balance of power, seriously affected by the great force of the first decision, will be more fully shown hereafter. Here it is sufficient to show that a complete concentration of all available means in a moment of time is contradictory to the nature of War.

Now this, in itself, furnishes no ground for relaxing our efforts to accumulate strength to gain the first result, because an unfavourable issue is always a disadvantage to which no one would purposely expose himself, and also because the first decision, although not the only one, still will have the more influence on subsequent events, the greater it is in itself.

But the possibility of gaining a later result causes men to take refuge in that expectation, owing to the repugnance in the human mind to making excessive efforts; and therefore forces are not concentrated and measures are not taken for the first decision with that energy which would otherwise be used. Whatever one belligerent omits from weakness, becomes to the other a real objective ground for limiting his own efforts, and thus again, through this reciprocal action, extreme tendencies are brought down to efforts on a limited scale.

9. The Result in War Is Never Absolute

Lastly, even the final decision of a whole War is not always to be regarded as absolute. The conquered State often sees in it only a passing evil, which may be repaired in after times by means of political combinations. How much this must modify the degree of tension, and the vigour of the efforts made, is evident in itself.

10. The Probabilities of Real Life Take the Place of the Conceptions of the Extreme and the Absolute

In this manner, the whole act of War is removed from the rigorous law of forces exerted to the utmost. If the extreme is no longer to be apprehended, and no longer to be sought for, it is left to the judgment to determine the limits for the efforts to be made in place of it, and this can only be done on the data furnished by the facts of the real world by the *Laws of Probability*. Once the belligerents are no longer mere conceptions, but individual States and Governments, once the War is no longer an ideal, but a definite substantial procedure, then the reality will furnish the data to compute the unknown quantities which are required to be found.

From the character, the measures, the situation of the adversary, and the relations with which he is surrounded, each side will draw conclusions by the law of probability as to the designs of the other, and act accordingly.

11. The Political Object Now Reappears

Here the question which we had laid aside forces itself again into consideration, viz., the political object of the War. The law of the extreme, the view to disarm the adversary, to overthrow him, has hitherto to a certain extent usurped the place of this end or object. Just as this law loses its force, the political must again come forward. If the whole consideration is a calculation of probability based on definite persons and relations, then the political object, being the original motive, must be an essential factor in the product. The smaller the sacrifice we demand from ours, the smaller, it may be expected, will be the means of resistance which he will employ; but the smaller his preparation, the smaller will ours require to be. Further, the smaller our political object, the less value shall we set upon it, and the more easily shall we be induced to give it up altogether.

Thus, therefore, the political object, as the original motive of the War, will be the standard for determining both the aim of the military force and also the amount of effort to be made. This it cannot be in itself, but it is so in relation to both the belligerent States, because we are concerned with realities, not with mere abstractions. One and the same political object may produce totally different effects upon different people, or even upon the same people at different times; we can, therefore, only admit the political object as the measure, by considering it in its effects upon those masses which it is to move, and consequently the nature of those masses also comes into consideration. It is easy to see that thus the result may be very different according as these masses are animated with a spirit which will infuse vigour into the action or otherwise. It is quite possible for such a state of feeling to exist between two States that a very trifling political motive for War may produce an effect quite disproportionate—in fact, a perfect explosion.

This applies to the efforts which the political object will call forth in the two States, and to the aim which the military action shall prescribe for itself. At times it may itself be that aim, as, for example, the conquest of a province. At other times the political object itself is not suitable for the aim of military action; then such a one must be chosen as will be an equivalent for it, and stand in its place as regards the conclusion of peace. But also, in this, due attention to the peculiar character of the States concerned is always supposed. There are circumstances in which the equivalent must be much greater than the political object, in order to secure the latter. The political object will be so much the more the standard of aim and effort, and have more influence in itself, the more the masses are indifferent, the less that any mutual feeling of hostility prevails in the two States from other causes, and therefore there are cases where the political object almost alone will be decisive.

If the aim of the military action is an equivalent for the political object, that action will in general diminish as the political object diminishes, and in a greater degree the more the political object dominates. Thus it is explained how, without any contradiction in itself, there may be Wars of all degrees of importance and energy, from a War of extermination down to the mere use of an army of observation. This, however, leads to a question of another kind which we have hereafter to develop and answer.

12. A Suspension in the Action of War Unexplained by Anything Said as Yet

However insignificant the political claims mutually advanced, however weak the means put forth, however small the aim to which military action is directed, can this action be suspended even for a moment? This is a question which penetrates deeply into the nature of the subject.

Every transaction requires for its accomplishment a certain time which we call its duration. This may be longer or shorter, according as the person acting throws more or less despatch into his movements.

About this more or less we shall not trouble ourselves here. Each person acts in his own fashion; but the slow person does not protract the thing because he wishes to spend more time about it, but because by his nature he requires more time, and if he made more haste would not do the thing so well. This time, therefore, depends on subjective causes, and belongs to the length, so called, of the action.

If we allow now to every action in War this, its length, then we must assume, at first sight at least, that any expenditure of time beyond this length, that is, every suspension of hostile action, appears an absurdity; with respect to this it must not be forgotten that we now speak not of the progress of one or other of the two opponents, but of the general progress of the whole action of the War.

13. There Is Only One Cause Which Can Suspend the Action, and this Seems to Be Only Possible on One Side in Any Case

If two parties have armed themselves for strife, then a feeling of animosity must have moved them to it; as long now as they continue armed, that is, do not come to terms of peace, this feeling must exist; and it can only be brought to a standstill by either side by one single motive alone, which is, *That he waits for a more favourable moment for action.* Now, at first sight, it appears that this motive can never exist except on one side, because it, eo ipso, must be prejudicial to the other. If the one has an interest in acting, then the other must have an interest in waiting.

A complete equilibrium of forces can never produce a suspension of action, for during this suspension he who has the positive object (that is, the assailant) must continue progressing; for if we should imagine an equilibrium in this way, that he who has the positive object, therefore the strongest motive, can at the same time only command the lesser means, so that the equation is made up by the product of the motive and the power, then we must say, if no alteration in this condition of equilibrium is to be expected, the two parties must make peace; but if an alteration is to be expected, then it can only be favourable to one side, and therefore the other has a manifest interest to act without delay. We see that the conception of an equilibrium cannot explain a suspension of arms, but that it ends in the question of the *expectation of a more favourable moment.*

Let us suppose, therefore, that one of two States has a positive object, as, for instance, the conquest of one of the enemy's provinces—which is to be utilised in the settlement of peace. After this conquest, his political object is accomplished, the necessity for action ceases, and for him a pause ensues. If the adversary is also contented with this solution, he will make peace; if not, he must act. Now, if we suppose that in four weeks he will be in a better condition to act, then he has sufficient grounds for putting off the time of action.

But from that moment the logical course for the enemy appears to be to act that he may not give the conquered party *the desired* time. Of course, in this mode of reasoning a complete insight into the state of circumstances on both sides is supposed.

14. Thus a Continuance of Action Will Ensue Which Will Advance Towards a Climax

If this unbroken continuity of hostile operations really existed, the effect would be that everything would again be driven towards the extreme; for, irrespective of the effect of such incessant activity in inflaming the feelings, and infusing into the whole a greater degree of passion, a greater elementary force, there would also follow from this continuance of action a stricter continuity, a closer connection between cause and effect, and thus every single action would become of more importance, and consequently more replete with danger.

But we know that the course of action in War has seldom or never this unbroken continuity, and that there have been many Wars in which action occupied by far the smallest portion of time employed, the whole of the rest being consumed in inaction. It is impossible that this should be always an anomaly; suspension of action in War must therefore be possible, that is no contradiction in itself. We now proceed to show how this is.

15. Here, Therefore, the Principle of Polarity Is Brought into Requisition

As we have supposed the interests of one Commander to be always antagonistic to those of the other, we have assumed a true *polarity*. We reserve a fuller explanation of this for another chapter, merely making the following observation on it at present.

The principle of polarity is only valid when it can be conceived in one and the same thing, where the positive and its opposite the negative completely destroy each other. In a battle both sides strive to conquer; that is true polarity, for the victory of the one side destroys that of the other. But when we speak of two different things which have a common relation external to themselves, then it is not the things but their relations which have the polarity.

16. Attack and Defence Are Things Differing in Kind and of Unequal Force. Polarity Is, Therefore, Not Applicable to Them

If there was only one form of War, to wit, the attack of the enemy, therefore no defence; or, in other words, if the attack was distinguished from the defence merely by the positive motive, which the one has and the other has not, but the methods of each were precisely one and the same: then in this sort of fight every advantage gained on the one side would be a corresponding disadvantage on the other, and true polarity would exist.

But action in War is divided into two forms, attack and defence, which, as we shall hereafter explain more particularly, are very different and of unequal strength. Polarity therefore lies in that to which both bear a relation, in the decision, but not in the attack or defence itself.

If the one Commander wishes the solution put off, the other must wish to hasten it, but only by the same form of action. If it is A's interest not to attack his enemy at present, but four weeks hence, then it is B's interest to be attacked, not four weeks hence, but at the present moment. This is the direct antagonism of interests, but it by no means follows that it would be for B's interest to attack A at once. That is plainly something totally different.

17. The Effect of Polarity Is Often Destroyed by the Superiority of the Defence over the Attack, and Thus the Suspension of Action in War Is Explained

If the form of defence is stronger than that of offence, as we shall hereafter show, the question arises, Is the advantage of a deferred decision as great on the one side as the advantage of the defensive form on the other? If it is not, then it cannot by its counter-weight over-balance the latter, and thus influence the progress of the action of the War. We see, therefore, that the impulsive force existing in the polarity of interests may be lost in the difference between the strength of the offensive and the defensive, and thereby become ineffectual.

If, therefore, that side for which the present is favourable, is too weak to be able to dispense with the advantage of the defensive, he must put up with the unfavourable prospects which the future holds out; for it may still be better to fight a defensive battle in the unpromising future than to assume the offensive or make peace at present. Now, being convinced that the superiority of the defensive [It must be remembered that all this antedates by some years the introduction of long-range weapons.] (rightly understood) is very great, and much greater than may appear at first sight, we conceive that the greater number of those periods of inaction which occur in war are thus explained without involving any contradiction. The weaker the motives to action are, the more will those motives be absorbed and neutralised by this difference between attack and defence, the more frequently, therefore, will action in warfare be stopped, as indeed experience teaches.

18 A Second Ground Consists in the Imperfect Knowledge of Circumstances

But there is still another cause which may stop action in War, viz., an incomplete view of the situation. Each Commander can only fully know his own position; that of his opponent can only be known to him by reports, which are uncertain; he may, therefore, form a wrong judgment with respect to it upon data of this description, and, in consequence of that error, he may suppose that the power of taking the initiative rests with his adversary when it lies really with himself. This want of perfect insight might certainly just as often occasion an untimely action as untimely inaction, and hence it would in itself no more contribute to delay than to accelerate action in War. Still, it must always be regarded as one of the natural causes which may bring action in War to a standstill without involving a contradiction. But if we reflect how much more we are inclined and induced to estimate the power of our opponents too high than too low, because it lies in human nature to do so, we shall admit that our imperfect insight into facts in general must contribute very much to delay action in War, and to modify the application of the principles pending our conduct.

The possibility of a standstill brings into the action of War a new modification, inasmuch as it dilutes that action with the element of time, checks the influence or sense of danger in its course, and increases the means of reinstating a lost balance of force. The greater the tension of feelings from which the War springs, the greater therefore the energy with which it is carried on, so much the shorter will be the periods of inaction; on the other hand, the weaker the principle of warlike activity, the longer will be these periods: for powerful motives increase the force of the will, and this, as we know, is always a factor in the product of force.

19. Frequent Periods of Inaction in War Remove it Further from the Absolute, and Make it Still More a Calculation of Probabilities

But the slower the action proceeds in War, the more frequent and longer the periods of inaction, so much the more easily can an error be repaired; therefore, so much the bolder a General will be in his calculations, so much the more readily will he keep them below the line of the absolute, and build everything upon probabilities and conjecture. Thus, according as the course of the War is more or less slow, more or less time will be allowed for that which the nature of a concrete case particularly requires, calculation of probability based on given circumstances.

20. Therefore, the Element of Chance Only Is Wanting to Make of War a Game, and in That Element it Is Least of All Deficient

We see from the foregoing how much the objective nature of War makes it a calculation of probabilities; now there is only one single element still wanting to make it a game, and that element it certainly is not without: it is chance. There is no human affair which stands so constantly and so generally in close connection with chance as War. But together with chance, the accidental, and along with it good luck, occupy a great place in War.

21. War Is a Game Both Objectively and Subjectively

If we now take a look at the subjective nature of War, that is to say, at those conditions under which it is carried on, it will appear to us still more like a game. Primarily the element in which the operations of War are carried on is danger; but which of all the moral qualities is the first in danger? *Courage.* Now certainly courage is quite compatible with prudent calculation, but still they are things of quite a different kind, essentially different qualities of the mind; on the other hand, daring reliance on good fortune, boldness, rashness, are only expressions of courage, and all these propensities of the mind look for the fortuitous (or accidental), because it is their element.

We see, therefore, how, from the commencement, the absolute, the mathematical as it is called, nowhere finds any sure basis in the calculations in the Art of War; and that from the outset there is a play of possibilities, probabilities, good and bad luck, which spreads about with all the coarse and fine threads of its web, and makes War of all branches of human activity the most like a gambling game.

22. How this Accords Best with the Human Mind in General

Although our intellect always feels itself urged towards clearness and certainty, still our mind often feels itself attracted by uncertainty. Instead of threading its way with the understanding along the narrow path of philosophical investigations and logical conclusions, in order, almost unconscious of itself, to arrive in spaces where it feels itself a stranger, and where it seems to part from all well-known objects, it prefers to remain with the imagination in the realms of chance and luck. Instead of living yonder on poor necessity, it revels here in the wealth of possibilities; animated thereby, courage then takes wings to itself, and daring and danger make the element into which it launches itself as a fearless swimmer plunges into the stream.

Shall theory leave it here, and move on, self-satisfied with absolute conclusions and rules? Then it is of no practical use. Theory must also take into account the human element; it must accord a place to courage, to boldness, even to rashness. The Art of War has to deal with living and with moral forces, the consequence of which is that it can never attain the absolute and positive. There is therefore everywhere a margin for the accidental, and just as much in the greatest things as in the smallest. As there is room for this accidental on the one hand, so on the other there must be courage and self-reliance in proportion to the room available. If these qualities are forthcoming in a high degree, the margin left may likewise be great. Courage and self-reliance are, therefore, principles quite essential to War; consequently, theory must only set up such rules as allow ample scope for all degrees and varieties of these necessary and noblest of military virtues. In daring there may still be wisdom, and prudence as well, only they are estimated by a different standard of value.

23. War Is Always a Serious Means for a Serious Object. Its More Particular Definition

Such is War; such the Commander who conducts it; such the theory which rules it. But War is no pastime; no mere passion for venturing and winning; no work of a free enthusiasm: it is a serious means for a serious object. All that appearance which it wears from the varying hues of fortune, all that it assimilates into itself of the oscillations of passion, of courage, of imagination, of enthusiasm, are only particular properties of this means.

The War of a community—of whole Nations, and particularly of civilised Nations—always starts from a political condition, and is called forth by a political motive. It is, therefore, a political act. Now if it was a perfect, unrestrained, and absolute expression of force, as we had to deduct it from its mere conception, then the moment it is called forth by policy it would step into the place of policy, and as something quite independent of it would set it aside, and only follow its own laws, just as a mine at the moment of explosion cannot be guided into any other direction than that which has been given to it by preparatory arrangements. This is how the thing has really been viewed hitherto, whenever a want of harmony between policy and the conduct of a War has led to theoretical distinctions of the kind. But it is not so, and the idea is radically false. War in the real world, as we have already seen, is not an extreme thing which expends itself at one single discharge; it is the operation of powers which do not develop themselves completely in the same manner and in the same measure, but which at one time expand sufficiently to overcome the resistance opposed by inertia or friction, while at another they are too weak to produce an effect; it is therefore, in a certain measure, a pulsation of violent force more or less vehement, consequently making its discharges and exhausting its powers more or less quickly—in other words, conducting more or less quickly to the aim, but always lasting long enough to admit of influence being exerted on it in its course, so as to give it this or that direction, in short, to be subject to the will of a guiding intelligence., if we reflect that War has its root in a political object, then naturally this original motive which called it into existence should also continue the first and highest consideration in its conduct. Still, the political object is no despotic lawgiver on that account; it must accommodate itself to the nature of the means, and though changes in these means may involve modification in the political objective, the latter always retains a prior right to consideration. Policy, therefore, is interwoven with the whole action of War, and must exercise a continuous influence upon it, as far as the nature of the forces liberated by it will permit.

24. War Is a Mere Continuation of Policy by Other Means

We see, therefore, that War is not merely a political act, but also a real political instrument, a continuation of political commerce, a carrying out of the same by other means. All beyond this which is strictly peculiar to War relates merely to the peculiar nature of the means which it uses. That the tendencies and views of policy shall not be incompatible with these means, the Art of War in general and the Commander in each particular case may demand, and this claim is truly not a trifling one. But however powerfully this may react on political views in particular cases, still it must always be regarded as only a modification of them; for the political view is the object, War is the means, and the means must always include the object in our conception.

25. Diversity in the Nature of Wars

The greater and the more powerful the motives of a War, the more it affects the whole existence of a people. The more violent the excitement which precedes the War, by so much the nearer will the War approach to its abstract form, so much the more will it be directed to the destruction of the enemy, so much the nearer will the military and political ends coincide, so much the more purely military and less political the War appears to be; but the weaker the motives and the tensions, so much the less will the natural direction of the military element—that is, force—be coincident with the direction which the political element indicates; so much the more must, therefore, the War become diverted from its natural direction, the political object diverge from the aim of an ideal War, and the War appear to become political.

But, that the reader may not form any false conceptions, we must here observe that by this natural tendency of War we only mean the philosophical, the strictly logical, and by no means the tendency of forces actually engaged in conflict, by which would be supposed to be included all the emotions and passions of the combatants. No doubt in some cases these also might be excited to such a degree as to be with difficulty restrained and confined to the political road; but in most cases such a contradiction will not arise, because by the existence of such strenuous exertions a great plan in harmony therewith would be implied. If the plan is directed only upon a small object, then the impulses of feeling amongst the masses will be also so weak that these masses will require to be stimulated rather than repressed.

26. They May All Be Regarded as Political Acts

Returning now to the main subject, although it is true that in one kind of War the political element seems almost to disappear, whilst in another kind it occupies a very prominent place, we may still affirm that the one is as political as the other; for if we regard the State policy as the intelligence of the personified State, then amongst all the constellations in the political sky whose movements it has to compute, those must be included which arise when the nature of its relations imposes the necessity of a great War. It is only if we understand by policy not a true appreciation of affairs in general, but the conventional conception of a cautious, subtle, also dishonest craftiness, averse from violence, that the latter kind of War may belong more to policy than the first.

27. Influence of this View on the Right Understanding of Military History, and on the Foundations of Theory

We see, therefore, in the first place, that under all circumstances War is to be regarded not as an independent thing, but as a political instrument; and it is only by taking this point of view that we can avoid finding ourselves in opposition to all military history. This is the only means of unlocking the great book and making it intelligible. Secondly, this view shows us how Wars must differ in character according to the nature of the motives and circumstances from which they proceed.

Now, the first, the grandest, and most decisive act of judgment which the Statesman and General exercises is rightly to understand in this respect the War in which he engages, not to take it for something, or to wish to make of it something, which by the nature of its relations it is impossible for it to be.

This is, therefore, the first, the most comprehensive, of all strategical questions. We shall enter into this more fully in treating of the plan of a War.

For the present we content ourselves with having brought the subject up to this point, and having thereby fixed the chief point of view from which War and its theory are to be studied.

28. Result for Theory

War is, therefore, not only chameleon-like in character, because it changes its colour in some degree in each particular case, but it is also, as a whole, in relation to the predominant tendencies which are in it, a wonderful trinity, composed of the original violence of its elements, hatred and animosity, which may be looked upon as blind instinct; of the play of probabilities and chance, which make it a free activity of the soul; and of the subordinate nature of a political instrument, by which it belongs purely to the reason.

The first of these three phases concerns more the people the second, more the General and his Army; the third, more the Government. The passions which break forth in War must already have a latent existence in the peoples. The range which the display of courage and talents shall get in the realm of probabilities and of chance depends on the particular characteristics of the General and his Army, but the political objects belong to the Government alone.

These three tendencies, which appear like so many different law-givers, are deeply rooted in the nature of the subject, and at the same time variable in degree. A theory which would leave any one of them out of account, or set up any arbitrary relation between them, would immediately become involved in such a contradiction with the reality, that it might be regarded as destroyed at once by that alone.

The problem is, therefore, that theory shall keep itself poised in a manner between these three tendencies, as between three points of attraction.

The way in which alone this difficult problem can be solved we shall examine in the book on the "Theory of War." In every case the conception of War, as here defined, will be the first ray of light which shows us the true foundation of theory, and which first separates the great masses and allows us to distinguish them from one another.

End and Means in War

Having in the foregoing chapter ascertained the complicated and variable nature of War, we shall now occupy ourselves in examining into the influence which this nature has upon the end and means in War.

If we ask, first of all, for the object upon which the whole effort of War is to be directed, in order that it may suffice for the attainment of the political object, we shall find that it is just as variable as are the political object and the particular circumstances of the War.

If, in the next place, we keep once more to the pure conception of War, then we must say that the political object properly lies out of its province, for if War is an act of violence to compel the enemy to fulfil our will, then in every case all depends on our overthrowing the enemy, that is, disarming him, and on that alone. This object, developed from abstract conceptions, but which is also the one aimed at in a great many cases in reality, we shall, in the first place, examine in this reality.

In connection with the plan of a campaign we shall hereafter examine more closely into the meaning of disarming a nation, but here we must at once draw a distinction between three things, which, as three general objects, comprise everything else within them. They are the *Military power, the country, and the will of the enemy.*

The military power must be destroyed, that is, reduced to such a state as not to be able to prosecute the War. This is the sense in which we wish to be understood hereafter, whenever we use the expression "destruction of the enemy's military power."

The country must be conquered, for out of the country a new military force may be formed.

But even when both these things are done, still the War, that is, the hostile feeling and action of hostile agencies, cannot be considered as at an end as long as the will of the enemy is not subdued also; that is, its Government and its Allies must be forced into signing a peace, or the people into submission; for whilst we are in full occupation of the country, the War may break out afresh, either in the interior or through assistance given by Allies. No doubt, this may also take place after a peace, but that shows nothing more than that every War does not carry in itself the elements for a complete decision and final settlement.

But even if this is the case, still with the conclusion of peace a number of sparks are always extinguished which would have smouldered on quietly, and the excitement of the passions abates, because all those whose minds are disposed to peace, of which in all nations and under all circumstances there is always a great number, turn themselves away completely from the road to resistance. Whatever may take place subsequently, we must always look upon the object as attained, and the business of War as ended, by a peace.

As protection of the country is the primary object for which the military force exists, therefore the natural order is, that first of all this force should be destroyed, then the country subdued; and through the effect of these two results, as well as the position we then hold, the enemy should be forced to make peace. Generally the destruction of the enemy's force is done by degrees, and in just the same measure the conquest of the country follows immediately. The two likewise usually react upon each other, because the loss of provinces occasions a diminution of military force. But this order is by no means necessary, and on that account it also does not always take place. The enemy's Army, before it is sensibly weakened, may retreat to the opposite side of the country, or even quite outside of it. In this case, therefore, the greater part or the whole of the country is conquered.

But this object of War in the abstract, this final means of attaining the political object in which all others are combined, the *disarming the enemy*, is rarely attained in practice and is not a condition necessary to peace. Therefore it can in no wise be set up in theory as a law. There are innumerable instances of treaties in which peace has been settled before either party could be looked upon as disarmed; indeed, even before the balance of power had undergone any sensible alteration. Nay, further, if we look at the case in the concrete, then we must say that in a whole class of cases, the idea of a complete defeat of the enemy would be a mere imaginative flight, especially when the enemy is considerably superior.

The reason why the object deduced from the conception of War is not adapted in general to real War lies in the difference between the two, which is discussed in the preceding chapter. If it was as pure theory gives it, then a War between two States of very unequal military strength would appear an absurdity; therefore impossible. At most, the inequality between the physical forces might be such that it could be balanced by the moral forces, and that would not go far with our present social condition in Europe. Therefore, if we have seen Wars take place between States of very unequal power, that has been the case because there is a wide difference between War in reality and its original conception.

There are two considerations which as motives may practically take the place of inability to continue the contest. The first is the improbability, the second is the excessive price, of success.

According to what we have seen in the foregoing chapter, War must always set itself free from the strict law of logical necessity, and seek aid from the calculation of probabilities; and as this is so much the more the case, the more the War has a bias that way, from the circumstances out of which it has arisen—the smaller its motives are, and the excitement it has raised—so it is also conceivable how out of this calculation of probabilities even motives to peace may arise. War does not, therefore, always require to be fought out until one party is overthrown; and we may suppose that, when the motives and passions are slight, a weak probability will suffice to move that side to which it is unfavourable to give way. Now, were the other side convinced of this beforehand, it is natural that he would strive for this probability only, instead of first wasting time and effort in the attempt to achieve the total destruction of the enemy's Army.

Still more general in its influence on the resolution to peace is the consideration of the expenditure of force already made, and further required. As War is no act of blind passion, but is dominated by the political object, therefore the value of that object determines the measure of the sacrifices by which it is to be purchased. This will be the case, not only as regards extent, but also as regards duration. As soon, therefore, as the required outlay becomes so great that the political object is no longer equal in value, the object must be given up, and peace will be the result.

We see, therefore, that in Wars where one side cannot completely disarm the other, the motives to peace on both sides will rise or fall on each side according to the probability of future success and the required outlay. If these motives were equally strong on both sides, they would meet in the centre of their political difference. Where they are strong on one side, they might be weak on the other. If their amount is only sufficient, peace will follow, but naturally to the advantage of that side which has the weakest motive for its conclusion. We purposely pass over here the difference which the *positive* and *negative* character of the political end must necessarily produce practically; for although that is, as we shall hereafter show, of the highest importance, still we are obliged to keep here to a more general point of view, because the original political views in the course of the War change very much, and at last may become totally different, *Just because they are determined by results and probable events.*

Now comes the question how to influence the probability of success. In the first place, naturally by the same means which we use when the object is the subjugation of the enemy, by the destruction of his military force and the conquest of his provinces; but these two means are not exactly of the same import here as they would be in reference to that object. If we attack the enemy's Army, it is a very different thing whether we intend to follow up the first blow with a succession of others, until the whole force is destroyed, or whether we mean to content ourselves with a victory to shake the enemy's feeling of security, to convince him of our superiority, and to instil into him a feeling of apprehension about the future. If this is our object, we only go so far in the destruction of his forces as is sufficient. In like manner, the conquest, of the enemy's provinces is quite a different measure if the object is not the destruction of the enemy's Army. In the latter case the destruction of the Army is the real effectual action, and the taking of the provinces only a consequence of it; to take them before the Army had been defeated would always be looked upon only as a necessary evil. On the other hand, if our views are not directed upon the complete destruction of the enemy's force, and if we are sure that the enemy does not seek but fears to bring matters to a bloody decision, the taking possession of a weak or defenceless province is an advantage in itself, and if this advantage is of sufficient importance to make the enemy apprehensive about the general result, then it may also be regarded as a shorter road to peace.

But now we come upon a peculiar means of influencing the probability of the result without destroying the enemy's Army, namely, upon the expeditions which have a direct connection with political views. If there are any enterprises which are particularly likely to break up the enemy's alliances or make them inoperative, to gain new alliances for ourselves, to raise political powers in our own favour, etc. etc., then it is easy to conceive how much these may increase the probability of success, and become a shorter way towards our object than the routing of the enemy's forces.

The second question is how to act upon the enemy's expenditure in strength, that is, to raise the price of success.

The enemy's outlay in strength lies in the *wear and tear* of his forces, consequently in the *destruction* of them on our part, and in the *loss* of *provinces*, consequently the *Conquest* of them by us.

Here, again, on account of the various significations of these means, so likewise it will be found that neither of them will be identical in its signification in all cases if the objects are different. The smallness in general of this difference must not cause us perplexity, for in reality the weakest motives, the finest shades of difference, often decide in favour of this or that method of applying force. Our only business here is to show that, certain conditions being supposed, the possibility of attaining our purpose in different ways is no contradiction, absurdity, nor even error.

Besides these two means, there are three other peculiar ways of directly increasing the waste of the enemy's force. The first is *invasion*, that is *The occupation of the enemy's territory, not with a view to keeping it*, but in order to levy contributions upon it, or to devastate it.

The immediate object here is neither the conquest of the enemy's territory nor the defeat of his armed force, but merely to *do him damage in a general way*. The second way is to select for the object of our enterprises those points at which we can do the enemy most harm. Nothing is easier to conceive than two different directions in which our force may be employed, the first of which is to be preferred if our object is to defeat the enemy's Army, while the other is more advantageous if the defeat of the enemy is out of the question. According to the usual mode of speaking, we should say that the first is primarily military, the other more political. But if we take our view from the highest point, both are equally military, and neither the one nor the other can be eligible unless it suits the circumstances of the case. The third, by far the most important, from the great number of cases which it embraces, is the *wearing out* of the enemy. We choose this expression not only to explain our meaning in few words, but because it represents the thing exactly, and is not so figurative as may at first appear. The idea of wearing out in a struggle amounts in practice to *a gradual exhaustion of the physical powers and of the will by the long continuance of exertion*.

Now, if we want to overcome the enemy by the duration of the contest, we must content ourselves with as small objects as possible, for it is in the nature of the thing that a great end requires a greater expenditure of force than a small one; but the smallest object that we can propose to ourselves is simple passive resistance, that is a combat without any positive view. In this way, therefore, our means attain their greatest relative value, and therefore the result is best secured. How far now can this negative mode of proceeding be carried? Plainly not to absolute passivity, for mere endurance would not be fighting; and the defensive is an activity by which so much of the enemy's power must be destroyed that he must give up his object. That alone is what we aim at in each single act, and therein consists the negative nature of our object.

No doubt this negative object in its single act is not so effective as the positive object in the same direction would be, supposing it successful; but there is this difference in its favour, that it succeeds more easily than the positive, and therefore it holds out greater certainty of success; what is wanting in the efficacy of its single act must be gained through time, that is, through the duration of the contest, and therefore this negative intention, which constitutes the principle of the pure defensive, is also the natural means of overcoming the enemy by the duration of the combat, that is of wearing him out.

Here lies the origin of that difference of *offensive* and *Defensive*, the influence of which prevails throughout the whole province of War. We cannot at present pursue this subject further than to observe that from this negative intention are to be deduced all the advantages and all the stronger forms of combat which are on the side of the Defensive, and in which that philosophical-dynamic law which exists between the greatness and the certainty of success is realised. We shall resume the consideration of all this hereafter.

If then the negative purpose, that is the concentration of all the means into a state of pure resistance, affords a superiority in the contest, and if this advantage is sufficient to *balance* whatever superiority in numbers the adversary may have, then the mere *Duration* of the contest will suffice gradually to bring the loss of force on the part of the adversary to a point at which the political object can no longer be an equivalent, a point at which, therefore, he must give up the contest. We see then that this class of means, the wearing out of the enemy, includes the great number of cases in which the weaker resists the stronger.

Frederick the Great, during the Seven Years' War, was never strong enough to overthrow the Austrian monarchy; and if he had tried to do so after the fashion of Charles the Twelfth, he would inevitably have had to succumb himself. But after his skilful application of the system of husbanding his resources had shown the powers allied against him, through a seven years' struggle, that the actual expenditure of strength far exceeded what they had at first anticipated, they made peace.

We see then that there are many ways to one's object in War; that the complete subjugation of the enemy is not essential in every case; that the destruction of the enemy's military force, the conquest of the enemy's provinces, the mere occupation of them, the mere invasion of them—enterprises which are aimed directly at political objects—lastly, a passive expectation of the enemy's blow, are all means which, each in itself, may be used to force the enemy's will according as the peculiar circumstances of the case lead us to expect more from the one or the other. We could still add to these a whole category of shorter methods of gaining the end, which might be called arguments ad hominem. What branch of human affairs is there in which these sparks of individual spirit have not made their appearance, surmounting all formal considerations? And least of all can they fail to appear in War, where the personal character of the combatants plays such an important part, both in the cabinet and in the field. We limit ourselves to pointing this out, as it would be pedantry to attempt to reduce such influences into classes. Including these, we may say that the number of possible ways of reaching the object rises to infinity.

To avoid under-estimating these different short roads to one's purpose, either estimating them only as rare exceptions, or holding the difference which they cause in the conduct of War as insignificant, we must bear in mind the diversity of political objects which may cause a War—measure at a glance the distance which there is between a death struggle for political existence and a War which a forced or tottering alliance makes a matter of disagreeable duty. Between the two innumerable gradations occur in practice. If we reject one of these gradations in theory, we might with equal right reject the whole, which would be tantamount to shutting the real world completely out of sight.

These are the circumstances in general connected with the aim which we have to pursue in War; let us now turn to the means.

There is only one single means, it is the *fight*. However diversified this may be in form, however widely it may differ from a rough vent of hatred and animosity in a hand-to-hand encounter, whatever number of things may introduce themselves which are not actual fighting, still it is always implied in the conception of War that all the effects manifested have their roots in the combat.

That this must always be so in the greatest diversity and complication of the reality is proved in a very simple manner. All that takes place in War takes place through armed forces, but where the forces of War, i.e., armed men, are applied, there the idea of fighting must of necessity be at the foundation.

All, therefore, that relates to forces of War—all that is connected with their creation, maintenance, and application—belongs to military activity.

Creation and maintenance are obviously only the means, whilst application is the object.

The contest in War is not a contest of individual against individual, but an organised whole, consisting of manifold parts; in this great whole we may distinguish units of two kinds, the one determined by the subject, the other by the object. In an Army the mass of combatants ranges itself always into an order of new units, which again form members of a higher order. The combat of each of these members forms, therefore, also a more or less distinct unit. Further, the motive of the fight; therefore its object forms its unit.

Now, to each of these units which we distinguish in the contest we attach the name of combat.

If the idea of combat lies at the foundation of every application of armed power, then also the application of armed force in general is nothing more than the determining and arranging a certain number of combats.

Every activity in War, therefore, necessarily relates to the combat either directly or indirectly. The soldier is levied, clothed, armed, exercised, he sleeps, eats, drinks, and marches, all *merely to fight at the right time and place.*

If, therefore, all the threads of military activity terminate in the combat, we shall grasp them all when we settle the order of the combats. Only from this order and its execution proceed the effects, never directly from the conditions preceding them. Now, in the combat all the action is directed to the *destruction* of the enemy, or rather of *his fighting powers,* for this lies in the conception of combat. The destruction of the enemy's fighting power is, therefore, always the means to attain the object of the combat.

This object may likewise be the mere destruction of the enemy's armed force; but that is not by any means necessary, and it may be something quite different. Whenever, for instance, as we have shown, the defeat of the enemy is not the only means to attain the political object, whenever there are other objects which may be pursued as the aim in a War, then it follows of itself that such other objects may become the object of particular acts of Warfare, and therefore also the object of combats.

But even those combats which, as subordinate acts, are in the strict sense devoted to the destruction of the enemy's fighting force need not have that destruction itself as their first object.

If we think of the manifold parts of a great armed force, of the number of circumstances which come into activity when it is employed, then it is clear that the combat of such a force must also require a manifold organisation, a subordinating of parts and formation. There may and must naturally arise for particular parts a number of objects which are not themselves the destruction of the enemy's armed force, and which, while they certainly contribute to increase that destruction, do so only in an indirect manner. If a battalion is ordered to drive the enemy from a rising ground, or a bridge, etc., then properly the occupation of any such locality is the real object, the destruction of the enemy's armed force which takes place only the means or secondary matter. If the enemy can be driven away merely by a demonstration, the object is attained all the same; but this hill or bridge is, in point of fact, only required as a means of increasing the gross amount of loss inflicted on the enemy's armed force. It is the case on the field of battle, much more must it be so on the whole theatre of war, where not only one Army is opposed to another, but one State, one Nation, one whole country to another. Here the number of possible relations, and consequently possible combinations, is much greater, the diversity of measures increased, and by the gradation of objects, each subordinate to another the first means employed is further apart from the ultimate object.

It is therefore for many reasons possible that the object of a combat is not the destruction of the enemy's force, that is, of the force immediately opposed to us, but that this only appears as a means. But in all such cases it is no longer a question of complete destruction, for the combat is here nothing else but a measure of strength—has in itself no value except only that of the present result, that is, of its decision.

But a measuring of strength may be effected in cases where the opposing sides are very unequal by a mere comparative estimate. In such cases no fighting will take place, and the weaker will immediately give way.

If the object of a combat is not always the destruction of the enemy's forces therein engaged—and if its object can often be attained as well without the combat taking place at all, by merely making a resolve to fight, and by the circumstances to which this resolution gives rise—then that explains how a whole campaign may be carried on with great activity without the actual combat playing any notable part in it.

That this may be so military history proves by a hundred examples. How many of those cases can be justified, that is, without involving a contradiction and whether some of the celebrities who rose out of them would stand criticism, we shall leave undecided, for all we have to do with the matter is to show the possibility of such a course of events in War.

We have only one means in War—the battle; but this means, by the infinite variety of paths in which it may be applied, leads us into all the different ways which the multiplicity of objects allows of, so that we seem to have gained nothing; but that is not the case, for from this unity of means proceeds a thread which assists the study of the subject, as it runs through the whole web of military activity and holds it together.

But we have considered the destruction of the enemy's force as one of the objects which maybe pursued in War, and left undecided what relative importance should be given to it amongst other objects. In certain cases it will depend on circumstances, and as a general question we have left its value undetermined. We are once more brought back upon it, and we shall be able to get an insight into the value which must necessarily be accorded to it.

The combat is the single activity in War; in the combat the destruction of the enemy opposed to us is the means to the end; it is so even when the combat does not actually take place, because in that case there lies at the root of the decision the supposition at all events that this destruction is to be regarded as beyond doubt. It follows, therefore, that the destruction of the enemy's military force is the foundation-stone of all action in War, the great support of all combinations, which rest upon it like the arch on its abutments. All action, therefore, takes place on the supposition that if the solution by force of arms which lies at its foundation should be realised, it will be a favourable one. The decision by arms is, for all operations in War, great and small, what cash payment is in bill transactions. However remote from each other these relations, however seldom the realisation may take place, still it can never entirely fail to occur.

If the decision by arms lies at the foundation of all combinations, then it follows that the enemy can defeat each of them by gaining a victory on the field, not merely in the one on which our combination directly depends, but also in any other encounter, if it is only important enough; for every important decision by arms—that is, destruction of the enemy's forces—reacts upon all preceding it, because, like a liquid element, they tend to bring themselves to a level.

Thus, the destruction of the enemy's armed force appears, therefore, always as the superior and more effectual means, to which all others must give way.

It is, however, only when there is a supposed equality in all other conditions that we can ascribe to the destruction of the enemy's armed force the greater efficacy. It would, therefore, be a great mistake to draw the conclusion that a blind dash must always gain the victory over skill and caution. An unskilful attack would lead to the destruction of our own and not of the enemy's force, and therefore is not what is here meant. The superior efficacy belongs not to the *means* but to the *end*, and we are only comparing the effect of one realised purpose with the other.

If we speak of the destruction of the enemy's armed force, we must expressly point out that nothing obliges us to confine this idea to the mere physical force; on the contrary, the moral is necessarily implied as well, because both in fact are interwoven with each other, even in the most minute details, and therefore cannot be separated. But it is just in connection with the inevitable effect which has been referred to, of a great act of destruction (a great victory) upon all other decisions by arms, that this moral element is most fluid, if we may use that expression, and therefore distributes itself the most easily through all the parts.

Against the far superior worth which the destruction of the enemy's armed force has over all other means stands the expense and risk of this means, and it is only to avoid these that any other means are taken. That these must be costly stands to reason, for the waste of our own military forces must, ceteris paribus, always be greater the more our aim is directed upon the destruction of the enemy's power.

The danger lies in this, that the greater efficacy which we seek recoils on ourselves, and therefore has worse consequences in case we fail of success.

Other methods are, therefore, less costly when they succeed, less dangerous when they fail; but in this is necessarily lodged the condition that they are only opposed to similar ones, that is, that the enemy acts on the same principle; for if the enemy should choose the way of a great decision by arms, *Our Means must on That Account Be Changed Against Our Will, in Order to Correspond with His.* Then all depends on the issue of the act of destruction; but of course it is evident that, ceteris paribus, in this act we must be at a disadvantage in all respects because our views and our means had been directed in part upon other objects, which is not the case with the enemy. Two different objects of which one is not part, the other exclude each other, and therefore a force which may be applicable for the one may not serve for the other. If, therefore, one of two belligerents is determined to seek the great decision by arms, then he has a high probability of success, as soon as he is certain his opponent will not take that way, but follows a different object; and every one who sets before himself any such other aim only does so in a reasonable manner, provided he acts on the supposition that his adversary has as little intention as he has of resorting to the great decision by arms.

But what we have here said of another direction of views and forces relates only to Other *Positive Objects*, which we may propose to ourselves in War, besides the destruction of the enemy's force, not by any means to the pure defensive, which may be adopted with a view thereby to exhaust the enemy's forces. In the pure defensive the positive object is wanting, and therefore, while on the defensive, our forces cannot at the same time be directed on other objects; they can only be employed to defeat the intentions of the enemy.

We have now to consider the opposite of the destruction of the enemy's armed force, that is to say, the preservation of our own. These two efforts always go together, as they mutually act and react on each other; they are integral parts of one and the same view, and we have only to ascertain what effect is produced when one or the other has the predominance. The endeavour to destroy the enemy's force has a positive object, and leads to positive results, of which the final aim is the conquest of the enemy. The preservation of our own forces has a negative object, leads therefore to the defeat of the enemy's intentions, that is to pure resistance, of which the final aim can be nothing more than to prolong the duration of the contest, so that the enemy shall exhaust himself in it.

The effort with a positive object calls into existence the act of destruction; the effort with the negative object awaits it.

How far this state of expectation should and may be carried we shall enter into more particularly in the theory of attack and defence, at the origin of which we again find ourselves. Here we shall content ourselves with saying that the awaiting must be no absolute endurance, and that in the action bound up with it the destruction of the enemy's armed force engaged in this conflict may be the aim just as well as anything else. It would therefore be a great error in the fundamental idea to suppose that the consequence of the negative course is that we are precluded from choosing the destruction of the enemy's military force as our object, and must prefer a bloodless solution. The advantage which the negative effort gives may certainly lead to that, but only at the risk of its not being the most advisable method, as that question is dependent on totally different conditions, resting not with ourselves but with our opponents. This other bloodless way cannot, therefore, be looked upon at all as the natural means of satisfying our great anxiety to spare our forces; on the contrary, when circumstances are not favourable, it would be the means of completely ruining them. Very many Generals have fallen into this error, and been ruined by it. The only necessary effect resulting from the superiority of the negative effort is the delay of the decision, so that the party acting takes refuge in that way, as it were, in the expectation of the decisive moment. The consequence of that is generally *The postponement of the action* as much as possible in time, and also in space, in so far as space is in connection with it. If the moment has arrived in which this can no longer be done without ruinous disadvantage, then the advantage of the negative must be considered as exhausted, and then comes forward unchanged the effort for the destruction of the enemy's force, which was kept back by a counterpoise, but never discarded.

We have seen, therefore, in the foregoing reflections, that there are many ways to the aim, that is, to the attainment of the political object; but that the only means is the combat, and that consequently everything is subject to a supreme law: which is the *decision by arms*; that where this is really demanded by one, it is a redress which cannot be refused by the other; that, therefore, a belligerent who takes any other way must make sure that his opponent will not take this means of redress, or his cause may be lost in that supreme court; hence therefore the destruction of the enemy's armed force, amongst all the objects which can be pursued in War, appears always as the one which overrules all others.

What may be achieved by combinations of another kind in War we shall only learn in the sequel, and naturally only by degrees. We content ourselves here with acknowledging in general their possibility, as something pointing to the difference between the reality and the conception, and to the influence of particular circumstances. But we could not avoid showing at once that the *bloody solution of the crisis*, the effort for the destruction of the enemy's force, is the firstborn son of War. If when political objects are unimportant, motives weak, the excitement of forces small, a cautious commander tries in all kinds of ways, without great crises and bloody solutions, to twist himself skilfully into a peace through the characteristic weaknesses of his enemy in the field and in the Cabinet, we have no right to find fault with him, if the premises on which he acts are well founded and justified by success; still we must require him to remember that he only travels on forbidden tracks, where the God of War may surprise him; that he ought always to keep his eye on the enemy, in order that he may not have to defend himself with a dress rapier if the enemy takes up a sharp sword.

The consequences of the nature of War, how ends and means act in it, how in the modifications of reality it deviates sometimes more, sometimes less, from its strict original conception, fluctuating backwards and forwards, yet always remaining under that strict conception as under a supreme law: all this we must retain before us, and bear constantly in mind in the consideration of each of the succeeding subjects, if we would rightly comprehend their true relations and proper importance, and not become involved incessantly in the most glaring contradictions with the reality, and at last with our own selves.

The Genius for War

Every special calling in life, if it is to be followed with success, requires peculiar qualifications of understanding and soul. Where these are of a high order, and manifest themselves by extraordinary achievements, the mind to which they belong is termed *Genius*.

We know very well that this word is used in many significations which are very different both in extent and nature, and that with many of these significations it is a very difficult task to define the essence of Genius; but as we neither profess to be philosopher nor grammarian, we must be allowed to keep to the meaning usual in ordinary language, and to understand by "genius" a very high mental capacity for certain employments.

We wish to stop for a moment over this faculty and dignity of the mind, in order to vindicate its title, and to explain more fully the meaning of the conception. But we shall not dwell on that (genius) which has obtained its title through a very great talent, on genius properly so called, that is a conception which has no defined limits. What we have to do is to bring under consideration every common tendency of the powers of the mind and soul towards the business of War, the whole of which common tendencies we may look upon as the *essence of military genius*. We say "common," for just therein consists military genius, that it is not one single quality bearing upon War, as, for instance, courage, while other qualities of mind and soul are wanting or have a direction which is unserviceable for War, but that it is *an harmonious association of powers*, in which one or other may predominate, but none must be in opposition.

If every combatant required to be more or less endowed with military genius, then our armies would be very weak; for as it implies a peculiar bent of the intelligent powers, therefore it can only rarely be found where the mental powers of a people are called into requisition and trained in many different ways. The fewer the employments followed by a Nation, the more that of arms predominates, so much the more prevalent will military genius also be found. But this merely applies to its prevalence, by no means to its degree, for that depends on the general state of intellectual culture in the country. If we look at a wild, warlike race, then we find a warlike spirit in individuals much more common than in a civilised people; for in the former almost every warrior possesses it, whilst in the civilised whole, masses are only carried away by it from necessity, never by inclination. But amongst uncivilised people we never find a really great General, and very seldom what we can properly call a military genius, because that requires a development of the intelligent powers which cannot be found in an uncivilised state. That a civilised people may also have a warlike tendency and development is a matter of course; and the more this is general, the more frequently also will military spirit be found in individuals in their armies. Now as this coincides in such case with the higher degree of civilisation, therefore from such nations have issued forth the most brilliant military exploits, as the Romans and the French have exemplified. The greatest names in these and in all other nations that have been renowned in War belong strictly to epochs of higher culture.

From this we may infer how great a share the intelligent powers have in superior military genius. We shall now look more closely into this point.

War is the province of danger, and therefore courage above all things is the first quality of a warrior.

Courage is of two kinds: first, physical courage, or courage in presence of danger to the person; and next, moral courage, or courage before responsibility, whether it be before the judgment-seat of external authority, or of the inner power, the conscience. We only speak here of the first.

Courage before danger to the person, again, is of two kinds. First, it may be indifference to danger, whether proceeding from the organism of the individual, contempt of death, or habit: in any of these cases it is to be regarded as a permanent condition.

Secondly, courage may proceed from positive motives, such as personal pride, patriotism, enthusiasm of any kind. In this case courage is not so much a normal condition as an impulse.

We may conceive that the two kinds act differently. The first kind is more certain, because it has become a second nature, never forsakes the man; the second often leads him farther. In the first there is more of firmness, in the second, of boldness. The first leaves the judgment cooler, the second raises its power at times, but often bewilders it. The two combined make up the most perfect kind of courage.

War is the province of physical exertion and suffering. In order not to be completely overcome by them, a certain strength of body and mind is required, which, either natural or acquired, produces indifference to them. With these qualifications, under the guidance of simply a sound understanding, a man is at once a proper instrument for War; and these are the qualifications so generally to be met with amongst wild and half-civilised tribes. If we go further in the demands which War makes on it, then we find the powers of the understanding predominating. War is the province of uncertainty: three-fourths of those things upon which action in War must be calculated, are hidden more or less in the clouds of great uncertainty. Here, then, above all a fine and penetrating mind is called for, to search out the truth by the tact of its judgment.

An average intellect may, at one time, perhaps hit upon this truth by accident; an extraordinary courage, at another, may compensate for the want of this tact; but in the majority of cases the average result will always bring to light the deficient understanding.

War is the province of chance. In no sphere of human activity is such a margin to be left for this intruder, because none is so much in constant contact with him on all sides. He increases the uncertainty of every circumstance, and deranges the course of events.

From this uncertainty of all intelligence and suppositions, this continual interposition of chance, the actor in War constantly finds things different from his expectations; and this cannot fail to have an influence on his plans, or at least on the presumptions connected with these plans. If this influence is so great as to render the pre-determined plan completely nugatory, then, as a rule, a new one must be substituted in its place; but at the moment the necessary data are often wanting for this, because in the course of action circumstances press for immediate decision, and allow no time to look about for fresh data, often not enough for mature consideration.

But it more often happens that the correction of one premise, and the knowledge of chance events which have arisen, are not sufficient to overthrow our plans completely, but only suffice to produce hesitation. Our knowledge of circumstances has increased, but our uncertainty, instead of having diminished, has only increased. The reason of this is, that we do not gain all our experience at once, but by degrees; thus our determinations continue to be assailed incessantly by fresh experience; and the mind, if we may use the expression, must always be "under arms."

Now, if it is to get safely through this perpetual conflict with the unexpected, two qualities are indispensable: in the first place an intellect which, even in the midst of this intense obscurity, is not without some traces of inner light, which lead to the truth, and then the courage to follow this faint light. The first is figuratively expressed by the French phrase coup d'oeil. The other is resolution. As the battle is the feature in War to which attention was originally chiefly directed, and as time and space are important elements in it, more particularly when cavalry with their rapid decisions were the chief arm, the idea of rapid and correct decision related in the first instance to the estimation of these two elements, and to denote the idea an expression was adopted which actually only points to a correct judgment by eye. Many teachers of the Art of War then gave this limited signification as the definition of coup d'oeil. But it is undeniable that all able decisions formed in the moment of action soon came to be understood by the expression, as, for instance, the hitting upon the right point of attack, etc. It is, therefore, not only the physical, but more frequently the mental eye which is meant in coup d'oeil. Naturally, the expression, like the thing, is always more in its place in the field of tactics: still, it must not be wanting in strategy, inasmuch as in it rapid decisions are often necessary. If we strip this conception of that which the expression has given it of the over-figurative and restricted, then it amounts simply to the rapid discovery of a truth which to the ordinary mind is either not visible at all or only becomes so after long examination and reflection.

Resolution is an act of courage in single instances, and if it becomes a characteristic trait, it is a habit of the mind. But here we do not mean courage in face of bodily danger, but in face of responsibility, therefore, to a certain extent against moral danger. This has been often called courage d'esprit, on the ground that it springs from the understanding; nevertheless, it is no act of the understanding on that account; it is an act of feeling. Mere intelligence is still not courage, for we often see the cleverest people devoid of resolution. The mind must, therefore, first awaken the feeling of courage, and then be guided and supported by it, because in momentary emergencies the man is swayed more by his feelings than his thoughts.

We have assigned to resolution the office of removing the torments of doubt, and the dangers of delay, when there are no sufficient motives for guidance. Through the unscrupulous use of language which is prevalent, this term is often applied to the mere propensity to daring, to bravery, boldness, or temerity. But, when there are *sufficient motives* in the man, let them be objective or subjective, true or false, we have no right to speak of his resolution; for, when we do so, we put ourselves in his place, and we throw into the scale doubts which did not exist with him.

Here there is no question of anything but of strength and weakness. We are not pedantic enough to dispute with the use of language about this little misapplication, our observation is only intended to remove wrong objections.

This resolution now, which overcomes the state of doubting, can only be called forth by the intellect, and, in fact, by a peculiar tendency of the same. We maintain that the mere union of a superior understanding and the necessary feelings are not sufficient to make up resolution. There are persons who possess the keenest perception for the most difficult problems, who are also not fearful of responsibility, and yet in cases of difficulty cannot come to a resolution. Their courage and their sagacity operate independently of each other, do not give each other a hand, and on that account do not produce resolution as a result. The forerunner of resolution is an act of the mind making evident the necessity of venturing, and thus influencing the will. This quite peculiar direction of the mind, which conquers every other fear in man by the fear of wavering or doubting, is what makes up resolution in strong minds; therefore, in our opinion, men who have little intelligence can never be resolute. They may act without hesitation under perplexing circumstances, but then they act without reflection. Now, of course, when a man acts without reflection he cannot be at variance with himself by doubts, and such a mode of action may now and then lead to the right point; but we say now as before, it is the average result which indicates the existence of military genius. Should our assertion appear extraordinary to any one, because he knows many a resolute hussar officer who is no deep thinker, we must remind him that the question here is about a peculiar direction of the mind, and not about great thinking powers.

We believe, therefore, that resolution is indebted to a special direction of the mind for its existence, a direction which belongs to a strong head rather than to a brilliant one. In corroboration of this genealogy of resolution we may add that there have been many instances of men who have shown the greatest resolution in an inferior rank, and have lost it in a higher position. While, on the one hand, they are obliged to resolve, on the other they see the dangers of a wrong decision, and as they are surrounded with things new to them, their understanding loses its original force, and they become only the more timid the more they become aware of the danger of the irresolution into which they have fallen, and the more they have formerly been in the habit of acting on the spur of the moment.

From the coup d'oeil and resolution we are naturally to speak of its kindred quality, *presence of mind*, which in a region of the unexpected like War must act a great part, for it is indeed nothing but a great conquest over the unexpected. As we admire presence of mind in a pithy answer to anything said unexpectedly, so we admire it in a ready expedient on sudden danger. Neither the answer nor the expedient need be in themselves extraordinary, if they only hit the point; for that which as the result of mature reflection would be nothing unusual, therefore insignificant in its impression on us, may as an instantaneous act of the mind produce a pleasing impression. The expression "presence of mind" certainly denotes very fitly the readiness and rapidity of the help rendered by the mind.

Whether this noble quality of a man is to be ascribed more to the peculiarity of his mind or to the equanimity of his feelings, depends on the nature of the case, although neither of the two can be entirely wanting. A telling repartee bespeaks rather a ready wit, a ready expedient on sudden danger implies more particularly a well-balanced mind.

If we take a general view of the four elements composing the atmosphere in which War moves, of *danger, physical effort, uncertainty*, and *chance*, it is easy to conceive that a great force of mind and understanding is requisite to be able to make way with safety and success amongst such opposing elements, a force which, according to the different modifications arising out of circumstances, we find termed by military writers and annalists as *energy, firmness, staunchness, strength of mind and character*. All these manifestations of the heroic nature might be regarded as one and the same power of volition, modified according to circumstances; but nearly related as these things are to each other, still they are not one and the same, and it is desirable for us to distinguish here a little more closely at least the action of the powers of the soul in relation to them.

In the first place, to make the conception clear, it is essential to observe that the weight, burden, resistance, or whatever it may be called, by which that force of the soul in the General is brought to light, is only in a very small measure the enemy's activity, the enemy's resistance, the enemy's action directly. The enemy's activity only affects the General directly in the first place in relation to his person, without disturbing his action as Commander. If the enemy, instead of two hours, resists for four, the Commander instead of two hours is four hours in danger; this is a quantity which plainly diminishes the higher the rank of the Commander. What is it for one in the post of Commander-in-Chief? It is nothing.

Secondly, although the opposition offered by the enemy has a direct effect on the Commander through the loss of means arising from prolonged resistance, and the responsibility connected with that loss, and his force of will is first tested and called forth by these anxious considerations, still we maintain that this is not the heaviest burden by far which he has to bear, because he has only himself to settle with. All the other effects of the enemy's resistance act directly upon the combatants under his command, and through them react upon him.

As long as his men full of good courage fight with zeal and spirit, it is seldom necessary for the Chief to show great energy of purpose in the pursuit of his object. But as soon as difficulties arise—and that must always happen when great results are at stake—then things no longer move on of themselves like a well-oiled machine, the machine itself then begins to offer resistance, and to overcome this the Commander must have a great force of will. By this resistance we must not exactly suppose disobedience and murmurs, although these are frequent enough with particular individuals; it is the whole feeling of the dissolution of all physical and moral power, it is the heartrending sight of the bloody sacrifice which the Commander has to contend with in himself, and then in all others who directly or indirectly transfer to him their impressions, feelings, anxieties, and desires. As the forces in one individual after another become prostrated, and can no longer be excited and supported by an effort of his own will, the whole inertia of the mass gradually rests its weight on the Will of the Commander: by the spark in his breast, by the light of his spirit, the spark of purpose, the light of hope, must be kindled afresh in others: in so far only as he is equal to this, he stands above the masses and continues to be their master; whenever that influence ceases, and his own spirit is no longer strong enough to revive the spirit of all others, the masses drawing him down with them sink into the lower region of animal nature, which shrinks from danger and knows not shame. These are the weights which the courage and intelligent faculties of the military Commander have to overcome if he is to make his name illustrious. They increase with the masses, and therefore, if the forces in question are to continue equal to the burden, they must rise in proportion to the height of the station.

Energy in action expresses the strength of the motive through which the action is excited, let the motive have its origin in a conviction of the understanding, or in an impulse. But the latter can hardly ever be wanting where great force is to show itself.

Of all the noble feelings which fill the human heart in the exciting tumult of battle, none, we must admit, are so powerful and constant as the soul's thirst for honour and renown, which the German language treats so unfairly and tends to depreciate by the unworthy associations in the words Ehrgeiz (greed of honour) and Ruhmsucht (hankering after glory). No doubt it is just in War that the abuse of these proud aspirations of the soul must bring upon the human race the most shocking outrages, but by their origin they are certainly to be counted amongst the noblest feelings which belong to human nature, and in War they are the vivifying principle which gives the enormous body a spirit. Although other feelings may be more general in their influence, and many of them—such as love of country, fanaticism, revenge, enthusiasm of every kind—may seem to stand higher, the thirst for honour and renown still remains indispensable. Those other feelings may rouse the great masses in general, and excite them more powerfully, but they do not give the Leader a desire to will more than others, which is an essential requisite in his position if he is to make himself distinguished in it. They do not, like a thirst for honour, make the military act specially the property of the Leader, which he strives to turn to the best account; where he ploughs with toil, sows with care, that he may reap plentifully. It is through these aspirations we have been speaking of in Commanders, from the highest to the lowest, this sort of energy, this spirit of emulation, these incentives, that the action of armies is chiefly animated and made successful. And now as to that which specially concerns the head of all, we ask, Has there ever been a great Commander destitute of the love of honour, or is such a character even conceivable?

Firmness denotes the resistance of the will in relation to the force of a single blow, STAUNCHNESS in relation to a continuance of blows. Close as is the analogy between the two, and often as the one is used in place of the other, still there is a notable difference between them which cannot be mistaken, inasmuch as firmness against a single powerful impression may have its root in the mere strength of a feeling, but staunchness must be supported rather by the understanding, for the greater the duration of an action the more systematic deliberation is connected with it, and from this staunchness partly derives its power.

If we now turn to *Strength of mind or soul*, then the first question is, What are we to understand thereby?

Plainly it is not vehement expressions of feeling, nor easily excited passions, for that would be contrary to all the usage of language, but the power of listening to reason in the midst of the most intense excitement, in the storm of the most violent passions. Should this power depend on strength of understanding alone? We doubt it. The fact that there are men of the greatest intellect who cannot command themselves certainly proves nothing to the contrary, for we might say that it perhaps requires an understanding of a powerful rather than of a comprehensive nature; but we believe we shall be nearer the truth if we assume that the power of submitting oneself to the control of the understanding, even in moments of the most violent excitement of the feelings, that power which we call *self-command*, has its root in the heart itself. It is, in point of fact, another feeling, which in strong minds balances the excited passions without destroying them; and it is only through this equilibrium that the mastery of the understanding is secured. This counterpoise is nothing but a sense of the dignity of man, that noblest pride, that deeply-seated desire of the soul always to act as a being endued with understanding and reason. We may therefore say that a strong mind is one which does not lose its balance even under the most violent excitement.

If we cast a glance at the variety to be observed in the human character in respect to feeling, we find, first, some people who have very little excitability, who are called phlegmatic or indolent.

Secondly, some very excitable, but whose feelings still never overstep certain limits, and who are therefore known as men full of feeling, but sober-minded.

Thirdly, those who are very easily roused, whose feelings blaze up quickly and violently like gunpowder, but do not last.

Fourthly, and lastly, those who cannot be moved by slight causes, and who generally are not to be roused suddenly, but only gradually; but whose feelings become very powerful and are much more lasting. These are men with strong passions, lying deep and latent.

This difference of character lies probably close on the confines of the physical powers which move the human organism, and belongs to that amphibious organisation which we call the nervous system, which appears to be partly material, partly spiritual. With our weak philosophy, we shall not proceed further in this mysterious field. But it is important for us to spend a moment over the effects which these different natures have on, action in War, and to see how far a great strength of mind is to be expected from them.

Indolent men cannot easily be thrown out of their equanimity, but we cannot certainly say there is strength of mind where there is a want of all manifestation of power.

At the same time, it is not to be denied that such men have a certain peculiar aptitude for War, on account of their constant equanimity. They often want the positive motive to action, impulse, and consequently activity, but they are not apt to throw things into disorder.

The peculiarity of the second class is that they are easily excited to act on trifling grounds, but in great matters they are easily overwhelmed. Men of this kind show great activity in helping an unfortunate individual, but by the distress of a whole Nation they are only inclined to despond, not roused to action.

Such people are not deficient in either activity or equanimity in War; but they will never accomplish anything great unless a great intellectual force furnishes the motive, and it is very seldom that a strong, independent mind is combined with such a character.

Excitable, inflammable feelings are in themselves little suited for practical life, and therefore they are not very fit for War. They have certainly the advantage of strong impulses, but that cannot long sustain them. At the same time, if the excitability in such men takes the direction of courage, or a sense of honour, they may often be very useful in inferior positions in War, because the action in War over which commanders in inferior positions have control is generally of shorter duration. Here one courageous resolution, one effervescence of the forces of the soul, will often suffice. A brave attack, a soul-stirring hurrah, is the work of a few moments, whilst a brave contest on the battle-field is the work of a day, and a campaign the work of a year.

Owing to the rapid movement of their feelings, it is doubly difficult for men of this description to preserve equilibrium of the mind; therefore they frequently lose head, and that is the worst phase in their nature as respects the conduct of War. But it would be contrary to experience to maintain that very excitable spirits can never preserve a steady equilibrium—that is to say, that they cannot do so even under the strongest excitement. Why should they not have the sentiment of self-respect, for, as a rule, they are men of a noble nature? This feeling is seldom wanting in them, but it has not time to produce an effect. After an outburst they suffer most from a feeling of inward humiliation. If through education, self-observance, and experience of life, they have learned, sooner or later, the means of being on their guard, so that at the moment of powerful excitement they are conscious betimes of the counteracting force within their own breasts, then even such men may have great strength of mind.

Lastly, those who are difficult to move, but on that account susceptible of very deep feelings, men who stand in the same relation to the preceding as red heat to a flame, are the best adapted by means of their Titanic strength to roll away the enormous masses by which we may figuratively represent the difficulties which beset command in War. The effect of their feelings is like the movement of a great body, slower, but more irresistible.

Although such men are not so likely to be suddenly surprised by their feelings and carried away so as to be afterwards ashamed of themselves, like the preceding, still it would be contrary to experience to believe that they can never lose their equanimity, or be overcome by blind passion; on the contrary, this must always happen whenever the noble pride of self-control is wanting, or as often as it has not sufficient weight. We see examples of this most frequently in men of noble minds belonging to savage nations, where the low degree of mental cultivation favours always the dominance of the passions. But even amongst the most civilised classes in civilised States, life is full of examples of this kind—of men carried away by the violence of their passions, like the poacher of old chained to the stag in the forest.

We therefore say once more a strong mind is not one that is merely susceptible of strong excitement, but one which can maintain its serenity under the most powerful excitement, so that, in spite of the storm in the breast, the perception and judgment can act with perfect freedom, like the needle of the compass in the storm-tossed ship.

By the term *strength of character*, or simply *character*, is denoted tenacity of conviction, let it be the result of our own or of others' views, and whether they are principles, opinions, momentary inspirations, or any kind of emanations of the understanding; but this kind of firmness certainly cannot manifest itself if the views themselves are subject to frequent change. This frequent change need not be the consequence of external influences; it may proceed from the continuous activity of our own mind, in which case it indicates a characteristic unsteadiness of mind. Evidently we should not say of a man who changes his views every moment, however much the motives of change may originate with himself, that he has character. Only those men, therefore, can be said to have this quality whose conviction is very constant, either because it is deeply rooted and clear in itself, little liable to alteration, or because, as in the case of indolent men, there is a want of mental activity, and therefore a want of motives to change; or lastly, because an explicit act of the will, derived from an imperative maxim of the understanding, refuses any change of opinion up to a certain point.

Now in War, owing to the many and powerful impressions to which the mind is exposed, and in the uncertainty of all knowledge and of all science, more things occur to distract a man from the road he has entered upon, to make him doubt himself and others, than in any other human activity.

The harrowing sight of danger and suffering easily leads to the feelings gaining ascendency over the conviction of the understanding; and in the twilight which surrounds everything a deep clear view is so difficult that a change of opinion is more conceivable and more pardonable. It is, at all times, only conjecture or guesses at truth which we have to act upon. This is why differences of opinion are nowhere so great as in War, and the stream of impressions acting counter to one's own convictions never ceases to flow. Even the greatest impassibility of mind is hardly proof against them, because the impressions are powerful in their nature, and always act at the same time upon the feelings.

When the discernment is clear and deep, none but general principles and views of action from a high standpoint can be the result; and on these principles the opinion in each particular case immediately under consideration lies, as it were, at anchor. But to keep to these results of bygone reflection, in opposition to the stream of opinions and phenomena which the present brings with it, is just the difficulty. Between the particular case and the principle there is often a wide space which cannot always be traversed on a visible chain of conclusions, and where a certain faith in self is necessary and a certain amount of scepticism is serviceable. Here often nothing else will help us but an imperative maxim which, independent of reflection, at once controls it: that maxim is, in all doubtful cases to adhere to the first opinion, and not to give it up until a clear conviction forces us to do so. We must firmly believe in the superior authority of well-tried maxims, and under the dazzling influence of momentary events not forget that their value is of an inferior stamp. By this preference which in doubtful cases we give to first convictions, by adherence to the same our actions acquire that stability and consistency which make up what is called character.

It is easy to see how essential a well-balanced mind is to strength of character; therefore men of strong minds generally have a great deal of character.

Force of character leads us to a spurious variety of it—*obstinacy*.

It is often very difficult in concrete cases to say where the one ends and the other begins; on the other hand, it does not seem difficult to determine the difference in idea.

Obstinacy is no fault of the understanding; we use the term as denoting a resistance against our better judgment, and it would be inconsistent to charge that to the understanding, as the understanding is the power of judgment. Obstinacy is *a fault of the feelings* or heart. This inflexibility of will, this impatience of contradiction, have their origin only in a particular kind of egotism, which sets above every other pleasure that of governing both self and others by its own mind alone. We should call it a kind of vanity, were it not decidedly something better. Vanity is satisfied with mere show, but obstinacy rests upon the enjoyment of the thing.

We say, therefore, force of character degenerates into obstinacy whenever the resistance to opposing judgments proceeds not from better convictions or a reliance upon a trustworthy maxim, but from a feeling of opposition. If this definition, as we have already admitted, is of little assistance practically, still it will prevent obstinacy from being considered merely force of character intensified, whilst it is something essentially different—something which certainly lies close to it and is cognate to it, but is at the same time so little an intensification of it that there are very obstinate men who from want of understanding have very little force of character.

Having in these high attributes of a great military Commander made ourselves acquainted with those qualities in which heart and head co-operate, we now come to a speciality of military activity which perhaps may be looked upon as the most marked if it is not the most important, and which only makes a demand on the power of the mind without regard to the forces of feelings. It is the connection which exists between War and country or ground.

This connection is, in the first place, a permanent condition of War, for it is impossible to imagine our organised Armies effecting any operation otherwise than in some given space; it is, secondly, of the most decisive importance, because it modifies, at times completely alters, the action of all forces; thirdly, while on the one hand it often concerns the most minute features of locality, on the other it may apply to immense tracts of country.

In this manner a great peculiarity is given to the effect of this connection of War with country and ground. If we think of other occupations of man which have a relation to these objects, on horticulture, agriculture, on building houses and hydraulic works, on mining, on the chase, and forestry, they are all confined within very limited spaces which may be soon explored with sufficient exactness. But the Commander in War must commit the business he has in hand to a corresponding space which his eye cannot survey, which the keenest zeal cannot always explore, and with which, owing to the constant changes taking place, he can also seldom become properly acquainted. Certainly the enemy generally is in the same situation; still, in the first place, the difficulty, although common to both, is not the less a difficulty, and he who by talent and practice overcomes it will have a great advantage on his side; secondly, this equality of the difficulty on both sides is merely an abstract supposition which is rarely realised in the particular case, as one of the two opponents (the defensive) usually knows much more of the locality than his adversary.

This very peculiar difficulty must be overcome by a natural mental gift of a special kind which is known by the—too restricted—term of Orisinn sense of locality. It is the power of quickly forming a correct geometrical idea of any portion of country, and consequently of being able to find one's place in it exactly at any time. This is plainly an act of the imagination. The perception no doubt is formed partly by means of the physical eye, partly by the mind, which fills up what is wanting with ideas derived from knowledge and experience, and out of the fragments visible to the physical eye forms a whole; but that this whole should present itself vividly to the reason, should become a picture, a mentally drawn map, that this picture should be fixed, that the details should never again separate themselves—all that can only be effected by the mental faculty which we call imagination. If some great poet or painter should feel hurt that we require from his goddess such an office; if he shrugs his shoulders at the notion that a sharp gamekeeper must necessarily excel in imagination, we readily grant that we only speak here of imagination in a limited sense, of its service in a really menial capacity. But, however slight this service, still it must be the work of that natural gift, for if that gift is wanting, it would be difficult to imagine things plainly in all the completeness of the visible. That a good memory is a great assistance we freely allow, but whether memory is to be considered as an independent faculty of the mind in this case, or whether it is just that power of imagination which here fixes these things better on the memory, we leave undecided, as in many respects it seems difficult upon the whole to conceive these two mental powers apart from each other.

That practice and mental acuteness have much to do with it is not to be denied. Puysegur, the celebrated Quartermaster-General of the famous Luxemburg, used to say that he had very little confidence in himself in this respect at first, because if he had to fetch the parole from a distance he always lost his way.

It is natural that scope for the exercise of this talent should increase along with rank. If the hussar and rifleman in command of a patrol must know well all the highways and byways, and if for that a few marks, a few limited powers of observation, are sufficient, the Chief of an Army must make himself familiar with the general geographical features of a province and of a country; must always have vividly before his eyes the direction of the roads, rivers, and hills, without at the same time being able to dispense with the narrower "sense of locality" Orisinn. No doubt, information of various kinds as to objects in general, maps, books, memoirs, and for details the assistance of his Staff, are a great help to him; but it is nevertheless certain that if he has himself a talent for forming an ideal picture of a country quickly and distinctly, it lends to his action an easier and firmer step, saves him from a certain mental helplessness, and makes him less dependent on others.

If this talent then is to be ascribed to imagination, it is also almost the only service which military activity requires from that erratic goddess, whose influence is more hurtful than useful in other respects.

We think we have now passed in review those manifestations of the powers of mind and soul which military activity requires from human nature. Everywhere intellect appears as an essential co-operative force; and thus we can understand how the work of War, although so plain and simple in its effects, can never be conducted with distinguished success by people without distinguished powers of the understanding.

When we have reached this view, then we need no longer look upon such a natural idea as the turning an enemy's position, which has been done a thousand times, and a hundred other similar conceptions, as the result of a great effort of genius.

Certainly one is accustomed to regard the plain honest soldier as the very opposite of the man of reflection, full of inventions and ideas, or of the brilliant spirit shining in the ornaments of refined education of every kind. This antithesis is also by no means devoid of truth; but it does not show that the efficiency of the soldier consists only in his courage, and that there is no particular energy and capacity of the brain required in addition to make a man merely what is called a true soldier. We must again repeat that there is nothing more common than to hear of men losing their energy on being raised to a higher position, to which they do not feel themselves equal; but we must also remind our readers that we are speaking of pre-eminent services, of such as give renown in the branch of activity to which they belong. Each grade of command in War therefore forms its own stratum of requisite capacity of fame and honour.

An immense space lies between a General—that is, one at the head of a whole War, or of a theatre of War—and his Second in Command, for the simple reason that the latter is in more immediate subordination to a superior authority and supervision, consequently is restricted to a more limited sphere of independent thought. This is why common opinion sees no room for the exercise of high talent except in high places, and looks upon an ordinary capacity as sufficient for all beneath: this is why people are rather inclined to look upon a subordinate General grown grey in the service, and in whom constant discharge of routine duties has produced a decided poverty of mind, as a man of failing intellect, and, with all respect for his bravery, to laugh at his simplicity. It is not our object to gain for these brave men a better lot—that would contribute nothing to their efficiency, and little to their happiness; we only wish to represent things as they are, and to expose the error of believing that a mere bravo without intellect can make himself distinguished in War.

As we consider distinguished talents requisite for those who are to attain distinction, even in inferior positions, it naturally follows that we think highly of those who fill with renown the place of Second in Command of an Army; and their seeming simplicity of character as compared with a polyhistor, with ready men of business, or with councillors of state, must not lead us astray as to the superior nature of their intellectual activity. It happens sometimes that men import the fame gained in an inferior position into a higher one, without in reality deserving it in the new position; and then if they are not much employed, and therefore not much exposed to the risk of showing their weak points, the judgment does not distinguish very exactly what degree of fame is really due to them; and thus such men are often the occasion of too low an estimate being formed of the characteristics required to shine in certain situations.

For each station, from the lowest upwards, to render distinguished services in War, there must be a particular genius. But the title of genius, history and the judgment of posterity only confer, in general, on those minds which have shone in the highest rank, that of Commanders-in-Chief. The reason is that here, in point of fact, the demand on the reasoning and intellectual powers generally is much greater.

To conduct a whole War, or its great acts, which we call campaigns, to a successful termination, there must be an intimate knowledge of State policy in its higher relations. The conduct of the War and the policy of the State here coincide, and the General becomes at the same time the Statesman.

We do not give Charles XII. the name of a great genius, because he could not make the power of his sword subservient to a higher judgment and philosophy—could not attain by it to a glorious object. We do not give that title to Henry IV. (of France), because he did not live long enough to set at rest the relations of different States by his military activity, and to occupy himself in that higher field where noble feelings and a chivalrous disposition have less to do in mastering the enemy than in overcoming internal dissension.

In order that the reader may appreciate all that must be comprehended and judged of correctly at a glance by a General, we refer to the first chapter. We say the General becomes a Statesman, but he must not cease to be the General. He takes into view all the relations of the State on the one hand; on the other, he must know exactly what he can do with the means at his disposal.

As the diversity, and undefined limits, of all the circumstances bring a great number of factors into consideration in War, as the most of these factors can only be estimated according to probability, therefore, if the Chief of an Army does not bring to bear upon them a mind with an intuitive perception of the truth, a confusion of ideas and views must take place, in the midst of which the judgment will become bewildered. In this sense, Buonaparte was right when he said that many of the questions which come before a General for decision would make problems for a mathematical calculation not unworthy of the powers of Newton or Euler.

What is here required from the higher powers of the mind is a sense of unity, and a judgment raised to such a compass as to give the mind an extraordinary faculty of vision which in its range allays and sets aside a thousand dim notions which an ordinary understanding could only bring to light with great effort, and over which it would exhaust itself. But this higher activity of the mind, this glance of genius, would still not become matter of history if the qualities of temperament and character of which we have treated did not give it their support.

Truth alone is but a weak motive of action with men, and hence there is always a great difference between knowing and action, between science and art. The man receives the strongest impulse to action through the feelings, and the most powerful succour, if we may use the expression, through those faculties of heart and mind which we have considered under the terms of resolution, firmness, perseverance, and force of character.

If, however, this elevated condition of heart and mind in the General did not manifest itself in the general effects resulting from it, and could only be accepted on trust and faith, then it would rarely become matter of history.

All that becomes known of the course of events in War is usually very simple, and has a great sameness in appearance; no one on the mere relation of such events perceives the difficulties connected with them which had to be overcome. It is only now and again, in the memoirs of Generals or of those in their confidence, or by reason of some special historical inquiry directed to a particular circumstance, that a portion of the many threads composing the whole web is brought to light. The reflections, mental doubts, and conflicts which precede the execution of great acts are purposely concealed because they affect political interests, or the recollection of them is accidentally lost because they have been looked upon as mere scaffolding which had to be removed on the completion of the building.

If, now, in conclusion, without venturing upon a closer definition of the higher powers of the soul, we should admit a distinction in the intelligent faculties themselves according to the common ideas established by language, and ask ourselves what kind of mind comes closest to military genius, then a look at the subject as well as at experience will tell us that searching rather than inventive minds, comprehensive minds rather than such as have a special bent, cool rather than fiery heads, are those to which in time of War we should prefer to trust the welfare of our women and children, the honour and the safety of our fatherland.

Of Danger in War

Usually before we have learnt what danger really is, we form an idea of it which is rather attractive than repulsive. In the intoxication of enthusiasm, to fall upon the enemy at the charge—who cares then about bullets and men falling? To throw oneself, blinded by excitement for a moment, against cold death, uncertain whether we or another shall escape him, and all this close to the golden gate of victory, close to the rich fruit which ambition thirsts for—can this be difficult? It will not be difficult, and still less will it appear so. But such moments, which, however, are not the work of a single pulse-beat, as is supposed, but rather like doctors' draughts, must be taken diluted and spoilt by mixture with time—such moments, we say, are but few.

Let us accompany the novice to the battle-field. As we approach, the thunder of the cannon becoming plainer and plainer is soon followed by the howling of shot, which attracts the attention of the inexperienced. Balls begin to strike the ground close to us, before and behind. We hasten to the hill where stands the General and his numerous Staff. Here the close striking of the cannon balls and the bursting of shells is so frequent that the seriousness of life makes itself visible through the youthful picture of imagination. Suddenly some one known to us falls—a shell strikes amongst the crowd and causes some involuntary movements—we begin to feel that we are no longer perfectly at ease and collected; even the bravest is at least to some degree confused. Now, a step farther into the battle which is raging before us like a scene in a theatre, we get to the nearest General of Division; here ball follows ball, and the noise of our own guns increases the confusion. From the General of Division to the Brigadier. He, a man of acknowledged bravery, keeps carefully behind a rising ground, a house, or a tree—a sure sign of increasing danger. Grape rattles on the roofs of the houses and in the fields; cannon balls howl over us, and plough the air in all directions, and soon there is a frequent whistling of musket balls. A step farther towards the troops, to that sturdy infantry which for hours has maintained its firmness under this heavy fire; here the air is filled with the hissing of balls which announce their proximity by a short sharp noise as they pass within an inch of the ear, the head, or the breast.

To add to all this, compassion strikes the beating heart with pity at the sight of the maimed and fallen. The young soldier cannot reach any of these different strata of danger without feeling that the light of reason does not move here in the same medium, that it is not refracted in the same manner as in speculative contemplation. Indeed, he must be a very extraordinary man who, under these impressions for the first time, does not lose the power of making any instantaneous decisions. It is true that habit soon blunts such impressions; in half in hour we begin to be more or less indifferent to all that is going on around us: but an ordinary character never attains to complete coolness and the natural elasticity of mind; and so we perceive that here again ordinary qualities will not suffice—a thing which gains truth, the wider the sphere of activity which is to be filled. Enthusiastic, stoical, natural bravery, great ambition, or also long familiarity with danger—much of all this there must be if all the effects produced in this resistant medium are not to fall far short of that which in the student's chamber may appear only the ordinary standard.

Danger in War belongs to its friction; a correct idea of its influence is necessary for truth of perception, and therefore it is brought under notice here.

Of Bodily Exertion in War

If no one were allowed to pass an opinion on the events of War, except at a moment when he is benumbed by frost, sinking from heat and thirst, or dying with hunger and fatigue, we should certainly have fewer judgments correct objectively; but they would be so, *subjectively*, at least; that is, they would contain in themselves the exact relation between the person giving the judgment and the object. We can perceive this by observing how modestly subdued, even spiritless and desponding, is the opinion passed upon the results of untoward events by those who have been eye-witnesses, but especially if they have been parties concerned. This is, according to our view, a criterion of the influence which bodily fatigue exercises, and of the allowance to be made for it in matters of opinion.

Amongst the many things in War for which no tariff can be fixed, bodily effort may be specially reckoned. Provided there is no waste, it is a coefficient of all the forces, and no one can tell exactly to what extent it may be carried. But what is remarkable is, that just as only a strong arm enables the archer to stretch the bowstring to the utmost extent, so also in War it is only by means of a great directing spirit that we can expect the full power latent in the troops to be developed. For it is one thing if an Army, in consequence of great misfortunes, surrounded with danger, falls all to pieces like a wall that has been thrown down, and can only find safety in the utmost exertion of its bodily strength; it is another thing entirely when a victorious Army, drawn on by proud feelings only, is conducted at the will of its Chief. The same effort which in the one case might at most excite our pity must in the other call forth our admiration, because it is much more difficult to sustain.

By this comes to light for the inexperienced eye one of those things which put fetters in the dark, as it were, on the action of the mind, and wear out in secret the powers of the soul.

Although here the question is strictly only respecting the extreme effort required by a Commander from his Army, by a leader from his followers, therefore of the spirit to demand it and of the art of getting it, still the personal physical exertion of Generals and of the Chief Commander must not be overlooked. Having brought the analysis of War conscientiously up to this point, we could not but take account also of the weight of this small remaining residue.

We have spoken here of bodily effort, chiefly because, like danger, it belongs to the fundamental causes of friction, and because its indefinite quantity makes it like an elastic body, the friction of which is well known to be difficult to calculate.

To check the abuse of these considerations, of such a survey of things which aggravate the difficulties of War, nature has given our judgment a guide in our sensibilities, just as an individual cannot with advantage refer to his personal deficiencies if he is insulted and ill-treated, but may well do so if he has successfully repelled the affront, or has fully revenged it, so no Commander or Army will lessen the impression of a disgraceful defeat by depicting the danger, the distress, the exertions, things which would immensely enhance the glory of a victory. Thus our feeling, which after all is only a higher kind of judgment, forbids us to do what seems an act of justice to which our judgment would be inclined.

Information in War

By the word "information" we denote all the knowledge which we have of the enemy and his country; therefore, in fact, the foundation of all our ideas and actions. Let us just consider the nature of this foundation, its want of trustworthiness, its changefulness, and we shall soon feel what a dangerous edifice War is, how easily it may fall to pieces and bury us in its ruins. For although it is a maxim in all books that we should trust only certain information, that we must be always suspicious, that is only a miserable book comfort, belonging to that description of knowledge in which writers of systems and compendiums take refuge for want of anything better to say.

Great part of the information obtained in War is contradictory, a still greater part is false, and by far the greatest part is of a doubtful character. What is required of an officer is a certain power of discrimination, which only knowledge of men and things and good judgment can give. The law of probability must be his guide. This is not a trifling difficulty even in respect of the first plans, which can be formed in the chamber outside the real sphere of War, but it is enormously increased when in the thick of War itself one report follows hard upon the heels of another; it is then fortunate if these reports in contradicting each other show a certain balance of probability, and thus themselves call forth a scrutiny. It is much worse for the inexperienced when accident does not render him this service, but one report supports another, confirms it, magnifies it, finishes off the picture with fresh touches of colour, until necessity in urgent haste forces from us a resolution which will soon be discovered to be folly, all those reports having been lies, exaggerations, errors, etc. etc. In a few words, most reports are false, and the timidity of men acts as a multiplier of lies and untruths. As a general rule, every one is more inclined to lend credence to the bad than the good. Every one is inclined to magnify the bad in some measure, and although the alarms which are thus propagated like the waves of the sea subside into themselves, still, like them, without any apparent cause they rise again. Firm in reliance on his own better convictions, the Chief must stand like a rock against which the sea breaks its fury in vain. The role is not easy; he who is not by nature of a buoyant disposition, or trained by experience in War, and matured in judgment, may let it be his rule to do violence to his own natural conviction by inclining from the side of fear to that of hope; only by that means will he be able to preserve his balance. This difficulty of seeing things correctly, which is one of the greatest sources of friction in War, makes things appear quite different from what was expected. The impression of the senses is stronger than the force of the ideas resulting from methodical reflection, and this goes so far that no important undertaking was ever yet carried out without the Commander having to subdue new doubts in himself at the time of commencing the execution of his work. Ordinary men who follow the suggestions of others become, therefore, generally undecided on the spot; they think that they have found circumstances different from what they had expected, and this view gains strength by their again yielding to the suggestions of others. But even the man who has made his own plans, when he comes to see things with his own eyes will often think he has done wrong. Firm reliance on self must make him proof against the seeming pressure of the moment; his first conviction will in the end prove true, when the foreground scenery which fate has pushed on to the stage of War, with its accompaniments of terrific objects, is drawn aside and the horizon extended. This is one of the great chasms which separate *conception* from *execution*.

Friction in War

As long as we have no personal knowledge of War, we cannot conceive where those difficulties lie of which so much is said, and what that genius and those extraordinary mental powers required in a General have really to do. All appears so simple, all the requisite branches of knowledge appear so plain, all the combinations so unimportant, that in comparison with them the easiest problem in higher mathematics impresses us with a certain scientific dignity. But if we have seen War, all becomes intelligible; and still, after all, it is extremely difficult to describe what it is which brings about this change, to specify this invisible and completely efficient factor.

Everything is very simple in War, but the simplest thing is difficult. These difficulties accumulate and produce a friction which no man can imagine exactly who has not seen War, Suppose now a traveller, who towards evening expects to accomplish the two stages at the end of his day's journey, four or five leagues, with post-horses, on the high road—it is nothing. He arrives now at the last station but one, finds no horses, or very bad ones; then a hilly country, bad roads; it is a dark night, and he is glad when, after a great deal of trouble, he reaches the next station, and finds there some miserable accommodation. So in War, through the influence of an infinity of petty circumstances, which cannot properly be described on paper, things disappoint us, and we fall short of the mark. A powerful iron will overcomes this friction; it crushes the obstacles, but certainly the machine along with them. We shall often meet with this result. Like an obelisk towards which the principal streets of a town converge, the strong will of a proud spirit stands prominent and commanding in the middle of the Art of War.

Friction is the only conception which in a general way corresponds to that which distinguishes real War from War on paper. The military machine, the Army and all belonging to it, is in fact simple, and appears on this account easy to manage. But let us reflect that no part of it is in one piece, that it is composed entirely of individuals, each of which keeps up its own friction in all directions. Theoretically all sounds very well: the commander of a battalion is responsible for the execution of the order given; and as the battalion by its discipline is glued together into one piece, and the chief must be a man of acknowledged zeal, the beam turns on an iron pin with little friction. But it is not so in reality, and all that is exaggerated and false in such a conception manifests itself at once in War. The battalion always remains composed of a number of men, of whom, if chance so wills, the most insignificant is able to occasion delay and even irregularity. The danger which War brings with it, the bodily exertions which it requires, augment this evil so much that they may be regarded as the greatest causes of it.

This enormous friction, which is not concentrated, as in mechanics, at a few points, is therefore everywhere brought into contact with chance, and thus incidents take place upon which it was impossible to calculate, their chief origin being chance. As an instance of one such chance: the weather. Here the fog prevents the enemy from being discovered in time, a battery from firing at the right moment, a report from reaching the General; there the rain prevents a battalion from arriving at the right time, because instead of for three it had to march perhaps eight hours; the cavalry from charging effectively because it is stuck fast in heavy ground.

These are only a few incidents of detail by way of elucidation, that the reader may be able to follow the author, for whole volumes might be written on these difficulties. To avoid this, and still to give a clear conception of the host of small difficulties to be contended with in War, we might go on heaping up illustrations, if we were not afraid of being tiresome. But those who have already comprehended us will permit us to add a few more.

Activity in War is movement in a resistant medium. Just as a man immersed in water is unable to perform with ease and regularity the most natural and simplest movement, that of walking, so in War, with ordinary powers, one cannot keep even the line of mediocrity. This is the reason that the correct theorist is like a swimming master, who teaches on dry land movements which are required in the water, which must appear grotesque and ludicrous to those who forget about the water. This is also why theorists, who have never plunged in themselves, or who cannot deduce any generalities from their experience, are unpractical and even absurd, because they only teach what every one knows—how to walk.

Further, every War is rich in particular facts, while at the same time each is an unexplored sea, full of rocks which the General may have a suspicion of, but which he has never seen with his eye, and round which, moreover, he must steer in the night. If a contrary wind also springs up, that is, if any great accidental event declares itself adverse to him, then the most consummate skill, presence of mind, and energy are required, whilst to those who only look on from a distance all seems to proceed with the utmost ease. The knowledge of this friction is a chief part of that so often talked of, experience in War, which is required in a good General. Certainly he is not the best General in whose mind it assumes the greatest dimensions, who is the most over-awed by it (this includes that class of over-anxious Generals, of whom there are so many amongst the experienced); but a General must be aware of it that he may overcome it, where that is possible, and that he may not expect a degree of precision in results which is impossible on account of this very friction. Besides, it can never be learnt theoretically; and if it could, there would still be wanting that experience of judgment which is called tact, and which is always more necessary in a field full of innumerable small and diversified objects than in great and decisive cases, when one's own judgment may be aided by consultation with others. Just as the man of the world, through tact of judgment which has become habit, speaks, acts, and moves only as suits the occasion, so the officer experienced in War will always, in great and small matters, at every pulsation of War as we may say, decide and determine suitably to the occasion. Through this experience and practice the idea comes to his mind of itself that so and so will not suit. And thus he will not easily place himself in a position by which he is compromised, which, if it often occurs in War, shakes all the foundations of confidence and becomes extremely dangerous.

It is therefore this friction, or what is so termed here, which makes that which appears easy in War difficult in reality. As we proceed, we shall often meet with this subject again, and it will hereafter become plain that besides experience and a strong will, there are still many other rare qualities of the mind required to make a man a consummate General.

Concluding Remarks, Book I

Those things which as elements meet together in the atmosphere of War and make it a resistant medium for every activity we have designated under the terms danger, bodily effort (exertion), information, and friction. In their impedient effects they may therefore be comprehended again in the collective notion of a general friction. Now is there, then, no kind of oil which is capable of diminishing this friction? Only one, and that one is not always available at the will of the Commander or his Army. It is the habituation of an Army to War.

Habit gives strength to the body in great exertion, to the mind in great danger, to the judgment against first impressions. By it a valuable circumspection is generally gained throughout every rank, from the hussar and rifleman up to the General of Division, which facilitates the work of the Chief Commander.

As the human eye in a dark room dilates its pupil, draws in the little light that there is, partially distinguishes objects by degrees, and at last knows them quite well, so it is in War with the experienced soldier, whilst the novice is only met by pitch dark night.

Habituation to War no General can give his Army at once, and the camps of manoeuvre (peace exercises) furnish but a weak substitute for it, weak in comparison with real experience in War, but not weak in relation to other Armies in which the training is limited to mere mechanical exercises of routine. So to regulate the exercises in peace time as to include some of these causes of friction, that the judgment, circumspection, even resolution of the separate leaders may be brought into exercise, is of much greater consequence than those believe who do not know the thing by experience. It is of immense importance that the soldier, high or low, whatever rank he has, should not have to encounter in War those things which, when seen for the first time, set him in astonishment and perplexity; if he has only met with them one single time before, even by that he is half acquainted with them. This relates even to bodily fatigues. They should be practised less to accustom the body to them than the mind. In War the young soldier is very apt to regard unusual fatigues as the consequence of faults, mistakes, and embarrassment in the conduct of the whole, and to become distressed and despondent as a consequence. This would not happen if he had been prepared for this beforehand by exercises in peace.

Another less comprehensive but still very important means of gaining habituation to War in time of peace is to invite into the service officers of foreign armies who have had experience in War. Peace seldom reigns over all Europe, and never in all quarters of the world. A State which has been long at peace should, therefore, always seek to procure some officers who have done good service at the different scenes of Warfare, or to send there some of its own, that they may get a lesson in War.

However small the number of officers of this description may appear in proportion to the mass, still their influence is very sensibly felt. [The War of 1870 furnishes a marked illustration. Von Moltke and von Goeben, not to mention many others, had both seen service in this manner, the former in Turkey and Syria, the latter in Spain—Editor] Their experience, the bent of their genius, the stamp of their character, influence their subordinates and comrades; and besides that, if they cannot be placed in positions of superior command, they may always be regarded as men acquainted with the country, who may be questioned on many special occasions.

Branches of the Art of War

War in its literal meaning is fighting, for fighting alone is the efficient principle in the manifold activity which in a wide sense is called War. But fighting is a trial of strength of the moral and physical forces by means of the latter. That the moral cannot be omitted is evident of itself, for the condition of the mind has always the most decisive influence on the forces employed in War.

The necessity of fighting very soon led men to special inventions to turn the advantage in it in their own favour: in consequence of these the mode of fighting has undergone great alterations; but in whatever way it is conducted its conception remains unaltered, and fighting is that which constitutes War.

The inventions have been from the first weapons and equipments for the individual combatants. These have to be provided and the use of them learnt before the War begins. They are made suitable to the nature of the fighting, consequently are ruled by it; but plainly the activity engaged in these appliances is a different thing from the fight itself; it is only the preparation for the combat, not the conduct of the same. That arming and equipping are not essential to the conception of fighting is plain, because mere wrestling is also fighting.

Fighting has determined everything appertaining to arms and equipment, and these in turn modify the mode of fighting; there is, therefore, a reciprocity of action between the two.

Nevertheless, the fight itself remains still an entirely special activity, more particularly because it moves in an entirely special element, namely, in the element of danger.

If, then, there is anywhere a necessity for drawing a line between two different activities, it is here; and in order to see clearly the importance of this idea, we need only just to call to mind how often eminent personal fitness in one field has turned out nothing but the most useless pedantry in the other.

It is also in no way difficult to separate in idea the one activity from the other, if we look at the combatant forces fully armed and equipped as a given means, the profitable use of which requires nothing more than a knowledge of their general results.

The Art of War is therefore, in its proper sense, the art of making use of the given means in fighting, and we cannot give it a better name than the "Conduct of War." On the other hand, in a wider sense all activities which have their existence on account of War, therefore the whole creation of troops, that is levying them, arming, equipping, and exercising them, belong to the Art of War.

To make a sound theory it is most essential to separate these two activities, for it is easy to see that if every act of War is to begin with the preparation of military forces, and to presuppose forces so organised as a primary condition for conducting War, that theory will only be applicable in the few cases to which the force available happens to be exactly suited. If, on the other hand, we wish to have a theory which shall suit most cases, and will not be wholly useless in any case, it must be founded on those means which are in most general use, and in respect to these only on the actual results springing from them.

The conduct of War is, therefore, the formation and conduct of the fighting. If this fighting was a single act, there would be no necessity for any further subdivision, but the fight is composed of a greater or less number of single acts, complete in themselves, which we call combats, as we have shown in the first chapter of the first book, and which form new units. From this arises the totally different activities, that of the *formation* and *conduct* of these single combats in themselves, and the *Combination* of them with one another, with a view to the ultimate object of the War. The first is called *tactics*, the other *strategy*.

This division into tactics and strategy is now in almost general use, and every one knows tolerably well under which head to place any single fact, without knowing very distinctly the grounds on which the classification is founded. But when such divisions are blindly adhered to in practice, they must have some deep root. We have searched for this root, and we might say that it is just the usage of the majority which has brought us to it. On the other hand, we look upon the arbitrary, unnatural definitions of these conceptions sought to be established by some writers as not in accordance with the general usage of the terms.

According to our classification, therefore, tactics *is the theory of the use of military forces in combat.* Strategy *is the theory of the use of combats for the object of the war.*

The way in which the conception of a single, or independent combat, is more closely determined, the conditions to which this unit is attached, we shall only be able to explain clearly when we consider the combat; we must content ourselves for the present with saying that in relation to space, therefore in combats taking place at the same time, the unit reaches just as far as *personal command* reaches; but in regard to time, and therefore in relation to combats which follow each other in close succession, it reaches to the moment when the crisis which takes place in every combat is entirely passed.

That doubtful cases may occur, cases, for instance, in which several combats may perhaps be regarded also as a single one, will not overthrow the ground of distinction we have adopted, for the same is the case with all grounds of distinction of real things which are differentiated by a gradually diminishing scale. There may, therefore, certainly be acts of activity in War which, without any alteration in the point of view, may just as well be counted strategic as tactical; for example, very extended positions resembling a chain of posts, the preparations for the passage of a river at several points, etc.

Our classification reaches and covers only the *use of the military force*. But now there are in War a number of activities which are subservient to it, and still are quite different from it; sometimes closely allied, sometimes less near in their affinity. All these activities relate to the *maintenance of the military force*. In the same way as its creation and training precede its use, so its maintenance is always a necessary condition. But, strictly viewed, all activities thus connected with it are always to be regarded only as preparations for fighting; they are certainly nothing more than activities which are very close to the action, so that they run through the hostile act alternate in importance with the use of the forces. We have therefore a right to exclude them as well as the other preparatory activities from the Art of War in its restricted sense, from the conduct of War properly so called; and we are obliged to do so if we would comply with the first principle of all theory, the elimination of all heterogeneous elements. Who would include in the real "conduct of War" the whole litany of subsistence and administration, because it is admitted to stand in constant reciprocal action with the use of the troops, but is something essentially different from it?

We have said, in the third chapter of our first book, that as the fight or combat is the only directly effective activity, therefore the threads of all others, as they end in it, are included in it. By this we meant to say that to all others an object was thereby appointed which, in accordance with the laws peculiar to themselves, they must seek to attain. Here we must go a little closer into this subject.

The subjects which constitute the activities outside of the combat are of various kinds.

The one part belongs, in one respect, to the combat itself, is identical with it, whilst it serves in another respect for the maintenance of the military force. The other part belongs purely to the subsistence, and has only, in consequence of the reciprocal action, a limited influence on the combats by its results. The subjects which in one respect belong to the fighting itself are *marches, camps,* and *cantonments,* for they suppose so many different situations of troops, and where troops are supposed there the idea of the combat must always be present.

The other subjects, which only belong to the maintenance, are *Subsistence, care of the sick,* the *supply and repair of arms and equipment.*

Marches are quite identical with the use of the troops. The act of marching in the combat, generally called manoeuvring, certainly does not necessarily include the use of weapons, but it is so completely and necessarily combined with it that it forms an integral part of that which we call a combat. But the march outside the combat is nothing but the execution of a strategic measure. By the strategic plan is settled *when, where,* and *with what forces* a battle is to be delivered—and to carry that into execution the march is the only means.

The march outside of the combat is therefore an instrument of strategy, but not on that account exclusively a subject of strategy, for as the armed force which executes it may be involved in a possible combat at any moment, therefore its execution stands also under tactical as well as strategic rules. If we prescribe to a column its route on a particular side of a river or of a branch of a mountain, then that is a strategic measure, for it contains the intention of fighting on that particular side of the hill or river in preference to the other, in case a combat should be necessary during the march.

But if a column, instead of following the road through a valley, marches along the parallel ridge of heights, or for the convenience of marching divides itself into several columns, then these are tactical arrangements, for they relate to the manner in which we shall use the troops in the anticipated combat.

The particular order of march is in constant relation with readiness for combat, is therefore tactical in its nature, for it is nothing more than the first or preliminary disposition for the battle which may possibly take place.

As the march is the instrument by which strategy apportions its active elements, the combats, but these last often only appear by their results and not in the details of their real course, it could not fail to happen that in theory the instrument has often been substituted for the efficient principle. Thus we hear of a decisive skilful march, allusion being thereby made to those combat-combinations to which these marches led. This substitution of ideas is too natural and conciseness of expression too desirable to call for alteration, but still it is only a condensed chain of ideas in regard to which we must never omit to bear in mind the full meaning, if we would avoid falling into error.

We fall into an error of this description if we attribute to strategical combinations a power independent of tactical results. We read of marches and manoeuvres combined, the object attained, and at the same time not a word about combat, from which the conclusion is drawn that there are means in War of conquering an enemy without fighting. The prolific nature of this error we cannot show until hereafter.

But although a march can be regarded absolutely as an integral part of the combat, still there are in it certain relations which do not belong to the combat, and therefore are neither tactical nor strategic. To these belong all arrangements which concern only the accommodation of the troops, the construction of bridges, roads, etc. These are only conditions; under many circumstances they are in very close connection, and may almost identify themselves with the troops, as in building a bridge in presence of the enemy; but in themselves they are always activities, the theory of which does not form part of the theory of the conduct of War.

Camps, by which we mean every disposition of troops in concentrated, therefore in battle order, in contradistinction to cantonments or quarters, are a state of rest, therefore of restoration; but they are at the same time also the strategic appointment of a battle on the spot, chosen; and by the manner in which they are taken up they contain the fundamental lines of the battle, a condition from which every defensive battle starts; they are therefore essential parts of both strategy and tactics.

Cantonments take the place of camps for the better refreshment of the troops. They are therefore, like camps, strategic subjects as regards position and extent; tactical subjects as regards internal organisation, with a view to readiness to fight.

The occupation of camps and cantonments no doubt usually combines with the recuperation of the troops another object also, for example, the covering a district of country, the holding a position; but it can very well be only the first. We remind our readers that strategy may follow a great diversity of objects, for everything which appears an advantage may be the object of a combat, and the preservation of the instrument with which War is made must necessarily very often become the object of its partial combinations.

If, therefore, in such a case strategy ministers only to the maintenance of the troops, we are not on that account out of the field of strategy, for we are still engaged with the use of the military force, because every disposition of that force upon any point Whatever of the theatre of War is such a use.

But if the maintenance of the troops in camp or quarters calls forth activities which are no employment of the armed force, such as the construction of huts, pitching of tents, subsistence and sanitary services in camps or quarters, then such belong neither to strategy nor tactics.

Even entrenchments, the site and preparation of which are plainly part of the order of battle, therefore tactical subjects, do not belong to the theory of the conduct of War so far as respects the execution of their construction the knowledge and skill required for such work being, in point of fact, qualities inherent in the nature of an organised Army; the theory of the combat takes them for granted.

Amongst the subjects which belong to the mere keeping up of an armed force, because none of the parts are identified with the combat, the victualling of the troops themselves comes first, as it must be done almost daily and for each individual. Thus it is that it completely permeates military action in the parts constituting strategy—we say parts constituting strategy, because during a battle the subsistence of troops will rarely have any influence in modifying the plan, although the thing is conceivable enough. The care for the subsistence of the troops comes therefore into reciprocal action chiefly with strategy, and there is nothing more common than for the leading strategic features of a campaign and War to be traced out in connection with a view to this supply. But however frequent and however important these views of supply may be, the subsistence of the troops always remains a completely different activity from the use of the troops, and the former has only an influence on the latter by its results.

The other branches of administrative activity which we have mentioned stand much farther apart from the use of the troops. The care of sick and wounded, highly important as it is for the good of an Army, directly affects it only in a small portion of the individuals composing it, and therefore has only a weak and indirect influence upon the use of the rest. The completing and replacing articles of arms and equipment, except so far as by the organism of the forces it constitutes a continuous activity inherent in them—takes place only periodically, and therefore seldom affects strategic plans.

We must, however, here guard ourselves against a mistake. In certain cases these subjects may be really of decisive importance. The distance of hospitals and depôts of munitions may very easily be imagined as the sole cause of very important strategic decisions. We do not wish either to contest that point or to throw it into the shade. But we are at present occupied not with the particular facts of a concrete case, but with abstract theory; and our assertion therefore is that such an influence is too rare to give the theory of sanitary measures and the supply of munitions and arms an importance in theory of the conduct of War such as to make it worth while to include in the theory of the conduct of War the consideration of the different ways and systems which the above theories may furnish, in the same way as is certainly necessary in regard to victualling troops.

If we have clearly understood the results of our reflections, then the activities belonging to War divide themselves into two principal classes, into such as are only "preparations for War" and into the "War itself." This division must therefore also be made in theory.

The knowledge and applications of skill in the preparations for War are engaged in the creation, discipline, and maintenance of all the military forces; what general names should be given to them we do not enter into, but we see that artillery, fortification, elementary tactics, as they are called, the whole organisation and administration of the various armed forces, and all such things are included. But the theory of War itself occupies itself with the use of these prepared means for the object of the war. It needs of the first only the results, that is, the knowledge of the principal properties of the means taken in hand for use. This we call "The Art of War" in a limited sense, or "Theory of the Conduct of War," or "Theory of the Employment of Armed Forces," all of them denoting for us the same thing.

The present theory will therefore treat the combat as the real contest, marches, camps, and cantonments as circumstances which are more or less identical with it. The subsistence of the troops will only come into consideration like *other given circumstances* in respect of its results, not as an activity belonging to the combat.

The Art of War thus viewed in its limited sense divides itself again into tactics and strategy. The former occupies itself with the form of the separate combat, the latter with its use. Both connect themselves with the circumstances of marches, camps, cantonments only through the combat, and these circumstances are tactical or strategic according as they relate to the form or to the signification of the battle.

No doubt there will be many readers who will consider superfluous this careful separation of two things lying so close together as tactics and strategy, because it has no direct effect on the conduct itself of War. We admit, certainly that it would be pedantry to look for direct effects on the field of battle from a theoretical distinction.

But the first business of every theory is to clear up conceptions and ideas which have been jumbled together, and, we may say, entangled and confused; and only when a right understanding is established, as to names and conceptions, can we hope to progress with clearness and facility, and be certain that author and reader will always see things from the same point of view. Tactics and strategy are two activities mutually permeating each other in time and space, at the same time essentially different activities, the inner laws and mutual relations of which cannot be intelligible at all to the mind until a clear conception of the nature of each activity is established.

He to whom all this is nothing, must either repudiate all theoretical consideration, *or his understanding has not as yet been pained* by the confused and perplexing ideas resting on no fixed point of view, leading to no satisfactory result, sometimes dull, sometimes fantastic, sometimes floating in vague generalities, which we are often obliged to hear and read on the conduct of War, owing to the spirit of scientific investigation having hitherto been little directed to these subjects.

On the Theory of War

1. The First Conception of the "Art of War" Was Merely the Preparation of the Armed Forces

Formerly by the term "Art of War," or "Science of War," nothing was understood but the totality of those branches of knowledge and those appliances of skill occupied with material things. The pattern and preparation and the mode of using arms, the construction of fortifications and entrenchments, the organism of an army and the mechanism of its movements, were the subject; these branches of knowledge and skill above referred to, and the end and aim of them all was the establishment of an armed force fit for use in War. All this concerned merely things belonging to the material world and a one-sided activity only, and it was in fact nothing but an activity advancing by gradations from the lower occupations to a finer kind of mechanical art. The relation of all this to War itself was very much the same as the relation of the art of the sword cutler to the art of using the sword. The employment in the moment of danger and in a state of constant reciprocal action of the particular energies of mind and spirit in the direction proposed to them was not yet even mooted.

2. True War First Appears in the Art of Sieges

In the art of sieges we first perceive a certain degree of guidance of the combat, something of the action of the intellectual faculties upon the material forces placed under their control, but generally only so far that it very soon embodied itself again in new material forms, such as approaches, trenches, counter-approaches, batteries, etc., and every step which this action of the higher faculties took was marked by some such result; it was only the thread that was required on which to string these material inventions in order. As the intellect can hardly manifest itself in this kind of War, except in such things, so therefore nearly all that was necessary was done in that way.

3. Then Tactics Tried to Find its Way in the Same Direction

Afterwards tactics attempted to give to the mechanism of its joints the character of a general disposition, built upon the peculiar properties of the instrument, which character leads indeed to the battle-field, but instead of leading to the free activity of mind, leads to an Army made like an automaton by its rigid formations and orders of battle, which, movable only by the word of command, is intended to unwind its activities like a piece of clockwork.

4. The Real Conduct of War Only Made its Appearance Incidentally and Incognito

The conduct of War properly so called, that is, a use of the prepared means adapted to the most special requirements, was not considered as any suitable subject for theory, but one which should be left to natural talents alone. By degrees, as War passed from the hand-to-hand encounters of the middle ages into a more regular and systematic form, stray reflections on this point also forced themselves into men's minds, but they mostly appeared only incidentally in memoirs and narratives, and in a certain measure incognito.

5. Reflections on Military Events Brought about the Want of a Theory

As contemplation on War continually increased, and its history every day assumed more of a critical character, the urgent want appeared of the support of fixed maxims and rules, in order that in the controversies naturally arising about military events the war of opinions might be brought to some one point. This whirl of opinions, which neither revolved on any central pivot nor according to any appreciable laws, could not but be very distasteful to people's minds.

6. Endeavours to Establish a Positive Theory

There arose, therefore, an endeavour to establish maxims, rules, and even systems for the conduct of War. By this the attainment of a positive object was proposed, without taking into view the endless difficulties which the conduct of War presents in that respect. The conduct of War, as we have shown, has no definite limits in any direction, while every system has the circumscribing nature of a synthesis, from which results an irreconcileable opposition between such a theory and practice.

7. Limitation to Material Objects

Writers on theory felt the difficulty of the subject soon enough, and thought themselves entitled to get rid of it by directing their maxims and systems only upon material things and a one-sided activity. Their aim was to reach results, as in the science for the preparation for War, entirely certain and positive, and therefore only to take into consideration that which could be made matter of calculation.

8. Superiority of Numbers

The superiority in numbers being a material condition, it was chosen from amongst all the factors required to produce victory, because it could be brought under mathematical laws through combinations of time and space. It was thought possible to leave out of sight all other circumstances, by supposing them to be equal on each side, and therefore to neutralise one another. This would have been very well if it had been done to gain a preliminary knowledge of this one factor, according to its relations, but to make it a rule for ever to consider superiority of numbers as the sole law; to see the whole secret of the Art of War in the formula, *In a certain time, at a certain point, to bring up superior masses*—was a restriction overruled by the force of realities.

9. Victualling of Troops

By one theoretical school an attempt was made to systematise another material element also, by making the subsistence of troops, according to a previously established organism of the Army, the supreme legislator in the higher conduct of War. In this way certainly they arrived at definite figures, but at figures which rested on a number of arbitrary calculations, and which therefore could not stand the test of practical application.

10. Base

An ingenious author tried to concentrate in a single conception, that of a *base*, a whole host of objects amongst which sundry relations even with immaterial forces found their way in as well. The list comprised the subsistence of the troops, the keeping them complete in numbers and equipment, the security of communications with the home country, lastly, the security of retreat in case it became necessary; and, first of all, he proposed to substitute this conception of a base for all these things; then for the base itself to substitute its own length (extent); and, last of all, to substitute the angle formed by the army with this base: all this was done to obtain a pure geometrical result utterly useless. This last is, in fact, unavoidable, if we reflect that none of these substitutions could be made without violating truth and leaving out some of the things contained in the original conception. The idea of a base is a real necessity for strategy, and to have conceived it is meritorious; but to make such a use of it as we have depicted is completely inadmissible, and could not but lead to partial conclusions which have forced these theorists into a direction opposed to common sense, namely, to a belief in the decisive effect of the enveloping form of attack.

11. Interior Lines

As a reaction against this false direction, another geometrical principle, that of the so-called interior lines, was then elevated to the throne. Although this principle rests on a sound foundation, on the truth that the combat is the only effectual means in War, still it is, just on account of its purely geometrical nature, nothing but another case of one-sided theory which can never gain ascendency in the real world.

12. All These Attempts Are Open to Objection

All these attempts at theory are only to be considered in their analytical part as progress in the province of truth, but in their synthetical part, in their precepts and rules, they are quite unserviceable.

They strive after determinate quantities, whilst in War all is undetermined, and the calculation has always to be made with varying quantities.

They direct the attention only upon material forces, while the whole military action is penetrated throughout by intelligent forces and their effects.

They only pay regard to activity on one side, whilst War is a constant state of reciprocal action, the effects of which are mutual.

13. As a Rule They Exclude Genius

All that was not attainable by such miserable philosophy, the offspring of partial views, lay outside the precincts of science—and was the field of genius, which *raises itself above rules*.

Pity the warrior who is contented to crawl about in this beggardom of rules, which are too bad for genius, over which it can set itself superior, over which it can perchance make merry! What genius does must be the best of all rules, and theory cannot do better than to show how and why it is so.

Pity the theory which sets itself in opposition to the mind! It cannot repair this contradiction by any humility, and the humbler it is so much the sooner will ridicule and contempt drive it out of real life.

14. The Difficulty of Theory as Soon as Moral Quantities Come into Consideration.

Every theory becomes infinitely more difficult from the moment that it touches on the province of moral quantities. Architecture and painting know quite well what they are about as long as they have only to do with matter; there is no dispute about mechanical or optical construction. But as soon as the moral activities begin their work, as soon as moral impressions and feelings are produced, the whole set of rules dissolves into vague ideas.

The science of medicine is chiefly engaged with bodily phenomena only; its business is with the animal organism, which, liable to perpetual change, is never exactly the same for two moments. This makes its practice very difficult, and places the judgment of the physician above his science; but how much more difficult is the case if a moral effect is added, and how much higher must we place the physician of the mind?

15. The Moral Quantities must Not Be Excluded in War

But now the activity in War is never directed solely against matter; it is always at the same time directed against the intelligent force which gives life to this matter, and to separate the two from each other is impossible.

But the intelligent forces are only visible to the inner eye, and this is different in each person, and often different in the same person at different times.

As danger is the general element in which everything moves in War, it is also chiefly by courage, the feeling of one's own power, that the judgment is differently influenced. It is to a certain extent the crystalline lens through which all appearances pass before reaching the understanding.

And yet we cannot doubt that these things acquire a certain objective value simply through experience.

Every one knows the moral effect of a surprise, of an attack in flank or rear. Every one thinks less of the enemy's courage as soon as he turns his back, and ventures much more in pursuit than when pursued. Every one judges of the enemy's General by his reputed talents, by his age and experience, and shapes his course accordingly. Every one casts a scrutinising glance at the spirit and feeling of his own and the enemy's troops. All these and similar effects in the province of the moral nature of man have established themselves by experience, are perpetually recurring, and therefore warrant our reckoning them as real quantities of their kind. What could we do with any theory which should leave them out of consideration?

Certainly experience is an indispensable title for these truths. With psychological and philosophical sophistries no theory, no General, should meddle.

16. Principal Difficulty of a Theory for the Conduct of War

In order to comprehend clearly the difficulty of the proposition which is contained in a theory for the conduct of War, and thence to deduce the necessary characteristics of such a theory, we must take a closer view of the chief particulars which make up the nature of activity in War.

17. First Speciality.—Moral Forces and Their Effects (Hostile Feeling)

The first of these specialities consists in the moral forces and effects.

The combat is, in its origin, the expression of *hostile feeling*, but in our great combats, which we call Wars, the hostile feeling frequently resolves itself into merely a hostile *view*, and there is usually no innate hostile feeling residing in individual against individual. Nevertheless, the combat never passes off without such feelings being brought into activity. National hatred, which is seldom wanting in our Wars, is a substitute for personal hostility in the breast of individual opposed to individual. But where this also is wanting, and at first no animosity of feeling subsists, a hostile feeling is kindled by the combat itself; for an act of violence which any one commits upon us by order of his superior, will excite in us a desire to retaliate and be revenged on him, sooner than on the superior power at whose command the act was done. This is human, or animal if we will; still it is so. We are very apt to regard the combat in theory as an abstract trial of strength, without any participation on the part of the feelings, and that is one of the thousand errors which theorists deliberately commit, because they do not see its consequences.

Besides that excitation of feelings naturally arising from the combat itself, there are others also which do not essentially belong to it, but which, on account of their relationship, easily unite with it—ambition, love of power, enthusiasm of every kind, etc. etc.

18. The Impressions of Danger (Courage)

Finally, the combat begets the element of danger, in which all the activities of War must live and move, like the bird in the air or the fish in the water. But the influences of danger all pass into the feelings, either directly—that is, instinctively—or through the medium of the understanding. The effect in the first case would be a desire to escape from the danger, and, if that cannot be done, fright and anxiety. If this effect does not take place, then it is COURAGE, which is a counterpoise to that instinct. Courage is, however, by no means an act of the understanding, but likewise a feeling, like fear; the latter looks to the physical preservation, courage to the moral preservation. Courage, then, is a nobler instinct. But because it is so, it will not allow itself to be used as a lifeless instrument, which produces its effects exactly according to prescribed measure. Courage is therefore no mere counterpoise to danger in order to neutralise the latter in its effects, but a peculiar power in itself.

19. Extent of the Influence of Danger

But to estimate exactly the influence of danger upon the principal actors in War, we must not limit its sphere to the physical danger of the moment. It dominates over the actor, not only by threatening him, but also by threatening all entrusted to him, not only at the moment in which it is actually present, but also through the imagination at all other moments, which have a connection with the present; lastly, not only directly by itself, but also indirectly by the responsibility which makes it bear with tenfold weight on the mind of the chief actor. Who could advise, or resolve upon a great battle, without feeling his mind more or less wrought up, or perplexed by, the danger and responsibility which such a great act of decision carries in itself? We may say that action in War, in so far as it is real action, not a mere condition, is never out of the sphere of danger.

20. Other Powers of Feeling

If we look upon these affections which are excited by hostility and danger as peculiarly belonging to War, we do not, therefore, exclude from it all others accompanying man in his life's journey. They will also find room here frequently enough. Certainly we may say that many a petty action of the passions is silenced in this serious business of life; but that holds good only in respect to those acting in a lower sphere, who, hurried on from one state of danger and exertion to another, lose sight of the rest of the things of Life, *Become Unused to Deceit*, because it is of no avail with death, and so attain to that soldierly simplicity of character which has always been the best representative of the military profession. In higher regions it is otherwise, for the higher a man's rank, the more he must look around him; then arise interests on every side, and a manifold activity of the passions of good and bad. Envy and generosity, pride and humility, fierceness and tenderness, all may appear as active powers in this great drama.

21. Peculiarity of Mind

The peculiar characteristics of mind in the chief actor have, as well as those of the feelings, a high importance. From an imaginative, flighty, inexperienced head, and from a calm, sagacious understanding, different things are to be expected.

22. From the Diversity in Mental Individualities Arises the Diversity of Ways Leading to the End

It is this great diversity in mental individuality, the influence of which is to be supposed as chiefly felt in the higher ranks, because it increases as we progress upwards, which chiefly produces the diversity of ways leading to the end noticed by us in the first book, and which gives, to the play of probabilities and chance, such an unequal share in determining the course of events.

23. Second Peculiarity.— Living Reaction

The second peculiarity in War is the living reaction, and the reciprocal action resulting therefrom. We do not here speak of the difficulty of estimating that reaction, for that is included in the difficulty before mentioned, of treating the moral powers as quantities; but of this, that reciprocal action, by its nature, opposes anything like a regular plan. The effect which any measure produces upon the enemy is the most distinct of all the data which action affords; but every theory must keep to classes (or groups) of phenomena, and can never take up the really individual case in itself: that must everywhere be left to judgment and talent. It is therefore natural that in a business such as War, which in its plan—built upon general circumstances—is so often thwarted by unexpected and singular accidents, more must generally be left to talent; and less use can be made of a *theoretical guide* than in any other.

24. Third Peculiarity.— Uncertainty of All Data

Lastly, the great uncertainty of all data in War is a peculiar difficulty, because all action must, to a certain extent, be planned in a mere twilight, which in addition not unfrequently—like the effect of a fog or moonshine—gives to things exaggerated dimensions and an unnatural appearance.

What this feeble light leaves indistinct to the sight talent must discover, or must be left to chance. It is therefore again talent, or the favour of fortune, on which reliance must be placed, for want of objective knowledge.

25. Positive Theory Is Impossible

With materials of this kind we can only say to ourselves that it is a sheer impossibility to construct for the Art of War a theory which, like a scaffolding, shall ensure to the chief actor an external support on all sides. In all those cases in which he is thrown upon his talent he would find himself away from this scaffolding of theory and in opposition to it, and, however many-sided it might be framed, the same result would ensue of which we spoke when we said that talent and genius act beyond the law, and theory is in opposition to reality.

26. Means Left by Which a Theory Is Possible (The Difficulties Are Not Everywhere Equally Great)

Two means present themselves of getting out of this difficulty. In the first place, what we have said of the nature of military action in general does not apply in the same manner to the action of every one, whatever may be his standing. In the lower ranks the spirit of self-sacrifice is called more into request, but the difficulties which the understanding and judgment meet with are infinitely less. The field of occurrences is more confined. Ends and means are fewer in number. Data more distinct; mostly also contained in the actually visible. But the higher we ascend the more the difficulties increase, until in the Commander-in-Chief they reach their climax, so that with him almost everything must be left to genius.

Further, according to a division of the subject in *agreement with its nature*, the difficulties are not everywhere the same, but diminish the more results manifest themselves in the material world, and increase the more they pass into the moral, and become motives which influence the will. Therefore it is easier to determine, by theoretical rules, the order and conduct of a battle, than the use to be made of the battle itself. Yonder physical weapons clash with each other, and although mind is not wanting therein, matter must have its rights. But in the effects to be produced by battles when the material results become motives, we have only to do with the moral nature. In a word, it is easier to make a theory for *tactics* than for *strategy*.

27. Theory must Be of the Nature of Observations Not of Doctrine

The second opening for the possibility of a theory lies in the point of view that it does not necessarily require to be a *direction* for action. As a general rule, whenever an *activity* is for the most part occupied with the same objects over and over again, with the same ends and means, although there may be trifling alterations and a corresponding number of varieties of combination, such things are capable of becoming a subject of study for the reasoning faculties. But such study is just the most essential part of every *theory*, and has a peculiar title to that name. It is an analytical investigation of the subject that leads to an exact knowledge; and if brought to bear on the results of experience, which in our case would be military history, to a thorough familiarity with it. The nearer theory attains the latter object, so much the more it passes over from the objective form of knowledge into the subjective one of skill in action; and so much the more, therefore, it will prove itself effective when circumstances allow of no other decision but that of personal talents; it will show its effects in that talent itself. If theory investigates the subjects which constitute War; if it separates more distinctly that which at first sight seems amalgamated; if it explains fully the properties of the means; if it shows their probable effects; if it makes evident the nature of objects; if it brings to bear all over the field of War the light of essentially critical investigation—then it has fulfilled the chief duties of its province. It becomes then a guide to him who wishes to make himself acquainted with War from books; it lights up the whole road for him, facilitates his progress, educates his judgment, and shields him from error.

If a man of expertness spends half his life in the endeavour to clear up an obscure subject thoroughly, he will probably know more about it than a person who seeks to master it in a short time. Theory is instituted that each person in succession may not have to go through the same labour of clearing the ground and toiling through his subject, but may find the thing in order, and light admitted on it. It should educate the mind of the future leader in War, or rather guide him in his self-instruction, but not accompany him to the field of battle; just as a sensible tutor forms and enlightens the opening mind of a youth without, therefore, keeping him in leading strings all through his life.

If maxims and rules result of themselves from the considerations which theory institutes, if the truth accretes itself into that form of crystal, then theory will not oppose this natural law of the mind; it will rather, if the arch ends in such a keystone, bring it prominently out; but so does this, only in order to satisfy the philosophical law of reason, in order to show distinctly the point to which the lines all converge, not in order to form out of it an algebraical formula for use upon the battle-field; for even these maxims and rules serve more to determine in the reflecting mind the leading outline of its habitual movements than as landmarks indicating to it the way in the act of execution.

28. By this Point of View Theory Becomes Possible, and Ceases to Be in Contradiction to Practice

Taking this point of view, there is a possibility afforded of a satisfactory, that is, of a useful, theory of the conduct of War, never coming into opposition with the reality, and it will only depend on rational treatment to bring it so far into harmony with action that between theory and practice there shall no longer be that absurd difference which an unreasonable theory, in defiance of common sense, has often produced, but which, just as often, narrow-mindedness and ignorance have used as a pretext for giving way to their natural incapacity.

29. Theory Therefore Considers the Nature of Ends and Means—ends and Means in Tactics

Theory has therefore to consider the nature of the means and ends.

In tactics the means are the disciplined armed forces which are to carry on the contest. The object is victory. The precise definition of this conception can be better explained hereafter in the consideration of the combat. Here we content ourselves by denoting the retirement of the enemy from the field of battle as the sign of victory. By means of this victory strategy gains the object for which it appointed the combat, and which constitutes its special signification. This signification has certainly some influence on the nature of the victory. A victory which is intended to weaken the enemy's armed forces is a different thing from one which is designed only to put us in possession of a position. The signification of a combat may therefore have a sensible influence on the preparation and conduct of it, consequently will be also a subject of consideration in tactics.

30. Circumstances Which Always Attend the Application of the Means

As there are certain circumstances which attend the combat throughout, and have more or less influence upon its result, therefore these must be taken into consideration in the application of the armed forces.

These circumstances are the locality of the combat (ground), the time of day, and the weather.

31. Locality

The locality, which we prefer leaving for solution, under the head of "Country and Ground," might, strictly speaking, be without any influence at all if the combat took place on a completely level and uncultivated plain.

In a country of steppes such a case may occur, but in the cultivated countries of Europe it is almost an imaginary idea. Therefore a combat between civilised nations, in which country and ground have no influence, is hardly conceivable.

32. Time of Day

The time of day influences the combat by the difference between day and night; but the influence naturally extends further than merely to the limits of these divisions, as every combat has a certain duration, and great battles last for several hours. In the preparations for a great battle, it makes an essential difference whether it begins in the morning or the evening. At the same time, certainly many battles may be fought in which the question of the time of day is quite immaterial, and in the generality of cases its influence is only trifling.

33. Weather

Still more rarely has the weather any decisive influence, and it is mostly only by fogs that it plays a part.

34. End and Means in Strategy

Strategy has in the first instance only the victory, that is, the tactical result, as a means to its object, and ultimately those things which lead directly to peace. The application of its means to this object is at the same time attended by circumstances which have an influence thereon more or less.

35. Circumstances Which Attend the Application of the Means of Strategy

These circumstances are country and ground, the former including the territory and inhabitants of the whole theatre of war; next the time of the day, and the time of the year as well; lastly, the weather, particularly any unusual state of the same, severe frost, etc.

36. These Form New Means

By bringing these things into combination with the results of a combat, strategy gives this result—and therefore the combat—a special signification, places before it a particular object. But when this object is not that which leads directly to peace, therefore a subordinate one, it is only to be looked upon as a means; and therefore in strategy we may look upon the results of combats or victories, in all their different significations, as means. The conquest of a position is such a result of a combat applied to ground. But not only are the different combats with special objects to be considered as means, but also every higher aim which we may have in view in the combination of battles directed on a common object is to be regarded as a means. A winter campaign is a combination of this kind applied to the season.

There remain, therefore, as objects, only those things which may be supposed as leading *directly* to peace, Theory investigates all these ends and means according to the nature of their effects and their mutual relations.

37. Strategy Deduces Only from Experience the Ends and Means to Be Examined

The first question is, How does strategy arrive at a complete list of these things? If there is to be a philosophical inquiry leading to an absolute result, it would become entangled in all those difficulties which the logical necessity of the conduct of War and its theory exclude. It therefore turns to experience, and directs its attention on those combinations which military history can furnish. In this manner, no doubt, nothing more than a limited theory can be obtained, which only suits circumstances such as are presented in history. But this incompleteness is unavoidable, because in any case theory must either have deduced from, or have compared with, history what it advances with respect to things. Besides, this incompleteness in every case is more theoretical than real.

One great advantage of this method is that theory cannot lose itself in abstruse disquisitions, subtleties, and chimeras, but must always remain practical.

38. How Far the Analysis of the Means Should Be Carried

Another question is, How far should theory go in its analysis of the means? Evidently only so far as the elements in a separate form present themselves for consideration in practice. The range and effect of different weapons is very important to tactics; their construction, although these effects result from it, is a matter of indifference; for the conduct of War is not making powder and cannon out of a given quantity of charcoal, sulphur, and saltpetre, of copper and tin: the given quantities for the conduct of War are arms in a finished state and their effects. Strategy makes use of maps without troubling itself about triangulations; it does not inquire how the country is subdivided into departments and provinces, and how the people are educated and governed, in order to attain the best military results; but it takes things as it finds them in the community of European States, and observes where very different conditions have a notable influence on War.

39. Great Simplification of the Knowledge Required

That in this manner the number of subjects for theory is much simplified, and the knowledge requisite for the conduct of War much reduced, is easy to perceive. The very great mass of knowledge and appliances of skill which minister to the action of War in general, and which are necessary before an army fully equipped can take the field, unite in a few great results before they are able to reach, in actual War, the final goal of their activity; just as the streams of a country unite themselves in rivers before they fall into the sea. Only those activities emptying themselves directly into the sea of War have to be studied by him who is to conduct its operations.

40. This Explains the Rapid Growth of Great Generals, and Why a General Is Not a Man of Learning

This result of our considerations is in fact so necessary, any other would have made us distrustful of their accuracy. Only thus is explained how so often men have made their appearance with great success in War, and indeed in the higher ranks even in supreme Command, whose pursuits had been previously of a totally different nature; indeed how, as a rule, the most distinguished Generals have never risen from the very learned or really erudite class of officers, but have been mostly men who, from the circumstances of their position, could not have attained to any great amount of knowledge. On that account those who have considered it necessary or even beneficial to commence the education of a future General by instruction in all details have always been ridiculed as absurd pedants. It would be easy to show the injurious tendency of such a course, because the human mind is trained by the knowledge imparted to it and the direction given to its ideas. Only what is great can make it great; the little can only make it little, if the mind itself does not reject it as something repugnant.

41. Former Contradictions

Because this simplicity of knowledge requisite in War was not attended to, but that knowledge was always jumbled up with the whole impedimenta of subordinate sciences and arts, therefore the palpable opposition to the events of real life which resulted could not be solved otherwise than by ascribing it all to genius, which requires no theory and for which no theory could be prescribed.

42. On this Account All Use of Knowledge Was Denied, and Everything Ascribed to Natural Talents

People with whom common sense had the upper hand felt sensible of the immense distance remaining to be filled up between a genius of the highest order and a learned pedant; and they became in a manner free-thinkers, rejected all belief in theory, and affirmed the conduct of War to be a natural function of man, which he performs more or less well according as he has brought with him into the world more or less talent in that direction. It cannot be denied that these were nearer to the truth than those who placed a value on false knowledge: at the same time it may easily be seen that such a view is itself but an exaggeration. No activity of the human understanding is possible without a certain stock of ideas; but these are, for the greater part at least, not innate but acquired, and constitute his knowledge. The only question therefore is, of what kind should these ideas be; and we think we have answered it if we say that they should be directed on those things which man has directly to deal with in War.

43. The Knowledge must Be Made Suitable to the Position

Inside this field itself of military activity, the knowledge required must be different according to the station of the Commander. It will be directed on smaller and more circumscribed objects if he holds an inferior, upon greater and more comprehensive ones if he holds a higher situation. There are Field Marshals who would not have shone at the head of a cavalry regiment, and vice versa.

44. The Knowledge in War Is Very Simple, but Not, at the Same Time, Very Easy

But although the knowledge in War is simple, that is to say directed to so few subjects, and taking up those only in their final results, the art of execution is not, on that account, easy. Of the difficulties to which activity in War is subject generally, we have already spoken in the first book; we here omit those things which can only be overcome by courage, and maintain also that the activity of mind, is only simple, and easy in inferior stations, but increases in difficulty with increase of rank, and in the highest position, in that of Commander-in-Chief, is to be reckoned among the most difficult which there is for the human mind.

45. Of the Nature of this Knowledge

The Commander of an Army neither requires to be a learned explorer of history nor a publicist, but he must be well versed in the higher affairs of State; he must know, and be able to judge correctly of traditional tendencies, interests at stake, the immediate questions at issue, and the characters of leading persons; he need not be a close observer of men, a sharp dissector of human character, but he must know the character, the feelings, the habits, the peculiar faults and inclinations of those whom he is to command. He need not understand anything about the make of a carriage, or the harness of a battery horse, but he must know how to calculate exactly the march of a column, under different circumstances, according to the time it requires. These are matters the knowledge of which cannot be forced out by an apparatus of scientific formula and machinery: they are only to be gained by the exercise of an accurate judgment in the observation of things and of men, aided by a special talent for the apprehension of both.

The necessary knowledge for a high position in military action is therefore distinguished by this, that by observation, therefore by study and reflection, it is only to be attained through a special talent which as an intellectual instinct understands how to extract from the phenomena of life only the essence or spirit, as bees do the honey from the flowers; and that it is also to be gained by experience of life as well as by study and reflection. Life will never bring forth a Newton or an Euler by its rich teachings, but it may bring forth great calculators in War, such as Conde' or Frederick.

It is therefore not necessary that, in order to vindicate the intellectual dignity of military activity, we should resort to untruth and silly pedantry. There never has been a great and distinguished Commander of contracted mind, but very numerous are the instances of men who, after serving with the greatest distinction in inferior positions, remained below mediocrity in the highest, from insufficiency of intellectual capacity. That even amongst those holding the post of Commander-in-Chief there may be a difference according to the degree of their plenitude of power is a matter of course.

46. Science must Become Art

Now we have yet to consider one condition which is more necessary for the knowledge of the conduct of War than for any other, which is, that it must pass completely into the mind and almost completely cease to be something objective. In almost all other arts and occupations of life the active agent can make use of truths which he has only learnt once, and in the spirit and sense of which he no longer lives, and which he extracts from dusty books. Even truths which he has in hand and uses daily may continue something external to himself, If the architect takes up a pen to settle the strength of a pier by a complicated calculation, the truth found as a result is no emanation from his own mind. He had first to find the data with labour, and then to submit these to an operation of the mind, the rule for which he did not discover, the necessity of which he is perhaps at the moment only partly conscious of, but which he applies, for the most part, as if by mechanical dexterity. But it is never so in War. The moral reaction, the ever-changeful form of things, makes it necessary for the chief actor to carry in himself the whole mental apparatus of his knowledge, that anywhere and at every pulse-beat he may be capable of giving the requisite decision from himself. Knowledge must, by this complete assimilation with his own mind and life, be converted into real power. This is the reason why everything seems so easy with men distinguished in War, and why everything is ascribed to natural talent. We say natural talent, in order thereby to distinguish it from that which is formed and matured by observation and study.

We think that by these reflections we have explained the problem of a theory of the conduct of War; and pointed out the way to its solution.

Of the two fields into which we have divided the conduct of War, tactics and strategy, the theory of the latter contains unquestionably, as before observed, the greatest difficulties, because the first is almost limited to a circumscribed field of objects, but the latter, in the direction of objects leading directly to peace, opens to itself an unlimited field of possibilities. Since for the most part the Commander-in-Chief has only to keep these objects steadily in view, therefore the part of strategy in which he moves is also that which is particularly subject to this difficulty.

Theory, therefore, especially where it comprehends the highest services, will stop much sooner in strategy than in tactics at the simple consideration of things, and content itself to assist the Commander to that insight into things which, blended with his whole thought, makes his course easier and surer, never forces him into opposition with himself in order to obey an objective truth.

Art or Science of War

1.—Usage Still Unsettled (Power and Knowledge. Science When Mere Knowing; Art, When Doing, Is the Object)

THE choice between these terms seems to be still unsettled, and no one seems to know rightly on what grounds it should be decided, and yet the thing is simple. We have already said elsewhere that "knowing" is something different from "doing." The two are so different that they should not easily be mistaken the one for the other. The "doing" cannot properly stand in any book, and therefore also Art should never be the title of a book. But because we have once accustomed ourselves to combine in conception, under the name of theory of Art, or simply Art, the branches of knowledge (which may be separately pure sciences) necessary for the practice of an Art, therefore it is consistent to continue this ground of distinction, and to call everything Art when the object is to carry out the "doing" (being able), as for example, Art of building; Science, when merely knowledge is the object; as Science of mathematics, of astronomy. That in every Art certain complete sciences may be included is intelligible of itself, and should not perplex us. But still it is worth observing that there is also no science without a mixture of Art. In mathematics, for instance, the use of figures and of algebra is an Art, but that is only one amongst many instances. The reason is, that however plain and palpable the difference is between knowledge and power in the composite results of human knowledge, yet it is difficult to trace out their line of separation in man himself.

2. Difficulty of Separating Perception from Judgment (Art of War)

All thinking is indeed Art. Where the logician draws the line, where the premises stop which are the result of cognition—where judgment begins, there Art begins. But more than this even the perception of the mind is judgment again, and consequently Art; and at last, even the perception by the senses as well. In a word, if it is impossible to imagine a human being possessing merely the faculty of cognition, devoid of judgment or the reverse, so also Art and Science can never be completely separated from each other.

The more these subtle elements of light embody themselves in the outward forms of the world, so much the more separate appear their domains; and now once more, where the object is creation and production, there is the province of Art; where the object is investigation and knowledge Science holds sway.—After all this it results of itself that it is more fitting to say Art of War than Science of War.

So much for this, because we cannot do without these conceptions. But now we come forward with the assertion that War is neither an Art nor a Science in the real signification, and that it is just the setting out from that starting-point of ideas which has led to a wrong direction being taken, which has caused War to be put on a par with other arts and sciences, and has led to a number of erroneous analogies.

This has indeed been felt before now, and on that it was maintained that War is a handicraft; but there was more lost than gained by that, for a handicraft is only an inferior art, and as such is also subject to definite and rigid laws. In reality the Art of War did go on for some time in the spirit of a handicraft—we allude to the times of the Condottieri—but then it received that direction, not from intrinsic but from external causes; and military history shows how little it was at that time in accordance with the nature of the thing.

3. War Is Part of the Intercourse of the Human Race

We say therefore War belongs not to the province of Arts and Sciences, but to the province of social life. It is a conflict of great interests which is settled by bloodshed, and only in that is it different from others. It would be better, instead of comparing it with any Art, to liken it to business competition, which is also a conflict of human interests and activities; and it is still more like State policy, which again, on its part, may be looked upon as a kind of business competition on a great scale. Besides, State policy is the womb in which War is developed, in which its outlines lie hidden in a rudimentary state, like the qualities of living creatures in their germs. [The analogy has become much closer since Clausewitz's time. Now that the first business of the State is regarded as the development of facilities for trade, War between great nations is only a question of time. No Hague Conferences can avert it—Editor]

4. Difference

The essential difference consists in this, that War is no activity of the will, which exerts itself upon inanimate matter like the mechanical Arts; or upon a living but still passive and yielding subject, like the human mind and the human feelings in the ideal Arts, but against a living and reacting force. How little the categories of Arts and Sciences are applicable to such an activity strikes us at once; and we can understand at the same time how that constant seeking and striving after laws like those which may be developed out of the dead material world could not but lead to constant errors. And yet it is just the mechanical Arts that some people would imitate in the Art of War. The imitation of the ideal Arts was quite out of the question, because these themselves dispense too much with laws and rules, and those hitherto tried, always acknowledged as insufficient and one-sided, are perpetually undermined and washed away by the current of opinions, feelings, and customs.

Whether such a conflict of the living, as takes place and is settled in War, is subject to general laws, and whether these are capable of indicating a useful line of action, will be partly investigated in this book; but so much is evident in itself, that this, like every other subject which does not surpass our powers of understanding, may be lighted up, and be made more or less plain in its inner relations by an inquiring mind, and that alone is sufficient to realise the idea of a *theory*.

Methodicism

In order to explain ourselves clearly as to the conception of method, and method of action, which play such an important part in War, we must be allowed to cast a hasty glance at the logical hierarchy through which, as through regularly constituted official functionaries, the world of action is governed.

Law, in the widest sense strictly applying to perception as well as action, has plainly something subjective and arbitrary in its literal meaning, and expresses just that on which we and those things external to us are dependent. As a subject of cognition, LAW is the relation of things and their effects to one another; as a subject of the will, it is a motive of action, and is then equivalent to *command* or *prohibition*.

Principle is likewise such a law for action, except that it has not the formal definite meaning, but is only the spirit and sense of law in order to leave the judgment more freedom of application when the diversity of the real world cannot be laid hold of under the definite form of a law. As the judgment must of itself suggest the cases in which the principle is not applicable, the latter therefore becomes in that way a real aid or guiding star for the person acting.

Principle is *objective* when it is the result of objective truth, and consequently of equal value for all men; it is *subjective*, and then generally called *Maxim* if there are subjective relations in it, and if it therefore has a certain value only for the person himself who makes it.

Rule is frequently taken in the sense of *law*, and then means the same as Principle, for we say "no rule without exceptions," but we do not say "no law without exceptions," a sign that with *rule* we retain to ourselves more freedom of application.

In another meaning *rule* is the means used of discerning a recondite truth in a particular sign lying close at hand, in order to attach to this particular sign the law of action directed upon the whole truth. Of this kind are all the rules of games of play, all abridged processes in mathematics, etc.

Directions and *instructions* are determinations of action which have an influence upon a number of minor circumstances too numerous and unimportant for general laws.

Lastly, *method, mode of acting*, is an always recurring proceeding selected out of several possible ones; and *methodicism (methodismus)* is that which is determined by methods instead of by general principles or particular prescriptions. By this the cases which are placed under such methods must necessarily be supposed alike in their essential parts. As they cannot all be this, then the point is that at least as many as possible should be; in other words, that Method should be calculated on the most probable cases. Methodicism is therefore not founded on determined particular premises, but on the average probability of cases one with another; and its ultimate tendency is to set up an average truth, the constant and uniform, application of which soon acquires something of the nature of a mechanical appliance, which in the end does that which is right almost unwittingly.

The conception of law in relation to perception is not necessary for the conduct of War, because the complex phenomena of War are not so regular, and the regular are not so complex, that we should gain anything more by this conception than by the simple truth. And where a simple conception and language is sufficient, to resort to the complex becomes affected and pedantic. The conception of law in relation to action cannot be used in the theory of the conduct of War, because owing to the variableness and diversity of the phenomena there is in it no determination of such a general nature as to deserve the name of law.

But principles, rules, prescriptions, and methods are conceptions indispensable to a theory of the conduct of War, in so far as that theory leads to positive doctrines, because in doctrines the truth can only crystallise itself in such forms.

As tactics is the branch of the conduct of War in which theory can attain the nearest to positive doctrine, therefore these conceptions will appear in it most frequently.

Not to use cavalry against unbroken infantry except in some case of special emergency, only to use firearms within effective range in the combat, to spare the forces as much as possible for the final struggle—these are tactical principles. None of them can be applied absolutely in every case, but they must always be present to the mind of the Chief, in order that the benefit of the truth contained in them may not be lost in cases where that truth can be of advantage.

If from the unusual cooking by an enemy's camp his movement is inferred, if the intentional exposure of troops in a combat indicates a false attack, then this way of discerning the truth is called rule, because from a single visible circumstance that conclusion is drawn which corresponds with the same.

If it is a rule to attack the enemy with renewed vigour, as soon as he begins to limber up his artillery in the combat, then on this particular fact depends a course of action which is aimed at the general situation of the enemy as inferred from the above fact, namely, that he is about to give up the fight, that he is commencing to draw off his troops, and is neither capable of making a serious stand while thus drawing off nor of making his retreat gradually in good order.

Regulations and *methods* bring preparatory theories into the conduct of War, in so far as disciplined troops are inoculated with them as active principles. The whole body of instructions for formations, drill, and field service are regulations and methods: in the drill instructions the first predominate, in the field service instructions the latter. To these things the real conduct of War attaches itself; it takes them over, therefore, as given modes of proceeding, and as such they must appear in the theory of the conduct of War.

But for those activities retaining freedom in the employment of these forces there cannot be regulations, that is, definite instructions, because they would do away with freedom of action. Methods, on the other hand, as a general way of executing duties as they arise, calculated, as we have said, on an average of probability, or as a dominating influence of principles and rules carried through to application, may certainly appear in the theory of the conduct of War, provided only they are not represented as something different from what they are, not as the absolute and necessary modes of action (systems), but as the best of general forms which may be used as shorter ways in place of a particular disposition for the occasion, at discretion.

But the frequent application of methods will be seen to be most essential and unavoidable in the conduct of War, if we reflect how much action proceeds on mere conjecture, or in complete uncertainty, because one side is prevented from learning all the circumstances which influence the dispositions of the other, or because, even if these circumstances which influence the decisions of the one were really known, there is not, owing to their extent and the dispositions they would entail, sufficient time for the other to carry out all necessary counteracting measures—that therefore measures in War must always be calculated on a certain number of possibilities; if we reflect how numberless are the trifling things belonging to any single event, and which therefore should be taken into account along with it, and that therefore there is no other means to suppose the one counteracted by the other, and to base our arrangements only upon what is of a general nature and probable; if we reflect lastly that, owing to the increasing number of officers as we descend the scale of rank, less must be left to the true discernment and ripe judgment of individuals the lower the sphere of action, and that when we reach those ranks where we can look for no other notions but those which the regulations of the service and experience afford, we must help them with the methodic forms bordering on those regulations. This will serve both as a support to their judgment and a barrier against those extravagant and erroneous views which are so especially to be dreaded in a sphere where experience is so costly.

Besides this absolute need of method in action, we must also acknowledge that it has a positive advantage, which is that, through the constant repetition of a formal exercise, a readiness, precision, and firmness is attained in the movement of troops which diminishes the natural friction, and makes the machine move easier.

Method will therefore be the more generally used, become the more indispensable, the farther down the scale of rank the position of the active agent; and on the other hand, its use will diminish upwards, until in the highest position it quite disappears. For this reason it is more in its place in tactics than in strategy.

War in its highest aspects consists not of an infinite number of little events, the diversities in which compensate each other, and which therefore by a better or worse method are better or worse governed, but of separate great decisive events which must be dealt with separately. It is not like a field of stalks, which, without any regard to the particular form of each stalk, will be mowed better or worse, according as the mowing instrument is good or bad, but rather as a group of large trees, to which the axe must be laid with judgment, according to the particular form and inclination of each separate trunk.

How high up in military activity the admissibility of method in action reaches naturally determines itself, not according to actual rank, but according to things; and it affects the highest positions in a less degree, only because these positions have the most comprehensive subjects of activity. A constant order of battle, a constant formation of advance guards and outposts, are methods by which a General ties not only his subordinates' hands, but also his own in certain cases. Certainly they may have been devised by himself, and may be applied by him according to circumstances, but they may also be a subject of theory, in so far as they are based on the general properties of troops and weapons. On the other hand, any method by which definite plans for wars or campaigns are to be given out all ready made as if from a machine are absolutely worthless.

As long as there exists no theory which can be sustained, that is, no enlightened treatise on the conduct of War, method in action cannot but encroach beyond its proper limits in high places, for men employed in these spheres of activity have not always had the opportunity of educating themselves, through study and through contact with the higher interests. In the impracticable and inconsistent disquisitions of theorists and critics they cannot find their way, their sound common sense rejects them, and as they bring with them no knowledge but that derived from experience, therefore in those cases which admit of, and require, a free individual treatment they readily make use of the means which experience gives them—that is, an imitation of the particular methods practised by great Generals, by which a method of action then arises of itself. If we see Frederick the Great's Generals always making their appearance in the so-called oblique order of battle, the Generals of the French Revolution always using turning movements with a long, extended line of battle, and Buonaparte's lieutenants rushing to the attack with the bloody energy of concentrated masses, then we recognise in the recurrence of the mode of proceeding evidently an adopted method, and see therefore that method of action can reach up to regions bordering on the highest. Should an improved theory facilitate the study of the conduct of War, form the mind and judgment of men who are rising to the highest commands, then also method in action will no longer reach so far, and so much of it as is to be considered indispensable will then at least be formed from theory itself, and not take place out of mere imitation. However pre-eminently a great Commander does things, there is always something subjective in the way he does them; and if he has a certain manner, a large share of his individuality is contained in it which does not always accord with the individuality of the person who copies his manner.

At the same time, it would neither be possible nor right to banish subjective methodicism or manner completely from the conduct of War: it is rather to be regarded as a manifestation of that influence which the general character of a War has upon its separate events, and to which satisfaction can only be done in that way if theory is not able to foresee this general character and include it in its considerations. What is more natural than that the War of the French Revolution had its own way of doing things? and what theory could ever have included that peculiar method? The evil is only that such a manner originating in a special case easily outlives itself, because it continues whilst circumstances imperceptibly change. This is what theory should prevent by lucid and rational criticism. When in the year 1806 the Prussian Generals, Prince Louis at Saalfeld, Tauentzien on the Dornberg near Jena, Grawert before and Ruechel behind Kappellendorf, all threw themselves into the open jaws of destruction in the oblique order of Frederick the Great, and managed to ruin Hohenlohe's Army in a way that no Army was ever ruined, even on the field of battle, all this was done through a manner which had outlived its day, together with the most downright stupidity to which methodicism ever led.

Criticism

The influence of theoretical principles upon real life is produced more through criticism than through doctrine, for as criticism is an application of abstract truth to real events, therefore it not only brings truth of this description nearer to life, but also accustoms the understanding more to such truths by the constant repetition of their application. We therefore think it necessary to fix the point of view for criticism next to that for theory.

From the simple narration of an historical occurrence which places events in chronological order, or at most only touches on their more immediate causes, we separate the *critical*.

In this *critical* three different operations of the mind may be observed.

First, the historical investigation and determining of doubtful facts. This is properly historical research, and has nothing in common with theory.

Secondly, the tracing of effects to causes. This is the *real critical inquiry*; it is indispensable to theory, for everything which in theory is to be established, supported, or even merely explained, by experience can only be settled in this way.

Thirdly, the testing of the means employed. This is criticism, properly speaking, in which praise and censure is contained. This is where theory helps history, or rather, the teaching to be derived from it.

In these two last strictly critical parts of historical study, all depends on tracing things to their primary elements, that is to say, up to undoubted truths, and not, as is so often done, resting half-way, that is, on some arbitrary assumption or supposition.

As respects the tracing of effect to cause, that is often attended with the insuperable difficulty that the real causes are not known. In none of the relations of life does this so frequently happen as in War, where events are seldom fully known, and still less motives, as the latter have been, perhaps purposely, concealed by the chief actor, or have been of such a transient and accidental character that they have been lost for history. For this reason critical narration must generally proceed hand in hand with historical investigation, and still such a want of connection between cause and effect will often present itself, that it does not seem justifiable to consider effects as the necessary results of known causes. Here, therefore must occur, that is, historical results which cannot be made use of for teaching. All that theory can demand is that the investigation should be rigidly conducted up to that point, and there leave off without drawing conclusions. A real evil springs up only if the known is made perforce to suffice as an explanation of effects, and thus a false importance is ascribed to it.

Besides this difficulty, critical inquiry also meets with another great and intrinsic one, which is that the progress of events in War seldom proceeds from one simple cause, but from several in common, and that it therefore is not sufficient to follow up a series of events to their origin in a candid and impartial spirit, but that it is then also necessary to apportion to each contributing cause its due weight. This leads, therefore, to a closer investigation of their nature, and thus a critical investigation may lead into what is the proper field of theory.

The critical *consideration*, that is, the testing of the means, leads to the question, Which are the effects peculiar to the means applied, and whether these effects were comprehended in the plans of the person directing?

The effects peculiar to the means lead to the investigation of their nature, and thus again into the field of theory.

We have already seen that in criticism all depends upon attaining to positive truth; therefore, that we must not stop at arbitrary propositions which are not allowed by others, and to which other perhaps equally arbitrary assertions may again be opposed, so that there is no end to pros and cons; the whole is without result, and therefore without instruction.

We have seen that both the search for causes and the examination of means lead into the field of theory; that is, into the field of universal truth, which does not proceed solely from the case immediately under examination. If there is a theory which can be used, then the critical consideration will appeal to the proofs there afforded, and the examination may there stop. But where no such theoretical truth is to be found, the inquiry must be pushed up to the original elements. If this necessity occurs often, it must lead the historian (according to a common expression) into a labyrinth of details. He then has his hands full, and it is impossible for him to stop to give the requisite attention everywhere; the consequence is, that in order to set bounds to his investigation, he adopts some arbitrary assumptions which, if they do not appear so to him, do so to others, as they are not evident in themselves or capable of proof.

A sound theory is therefore an essential foundation for criticism, and it is impossible for it, without the assistance of a sensible theory, to attain to that point at which it commences chiefly to be instructive, that is, where it becomes demonstration, both convincing and sans re'plique.

But it would be a visionary hope to believe in the possibility of a theory applicable to every abstract truth, leaving nothing for criticism to do but to place the case under its appropriate law: it would be ridiculous pedantry to lay down as a rule for criticism that it must always halt and turn round on reaching the boundaries of sacred theory. The same spirit of analytical inquiry which is the origin of theory must also guide the critic in his work; and it can and must therefore happen that he strays beyond the boundaries of the province of theory and elucidates those points with which he is more particularly concerned. It is more likely, on the contrary, that criticism would completely fail in its object if it degenerated into a mechanical application of theory. All positive results of theoretical inquiry, all principles, rules, and methods, are the more wanting in generality and positive truth the more they become positive doctrine. They exist to offer themselves for use as required, and it must always be left for judgment to decide whether they are suitable or not. Such results of theory must never be used in criticism as rules or norms for a standard, but in the same way as the person acting should use them, that is, merely as aids to judgment. If it is an acknowledged principle in tactics that in the usual order of battle cavalry should be placed behind infantry, not in line with it, still it would be folly on this account to condemn every deviation from this principle. Criticism must investigate the grounds of the deviation, and it is only in case these are insufficient that it has a right to appeal to principles laid down in theory. If it is further established in theory that a divided attack diminishes the probability of success, still it would be just as unreasonable, whenever there is a divided attack and an unsuccessful issue, to regard the latter as the result of the former, without further investigation into the connection between the two, as where a divided attack is successful to infer from it the fallacy of that theoretical principle. The spirit of investigation which belongs to criticism cannot allow either. Criticism therefore supports itself chiefly on the results of the analytical investigation of theory; what has been made out and determined by theory does not require to be demonstrated over again by criticism, and it is so determined by theory that criticism may find it ready demonstrated.

This office of criticism, of examining the effect produced by certain causes, and whether a means applied has answered its object, will be easy enough if cause and effect, means and end, are all near together.

If an Army is surprised, and therefore cannot make a regular and intelligent use of its powers and resources, then the effect of the surprise is not doubtful.—If theory has determined that in a battle the convergent form of attack is calculated to produce greater but less certain results, then the question is whether he who employs that convergent form had in view chiefly that greatness of result as his object; if so, the proper means were chosen. But if by this form he intended to make the result more certain, and that expectation was founded not on some exceptional circumstances (in this case), but on the general nature of the convergent form, as has happened a hundred times, then he mistook the nature of the means and committed an error.

Here the work of military investigation and criticism is easy, and it will always be so when confined to the immediate effects and objects. This can be done quite at option, if we abstract the connection of the parts with the whole, and only look at things in that relation.

But in War, as generally in the world, there is a connection between everything which belongs to a whole; and therefore, however small a cause may be in itself, its effects reach to the end of the act of warfare, and modify or influence the final result in some degree, let that degree be ever so small. In the same manner every means must be felt up to the ultimate object.

We can therefore trace the effects of a cause as long as events are worth noticing, and in the same way we must not stop at the testing of a means for the immediate object, but test also this object as a means to a higher one, and thus ascend the series of facts in succession, until we come to one so absolutely necessary in its nature as to require no examination or proof. In many cases, particularly in what concerns great and decisive measures, the investigation must be carried to the final aim, to that which leads immediately to peace.

It is evident that in thus ascending, at every new station which we reach a new point of view for the judgment is attained, so that the same means which appeared advisable at one station, when looked at from the next above it may have to be rejected.

The search for the causes of events and the comparison of means with ends must always go hand in hand in the critical review of an act, for the investigation of causes leads us first to the discovery of those things which are worth examining.

This following of the clue up and down is attended with considerable difficulty, for the farther from an event the cause lies which we are looking for, the greater must be the number of other causes which must at the same time be kept in view and allowed for in reference to the share which they have in the course of events, and then eliminated, because the higher the importance of a fact the greater will be the number of separate forces and circumstances by which it is conditioned. If we have unravelled the causes of a battle being lost, we have certainly also ascertained a part of the causes of the consequences which this defeat has upon the whole War, but only a part, because the effects of other causes, more or less according to circumstances, will flow into the final result.

The same multiplicity of circumstances is presented also in the examination of the means the higher our point of view, for the higher the object is situated, the greater must be the number of means employed to reach it. The ultimate object of the War is the object aimed at by all the Armies simultaneously, and it is therefore necessary that the consideration should embrace all that each has done or could have done.

It is obvious that this may sometimes lead to a wide field of inquiry, in which it is easy to wander and lose the way, and in which this difficulty prevails—that a number of assumptions or suppositions must be made about a variety of things which do not actually appear, but which in all probability did take place, and therefore cannot possibly be left out of consideration.

When Buonaparte, in 1797, [Compare Hinterlassene Werke, 2nd edition, vol. iv. p. 276 et seq.] at the head of the Army of Italy, advanced from the Tagliamento against the Archduke Charles, he did so with a view to force that General to a decisive action before the reinforcements expected from the Rhine had reached him. If we look, only at the immediate object, the means were well chosen and justified by the result, for the Archduke was so inferior in numbers that he only made a show of resistance on the Tagliamento, and when he saw his adversary so strong and resolute, yielded ground, and left open the passages, of the Norican Alps. Now to what use could Buonaparte turn this fortunate event? To penetrate into the heart of the Austrian empire itself, to facilitate the advance of the Rhine Armies under Moreau and Hoche, and open communication with them? This was the view taken by Buonaparte, and from this point of view he was right. But now, if criticism places itself at a higher point of view—namely, that of the French Directory, which body could see and know that the Armies on the Rhine could not commence the campaign for six weeks, then the advance of Buonaparte over the Norican Alps can only be regarded as an extremely hazardous measure; for if the Austrians had drawn largely on their Rhine Armies to reinforce their Army in Styria, so as to enable the Archduke to fall upon the Army of Italy, not only would that Army have been routed, but the whole campaign lost. This consideration, which attracted the serious attention of Buonaparte at Villach, no doubt induced him to sign the armistice of Leoben with so much readiness.

If criticism takes a still higher position, and if it knows that the Austrians had no reserves between the Army of the Archduke Charles and Vienna, then we see that Vienna became threatened by the advance of the Army of Italy.

Supposing that Buonaparte knew that the capital was thus uncovered, and knew that he still retained the same superiority in numbers over the Archduke as he had in Styria, then his advance against the heart of the

Austrian States was no longer without purpose, and its value depended on the value which the Austrians might place on preserving their capital. If that was so great that, rather than lose it, they would accept the conditions of peace which Buonaparte was ready to offer them, it became an object of the first importance to threaten Vienna. If Buonaparte had any reason to know this, then criticism may stop there, but if this point was only problematical, then criticism must take a still higher position, and ask what would have followed if the Austrians had resolved to abandon Vienna and retire farther into the vast dominions still left to them. But it is easy to see that this question cannot be answered without bringing into the consideration the probable movements of the Rhine Armies on both sides. Through the decided superiority of numbers on the side of the French—130,000 to 80,000—there could be little doubt of the result; but then next arises the question, What use would the Directory make of a victory; whether they would follow up their success to the opposite frontiers of the Austrian monarchy, therefore to the complete breaking up or overthrow of that power, or whether they would be satisfied with the conquest of a considerable portion to serve as a security for peace? The probable result in each case must be estimated, in order to come to a conclusion as to the probable determination of the Directory. Supposing the result of these considerations to be that the French forces were much too weak for the complete subjugation of the Austrian monarchy, so that the attempt might completely reverse the respective positions of the contending Armies, and that even the conquest and occupation of a considerable district of country would place the French Army in strategic relations to which they were not equal, then that result must naturally influence the estimate of the position of the Army of Italy, and compel it to lower its expectations. And this, it was no doubt which influenced Buonaparte, although fully aware of the helpless condition of the Archduke, still to sign the peace of Campo Formio, which imposed no greater sacrifices on the Austrians than the loss of provinces which, even if the campaign took the most favourable turn for them, they could not have reconquered. But the French could not have reckoned on even the moderate treaty of Campo Formio, and therefore it could not have been their object in making their bold advance if two considerations had not presented themselves to their view, the first of which consisted in the question, what degree of value the Austrians would attach to each of the above-mentioned results; whether, notwithstanding the probability of a satisfactory result in either of these cases, would it be worth while to make the sacrifices inseparable from a continuance of the War, when they could be spared those sacrifices by a peace on terms not too humiliating? The second consideration is the question whether the Austrian Government, instead of seriously weighing the possible results of a resistance pushed to extremities, would not prove completely disheartened by the impression of their present reverses.

The consideration which forms the subject of the first is no idle piece of subtle argument, but a consideration of such decidedly practical importance that it comes up whenever the plan of pushing War to the utmost extremity is mooted, and by its weight in most cases restrains the execution of such plans.

The second consideration is of equal importance, for we do not make War with an abstraction but with a reality, which we must always keep in view, and we may be sure that it was not overlooked by the bold Buonaparte—that is, that he was keenly alive to the terror which the appearance of his sword inspired. It was reliance on that which led him to Moscow. There it led him into a scrape. The terror of him had been weakened by the gigantic struggles in which he had been engaged; in the year 1797 it was still fresh, and the secret of a resistance pushed to extremities had not been discovered; nevertheless even in 1797 his boldness might have led to a negative result if, as already said, he had not with a sort of presentiment avoided it by signing the moderate peace of Campo Formio.

We must now bring these considerations to a close—they will suffice to show the wide sphere, the diversity and embarrassing nature of the subjects embraced in a critical examination carried to the fullest extent, that is, to those measures of a great and decisive class which must necessarily be included. It follows from them that besides a theoretical acquaintance with the subject, natural talent must also have a great influence on the value of critical examinations, for it rests chiefly with the latter to throw the requisite light on the interrelations of things, and to distinguish from amongst the endless connections of events those which are really essential.

But talent is also called into requisition in another way. Critical examination is not merely the appreciation of those means which have been actually employed, but also of all possible means, which therefore must be suggested in the first place—that is, must be discovered; and the use of any particular means is not fairly open to censure until a better is pointed out. Now, however small the number of possible combinations may be in most cases, still it must be admitted that to point out those which have not been used is not a mere analysis of actual things, but a spontaneous creation which cannot be prescribed, and depends on the fertility of genius.

We are far from seeing a field for great genius in a case which admits only of the application of a few simple combinations, and we think it exceedingly ridiculous to hold up, as is often done, the turning of a position as an invention showing the highest genius; still nevertheless this creative self-activity on the part of the critic is necessary, and it is one of the points which essentially determine the value of critical examination.

When Buonaparte on 30th July, 1796, [Compare Hinterlassene Werke, 2nd edition, vol. iv. p. 107 et seq.] determined to raise the siege of Mantua, in order to march with his whole force against the enemy, advancing in separate columns to the relief of the place, and to beat them in detail, this appeared the surest way to the attainment of brilliant victories. These victories actually followed, and were afterwards again repeated on a still more brilliant scale on the attempt to relieve the fortress being again renewed. We hear only one opinion on these achievements, that of unmixed admiration.

At the same time, Buonaparte could not have adopted this course on the 30th July without quite giving up the idea of the siege of Mantua, because it was impossible to save the siege train, and it could not be replaced by another in this campaign. In fact, the siege was converted into a blockade, and the town, which if the siege had continued must have very shortly fallen, held out for six months in spite of Buonaparte's victories in the open field.

Criticism has generally regarded this as an evil that was unavoidable, because critics have not been able to suggest any better course. Resistance to a relieving Army within lines of circumvallation had fallen into such disrepute and contempt that it appears to have entirely escaped consideration as a means. And yet in the reign of Louis XIV. that measure was so often used with success that we can only attribute to the force of fashion the fact that a hundred years later it never occurred to any one even to propose such a measure. If the practicability of such a plan had ever been entertained for a moment, a closer consideration of circumstances would have shown that 40,000 of the best infantry in the world under Buonaparte, behind strong lines of circumvallation round Mantua, had so little to fear from the 50,000 men coming to the relief under Wurmser, that it was very unlikely that any attempt even would be made upon their lines. We shall not seek here to establish this point, but we believe enough has been said to show that this means was one which had a right to a share of consideration. Whether Buonaparte himself ever thought of such a plan we leave undecided; neither in his memoirs nor in other sources is there any trace to be found of his having done so; in no critical works has it been touched upon, the measure being one which the mind had lost sight of. The merit of resuscitating the idea of this means is not great, for it suggests itself at once to any one who breaks loose from the trammels of fashion. Still it is necessary that it should suggest itself for us to bring it into consideration and compare it with the means which Buonaparte employed. Whatever may be the result of the comparison, it is one which should not be omitted by criticism.

When Buonaparte, in February, 1814, [Compare Hinterlassene Werks, 2nd edition. vol. vii. p. 193 et seq.] after gaining the battles at Etoges, Champ-Aubert, and Montmirail, left Bluecher's Army, and turning upon Schwartzenberg, beat his troops at Montereau and Mormant, every one was filled with admiration, because Buonaparte, by thus throwing his concentrated force first upon one opponent, then upon another, made a brilliant use of the mistakes which his adversaries had committed in dividing their forces. If these brilliant strokes in different directions failed to save him, it was generally considered to be no fault of his, at least. No one has yet asked the question, What would have been the result if, instead of turning from Bluecher upon Schwartzenberg, he had tried another blow at Bluecher, and pursued him to the Rhine? We are convinced that it would have completely changed the course of the campaign, and that the Army of the Allies, instead of marching to Paris, would have retired behind the Rhine. We do not ask others to share our conviction, but no one who understands the thing will doubt, at the mere mention of this alternative course, that it is one which should not be overlooked in criticism.

In this case the means of comparison lie much more on the surface than in the foregoing, but they have been equally overlooked, because one-sided views have prevailed, and there has been no freedom of judgment.

From the necessity of pointing out a better means which might have been used in place of those which are condemned has arisen the form of criticism almost exclusively in use, which contents itself with pointing out the better means without demonstrating in what the superiority consists. The consequence is that some are not convinced, that others start up and do the same thing, and that thus discussion arises which is without any fixed basis for the argument. Military literature abounds with matter of this sort.

The demonstration we require is always necessary when the superiority of the means propounded is not so evident as to leave no room for doubt, and it consists in the examination of each of the means on its own merits, and then of its comparison with the object desired. When once the thing is traced back to a simple truth, controversy must cease, or at all events a new result is obtained, whilst by the other plan the pros and cons go on for ever consuming each other.

Should we, for example, not rest content with assertion in the case before mentioned, and wish to prove that the persistent pursuit of Bluecher would have been more advantageous than the turning on Schwartzenberg, we should support the arguments on the following simple truths:

1. In general it is more advantageous to continue our blows in one and the same direction, because there is a loss of time in striking in different directions; and at a point where the moral power is already shaken by

considerable losses there is the more reason to expect fresh successes, therefore in that way no part of the preponderance already gained is left idle.

2. Because Bluecher, although weaker than Schwartzenberg, was, on account of his enterprising spirit, the more important adversary; in him, therefore, lay the centre of attraction which drew the others along in the same direction.

3. Because the losses which Bluecher had sustained almost amounted to a defeat, which gave Buonaparte such a preponderance over him as to make his retreat to the Rhine almost certain, and at the same time no reserves of any consequence awaited him there.

4. Because there was no other result which would be so terrific in its aspects, would appear to the imagination in such gigantic proportions, an immense advantage in dealing with a Staff so weak and irresolute as that of Schwartzenberg notoriously was at this time. What had happened to the Crown Prince of Wartemberg at Montereau, and to Count Wittgenstein at Mormant, Prince Schwartzenberg must have known well enough; but all the untoward events on Bluecher's distant and separate line from the Marne to the Rhine would only reach him by the avalanche of rumour. The desperate movements which Buonaparte made upon Vitry at the end of March, to see what the Allies would do if he threatened to turn them strategically, were evidently done on the principle of working on their fears; but it was done under far different circumstances, in consequence of his defeat at Laon and Arcis, and because Bluecher, with 100,000 men, was then in communication with Schwartzenberg.

There are people, no doubt, who will not be convinced on these arguments, but at all events they cannot retort by saying, that "whilst Buonaparte threatened Schwartzenberg's base by advancing to the Rhine, Schwartzenberg at the same time threatened Buonaparte's communications with Paris," because we have shown by the reasons above given that Schwartzenberg would never have thought of marching on Paris.

With respect to the example quoted by us from the campaign of 1796, we should say: Buonaparte looked upon the plan he adopted as the surest means of beating the Austrians; but admitting that it was so, still the object to be attained was only an empty victory, which could have hardly any sensible influence on the fall of Mantua. The way which we should have chosen would, in our opinion, have been much more certain to prevent the relief of Mantua; but even if we place ourselves in the position of the French General and assume that it was not so, and look upon the certainty of success to have been less, the question then amounts to a choice between a more certain but less useful, and therefore less important, victory on the one hand, and a somewhat less probable but far more decisive and important victory, on the other hand. Presented in this form, boldness must have declared for the second solution, which is the reverse of what took place, when the thing was only superficially viewed. Buonaparte certainly was anything but deficient in boldness, and we may be sure that he did not see the whole case and its consequences as fully and clearly as we can at the present time.

Naturally the critic, in treating of the means, must often appeal to military history, as experience is of more value in the Art of War than all philosophical truth. But this exemplification from history is subject to certain conditions, of which we shall treat in a special chapter and unfortunately these conditions are so seldom regarded that reference to history generally only serves to increase the confusion of ideas.

We have still a most important subject to consider, which is, How far criticism in passing judgments on particular events is permitted, or in duty bound, to make use of its wider view of things, and therefore also of that which is shown by results; or when and where it should leave out of sight these things in order to place itself, as far as possible, in the exact position of the chief actor?

If criticism dispenses praise or censure, it should seek to place itself as nearly as possible at the same point of view as the person acting, that is to say, to collect all he knew and all the motives on which he acted, and, on the other hand, to leave out of the consideration all that the person acting could not or did not know, and above all, the result. But this is only an object to aim at, which can never be reached because the state of circumstances from which an event proceeded can never be placed before the eye of the critic exactly as it lay before the eye of the person acting. A number of inferior circumstances, which must have influenced the result, are completely lost to sight, and many a subjective motive has never come to light.

The latter can only be learnt from the memoirs of the chief actor, or from his intimate friends; and in such things of this kind are often treated of in a very desultory manner, or purposely misrepresented. Criticism must, therefore, always forego much which was present in the minds of those whose acts are criticised.

On the other hand, it is much more difficult to leave out of sight that which criticism knows in excess. This is only easy as regards accidental circumstances, that is, circumstances which have been mixed up, but are in no way necessarily related. But it is very difficult, and, in fact, can never be completely done with regard to things really essential.

Let us take first, the result. If it has not proceeded from accidental circumstances, it is almost impossible that the knowledge of it should not have an effect on the judgment passed on events which have preceded it, for we see these things in the light of this result, and it is to a certain extent by it that we first become acquainted with them and appreciate them. Military history, with all its events, is a source of instruction for criticism itself, and it is only natural that criticism should throw that light on things which it has itself obtained from the consideration of the whole. If therefore it might wish in some cases to leave the result out of the consideration, it would be impossible to do so completely.

But it is not only in relation to the result, that is, with what takes place at the last, that this embarrassment arises; the same occurs in relation to preceding events, therefore with the data which furnished the motives to action. Criticism has before it, in most cases, more information on this point than the principal in the transaction. Now it may seem easy to dismiss from the consideration everything of this nature, but it is not so easy as we may think. The knowledge of preceding and concurrent events is founded not only on certain information, but on a number of conjectures and suppositions; indeed, there is hardly any of the information respecting things not purely accidental which has not been preceded by suppositions or conjectures destined to take the place of certain information in case such should never be supplied. Now is it conceivable that criticism in after times, which has before it as facts all the preceding and concurrent circumstances, should not allow itself to be thereby influenced when it asks itself the question, What portion of the circumstances, which at the moment of action were unknown, would it have held to be probable? We maintain that in this case, as in the case of the results, and for the same reason, it is impossible to disregard all these things completely.

If therefore the critic wishes to bestow praise or blame upon any single act, he can only succeed to a certain degree in placing himself in the position of the person whose act he has under review. In many cases he can do so sufficiently near for any practical purpose, but in many instances it is the very reverse, and this fact should never be overlooked.

But it is neither necessary nor desirable that criticism should completely identify itself with the person acting. In War, as in all matters of skill, there is a certain natural aptitude required which is called talent. This may be great or small. In the first case it may easily be superior to that of the critic, for what critic can pretend to the skill of a Frederick or a Buonaparte? Therefore, if criticism is not to abstain altogether from offering an opinion where eminent talent is concerned, it must be allowed to make use of the advantage which its enlarged horizon affords. Criticism must not, therefore, treat the solution of a problem by a great General like a sum in arithmetic; it is only through the results and through the exact coincidences of events that it can recognise with admiration how much is due to the exercise of genius, and that it first learns the essential combination which the glance of that genius devised.

But for every, even the smallest, act of genius it is necessary that criticism should take a higher point of view, so that, having at command many objective grounds of decision, it may be as little subjective as possible, and that the critic may not take the limited scope of his own mind as a standard.

This elevated position of criticism, its praise and blame pronounced with a full knowledge of all the circumstances, has in itself nothing which hurts our feelings; it only does so if the critic pushes himself forward, and speaks in a tone as if all the wisdom which he has obtained by an exhaustive examination of the event under consideration were really his own talent. Palpable as is this deception, it is one which people may easily fall into through vanity, and one which is naturally distasteful to others. It very often happens that although the critic has no such arrogant pretensions, they are imputed to him by the reader because he has not expressly disclaimed them, and then follows immediately a charge of a want of the power of critical judgment.

If therefore a critic points out an error made by a Frederick or a Buonaparte, that does not mean that he who makes the criticism would not have committed the same error; he may even be ready to grant that had he been in the place of these great Generals he might have made much greater mistakes; he merely sees this error from the chain of events, and he thinks that it should not have escaped the sagacity of the General.

This is, therefore, an opinion formed through the connection of events, and therefore through the *result*. But there is another quite different effect of the result itself upon the judgment, that is if it is used quite alone as an example for or against the soundness of a measure. This may be called *judgment according to the result*. Such a judgment appears at first sight inadmissible, and yet it is not.

When Buonaparte marched to Moscow in 1812, all depended upon whether the taking of the capital, and the events which preceded the capture, would force the Emperor Alexander to make peace, as he had been compelled to do after the battle of Friedland in 1807, and the Emperor Francis in 1805 and 1809 after Austerlitz and Wagram; for if Buonaparte did not obtain a peace at Moscow, there was no alternative but to return—that is, there was nothing for him but a strategic defeat. We shall leave out of the question what he did to get to Moscow, and whether in his advance he did not miss many opportunities of bringing the Emperor Alexander to peace; we shall also exclude all consideration of the disastrous circumstances which attended his retreat, and which perhaps had their origin in the general conduct of the campaign. Still the question remains the same, for however much more brilliant the course of the campaign up to Moscow might have been, still there was always an uncertainty whether the Emperor Alexander would be intimidated into making peace; and then, even if a retreat did not contain in itself the seeds of such disasters as did in fact occur, still it could never be anything else than a great strategic defeat. If the Emperor Alexander agreed to a peace which was disadvantageous to him, the campaign of 1812 would have ranked with those of Austerlitz, Friedland, and Wagram. But these campaigns also, if they had not led to peace, would in all probability have ended in similar catastrophes. Whatever, therefore, of genius, skill, and energy the Conqueror of the World applied to the task, this last question addressed to fate ["Frage an der Schicksal,"a familiar quotation from Schiller.—TR.] remained always the same. Shall we then discard the campaigns of 1805, 1807, 1809, and say on account of the campaign of 1812 that they were acts of imprudence; that the results were against the nature of things, and that in 1812 strategic justice at last found vent for itself in opposition to blind chance? That would be an unwarrantable conclusion, a most arbitrary judgment, a case only half proved, because no human, eye can trace the thread of the necessary connection of events up to the determination of the conquered Princes.

Still less can we say the campaign of 1812 merited the same success as the others, and that the reason why it turned out otherwise lies in something unnatural, for we cannot regard the firmness of Alexander as something unpredictable.

What can be more natural than to say that in the years 1805, 1807, 1809, Buonaparte judged his opponents correctly, and that in 1812 he erred in that point? On the former occasions, therefore, he was right, in the latter wrong, and in both cases we judge by the *result*.

All action in War, as we have already said, is directed on probable, not on certain, results. Whatever is wanting in certainty must always be left to fate, or chance, call it which you will. We may demand that what is so left should be as little as possible, but only in relation to the particular case—that is, as little as is possible in this one case, but not that the case in which the least is left to chance is always to be preferred. That would be an enormous error, as follows from all our theoretical views. There are cases in which the greatest daring is the greatest wisdom.

Now in everything which is left to chance by the chief actor, his personal merit, and therefore his responsibility as well, seems to be completely set aside; nevertheless we cannot suppress an inward feeling of satisfaction whenever expectation realises itself, and if it disappoints us our mind is dissatisfied; and more than this of right and wrong should not be meant by the judgment which we form from the mere result, or rather that we find there.

Nevertheless, it cannot be denied that the satisfaction which our mind experiences at success, the pain caused by failure, proceed from a sort of mysterious feeling; we suppose between that success ascribed to good fortune and the genius of the chief a fine connecting thread, invisible to the mind's eye, and the supposition gives pleasure. What tends to confirm this idea is that our sympathy increases, becomes more decided, if the successes and defeats of the principal actor are often repeated. Thus it becomes intelligible how good luck in War assumes a much nobler nature than good luck at play. In general, when a fortunate warrior does not otherwise lessen our interest in his behalf, we have a pleasure in accompanying him in his career.

Criticism, therefore, after having weighed all that comes within the sphere of human reason and conviction, will let the result speak for that part where the deep mysterious relations are not disclosed in any visible form, and will protect this silent sentence of a higher authority from the noise of crude opinions on the one hand, while on the other it prevents the gross abuse which might be made of this last tribunal.

This verdict of the result must therefore always bring forth that which human sagacity cannot discover; and it will be chiefly as regards the intellectual powers and operations that it will be called into requisition, partly because they can be estimated with the least certainty, partly because their close connection with the will is favourable to their exercising over it an important influence. When fear or bravery precipitates the decision, there is nothing objective intervening between them for our consideration, and consequently nothing by which sagacity and calculation might have met the probable result.

We must now be allowed to make a few observations on the instrument of criticism, that is, the language which it uses, because that is to a certain extent connected with the action in War; for the critical examination is nothing more than the deliberation which should precede action in War. We therefore think it very essential that the language used in criticism should have the same character as that which deliberation in War must have, for otherwise it would cease to be practical, and criticism could gain no admittance in actual life.

We have said in our observations on the theory of the conduct of War that it should educate the mind of the Commander for War, or that its teaching should guide his education; also that it is not intended to furnish him with positive doctrines and systems which he can use like mental appliances. But if the construction of scientific formulae is never required, or even allowable, in War to aid the decision on the case presented, if truth does not appear there in a systematic shape, if it is not found in an indirect way, but directly by the natural perception of the mind, then it must be the same also in a critical review.

It is true as we have seen that, wherever complete demonstration of the nature of things would be too tedious, criticism must support itself on those truths which theory has established on the point. But, just as in War the actor obeys these theoretical truths rather because his mind is imbued with them than because he regards them as objective inflexible laws, so criticism must also make use of them, not as an external law or an algebraic formula, of which fresh proof is not required each time they are applied, but it must always throw a light on this proof itself, leaving only to theory the more minute and circumstantial proof. Thus it avoids a mysterious, unintelligible phraseology, and makes its progress in plain language, that is, with a clear and always visible chain of ideas.

Certainly this cannot always be completely attained, but it must always be the aim in critical expositions. Such expositions must use complicated forms of science as sparingly as possible, and never resort to the construction of scientific aids as of a truth apparatus of its own, but always be guided by the natural and unbiassed impressions of the mind.

But this pious endeavour, if we may use the expression, has unfortunately seldom hitherto presided over critical examinations: the most of them have rather been emanations of a species of vanity—a wish to make a display of ideas.

The first evil which we constantly stumble upon is a lame, totally inadmissible application of certain one-sided systems as of a formal code of laws. But it is never difficult to show the one-sidedness of such systems, and this only requires to be done once to throw discredit for ever on critical judgments which are based on them. We have here to deal with a definite subject, and as the number of possible systems after all can be but small, therefore also they are themselves the lesser evil.

Much greater is the evil which lies in the pompous retinue of technical terms—scientific expressions and metaphors, which these systems carry in their train, and which like a rabble-like the baggage of an Army broken away from its Chief—hang about in all directions. Any critic who has not adopted a system, either because he has not found one to please him, or because he has not yet been able to make himself master of one, will at least occasionally make use of a piece of one, as one would use a ruler, to show the blunders committed by a General. The most of them are incapable of reasoning without using as a help here and there some shreds of scientific military theory. The smallest of these fragments, consisting in mere scientific words and metaphors, are often nothing more than ornamental flourishes of critical narration. Now it is in the nature of things that all technical and scientific expressions which belong to a system lose their propriety, if they ever had any, as soon as they are distorted, and used as general axioms, or as small crystalline talismans, which have more power of demonstration than simple speech.

Thus it has come to pass that our theoretical and critical books, instead of being straightforward, intelligible dissertations, in which the author always knows at least what he says and the reader what he reads, are brimful of these technical terms, which form dark points of interference where author and reader part company. But frequently they are something worse, being nothing but hollow shells without any kernel. The author himself has no clear perception of what he means, contents himself with vague ideas, which if expressed in plain language would be unsatisfactory even to himself.

A third fault in criticism is the *misuse of historical examples*, and a display of great reading or learning. What the history of the Art of War is we have already said, and we shall further explain our views on examples and on military history in general in special chapters. One fact merely touched upon in a very cursory manner may be used to support the most opposite views, and three or four such facts of the most heterogeneous description, brought together out of the most distant lands and remote times and heaped up, generally distract and bewilder the judgment and understanding without demonstrating anything; for when exposed to the light they turn out to be only trumpery rubbish, made use of to show off the author's learning.

But what can be gained for practical life by such obscure, partly false, confused arbitrary conceptions? So little is gained that theory on account of them has always been a true antithesis of practice, and frequently a subject of ridicule to those whose soldierly qualities in the field are above question.

But it is impossible that this could have been the case, if theory in simple language, and by natural treatment of those things which constitute the Art of making War, had merely sought to establish just so much as admits of being established; if, avoiding all false pretensions and irrelevant display of scientific forms and historical parallels, it had kept close to the subject, and gone hand in hand with those who must conduct affairs in the field by their own natural genius.

On Examples

Examples from history make everything clear, and furnish the best description of proof in the empirical sciences. This applies with more force to the Art of War than to any other. General Scharnhorst, whose handbook is the best ever written on actual War, pronounces historical examples to be of the first importance, and makes an admirable use of them himself. Had he survived the War in which he fell,[General Scharnhorst died in 1813, of a wound received in the battle of Bautzen or Grosz Gorchen—Editor]. the fourth part of his revised treatise on artillery would have given a still greater proof of the observing and enlightened spirit in which he sifted matters of experience.

But such use of historical examples is rarely made by theoretical writers; the way in which they more commonly make use of them is rather calculated to leave the mind unsatisfied, as well as to offend the understanding. We therefore think it important to bring specially into view the use and abuse of historical examples.

Unquestionably the branches of knowledge which lie at the foundation of the Art of War come under the denomination of empirical sciences; for although they are derived in a great measure from the nature of things, still we can only learn this very nature itself for the most part from experience; and besides that, the practical application is modified by so many circumstances that the effects can never be completely learnt from the mere nature of the means.

The effects of gunpowder, that great agent in our military activity, were only learnt by experience, and up to this hour experiments are continually in progress in order to investigate them more fully. That an iron ball to which powder has given a velocity of 1000 feet in a second, smashes every living thing which it touches in its course is intelligible in itself; experience is not required to tell us that; but in producing this effect how many hundred circumstances are concerned, some of which can only be learnt by experience! And the physical is not the only effect which we have to study, it is the moral which we are in search of, and that can only be ascertained by experience; and there is no other way of learning and appreciating it but by experience. In the middle ages, when firearms were first invented, their effect, owing to their rude make, was materially but trifling compared to what it now is, but their effect morally was much greater. One must have witnessed the firmness of one of those masses taught and led by Buonaparte, under the heaviest and most unintermittent cannonade, in order to understand what troops, hardened by long practice in the field of danger, can do, when by a career of victory they have reached the noble principle of demanding from themselves their utmost efforts. In pure conception no one would believe it. On the other hand, it is well known that there are troops in the service of European Powers at the present moment who would easily be dispersed by a few cannon shots.

But no empirical science, consequently also no theory of the Art of War, can always corroborate its truths by historical proof; it would also be, in some measure, difficult to support experience by single facts. If any means is once found efficacious in War, it is repeated; one nation copies another, the thing becomes the fashion, and in this manner it comes into use, supported by experience, and takes its place in theory, which contents itself with appealing to experience in general in order to show its origin, but not as a verification of its truth.

But it is quite otherwise if experience is to be used in order to overthrow some means in use, to confirm what is doubtful, or introduce something new; then particular examples from history must be quoted as proofs.

Now, if we consider closely the use of historical proofs, four points of view readily present themselves for the purpose.

First, they may be used merely as an *explanation* of an idea. In every abstract consideration it is very easy to be misunderstood, or not to be intelligible at all: when an author is afraid of this, an exemplification from history serves to throw the light which is wanted on his idea, and to ensure his being intelligible to his reader.

Secondly, it may serve as an *application* of an idea, because by means of an example there is an opportunity of showing the action of those minor circumstances which cannot all be comprehended and explained in any general expression of an idea; for in that consists, indeed, the difference between theory and experience. Both these cases belong to examples properly speaking, the two following belong to historical proofs.

Thirdly, a historical fact may be referred to particularly, in order to support what one has advanced. This is in all cases sufficient, if we have *only* to prove the *possibility* of a fact or effect.

Lastly, in the fourth place, from the circumstantial detail of a historical event, and by collecting together several of them, we may deduce some theory, which therefore has its true *proof* in this testimony itself.

For the first of these purposes all that is generally required is a cursory notice of the case, as it is only used partially. Historical correctness is a secondary consideration; a case invented might also serve the purpose as well, only historical ones are always to be preferred, because they bring the idea which they illustrate nearer to practical life.

The second use supposes a more circumstantial relation of events, but historical authenticity is again of secondary importance, and in respect to this point the same is to be said as in the first case.

For the third purpose the mere quotation of an undoubted fact is generally sufficient. If it is asserted that fortified positions may fulfil their object under certain conditions, it is only necessary to mention the position of Bunzelwitz [Frederick the Great's celebrated entrenched camp in 1761.] in support of the assertion.

But if, through the narrative of a case in history, an abstract truth is to be demonstrated, then everything in the case bearing on the demonstration must be analysed in the most searching and complete manner; it must, to a certain extent, develop itself carefully before the eyes of the reader. The less effectually this is done the weaker will be the proof, and the more necessary it will be to supply the demonstrative proof which is wanting in the single case by a number of cases, because we have a right to suppose that the more minute details which we are unable to give neutralise each other in their effects in a certain number of cases.

If we want to show by example derived from experience that cavalry are better placed behind than in a line with infantry; that it is very hazardous without a decided preponderance of numbers to attempt an enveloping movement, with widely separated columns, either on a field of battle or in the theatre of war—that is, either tactically or strategically—then in the first of these cases it would not be sufficient to specify some lost battles in which the cavalry was on the flanks and some gained in which the cavalry was in rear of the infantry; and in the tatter of these cases it is not sufficient to refer to the battles of Rivoli and Wagram, to the attack of the Austrians on the theatre of war in Italy, in 1796, or of the French upon the German theatre of war in the same year. The way in which these orders of battle or plans of attack essentially contributed to disastrous issues in those particular cases must be shown by closely tracing out circumstances and occurrences. Then it will appear how far such forms or measures are to be condemned, a point which it is very necessary to show, for a total condemnation would be inconsistent with truth.

It has been already said that when a circumstantial detail of facts is impossible, the demonstrative power which is deficient may to a certain extent be supplied by the number of cases quoted; but this is a very dangerous method of getting out of the difficulty, and one which has been much abused. Instead of one well-explained example, three or four are just touched upon, and thus a show is made of strong evidence. But there are matters where a whole dozen of cases brought forward would prove nothing, if, for instance, they are facts of frequent occurrence, and therefore a dozen other cases with an opposite result might just as easily be brought forward. If any one will instance a dozen lost battles in which the side beaten attacked in separate converging columns, we can instance a dozen that have been gained in which the same order was adopted. It is evident that in this way no result is to be obtained.

Upon carefully considering these different points, it will be seen how easily examples may be misapplied.

An occurrence which, instead of being carefully analysed in all its parts, is superficially noticed, is like an object seen at a great distance, presenting the same appearance on each side, and in which the details of its parts cannot be distinguished. Such examples have, in reality, served to support the most contradictory opinions. To some Daun's campaigns are models of prudence and skill. To others, they are nothing but examples of timidity and want of resolution. Buonaparte's passage across the Noric Alps in 1797 may be made to appear the noblest resolution, but also as an act of sheer temerity. His strategic defeat in 1812 may be represented as the consequence either of an excess, or of a deficiency, of energy. All these opinions have been broached, and it is easy to see that they might very well arise, because each person takes a different view of the connection of events. At the same time these antagonistic opinions cannot be reconciled with each other, and therefore one of the two must be wrong.

Much as we are obliged to the worthy Feuquieres for the numerous examples introduced in his memoirs—partly because a number of historical incidents have thus been preserved which might otherwise have been lost, and partly because he was one of the first to bring theoretical, that is, abstract, ideas into connection with the practical in war, in so far that the cases brought forward may be regarded as intended to exemplify and confirm what is theoretically asserted—yet, in the opinion of an impartial reader, he will hardly be allowed to have attained the object he proposed to himself, that of proving theoretical principles by historical examples. For although he sometimes relates occurrences with great minuteness, still he falls short very often of showing that the deductions drawn necessarily proceed from the inner relations of these events.

Another evil which comes from the superficial notice of historical events, is that some readers are either wholly ignorant of the events, or cannot call them to remembrance sufficiently to be able to grasp the author's meaning, so that there is no alternative between either accepting blindly what is said, or remaining unconvinced.

It is extremely difficult to put together or unfold historical events before the eyes of a reader in such a way as is necessary, in order to be able to use them as proofs; for the writer very often wants the means, and can neither afford the time nor the requisite space; but we maintain that, when the object is to establish a new or doubtful opinion, one single example, thoroughly analysed, is far more instructive than ten which are superficially treated. The great mischief of these superficial representations is not that the writer puts his story forward as a proof when it has only a false title, but that he has not made himself properly acquainted with the subject, and that from this sort of slovenly, shallow treatment of history, a hundred false views and attempts at the construction of theories arise, which would never have made their appearance if the writer had looked upon it as his duty to deduce from the strict connection of events everything new which he brought to market, and sought to prove from history.

When we are convinced of these difficulties in the use of historical examples, and at the same time of the necessity (of making use of such examples), then we shall also come to the conclusion that the latest military history is naturally the best field from which to draw them, inasmuch as it alone is sufficiently authentic and detailed.

In ancient times, circumstances connected with War, as well as the method of carrying it on, were different; therefore its events are of less use to us either theoretically or practically; in addition to which, military history, like every other, naturally loses in the course of time a number of small traits and lineaments originally to be seen, loses in colour and life, like a worn-out or darkened picture; so that perhaps at last only the large masses and leading features remain, which thus acquire undue proportions.

If we look at the present state of warfare, we should say that the Wars since that of the Austrian succession are almost the only ones which, at least as far as armament, have still a considerable similarity to the present, and which, notwithstanding the many important changes which have taken place both great and small, are still capable of affording much instruction. It is quite otherwise with the War of the Spanish succession, as the use of fire-arms had not then so far advanced towards perfection, and cavalry still continued the most important arm. The farther we go back, the less useful becomes military history, as it gets so much the more meagre and barren of detail. The most useless of all is that of the old world.

But this uselessness is not altogether absolute, it relates only to those subjects which depend on a knowledge of minute details, or on those things in which the method of conducting war has changed. Although we know very little about the tactics in the battles between the Swiss and the Austrians, the Burgundians and French, still we find in them unmistakable evidence that they were the first in which the superiority of a good infantry over the best cavalry was, displayed. A general glance at the time of the Condottieri teaches us how the whole method of conducting War is dependent on the instrument used; for at no period have the forces used in War had so much the characteristics of a special instrument, and been a class so totally distinct from the rest of the national community. The memorable way in which the Romans in the second Punic War attacked the Carthaginan possessions in Spain and Africa, while Hannibal still maintained himself in Italy, is a most instructive subject to study, as the general relations of the States and Armies concerned in this indirect act of defence are sufficiently well known.

But the more things descend into particulars and deviate in character from the most general relations, the less we can look for examples and lessons of experience from very remote periods, for we have neither the means of judging properly of corresponding events, nor can we apply them to our completely different method of War.

Unfortunately, however, it has always been the fashion with historical writers to talk about ancient times. We shall not say how far vanity and charlatanism may have had a share in this, but in general we fail to discover any honest intention and earnest endeavour to instruct and convince, and we can therefore only look upon such quotations and references as embellishments to fill up gaps and hide defects.

It would be an immense service to teach the Art of War entirely by historical examples, as Feuquieres proposed to do; but it would be full work for the whole life of a man, if we reflect that he who undertakes it must first qualify himself for the task by a long personal experience in actual War.

Whoever, stirred by ambition, undertakes such a task, let him prepare himself for his pious undertaking as for a long pilgrimage; let him give up his time, spare no sacrifice, fear no temporal rank or power, and rise above all feelings of personal vanity, of false shame, in order, according to the French code, to speak the truth, the whole truth, and nothing but the truth.

Strategy

In the second chapter of the second book, Strategy has been defined as "the employment of the battle as the means towards the attainment of the object of the War." Properly speaking it has to do with nothing but the battle, but its theory must include in this consideration the instrument of this real activity—the armed force—in itself and in its principal relations, for the battle is fought by it, and shows its effects upon it in turn. It must be well acquainted with the battle itself as far as relates to its possible results, and those mental and moral powers which are the most important in the use of the same.

Strategy is the employment of the battle to gain the end of the War; it must therefore give an aim to the whole military action, which must be in accordance with the object of the War; in other words, Strategy forms the plan of the War, and to this end it links together the series of acts which are to lead to the final decision, that, is to say, it makes the plans for the separate campaigns and regulates the combats to be fought in each. As these are all things which to a great extent can only be determined on conjectures some of which turn out incorrect, while a number of other arrangements pertaining to details cannot be made at all beforehand, it follows, as a matter of course, that Strategy must go with the Army to the field in order to arrange particulars on the spot, and to make the modifications in the general plan, which incessantly become necessary in War. Strategy can therefore never take its hand from the work for a moment.

That this, however, has not always been the view taken is evident from the former custom of keeping Strategy in the cabinet and not with the Army, a thing only allowable if the cabinet is so near to the Army that it can be taken for the chief head-quarters of the Army.

Theory will therefore attend on Strategy in the determination of its plans, or, as we may more properly say, it will throw a light on things in themselves, and on their relations to each other, and bring out prominently the little that there is of principle or rule.

If we recall to mind from the first chapter how many things of the highest importance War touches upon, we may conceive that a consideration of all requires a rare grasp of mind.

A Prince or General who knows exactly how to organise his War according to his object and means, who does neither too little nor too much, gives by that the greatest proof of his genius. But the effects of this talent are exhibited not so much by the invention of new modes of action, which might strike the eye immediately, as in the successful final result of the whole. It is the exact fulfilment of silent suppositions, it is the noiseless harmony of the whole action which we should admire, and which only makes itself known in the total result. Inquirer who, tracing back from the final result, does not perceive the signs of that harmony is one who is apt to seek for genius where it is not, and where it cannot be found.

The means and forms which Strategy uses are in fact so extremely simple, so well known by their constant repetition, that it only appears ridiculous to sound common sense when it hears critics so frequently speaking of them with high-flown emphasis. Turning a flank, which has been done a thousand times, is regarded here as a proof of the most brilliant genius, there as a proof of the most profound penetration, indeed even of the most comprehensive knowledge. Can there be in the book-world more absurd productions? [This paragraph refers to the works of Lloyd, Buelow, indeed to all the eighteenth-century writers, from whose influence we in England are not even yet free.—Ed.]

It is still more ridiculous if, in addition to this, we reflect that the same critic, in accordance with prevalent opinion, excludes all moral forces from theory, and will not allow it to be concerned with anything but the material forces, so that all must be confined to a few mathematical relations of equilibrium and preponderance, of time and space, and a few lines and angles. If it were nothing more than this, then out of such a miserable business there would not be a scientific problem for even a schoolboy.

But let us admit: there is no question here about scientific formulas and problems; the relations of material things are all very simple; the right comprehension of the moral forces which come into play is more difficult. Still, even in respect to them, it is only in the highest branches of Strategy that moral complications and a great diversity of quantities and relations are to be looked for, only at that point where Strategy borders on political science, or rather where the two become one, and there, as we have before observed, they have more influence on the "how much" and "how little" is to be done than on the form of execution. Where the latter is the principal question, as in the single acts both great and small in War, the moral quantities are already reduced to a very small number.

Thus, then, in Strategy everything is very simple, but not on that account very easy. Once it is determined from the relations of the State what should and may be done by War, then the way to it is easy to find; but to follow that way straightforward, to carry out the plan without being obliged to deviate from it a thousand times by a thousand varying influences, requires, besides great strength of character, great clearness and steadiness of mind, and out of a thousand men who are remarkable, some for mind, others for penetration, others again for boldness or strength of will, perhaps not one will combine in himself all those qualities which are required to raise a man above mediocrity in the career of a general.

It may sound strange, but for all who know War in this respect it is a fact beyond doubt, that much more strength of will is required to make an important decision in Strategy than in tactics. In the latter we are hurried on with the moment; a Commander feels himself borne along in a strong current, against which he durst not contend without the most destructive consequences, he suppresses the rising fears, and boldly ventures further. In Strategy, where all goes on at a slower rate, there is more room allowed for our own apprehensions and those of others, for objections and remonstrances, consequently also for unseasonable regrets; and as we do not see things in Strategy as we do at least half of them in tactics, with the living eye, but everything must be conjectured and assumed, the convictions produced are less powerful. The consequence is that most Generals, when they should act, remain stuck fast in bewildering doubts.

Now let us cast a glance at history—upon Frederick the Great's campaign of 1760, celebrated for its fine marches and manoeuvres: a perfect masterpiece of Strategic skill as critics tell us. Is there really anything to drive us out of our wits with admiration in the King's first trying to turn Daun's right flank, then his left, then again his right, etc.? Are we to see profound wisdom in this? No, that we cannot, if we are to decide naturally and without affectation. What we rather admire above all is the sagacity of the King in this respect, that while pursuing a great object with very limited means, he undertook nothing beyond his powers, and *just enough* to gain his object. This sagacity of the General is visible not only in this campaign, but throughout all the three Wars of the Great King!

To bring Silesia into the safe harbour of a well-guaranteed peace was his object.

At the head of a small State, which was like other States in most things, and only ahead of them in some branches of administration; he could not be an Alexander, and, as Charles XII, he would only, like him, have broken his head. We find, therefore, in the whole of his conduct of War, a controlled power, always well balanced, and never wanting in energy, which in the most critical moments rises to astonishing deeds, and the next moment oscillates quietly on again in subordination to the play of the most subtle political influences. Neither vanity, thirst for glory, nor vengeance could make him deviate from his course, and this course alone it is which brought him to a fortunate termination of the contest.

These few words do but scant justice to this phase of the genius of the great General; the eyes must be fixed carefully on the extraordinary issue of the struggle, and the causes which brought about that issue must be traced out, in order thoroughly to understand that nothing but the King's penetrating eye brought him safely out of all his dangers.

This is one feature in this great Commander which we admire in the campaign of 1760—and in all others, but in this especially—because in none did he keep the balance even against such a superior hostile force, with such a small sacrifice.

Another feature relates to the difficulty of execution. Marches to turn a flank, right or left, are easily combined; the idea of keeping a small force always well concentrated to be able to meet the enemy on equal terms at any point, to multiply a force by rapid movement, is as easily conceived as expressed; the mere contrivance in these points, therefore, cannot excite our admiration, and with respect to such simple things, there is nothing further than to admit that they are simple.

But let a General try to do these things like Frederick the Great. Long afterwards authors, who were eyewitnesses, have spoken of the danger, indeed of the imprudence, of the King's camps, and doubtless, at the time he pitched them, the danger appeared three times as great as afterwards.

It was the same with his marches, under the eyes, nay, often under the cannon of the enemy's Army; these camps were taken up, these marches made, not from want of prudence, but because in Daun's system, in his mode of drawing up his Army, in the responsibility which pressed upon him, and in his character, Frederick found that security which justified his camps and marches. But it required the King's boldness, determination, and strength of will to see things in this light, and not to be led astray and intimidated by the danger of which thirty years after people still wrote and spoke. Few Generals in this situation would have believed these simple strategic means to be practicable.

Again, another difficulty in execution lay in this, that the King's Army in this campaign was constantly in motion. Twice it marched by wretched cross-roads, from the Elbe into Silesia, in rear of Daun and pursued by Lascy (beginning of July, beginning of August). It required to be always ready for battle, and its marches had to be organised with a degree of skill which necessarily called forth a proportionate amount of exertion. Although attended and delayed by thousands of waggons, still its subsistence was extremely difficult. In Silesia, for eight days before the battle of Leignitz, it had constantly to march, defiling alternately right and left in front of the enemy:—this costs great fatigue, and entails great privations.

Is it to be supposed that all this could have been done without producing great friction in the machine? Can the mind of a Commander elaborate such movements with the same ease as the hand of a land surveyor uses the astrolabe? Does not the sight of the sufferings of their hungry, thirsty comrades pierce the hearts of the Commander and his Generals a thousand times? Must not the murmurs and doubts which these cause reach his ear? Has an ordinary man the courage to demand such sacrifices, and would not such efforts most certainly demoralise the Army, break up the bands of discipline, and, in short, undermine its military virtue, if firm reliance on the greatness and infallibility of the Commander did not compensate for all? Here, therefore, it is that we should pay respect; it is these miracles of execution which we should admire. But it is impossible to realise all this in its full force without a foretaste of it by experience. He who only knows War from books or the drill-ground cannot realise the whole effect of this counterpoise in action; *we beg him, therefore, to accept from us on faith and trust all that he is unable to supply from any personal experiences of his own.*

This illustration is intended to give more clearness to the course of our ideas, and in closing this chapter we will only briefly observe that in our exposition of Strategy we shall describe those separate subjects which appear to us the most important, whether of a moral or material nature; then proceed from the simple to the complex, and conclude with the inner connection of the whole act of War, in other words, with the plan for a War or campaign.

Observation

In an earlier manuscript of the second book are the following passages endorsed by the author himself to be used for the first Chapter of the second Book: the projected revision of that chapter not having been made, the passages referred to are introduced here in full.

By the mere assemblage of armed forces at a particular point, a battle there becomes possible, but does not always take place. Is that possibility now to be regarded as a reality and therefore an effective thing? Certainly, it is so by its results, and these effects, whatever they may be, can never fail.

1. Possible Combats Are on Account of Their Results to Be Looked upon as Real Ones.

If a detachment is sent away to cut off the retreat of a flying enemy, and the enemy surrenders in consequence without further resistance, still it is through the combat which is offered to him by this detachment sent after him that he is brought to his decision.

If a part of our Army occupies an enemy's province which was undefended, and thus deprives the enemy of very considerable means of keeping up the strength of his Army, it is entirely through the battle which our detached body gives the enemy to expect, in case he seeks to recover the lost province, that we remain in possession of the same.

In both cases, therefore, the mere possibility of a battle has produced results, and is therefore to be classed amongst actual events. Suppose that in these cases the enemy has opposed our troops with others superior in force, and thus forced ours to give up their object without a combat, then certainly our plan has failed, but the battle which we offered at (either of) those points has not on that account been without effect, for it attracted the enemy's forces to that point. And in case our whole undertaking has done us harm, it cannot be said that these positions, these possible battles, have been attended with no results; their effects, then, are similar to those of a lost battle.

In this manner we see that the destruction of the enemy's military forces, the overthrow of the enemy's power, is only to be done through the effect of a battle, whether it be that it actually takes place, or that it is merely offered, and not accepted.

2. Twofold Object of the Combat

But these effects are of two kinds, direct and indirect they are of the latter, if other things intrude themselves and become the object of the combat—things which cannot be regarded as the destruction of enemy's force, but only leading up to it, certainly by a circuitous road, but with so much the greater effect. The possession of provinces, towns, fortresses, roads, bridges, magazines, etc., may be the *immediate* object of a battle, but never the ultimate one. Things of this description can never be, looked upon otherwise than as means of gaining greater superiority, so as at last to offer battle to the enemy in such a way that it will be impossible for him to accept it. Therefore all these things must only be regarded as intermediate links, steps, as it were, leading up to the effectual principle, but never as that principle itself.

3. Example

In 1814, by the capture of Buonaparte's capital the object of the War was attained. The political divisions which had their roots in Paris came into active operation, and an enormous split left the power of the Emperor to collapse of itself. Nevertheless the point of view from which we must look at all this is, that through these causes the forces and defensive means of Buonaparte were suddenly very much diminished, the superiority of the Allies, therefore, just in the same measure increased, and any further resistance then became *impossible*. It was this impossibility which produced the peace with France. If we suppose the forces of the Allies at that moment diminished to a like extent through external causes;—if the superiority vanishes, then at the same time vanishes also all the effect and importance of the taking of Paris.

We have gone through this chain of argument in order to show that this is the natural and only true view of the thing from which it derives its importance. It leads always back to the question, What at any given moment of the War or campaign will be the probable result of the great or small combats which the two sides might offer to each other? In the consideration of a plan for a campaign, this question only is decisive as to the measures which are to be taken all through from the very commencement.

4. When this View Is Not Taken, Then a False Value Is Given to Other Things

If we do not accustom ourselves to look upon War, and the single campaigns in a War, as a chain which is all composed of battles strung together, one of which always brings on another; if we adopt the idea that the taking of a certain geographical point, the occupation of an undefended province, is in itself anything; then we are very likely to regard it as an acquisition which we may retain; and if we look at it so, and not as a term in the whole series of events, we do not ask ourselves whether this possession may not lead to greater disadvantages hereafter. How often we find this mistake recurring in military history.

We might say that, just as in commerce the merchant cannot set apart and place in security gains from one single transaction by itself, so in War a single advantage cannot be separated from the result of the whole. Just as the former must always operate with the whole bulk of his means, just so in War, only the sum total will decide on the advantage or disadvantage of each item.

If the mind's eye is always directed upon the series of combats, so far as they can be seen beforehand, then it is always looking in the right direction, and thereby the motion of the force acquires that rapidity, that is to say, willing and doing acquire that energy which is suitable to the matter, and which is not to be thwarted or turned aside by extraneous influences. [The whole of this chapter is directed against the theories of the Austrian Staff in

1814. It may be taken as the foundation of the modern teaching of the Prussian General Staff. See especially von Kammer.—Ed.]

Elements of Strategy

The causes which condition the use of the combat in Strategy may be easily divided into elements of different kinds, such as the moral, physical, mathematical, geographical and statistical elements.

The first class includes all that can be called forth by moral qualities and effects; to the second belong the whole mass of the military force, its organisation, the proportion of the three arms, etc. etc.; to the third, the angle of the lines of operation, the concentric and eccentric movements in as far as their geometrical nature has any value in the calculation; to the fourth, the influences of country, such as commanding points, hills, rivers, woods, roads, etc. etc.; lastly, to the fifth, all the means of supply. The separation of these things once for all in the mind does good in giving clearness and helping us to estimate at once, at a higher or lower value, the different classes as we pass onwards. For, in considering them separately, many lose of themselves their borrowed importance; one feels, for instance, quite plainly that the value of a base of operations, even if we look at nothing in it but its relative position to the line of operations, depends much less in that simple form on the geometrical element of the angle which they form with one another, than on the nature of the roads and the country through which they pass.

But to treat upon Strategy according to these elements would be the most unfortunate idea that could be conceived, for these elements are generally manifold, and intimately connected with each other in every single operation of War. We should lose ourselves in the most soulless analysis, and as if in a horrid dream, we should be for ever trying in vain to build up an arch to connect this base of abstractions with facts belonging to the real world. Heaven preserve every theorist from such an undertaking! We shall keep to the world of things in their totality, and not pursue our analysis further than is necessary from time to time to give distinctness to the idea which we wish to impart, and which has come to us, not by a speculative investigation, but through the impression made by the realities of War in their entirety.

Moral Forces

We must return again to this subject, which is touched upon in the third chapter of the second book, because the moral forces are amongst the most important subjects in War. They form the spirit which permeates the whole being of War. These forces fasten themselves soonest and with the greatest affinity on to the Will which puts in motion and guides the whole mass of powers, uniting with it as it were in one stream, because this is a moral force itself. Unfortunately they will escape from all book-analysis, for they will neither be brought into numbers nor into classes, and require to be both seen and felt.

The spirit and other moral qualities which animate an Army, a General, or Governments, public opinion in provinces in which a War is raging, the moral effect of a victory or of a defeat, are things which in themselves vary very much in their nature, and which also, according as they stand with regard to our object and our relations, may have an influence in different ways.

Although little or nothing can be said about these things in books, still they belong to the theory of the Art of War, as much as everything else which constitutes War. For I must here once more repeat that it is a miserable philosophy if, according to the old plan, we establish rules and principles wholly regardless of all moral forces, and then, as soon as these forces make their appearance, we begin to count exceptions which we thereby establish as it were theoretically, that is, make into rules; or if we resort to an appeal to genius, which is above all rules, thus giving out by implication, not only that rules were only made for fools, but also that they themselves are no better than folly.

Even if the theory of the Art of War does no more in reality than recall these things to remembrance, showing the necessity of allowing to the moral forces their full value, and of always taking them into consideration, by so doing it extends its borders over the region of immaterial forces, and by establishing that point of view, condemns beforehand every one who would endeavour to justify himself before its judgment seat by the mere physical relations of forces.

Further out of regard to all other so-called rules, theory cannot banish the moral forces beyond its frontier, because the effects of the physical forces and the moral are completely fused, and are not to be decomposed like a metal alloy by a chemical process. In every rule relating to the physical forces, theory must present to the mind at the same time the share which the moral powers will have in it, if it would not be led to categorical propositions, at one time too timid and contracted, at another too dogmatical and wide. Even the most matter-of-fact theories have, without knowing it, strayed over into this moral kingdom; for, as an example, the effects of a victory cannot in any way be explained without taking into consideration the moral impressions. And therefore the most of the subjects which we shall go through in this book are composed half of physical, half of moral causes and effects, and we might say the physical are almost no more than the wooden handle, whilst the moral are the noble metal, the real bright-polished weapon.

The value of the moral powers, and their frequently incredible influence, are best exemplified by history, and this is the most generous and the purest nourishment which the mind of the General can extract from it.—At the same time it is to be observed, that it is less demonstrations, critical examinations, and learned treatises, than sentiments, general impressions, and single flashing sparks of truth, which yield the seeds of knowledge that are to fertilise the mind.

We might go through the most important moral phenomena in War, and with all the care of a diligent professor try what we could impart about each, either good or bad. But as in such a method one slides too much into the commonplace and trite, whilst real mind quickly makes its escape in analysis, the end is that one gets imperceptibly to the relation of things which everybody knows. We prefer, therefore, to remain here more than usually incomplete and rhapsodical, content to have drawn attention to the importance of the subject in a general way, and to have pointed out the spirit in which the views given in this book have been conceived.

The Chief Moral Powers

These are The Talents of the Commander; The Military Virtue of the Army; Its National feeling. Which of these is the most important no one can tell in a general way, for it is very difficult to say anything in general of their strength, and still more difficult to compare the strength of one with that of another. The best plan is not to undervalue any of them, a fault which human judgment is prone to, sometimes on one side, sometimes on another, in its whimsical oscillations. It is better to satisfy ourselves of the undeniable efficacy of these three things by sufficient evidence from history.

It is true, however, that in modern times the Armies of European states have arrived very much at a par as regards discipline and fitness for service, and that the conduct of War has—as philosophers would say—naturally developed itself, thereby become a method, common as it were to all Armies, so that even from Commanders there is nothing further to be expected in the way of application of special means of Art, in the limited sense (such as Frederick the Second's oblique order). Hence it cannot be denied that, as matters now stand, greater scope is afforded for the influence of National spirit and habituation of an army to War. A long peace may again alter all this.[Written shortly after the Great Napoleonic campaigns]

The national spirit of an Army (enthusiasm, fanatical zeal, faith, opinion) displays itself most in mountain warfare, where every one down to the common soldier is left to himself. On this account, a mountainous country is the best campaigning ground for popular levies.

Expertness of an Army through training, and that well-tempered courage which holds the ranks together as if they had been cast in a mould, show their superiority in an open country.

The talent of a General has most room to display itself in a closely intersected, undulating country. In mountains he has too little command over the separate parts, and the direction of all is beyond his powers; in open plains it is simple and does not exceed those powers.

According to these undeniable elective affinities, plans should be regulated.

Military Virtue of an Army

This is distinguished from mere bravery, and still more from enthusiasm for the business of War. The first is certainly a necessary constituent part of it, but in the same way as bravery, which is a natural gift in some men, may arise in a soldier as a part of an Army from habit and custom, so with him it must also have a different direction from that which it has with others. It must lose that impulse to unbridled activity and exercise of force which is its characteristic in the individual, and submit itself to demands of a higher kind, to obedience, order, rule, and method. Enthusiasm for the profession gives life and greater fire to the military virtue of an Army, but does not necessarily constitute a part of it.

War is a special business, and however general its relations may be, and even if all the male population of a country, capable of bearing arms, exercise this calling, still it always continues to be different and separate from the other pursuits which occupy the life of man.—To be imbued with a sense of the spirit and nature of this business, to make use of, to rouse, to assimilate into the system the powers which should be active in it, to penetrate completely into the nature of the business with the understanding, through exercise to gain confidence and expertness in it, to be completely given up to it, to pass out of the man into the part which it is assigned to us to play in War, that is the military virtue of an Army in the individual.

However much pains may be taken to combine the soldier and the citizen in one and the same individual, whatever may be done to nationalise Wars, and however much we may imagine times have changed since the days of the old Condottieri, never will it be possible to do away with the individuality of the business; and if that cannot be done, then those who belong to it, as long as they belong to it, will always look upon themselves as a kind of guild, in the regulations, laws and customs in which the "Spirit of War" by preference finds its expression. And so it is in fact. Even with the most decided inclination to look at War from the highest point of view, it would be very wrong to look down upon this corporate spirit (e'sprit de corps) which may and should exist more or less in every Army. This corporate spirit forms the bond of union between the natural forces which are active in that which we have called military virtue. The crystals of military virtue have a greater affinity for the spirit of a corporate body than for anything else.

An Army which preserves its usual formations under the heaviest fire, which is never shaken by imaginary fears, and in the face of real danger disputes the ground inch by inch, which, proud in the feeling of its victories, never loses its sense of obedience, its respect for and confidence in its leaders, even under the depressing effects of defeat; an Army with all its physical powers, inured to privations and fatigue by exercise, like the muscles of an athlete; an Army which looks upon all its toils as the means to victory, not as a curse which hovers over its standards, and which is always reminded of its duties and virtues by the short catechism of one idea, namely the *honour of its arms*;—Such an Army is imbued with the true military spirit.

Soldiers may fight bravely like the Vende'ans, and do great things like the Swiss, the Americans, or Spaniards, without displaying this military virtue. A Commander may also be successful at the head of standing Armies, like Eugene and Marlborough, without enjoying the benefit of its assistance; we must not, therefore, say that a successful War without it cannot be imagined; and we draw especial attention to that point, in order the more to individualise the conception which is here brought forward, that the idea may not dissolve into a generalisation and that it may not be thought that military virtue is in the end everything. It is not so. Military virtue in an Army is a definite moral power which may be supposed wanting, and the influence of which may therefore be estimated—like any instrument the power of which may be calculated.

Having thus characterised it, we proceed to consider what can be predicated of its influence, and what are the means of gaining its assistance.

Military virtue is for the parts, what the genius of the Commander is for the whole. The General can only guide the whole, not each separate part, and where he cannot guide the part, there military virtue must be its leader. A General is chosen by the reputation of his superior talents, the chief leaders of large masses after careful probation; but this probation diminishes as we descend the scale of rank, and in just the same measure we may reckon less and less upon individual talents; but what is wanting in this respect military virtue should supply. The natural qualities of a warlike people play just this part: *bravery, aptitude, powers of endurance* and *enthusiasm*.

These properties may therefore supply the place of military virtue, and vice versa, from which the following may be deduced:

1. Military virtue is a quality of standing Armies only, but they require it the most. In national risings its place is supplied by natural qualities, which develop themselves there more rapidly.

2. Standing Armies opposed to standing Armies, can more easily dispense with it, than a standing Army opposed to a national insurrection, for in that case, the troops are more scattered, and the divisions left more to themselves.

But where an Army can be kept concentrated, the genius of the General takes a greater place, and supplies what is wanting in the spirit of the Army. Therefore generally military virtue becomes more necessary the more the theatre of operations and other circumstances make the War complicated, and cause the forces to be scattered.

From these truths the only lesson to be derived is this, that if an Army is deficient in this quality, every endeavour should be made to simplify the operations of the War as much as possible, or to introduce double efficiency in the organisation of the Army in some other respect, and not to expect from the mere name of a standing Army, that which only the veritable thing itself can give.

The military virtue of an Army is, therefore, one of the most important moral powers in War, and where it is wanting, we either see its place supplied by one of the others, such as the great superiority of generalship or popular enthusiasm, or we find the results not commensurate with the exertions made.—How much that is great, this spirit, this sterling worth of an army, this refining of ore into the polished metal, has already done, we see in the history of the Macedonians under Alexander, the Roman legions under Cesar, the Spanish infantry under Alexander Farnese, the Swedes under Gustavus Adolphus and Charles XII, the Prussians under Frederick the Great, and the French under Buonaparte. We must purposely shut our eyes against all historical proof, if we do not admit, that the astonishing successes of these Generals and their greatness in situations of extreme difficulty, were only possible with Armies possessing this virtue.

This spirit can only be generated from two sources, and only by these two conjointly; the first is a succession of campaigns and great victories; the other is, an activity of the Army carried sometimes to the highest pitch. Only by these, does the soldier learn to know his powers. The more a General is in the habit of demanding from his troops, the surer he will be that his demands will be answered. The soldier is as proud of overcoming toil, as he is of surmounting danger. Therefore it is only in the soil of incessant activity and exertion that the germ will thrive, but also only in the sunshine of victory. Once it becomes a *strong tree*, it will stand against the fiercest storms of misfortune and defeat, and even against the indolent inactivity of peace, at least for a time. It can therefore only be created in War, and under great Generals, but no doubt it may last at least for several generations, even under Generals of moderate capacity, and through considerable periods of peace.

With this generous and noble spirit of union in a line of veteran troops, covered with scars and thoroughly inured to War, we must not compare the self-esteem and vanity of a standing Army, [Clausewitz is, of course, thinking of the long-service standing armies of his own youth. Not of the short-service standing armies of to-day (Editor).] held together merely by the glue of service-regulations and a drill book; a certain plodding earnestness and strict discipline may keep up military virtue for a long time, but can never create it; these things therefore have a certain value, but must not be over-rated. Order, smartness, good will, also a certain degree of pride and high feeling, are qualities of an Army formed in time of peace which are to be prized, but cannot stand alone. The whole retains the whole, and as with glass too quickly cooled, a single crack breaks the whole mass. Above all, the highest spirit in the world changes only too easily at the first check into depression, and one might say into a kind of rhodomontade of alarm, the French sauve que peut.—Such an Army can only achieve something through its leader, never by itself. It must be led with double caution, until by degrees, in victory and hardships, the strength grows into the full armour. Beware then of confusing the *spirit* of an Army with its temper.

Boldness

The place and part which boldness takes in the dynamic system of powers, where it stands opposed to Foresight and prudence, has been stated in the chapter on the certainty of the result in order thereby to show, that theory has no right to restrict it by virtue of its legislative power.

But this noble impulse, with which the human soul raises itself above the most formidable dangers, is to be regarded as an active principle peculiarly belonging to War. In fact, in what branch of human activity should boldness have a right of citizenship if not in War?

From the transport-driver and the drummer up to the General, it is the noblest of virtues, the true steel which gives the weapon its edge and brilliancy.

Let us admit in fact it has in War even its own prerogatives. Over and above the result of the calculation of space, time, and quantity, we must allow a certain percentage which boldness derives from the weakness of others, whenever it gains the mastery. It is therefore, virtually, a creative power. This is not difficult to demonstrate philosophically. As often as boldness encounters hesitation, the probability of the result is of necessity in its favour, because the very state of hesitation implies a loss of equilibrium already. It is only when it encounters cautious foresight—which we may say is just as bold, at all events just as strong and powerful as itself—that it is at a disadvantage; such cases, however, rarely occur. Out of the whole multitude of prudent men in the world, the great majority are so from timidity.

Amongst large masses, boldness is a force, the special cultivation of which can never be to the detriment of other forces, because the great mass is bound to a higher will by the frame-work and joints of the order of battle and of the service, and therefore is guided by an intelligent power which is extraneous. Boldness is therefore here only like a spring held down until its action is required.

The higher the rank the more necessary it is that boldness should be accompanied by a reflective mind, that it may not be a mere blind outburst of passion to no purpose; for with increase of rank it becomes always less a matter of self-sacrifice and more a matter of the preservation of others, and the good of the whole. Where regulations of the service, as a kind of second nature, prescribe for the masses, reflection must be the guide of the General, and in his case individual boldness in action may easily become a fault. Still, at the same time, it is a fine failing, and must not be looked at in the same light as any other. Happy the Army in which an untimely boldness frequently manifests itself; it is an exuberant growth which shows a rich soil. Even foolhardiness, that is boldness without an object, is not to be despised; in point of fact it is the same energy of feeling, only exercised as a kind of passion without any co-operation of the intelligent faculties. It is only when it strikes at the root of obedience, when it treats with contempt the orders of superior authority, that it must be repressed as a dangerous evil, not on its own account but on account of the act of disobedience, for there is nothing in War which is of *greater importance than obedience.*

The reader will readily agree with us that, supposing an equal degree of discernment to be forthcoming in a certain number of cases, a thousand times as many of them will end in disaster through over-anxiety as through boldness.

One would suppose it natural that the interposition of a reasonable object should stimulate boldness, and therefore lessen its intrinsic merit, and yet the reverse is the case in reality.

The intervention of lucid thought or the general supremacy of mind deprives the emotional forces of a great part of their power. On that account *boldness becomes of rarer occurrence the higher we ascend the scale of rank*, for whether the discernment and the understanding do or do not increase with these ranks still the Commanders, in their several stations as they rise, are pressed upon more and more severely by objective things, by relations and claims from without, so that they become the more perplexed the lower the degree of their individual intelligence. This so far as regards War is the chief foundation of the truth of the French proverb:—

"Tel brille au second qui s' e'clipse an premier."

Almost all the Generals who are represented in history as merely having attained to mediocrity, and as wanting in decision when in supreme command, are men celebrated in their antecedent career for their boldness and decision. [Beaulieu, Benedek, Bazaine, Buller, Melas, Mack. etc. etc.]

In those motives to bold action which arise from the pressure of necessity we must make a distinction. Necessity has its degrees of intensity. If it lies near at hand, if the person acting is in the pursuit of his object driven into great dangers in order to escape others equally great, then we can only admire his resolution, which still has also its value. If a young man to show his skill in horsemanship leaps across a deep cleft, then he is bold; if he makes the same leap pursued by a troop of head-chopping Janissaries he is only resolute. But the farther off the necessity from the point of action, the greater the number of relations intervening which the mind has to traverse; in order to realise them, by so much the less does necessity take from boldness in action. If Frederick the Great, in the year 1756, saw that War was inevitable, and that he could only escape destruction by being beforehand with his enemies, it became necessary for him to commence the War himself, but at the same time it was certainly very bold: for few men in his position would have made up their minds to do so.

Although Strategy is only the province of Generals-in-Chief or Commanders in the higher positions, still boldness in all the other branches of an Army is as little a matter of indifference to it as their other military virtues. With an Army belonging to a bold race, and in which the spirit of boldness has been always nourished, very different things may be undertaken than with one in which this virtue, is unknown; for that reason we have considered it in connection with an Army. But our subject is specially the boldness of the General, and yet we have not much to say about it after having described this military virtue in a general way to the best of our ability.

The higher we rise in a position of command, the more of the mind, understanding, and penetration predominate in activity, the more therefore is boldness, which is a property of the feelings, kept in subjection, and for that reason we find it so rarely in the highest positions, but then, so much the more should it be admired. Boldness, directed by an overruling intelligence, is the stamp of the hero: this boldness does not consist in venturing directly against the nature of things, in a downright contempt of the laws of probability, but, if a choice is once made, in the rigorous adherence to that higher calculation which genius, the tact of judgment, has gone over with the speed of lightning. The more boldness lends wings to the mind and the discernment, so much the farther they will reach in their flight, so much the more comprehensive will be the view, the more exact the result, but certainly always only in the sense that with greater objects greater dangers are connected. The ordinary man, not to speak of the weak and irresolute, arrives at an exact result so far as such is possible without ocular demonstration, at most after diligent reflection in his chamber, at a distance from danger and responsibility. Let danger and responsibility draw close round him in every direction, then he loses the power of comprehensive vision, and if he retains this in any measure by the influence of others, still he will lose his power of *decision*, because in that point no one can help him.

We think then that it is impossible to imagine a distinguished General without boldness, that is to say, that no man can become one who is not born with this power of the soul, and we therefore look upon it as the first requisite for such a career. How much of this inborn power, developed and moderated through education and the circumstances of life, is left when the man has attained a high position, is the second question. The greater this power still is, the stronger will genius be on the wing, the higher will be its flight. The risks become always greater, but the purpose grows with them. Whether its lines proceed out of and get their direction from a distant necessity, or whether they converge to the keystone of a building which ambition has planned, whether Frederick or Alexander acts, is much the same as regards the critical view. If the one excites the imagination more because it is bolder, the other pleases the understanding most, because it has in it more absolute necessity.

We have still to advert to one very important circumstance.

The spirit of boldness can exist in an Army, either because it is in the people, or because it has been generated in a successful War conducted by able Generals. In the latter case it must of course be dispensed with at the commencement.

Now in our days there is hardly any other means of educating the spirit of a people in this respect, except by War, and that too under bold Generals. By it alone can that effeminacy of feeling be counteracted, that propensity to seek for the enjoyment of comfort, which cause degeneracy in a people rising in prosperity and immersed in an extremely busy commerce.

A Nation can hope to have a strong position in the political world only if its character and practice in actual War mutually support each other in constant reciprocal action.

Perseverance

The reader expects to hear of angles and lines, and finds, instead of these citizens of the scientific world, only people out of common life, such as he meets with every day in the street. And yet the author cannot make up his mind to become a hair's breadth more mathematical than the subject seems to him to require, and he is not alarmed at the surprise which the reader may show.

In War more than anywhere else in the world things happen differently to what we had expected, and look differently when near, to what they did at a distance. With what serenity the architect can watch his work gradually rising and growing into his plan. The doctor although much more at the mercy of mysterious agencies and chances than the architect, still knows enough of the forms and effects of his means. In War, on the other hand, the Commander of an immense whole finds himself in a constant whirlpool of false and true information, of mistakes committed through fear, through negligence, through precipitation, of contraventions of his authority, either from mistaken or correct motives, from ill will, true or false sense of duty, indolence or exhaustion, of accidents which no mortal could have foreseen. In short, he is the victim of a hundred thousand impressions, of which the most have an intimidating, the fewest an encouraging tendency. By long experience in War, the tact is acquired of readily appreciating the value of these incidents; high courage and stability of character stand proof against them, as the rock resists the beating of the waves. He who would yield to these impressions would never carry out an undertaking, and on that account *perseverance* in the proposed object, as long as there is no decided reason against it, is a most necessary counterpoise. Further, there is hardly any celebrated enterprise in War which was not achieved by endless exertion, pains, and privations; and as here the weakness of the physical and moral man is ever disposed to yield, only an immense force of will, which manifests itself in perseverance admired by present and future generations, can conduct to our goal.

Superiority of Numbers

This is in tactics, as well as in Strategy, the most general principle of victory, and shall be examined by us first in its generality, for which we may be permitted the following exposition:

Strategy fixes the point where, the time when, and the numerical force with which the battle is to be fought. By this triple determination it has therefore a very essential influence on the issue of the combat. If tactics has fought the battle, if the result is over, let it be victory or defeat, Strategy makes such use of it as can be made in accordance with the great object of the War. This object is naturally often a very distant one, seldom does it lie quite close at hand. A series of other objects subordinate themselves to it as means. These objects, which are at the same time means to a higher purpose, may be practically of various kinds; even the ultimate aim of the whole War may be a different one in every case. We shall make ourselves acquainted with these things according as we come to know the separate objects which they come, in contact with; and it is not our intention here to embrace the whole subject by a complete enumeration of them, even if that were possible. We therefore let the employment of the battle stand over for the present.

Even those things through which Strategy has an influence on the issue of the combat, inasmuch as it establishes the same, to a certain extent decrees them, are not so simple that they can be embraced in one single view. For as Strategy appoints time, place and force, it can do so in practice in many ways, each of which influences in a different manner the result of the combat as well as its consequences. Therefore we shall only get acquainted with this also by degrees, that is, through the subjects which more closely determine the application.

If we strip the combat of all modifications which it may undergo according to its immediate purpose and the circumstances from which it proceeds, lastly if we set aside the valour of the troops, because that is a given quantity, then there remains only the bare conception of the combat, that is a combat without form, in which we distinguish nothing but the number of the combatants.

This number will therefore determine victory. Now from the number of things above deducted to get to this point, it is shown that the superiority in numbers in a battle is only one of the factors employed to produce victory that therefore so far from having with the superiority in number obtained all, or even only the principal thing, we have perhaps got very little by it, according as the other circumstances which co-operate happen to vary.

But this superiority has degrees, it may be imagined as twofold, threefold or fourfold, and every one sees, that by increasing in this way, it must (at last) overpower everything else.

In such an aspect we grant, that the superiority in numbers is the most important factor in the result of a combat, only it must be sufficiently great to be a counterpoise to all the other co-operating circumstances. The direct result of this is, that the greatest possible number of troops should be brought into action at the decisive point.

Whether the troops thus brought are sufficient or not, we have then done in this respect all that our means allowed. This is the first principle in Strategy, therefore in general as now stated, it is just as well suited for Greeks and Persians, or for Englishmen and Mahrattas, as for French and Germans. But we shall take a glance at our relations in Europe, as respects War, in order to arrive at some more definite idea on this subject.

Here we find Armies much more alike in equipment, organisation, and practical skill of every kind. There only remains a difference in the military virtue of Armies, and in the talent of Generals which may fluctuate with time from side to side. If we go through the military history of modern Europe, we find no example of a Marathon.

Frederick the Great beat 80,000 Austrians at Leuthen with about 30,000 men, and at Rosbach with 25,000 some 50,000 allies; these are however the only instances of victories gained against an enemy double, or more than double in numbers. Charles XII, in the battle of Narva, we cannot well quote, for the Russians were at that time hardly to be regarded as Europeans, also the principal circumstances, even of the battle, are too little known. Buonaparte had at Dresden 120,000 against 220,000, therefore not the double. At Kollin, Frederick the Great did not succeed, with 30,000 against 50,000 Austrians, neither did Buonaparte in the desperate battle of Leipsic, where he was 160,000 strong, against 280,000.

From this we may infer, that it is very difficult in the present state of Europe, for the most talented General to gain a victory over an enemy double his strength. Now if we see double numbers prove such a weight in the scale against the greatest Generals, we may be sure, that in ordinary cases, in small as well as great combats, an important superiority of numbers, but which need not be over two to one, will be sufficient to ensure the victory, however disadvantageous other circumstances may be. Certainly, we may imagine a defile which even tenfold would not suffice to force, but in such a case it can be no question of a battle at all.

We think, therefore, that under our conditions, as well as in all similar ones, the superiority at the decisive point is a matter of capital importance, and that this subject, in the generality of cases, is decidedly the most important of all. The strength at the decisive point depends on the absolute strength of the Army, and on skill in making use of it.

The first rule is therefore to enter the field with an Army as strong as possible. This sounds very like a commonplace, but still it is really not so.

In order to show that for a long time the strength of forces was by no means regarded as a chief point, we need only observe, that in most, and even in the most detailed histories of the Wars in the eighteenth century, the strength of the Armies is either not given at all, or only incidentally, and in no case is any special value laid upon it. Tempelhof in his history of the Seven Years' War is the earliest writer who gives it regularly, but at the same time he does it only very superficially.

Even Massenbach, in his manifold critical observations on the Prussian campaigns of 1793-94 in the Vosges, talks a great deal about hills and valleys, roads and footpaths, but does not say a syllable about mutual strength.

Another proof lies in a wonderful notion which haunted the heads of many critical historians, according to which there was a certain size of an Army which was the best, a normal strength, beyond which the forces in excess were burdensome rather than serviceable. [Tempelhof and Montalembert are the first we recollect as examples—the first in a passage of his first part, page 148; the other in his correspondence relative to the plan of operations of the Russians in 1759.]

Lastly, there are a number of instances to be found, in which all the available forces were not really brought into the battle, [The Prussians at Jena, 1806. Wellington at Waterloo.] or into the War, because the superiority of numbers was not considered to have that importance which in the nature of things belongs to it.

If we are thoroughly penetrated with the conviction that with a considerable superiority of numbers everything possible is to be effected, then it cannot fail that this clear conviction reacts on the preparations for the War, so as to make us appear in the field with as many troops as possible, and either to give us ourselves the superiority, or at least to guard against the enemy obtaining it. So much for what concerns the absolute force with which the War is to be conducted.

The measure of this absolute force is determined by the Government; and although with this determination the real action of War commences, and it forms an essential part of the Strategy of the War, still in most cases the General who is to command these forces in the War must regard their absolute strength as a given quantity, whether it be that he has had no voice in fixing it, or that circumstances prevented a sufficient expansion being given to it.

There remains nothing, therefore, where an absolute superiority is not attainable, but to produce a relative one at the decisive point, by making skilful use of what we have.

The calculation of space and time appears as the most essential thing to this end—and this has caused that subject to be regarded as one which embraces nearly the whole art of using military forces. Indeed, some have gone so far as to ascribe to great strategists and tacticians a mental organ peculiarly adapted to this point.

But the calculation of time and space, although it lies universally at the foundation of Strategy, and is to a certain extent its daily bread, is still neither the most difficult, nor the most decisive one.

If we take an unprejudiced glance at military history, we shall find that the instances in which mistakes in such a calculation have proved the cause of serious losses are very rare, at least in Strategy. But if the conception of a skilful combination of time and space is fully to account for every instance of a resolute and active Commander beating several separate opponents with one and the same army (Frederick the Great, Buonaparte), then we perplex ourselves unnecessarily with conventional language. For the sake of clearness and the profitable use of conceptions, it is necessary that things should always be called by their right names.

The right appreciation of their opponents (Daun, Schwartzenberg), the audacity to leave for a short space of time a small force only before them, energy in forced marches, boldness in sudden attacks, the intensified activity which great souls acquire in the moment of danger, these are the grounds of such victories; and what have these to do with the ability to make an exact calculation of two such simple things as time and space?

But even this ricochetting play of forces, "when the victories at Rosbach and Montmirail give the impulse to victories at Leuthen and Montereau," to which great Generals on the defensive have often trusted, is still, if we would be clear and exact, only a rare occurrence in history.

Much more frequently the relative superiority—that is, the skilful assemblage of superior forces at the decisive point—has its foundation in the right appreciation of those points, in the judicious direction which by that means has been given to the forces from the very first, and in the resolution required to sacrifice the unimportant to the advantage of the important—that is, to keep the forces concentrated in an overpowering mass. In this, Frederick the Great and Buonaparte are particularly characteristic.

We think we have now allotted to the superiority in numbers the importance which belongs to it; it is to be regarded as the fundamental idea, always to be aimed at before all and as far as possible.

But to regard it on this account as a necessary condition of victory would be a complete misconception of our exposition; in the conclusion to be drawn from it there lies nothing more than the value which should attach to numerical strength in the combat. If that strength is made as great as possible, then the maxim is satisfied; a review of the total relations must then decide whether or not the combat is to be avoided for want of sufficient force. [Owing to our freedom from invasion, and to the condition which arise in our Colonial Wars, we have not yet, in England, arrived at a correct appreciation of the value of superior numbers in War, and still adhere to the idea of an Army just "big enough," which Clausewitz has so unsparingly ridiculed. (Editor.)]

The Surprise

From the subject of the foregoing chapter, the general endeavour to attain a relative superiority, there follows another endeavour which must consequently be just as general in its nature: this is the SURPRISE of the enemy. It lies more or less at the foundation of all undertakings, for without it the preponderance at the decisive point is not properly conceivable.

The surprise is, therefore, not only the means to the attainment of numerical superiority; but it is also to be regarded as a substantive principle in itself, on account of its moral effect. When it is successful in a high degree, confusion and broken courage in the enemy's ranks are the consequences; and of the degree to which these multiply a success, there are examples enough, great and small. We are not now speaking of the particular surprise which belongs to the attack, but of the endeavour by measures generally, and especially by the distribution of forces, to surprise the enemy, which can be imagined just as well in the defensive, and which in the tactical defence particularly is a chief point.

We say, surprise lies at the foundation of all undertakings without exception, only in very different degrees according to the nature of the undertaking and other circumstances.

This difference, indeed, originates in the properties or peculiarities of the Army and its Commander, in those even of the Government.

Secrecy and rapidity are the two factors in this product and these suppose in the Government and the Commander-in-Chief great energy, and on the part of the Army a high sense of military duty. With effeminacy and loose principles it is in vain to calculate upon a surprise. But so general, indeed so indispensable, as is this endeavour, and true as it is that it is never wholly unproductive of effect, still it is not the less true that it seldom succeeds to a *remarkable* degree, and this follows from the nature of the idea itself. We should form an erroneous conception if we believed that by this means chiefly there is much to be attained in War. In idea it promises a great deal; in the execution it generally sticks fast by the friction of the whole machine.

In tactics the surprise is much more at home, for the very natural reason that all times and spaces are on a smaller scale. It will, therefore, in Strategy be the more feasible in proportion as the measures lie nearer to the province of tactics, and more difficult the higher up they lie towards the province of policy.

The preparations for a War usually occupy several months; the assembly of an Army at its principal positions requires generally the formation of depôts and magazines, and long marches, the object of which can be guessed soon enough.

It therefore rarely happens that one State surprises another by a War, or by the direction which it gives the mass of its forces. In the seventeenth and eighteenth centuries, when War turned very much upon sieges, it was a frequent aim, and quite a peculiar and important chapter in the Art of War, to invest a strong place unexpectedly, but even that only rarely succeeded. [Railways, steamships, and telegraphs have, however, enormously modified the relative importance and practicability of surprise. (Editor.]

On the other hand, with things which can be done in a day or two, a surprise is much more conceivable, and, therefore, also it is often not difficult thus to gain a march upon the enemy, and thereby a position, a point of country, a road, etc. But it is evident that what surprise gains in this way in easy execution, it loses in the efficacy, as the greater the efficacy the greater always the difficulty of execution. Whoever thinks that with such surprises on a small scale, he may connect great results—as, for example, the gain of a battle, the capture of an important magazine—believes in something which it is certainly very possible to imagine, but for which there is no warrant in history; for there are upon the whole very few instances where anything great has resulted from such surprises; from which we may justly conclude that inherent difficulties lie in the way of their success.

Certainly, whoever would consult history on such points must not depend on sundry battle steeds of historical critics, on their wise dicta and self-complacent terminology, but look at facts with his own eyes. There is, for instance, a certain day in the campaign in Silesia, 1761, which, in this respect, has attained a kind of notoriety. It is the 22nd July, on which Frederick the Great gained on Laudon the march to Nossen, near Neisse, by which, as is said, the junction of the Austrian and Russian armies in Upper Silesia became impossible, and, therefore, a period of four weeks was gained by the King. Whoever reads over this occurrence carefully in the principal histories, [Tempelhof, The Veteran, Frederick the Great. Compare also (Clausewitz) "Hinterlassene Werke," vol. x., p. 158.] and considers it impartially, will, in the march of the 22nd July, never find this importance; and generally in the whole of the fashionable logic on this subject, he will see nothing but contradictions; but in the proceedings of Laudon, in this renowned period of manoeuvres, much that is unaccountable. How could one, with a thirst for truth, and clear conviction, accept such historical evidence?

When we promise ourselves great effects in a campaign from the principle of surprising, we think upon great activity, rapid resolutions, and forced marches, as the means of producing them; but that these things, even when forthcoming in a very high degree, will not always produce the desired effect, we see in examples given by Generals, who may be allowed to have had the greatest talent in the use of these means, Frederick the Great and Buonaparte. The first when he left Dresden so suddenly in July 1760, and falling upon Lascy, then turned against Dresden, gained nothing by the whole of that intermezzo, but rather placed his affairs in a condition notably worse, as the fortress Glatz fell in the meantime.

In 1813, Buonaparte turned suddenly from Dresden twice against Bluecher, to say nothing of his incursion into Bohemia from Upper Lusatia, and both times without in the least attaining his object. They were blows in the air which only cost him time and force, and might have placed him in a dangerous position in Dresden.

Therefore, even in this field, a surprise does not necessarily meet with great success through the mere activity, energy, and resolution of the Commander; it must be favoured by other circumstances. But we by no means deny that there can be success; we only connect with it a necessity of favourable circumstances, which, certainly do not occur very frequently, and which the Commander can seldom bring about himself.

Just those two Generals afford each a striking illustration of this. We take first Buonaparte in his famous enterprise against Bluecher's Army in February 1814, when it was separated from the Grand Army, and descending the Marne. It would not be easy to find a two days' march to surprise the enemy productive of greater results than this; Bluecher's Army, extended over a distance of three days' march, was beaten in detail, and suffered a loss nearly equal to that of defeat in a great battle. This was completely the effect of a surprise, for if Bluecher had thought of such a near possibility of an attack from Buonaparte [Bluecher believed his march to be covered by Pahlen's Cossacks, but these had been withdrawn without warning to him by the Grand Army Headquarters under Schwartzenberg.] he would have organised his march quite differently. To this mistake of Bluecher's the result is to be attributed. Buonaparte did not know all these circumstances, and so there was a piece of good fortune that mixed itself up in his favour.

It is the same with the battle of Liegnitz, 1760. Frederick the Great gained this fine victory through altering during the night a position which he had just before taken up. Laudon was through this completely surprised, and lost 70 pieces of artillery and 10,000 men. Although Frederick the Great had at this time adopted the principle of moving backwards and forwards in order to make a battle impossible, or at least to disconcert the enemy's plans, still the alteration of position on the night of the 14-15 was not made exactly with that intention, but as the King himself says, because the position of the 14th did not please him. Here, therefore, also chance was hard at work; without this happy conjunction of the attack and the change of position in the night, and the difficult nature of the country, the result would not have been the same.

Also in the higher and highest province of Strategy there are some instances of surprises fruitful in results. We shall only cite the brilliant marches of the Great Elector against the Swedes from Franconia to Pomerania and from the Mark (Brandenburg) to the Pregel in 1757, and the celebrated passage of the Alps by Buonaparte, 1800. In the latter case an Army gave up its whole theatre of war by a capitulation, and in 1757 another Army was very near giving up its theatre of war and itself as well. Lastly, as an instance of a War wholly unexpected, we may bring forward the invasion of Silesia by Frederick the Great. Great and powerful are here the results everywhere, but such events are not common in history if we do not confuse with them cases in which a State, for want of activity and energy (Saxony 1756, and Russia, 1812), has not completed its preparations in time.

Now there still remains an observation which concerns the essence of the thing. A surprise can only be effected by that party which gives the law to the other; and he who is in the right gives the law. If we surprise the adversary by a wrong measure, then instead of reaping good results, we may have to bear a sound blow in return; in any case the adversary need not trouble himself much about our surprise, he has in our mistake the means of turning off the evil. As the offensive includes in itself much more positive action than the defensive, so the surprise is certainly more in its place with the assailant, but by no means invariably, as we shall hereafter see. Mutual surprises by the offensive and defensive may therefore meet, and then that one will have the advantage who has hit the nail on the head the best.

So should it be, but practical life does not keep to this line so exactly, and that for a very simple reason. The moral effects which attend a surprise often convert the worst case into a good one for the side they favour, and do not allow the other to make any regular determination. We have here in view more than anywhere else not only the chief Commander, but each single one, because a surprise has the effect in particular of greatly loosening unity, so that the individuality of each separate leader easily comes to light.

Much depends here on the general relation in which the two parties stand to each other. If the one side through a general moral superiority can intimidate and outdo the other, then he can make use of the surprise with more success, and even reap good fruit where properly he should come to ruin.

Stratagem

Stratagem implies a concealed intention, and therefore is opposed to straightforward dealing, in the same way as wit is the opposite of direct proof. It has therefore nothing in common with means of persuasion, of self-interest, of force, but a great deal to do with deceit, because that likewise conceals its object. It is itself a deceit as well when it is done, but still it differs from what is commonly called deceit, in this respect that there is no direct breach of word. The deceiver by stratagem leaves it to the person himself whom he is deceiving to commit the errors of understanding which at last, flowing into *one* result, suddenly change the nature of things in his eyes. We may therefore say, as nit is a sleight of hand with ideas and conceptions, so stratagem is a sleight of hand with actions.

At first sight it appears as if Strategy had not improperly derived its name from stratagem; and that, with all the real and apparent changes which the whole character of War has undergone since the time of the Greeks, this term still points to its real nature.

If we leave to tactics the actual delivery of the blow, the battle itself, and look upon Strategy as the art of using this means with skill, then besides the forces of the character, such as burning ambition which always presses like a spring, a strong will which hardly bends etc. etc., there seems no subjective quality so suited to guide and inspire strategic activity as stratagem. The general tendency to surprise, treated of in the foregoing chapter, points to this conclusion, for there is a degree of stratagem, be it ever so small, which lies at the foundation of every attempt to surprise.

But however much we feel a desire to see the actors in War outdo each other in hidden activity, readiness, and stratagem, still we must admit that these qualities show themselves but little in history, and have rarely been able to work their way to the surface from amongst the mass of relations and circumstances.

The explanation of this is obvious, and it is almost identical with the subject matter of the preceding chapter.

Strategy knows no other activity than the regulating of combat with the measures which relate to it. It has no concern, like ordinary life, with transactions which consist merely of words—that is, in expressions, declarations, etc. But these, which are very inexpensive, are chiefly the means with which the wily one takes in those he practises upon.

That which there is like it in War, plans and orders given merely as make-believers, false reports sent on purpose to the enemy—is usually of so little effect in the strategic field that it is only resorted to in particular cases which offer of themselves, therefore cannot be regarded as spontaneous action which emanates from the leader.

But such measures as carrying out the arrangements for a battle, so far as to impose upon the enemy, require a considerable expenditure of time and power; of course, the greater the impression to be made, the greater the expenditure in these respects. And as this is usually not given for the purpose, very few demonstrations, so-called, in Strategy, effect the object for which they are designed. In fact, it is dangerous to detach large forces for any length of time merely for a trick, because there is always the risk of its being done in vain, and then these forces are wanted at the decisive point.

The chief actor in War is always thoroughly sensible of this sober truth, and therefore he has no desire to play at tricks of agility. The bitter earnestness of necessity presses so fully into direct action that there is no room for that game. In a word, the pieces on the strategical chess-board want that mobility which is the element of stratagem and subtility.

The conclusion which we draw, is that a correct and penetrating eye is a more necessary and more useful quality for a General than craftiness, although that also does no harm if it does not exist at the expense of necessary qualities of the heart, which is only too often the case.

But the weaker the forces become which are under the command of Strategy, so much the more they become adapted for stratagem, so that to the quite feeble and little, for whom no prudence, no sagacity is any longer sufficient at the point where all art seems to forsake him, stratagem offers itself as a last resource. The more helpless his situation, the more everything presses towards one single, desperate blow, the more readily stratagem comes to the aid of his boldness. Let loose from all further calculations, freed from all concern for the future, boldness and stratagem intensify each other, and thus collect at one point an infinitesimal glimmering of hope into a single ray, which may likewise serve to kindle a flame.

Assembly of Forces in Space

The best Strategy is *always to be very strong*, first generally then at the decisive point. Therefore, apart from the energy which creates the Army, a work which is not always done by the General, there is no more imperative and no simpler law for Strategy than to *keep the forces concentrated.* —No portion is to be separated from the main body unless called away by some urgent necessity. On this maxim we stand firm, and look upon it as a guide to be depended upon. What are the reasonable grounds on which a detachment of forces may be made we shall learn by degrees. Then we shall also see that this principle cannot have the same general effects in every War, but that these are different according to the means and end.

It seems incredible, and yet it has happened a hundred times, that troops have been divided and separated merely through a mysterious feeling of conventional manner, without any clear perception of the reason.

If the concentration of the whole force is acknowledged as the norm, and every division and separation as an exception which must be justified, then not only will that folly be completely avoided, but also many an erroneous ground for separating troops will be barred admission.

Assembly of Forces in Time

We have here to deal with a conception which in real life diffuses many kinds of illusory light. A clear definition and development of the idea is therefore necessary, and we hope to be allowed a short analysis.

War is the shock of two opposing forces in collision with each other, from which it follows as a matter of course that the stronger not only destroys the other, but carries it forward with it in its movement. This fundamentally admits of no successive action of powers, but makes the simultaneous application of all forces intended for the shock appear as a primordial law of War.

So it is in reality, but only so far as the struggle resembles also in practice a mechanical shock, but when it consists in a lasting, mutual action of destructive forces, then we can certainly imagine a successive action of forces. This is the case in tactics, principally because firearms form the basis of all tactics, but also for other reasons as well. If in a fire combat 1000 men are opposed to 500, then the gross loss is calculated from the amount of the enemy's force and our own; 1000 men fire twice as many shots as 500, but more shots will take effect on the 1000 than on the 500 because it is assumed that they stand in closer order than the other. If we were to suppose the number of hits to be double, then the losses on each side would be equal. From the 500 there would be for example 200 disabled, and out of the body of 1000 likewise the same; now if the 500 had kept another body of equal number quite out of fire, then both sides would have 800 effective men; but of these, on the one side there would be 500 men quite fresh, fully supplied with ammunition, and in their full vigour; on the other side only 800 all alike shaken in their order, in want of sufficient ammunition and weakened in physical force. The assumption that the 1000 men merely on account of their greater number would lose twice as many as 500 would have lost in their place, is certainly not correct; therefore the greater loss which the side suffers that has placed the half of its force in reserve, must be regarded as a disadvantage in that original formation; further it must be admitted, that in the generality of cases the 1000 men would have the advantage at the first commencement of being able to drive their opponent out of his position and force him to a retrograde movement; now, whether these two advantages are a counterpoise to the disadvantage of finding ourselves with 800 men to a certain extent disorganised by the combat, opposed to an enemy who is not materially weaker in numbers and who has 500 quite fresh troops, is one that cannot be decided by pursuing an analysis further, we must here rely upon experience, and there will scarcely be an officer experienced in War who will not in the generality of cases assign the advantage to that side which has the fresh troops.

In this way it becomes evident how the employment of too many forces in combat may be disadvantageous; for whatever advantages the superiority may give in the first moment, we may have to pay dearly for in the next.

But this danger only endures as long as the disorder, the state of confusion and weakness lasts, in a word, up to the crisis which every combat brings with it even for the conqueror. Within the duration of this relaxed state of exhaustion, the appearance of a proportionate number of fresh troops is decisive.

But when this disordering effect of victory stops, and therefore only the moral superiority remains which every victory gives, then it is no longer possible for fresh troops to restore the combat, they would only be carried along in the general movement; a beaten Army cannot be brought back to victory a day after by means of a strong reserve. Here we find ourselves at the source of a highly material difference between tactics and strategy.

The tactical results, the results within the four corners of the battle, and before its close, lie for the most part within the limits of that period of disorder and weakness. But the strategic result, that is to say, the result of the total combat, of the victories realised, let them be small or great, lies completely (beyond) outside of that period. It is only when the results of partial combats have bound themselves together into an independent whole, that the strategic result appears, but then, the state of crisis is over, the forces have resumed their original form, and are now only weakened to the extent of those actually destroyed (placed hors de combat).

The consequence of this difference is, that tactics can make a continued use of forces, Strategy only a simultaneous one.

If I cannot, in tactics, decide all by the first success, if I have to fear the next moment, it follows of itself that I employ only so much of my force for the success of the first moment as appears sufficient for that object, and keep the rest beyond the reach of fire or conflict of any kind, in order to be able to oppose fresh troops to fresh, or with such to overcome those that are exhausted. But it is not so in Strategy. Partly, as we have just shown, it has not so much reason to fear a reaction after a success realised, because with that success the crisis stops; partly all the forces strategically employed are not necessarily weakened. Only so much of them as have been tactically in conflict with the enemy's force, that is, engaged in partial combat, are weakened by it; consequently, only so much as was unavoidably necessary, but by no means all which was strategically in conflict with the enemy, unless tactics has expended them unnecessarily. Corps which, on account of the general superiority in numbers, have either been little or not at all engaged, whose presence alone has assisted in the result, are after the decision the same as they were before, and for new enterprises as efficient as if they had been entirely inactive. How greatly such corps which thus constitute our excess may contribute to the total success is evident in itself; indeed, it is not difficult to see how they may even diminish considerably the loss of the forces engaged in tactical, conflict on our side.

If, therefore, in Strategy the loss does not increase with the number of the troops employed, but is often diminished by it, and if, as a natural consequence, the decision in our favor is, by that means, the more certain, then it follows naturally that in Strategy we can never employ too many forces, and consequently also that they must be applied simultaneously to the immediate purpose.

But we must vindicate this proposition upon another ground. We have hitherto only spoken of the combat itself; it is the real activity in War, but men, time, and space, which appear as the elements of this activity, must, at the same time, be kept in view, and the results of their influence brought into consideration also.

Fatigue, exertion, and privation constitute in War a special principle of destruction, not essentially belonging to contest, but more or less inseparably bound up with it, and certainly one which especially belongs to Strategy. They no doubt exist in tactics as well, and perhaps there in the highest degree; but as the duration of the tactical acts is shorter, therefore the small effects of exertion and privation on them can come but little into consideration. But in Strategy on the other hand, where time and space, are on a larger scale, their influence is not only always very considerable, but often quite decisive. It is not at all uncommon for a victorious Army to lose many more by sickness than on the field of battle.

If, therefore, we look at this sphere of destruction in Strategy in the same manner as we have considered that of fire and close combat in tactics, then we may well imagine that everything which comes within its vortex will, at the end of the campaign or of any other strategic period, be reduced to a state of weakness, which makes the arrival of a fresh force decisive. We might therefore conclude that there is a motive in the one case as well as the other to strive for the first success with as few forces as possible, in order to keep up this fresh force for the last.

In order to estimate exactly this conclusion, which, in many cases in practice, will have a great appearance of truth, we must direct our attention to the separate ideas which it contains. In the first place, we must not confuse the notion of reinforcement with that of fresh unused troops. There are few campaigns at the end of which an increase of force is not earnestly desired by the conqueror as well as the conquered, and indeed should appear decisive; but that is not the point here, for that increase of force could not be necessary if the force had been so much larger at the first. But it would be contrary to all experience to suppose that an Army coming fresh into the field is to be esteemed higher in point of moral value than an Army already in the field, just as a tactical reserve is more to be esteemed than a body of troops which has been already severely handled in the fight. Just as much as an unfortunate campaign lowers the courage and moral powers of an Army, a successful one raises these elements in their value. In the generality of cases, therefore, these influences are compensated, and then there remains over and above as clear gain the habituation to War. We should besides look more here to successful than to unsuccessful campaigns, because when the greater probability of the latter may be seen beforehand, without doubt forces are wanted, and, therefore, the reserving a portion for future use is out of the question.

This point being settled, then the question is, Do the losses which a force sustains through fatigues and privations increase in proportion to the size of the force, as is the case in a combat? And to that we answer "No."

The fatigues of War result in a great measure from the dangers with which every moment of the act of War is more or less impregnated. To encounter these dangers at all points, to proceed onwards with security in the execution of one's plans, gives employment to a multitude of agencies which make up the tactical and strategic service of the Army. This service is more difficult the weaker an Army is, and easier as its numerical superiority over that of the enemy increases. Who can doubt this? A campaign against a much weaker enemy will therefore cost smaller efforts than against one just as strong or stronger.

So much for the fatigues. It is somewhat different with the privations; they consist chiefly of two things, the want of food, and the want of shelter for the troops, either in quarters or in suitable camps. Both these wants will no doubt be greater in proportion as the number of men on one spot is greater. But does not the superiority in force afford also the best means of spreading out and finding more room, and therefore more means of subsistence and shelter?

If Buonaparte, in his invasion of Russia in 1812, concentrated his Army in great masses upon one single road in a manner never heard of before, and thus caused privations equally unparalleled, we must ascribe it to his maxim *that it is impossible to be too strong at the decisive point.* Whether in this instance he did not strain the principle too far is a question which would be out of place here; but it is certain that, if he had made a point of avoiding the distress which was by that means brought about, he had only to advance on a greater breadth of front. Room was not wanted for the purpose in Russia, and in very few cases can it be wanted. Therefore, from this no ground can be deduced to prove that the simultaneous employment of very superior forces must produce greater weakening. But now, supposing that in spite of the general relief afforded by setting apart a portion of the Army, wind and weather and the toils of War had produced a diminution even on the part which as a spare force had been reserved for later use, still we must take a comprehensive general view of the whole, and therefore ask, Will this diminution of force suffice to counterbalance the gain in forces, which we, through our superiority in numbers, may be able to make in more ways than one?

But there still remains a most important point to be noticed. In a partial combat, the force required to obtain a great result can be approximately estimated without much difficulty, and, consequently, we can form an idea of what is superfluous. In Strategy this may be said to be impossible, because the strategic result has no such well-defined object and no such circumscribed limits as the tactical. Thus what can be looked upon in tactics as an excess of power, must be regarded in Strategy as a means to give expansion to success, if opportunity offers for it; with the magnitude of the success the gain in force increases at the same time, and in this way the superiority of numbers may soon reach a point which the most careful economy of forces could never have attained.

By means of his enormous numerical superiority, Buonaparte was enabled to reach Moscow in 1812, and to take that central capital. Had he by means of this superiority succeeded in completely defeating the Russian Army, he would, in all probability, have concluded a peace in Moscow which in any other way was much less attainable. This example is used to explain the idea, not to prove it, which would require a circumstantial demonstration, for which this is not the place.

All these reflections bear merely upon the idea of a successive employment of forces, and not upon the conception of a reserve properly so called, which they, no doubt, come in contact with throughout, but which, as we shall see in the following chapter, is connected with some other considerations.

What we desire to establish here is, that if in tactics the military force through the mere duration of actual employment suffers a diminution of power, if time, therefore, appears as a factor in the result, this is not the case in Strategy in a material degree. The destructive effects which are also produced upon the forces in Strategy by time, are partly diminished through their mass, partly made good in other ways, and, therefore, in Strategy it cannot be an object to make time an ally on its own account by bringing troops successively into action.

We say on "its own account," for the influence which time, on account of other circumstances which it brings about but which are different from itself can have, indeed must necessarily have, for one of the two parties, is quite another thing, is anything but indifferent or unimportant, and will be the subject of consideration hereafter.

The rule which we have been seeking to set forth is, therefore, that all forces which are available and destined for a strategic object should be *simultaneously* applied to it; and this application will be so much the more complete the more everything is compressed into one act and into one movement.

But still there is in Strategy a renewal of effort and a persistent action which, as a chief means towards the ultimate success, is more particularly not to be overlooked, it is the *continual development of new forces*. This is also the subject of another chapter, and we only refer to it here in order to prevent the reader from having something in view of which we have not been speaking.

We now turn to a subject very closely connected with our present considerations, which must be settled before full light can be thrown on the whole, we mean the *strategic reserve*.

Strategic Reserve

A reserve has two objects which are very distinct from each other, namely, first, the prolongation and renewal of the combat, and secondly, for use in case of unforeseen events. The first object implies the utility of a successive application of forces, and on that account cannot occur in Strategy. Cases in which a corps is sent to succour a point which is supposed to be about to fall are plainly to be placed in the category of the second object, as the resistance which has to be offered here could not have been sufficiently foreseen. But a corps which is destined expressly to prolong the combat, and with that object in view is placed in rear, would be only a corps placed out of reach of fire, but under the command and at the disposition of the General Commanding in the action, and accordingly would be a tactical and not a strategic reserve.

But the necessity for a force ready for unforeseen events may also take place in Strategy, and consequently there may also be a strategic reserve, but only where unforeseen events are imaginable. In tactics, where the enemy's measures are generally first ascertained by direct sight, and where they may be concealed by every wood, every fold of undulating ground, we must naturally always be alive, more or less, to the possibility of unforeseen events, in order to strengthen, subsequently, those points which appear too weak, and, in fact, to modify generally the disposition of our troops, so as to make it correspond better to that of the enemy.

Such cases must also happen in Strategy, because the strategic act is directly linked to the tactical. In Strategy also many a measure is first adopted in consequence of what is actually seen, or in consequence of uncertain reports arriving from day to day, or even from hour to hour, and lastly, from the actual results of the combats it is, therefore, an essential condition of strategic command that, according to the degree of uncertainty, forces must be kept in reserve against future contingencies.

In the defensive generally, but particularly in the defence of certain obstacles of ground, like rivers, hills, etc., such contingencies, as is well known, happen constantly.

But this uncertainty diminishes in proportion as the strategic activity has less of the tactical character, and ceases almost altogether in those regions where it borders on politics.

The direction in which the enemy leads his columns to the combat can be perceived by actual sight only; where he intends to pass a river is learnt from a few preparations which are made shortly before; the line by which he proposes to invade our country is usually announced by all the newspapers before a pistol shot has been fired. The greater the nature of the measure the less it will take the enemy by surprise. Time and space are so considerable, the circumstances out of which the action proceeds so public and little susceptible of alteration, that the coming event is either made known in good time, or can be discovered with reasonable certainty.

On the other hand the use of a reserve in this province of Strategy, even if one were available, will always be less efficacious the more the measure has a tendency towards being one of a general nature.

We have seen that the decision of a partial combat is nothing in itself, but that all partial combats only find their complete solution in the decision of the total combat.

But even this decision of the total combat has only a relative meaning of many different gradations, according as the force over which the victory has been gained forms a more or less great and important part of the whole. The lost battle of a corps may be repaired by the victory of the Army. Even the lost battle of an Army may not only be counterbalanced by the gain of a more important one, but converted into a fortunate event (the two days of Kulm, August 29 and 30, 1813 [Refers to the destruction of Vandamme's column, which had been sent unsupported to intercept the retreat of the Austrians and Prussians from Dresden—but was forgotten by Napoleon.—Editor.]). No one can doubt this; but it is just as clear that the weight of each victory (the successful issue of each total combat) is so much the more substantial the more important the part conquered, and that therefore the possibility of repairing the loss by subsequent events diminishes in the same proportion. In another place we shall have to examine this more in detail; it suffices for the present to have drawn attention to the indubitable existence of this progression.

If we now add lastly to these two considerations the third, which is, that if the persistent use of forces in tactics always shifts the great result to the end of the whole act, law of the simultaneous use of the forces in Strategy, on the contrary, lets the principal result (which need not be the final one) take place almost always at the commencement of the great (or whole) act, then in these three results we have grounds sufficient to find strategic reserves always more superfluous, always more useless, always more dangerous, the more general their destination.

The point where the idea of a strategic reserve begins to become inconsistent is not difficult to determine: it lies in the *supreme decision*. Employment must be given to all the forces within the space of the supreme decision, and every reserve (active force available) which is only intended for use after that decision is opposed to common sense.

If, therefore, tactics has in its reserves the means of not only meeting unforeseen dispositions on the part of the enemy, but also of repairing that which never can be foreseen, the result of the combat, should that be unfortunate; Strategy on the other hand must, at least as far as relates to the capital result, renounce the use of these means. As A rule, it can only repair the losses sustained at one point by advantages gained at another, in a few cases by moving troops from one point to another; the idea of preparing for such reverses by placing forces in reserve beforehand, can never be entertained in Strategy.

We have pointed out as an absurdity the idea of a strategic reserve which is not to co-operate in the capital result, and as it is so beyond a doubt, we should not have been led into such an analysis as we have made in these two chapters, were it not that, in the disguise of other ideas, it looks like something better, and frequently makes its appearance. One person sees in it the acme of strategic sagacity and foresight; another rejects it, and with it the idea of any reserve, consequently even of a tactical one. This confusion of ideas is transferred to real life, and if we would see a memorable instance of it we have only to call to mind that Prussia in 1806 left a reserve of 20,000 men cantoned in the Mark, under Prince Eugene of Wurtemberg, which could not possibly reach the Saale in time to be of any use, and that another force Of 25,000 men belonging to this power remained in East and South Prussia, destined only to be put on a war-footing afterwards as a reserve.

After these examples we cannot be accused of having been fighting with windmills.

Economy of Forces

The road of reason, as we have said, seldom allows itself to be reduced to a mathematical line by principles and opinions. There remains always a certain margin. But it is the same in all the practical arts of life. For the lines of beauty there are no abscissae and ordinates; circles and ellipses are not described by means of their algebraical formulae. The actor in War therefore soon finds he must trust himself to the delicate tact of judgment which, founded on natural quickness of perception, and educated by reflection, almost unconsciously seizes upon the right; he soon finds that at one time he must simplify the law (by reducing it) to some prominent characteristic points which form his rules; that at another the adopted method must become the staff on which he leans.

As one of these simplified characteristic points as a mental appliance, we look upon the principle of watching continually over the co-operation of all forces, or in other words, of keeping constantly in view that no part of them should ever be idle. Whoever has forces where the enemy does not give them sufficient employment, whoever has part of his forces on the march—that is, allows them to lie dead—while the enemy's are fighting, he is a bad manager of his forces. In this sense there is a waste of forces, which is even worse than their employment to no purpose. If there must be action, then the first point is that all parts act, because the most purposeless activity still keeps employed and destroys a portion of the enemy's force, whilst troops completely inactive are for the moment quite neutralised. Unmistakably this idea is bound up with the principles contained in the last three chapters, it is the same truth, but seen from a somewhat more comprehensive point of view and condensed into a single conception.

Geometrical Element

The length to which the geometrical element or form in the disposition of military force in War can become a predominant principle, we see in the art of fortification, where geometry looks after the great and the little. Also in tactics it plays a great part. It is the basis of elementary tactics, or of the theory of moving troops; but in field fortification, as well as in the theory of positions, and of their attack, its angles and lines rule like law givers who have to decide the contest. Many things here were at one time misapplied, and others were mere fribbles; still, however, in the tactics of the present day, in which in every combat the aim is to surround the enemy, the geometrical element has attained anew a great importance in a very simple, but constantly recurring application. Nevertheless, in tactics, where all is more movable, where the moral forces, individual traits, and chance are more influential than in a war of sieges, the geometrical element can never attain to the same degree of supremacy as in the latter. But less still is its influence in Strategy; certainly here, also, form in the disposition of troops, the shape of countries and states is of great importance; but the geometrical element is not decisive, as in fortification, and not nearly so important as in tactics.—The manner in which this influence exhibits itself, can only be shown by degrees at those places where it makes its appearance, and deserves notice. Here we wish more to direct attention to the difference which there is between tactics and Strategy in relation to it.

In tactics time and space quickly dwindle to their absolute minimum. If a body of troops is attacked in flank and rear by the enemy, it soon gets to a point where retreat no longer remains; such a position is very close to an absolute impossibility of continuing the fight; it must therefore extricate itself from it, or avoid getting into it. This gives to all combinations aiming at this from the first commencement a great efficiency, which chiefly consists in the disquietude which it causes the enemy as to consequences. This is why the geometrical disposition of the forces is such an important factor in the tactical product.

In Strategy this is only faintly reflected, on account of the greater space and time. We do not fire from one theatre of war upon another; and often weeks and months must pass before a strategic movement designed to surround the enemy can be executed. Further, the distances are so great that the probability of hitting the right point at last, even with the best arrangements, is but small.

In Strategy therefore the scope for such combinations, that is for those resting on the geometrical element, is much smaller, and for the same reason the effect of an advantage once actually gained at any point is much greater. Such advantage has time to bring all its effects to maturity before it is disturbed, or quite neutralised therein, by any counteracting apprehensions. We therefore do not hesitate to regard as an established truth, that in Strategy more depends on the number and the magnitude of the victorious combats, than on the form of the great lines by which they are connected.

A view just the reverse has been a favourite theme of modern theory, because a greater importance was supposed to be thus given to Strategy, and, as the higher functions of the mind were seen in Strategy, it was thought by that means to ennoble War, and, as it was said—through a new substitution of ideas—to make it more scientific. We hold it to be one of the principal uses of a complete theory openly to expose such vagaries, and as the geometrical element is the fundamental idea from which theory usually proceeds, therefore we have expressly brought out this point in strong relief.

On the Suspension of the Act in Warfare

If one considers War as an act of mutual destruction, we must of necessity imagine both parties as making some progress; but at the same time, as regards the existing moment, we must almost as necessarily suppose the one party in a state of expectation, and only the other actually advancing, for circumstances can never be actually the same on both sides, or continue so. In time a change must ensue, from which it follows that the present moment is more favourable to one side than the other. Now if we suppose that both commanders have a full knowledge of this circumstance, then the one has a motive for action, which at the same time is a motive for the other to wait; therefore, according to this it cannot be for the interest of both at the same time to advance, nor can waiting be for the interest of both at the same time. This opposition of interest as regards the object is not deduced here from the principle of general polarity, and therefore is not in opposition to the argument in the fifth chapter of the second book; it depends on the fact that here in reality the same thing is at once an incentive or motive to both commanders, namely the probability of improving or impairing their position by future action.

But even if we suppose the possibility of a perfect equality of circumstances in this respect, or if we take into account that through imperfect knowledge of their mutual position such an equality may appear to the two Commanders to subsist, still the difference of political objects does away with this possibility of suspension. One of the parties must of necessity be assumed politically to be the aggressor, because no War could take place from defensive intentions on both sides. But the aggressor has the positive object, the defender merely a negative one. To the first then belongs the positive action, for it is only by that means that he can attain the positive object; therefore, in cases where both parties are in precisely similar circumstances, the aggressor is called upon to act by virtue of his positive object.

Therefore, from this point of view, a suspension in the act of Warfare, strictly speaking, is in contradiction with the nature of the thing; because two Armies, being two incompatible elements, should destroy one another unremittingly, just as fire and water can never put themselves in equilibrium, but act and react upon one another, until one quite disappears. What would be said of two wrestlers who remained clasped round each other for hours without making a movement. Action in War, therefore, like that of a clock which is wound up, should go on running down in regular motion.—But wild as is the nature of War it still wears the chains of human weakness, and the contradiction we see here, viz., that man seeks and creates dangers which he fears at the same time will astonish no one.

If we cast a glance at military history in general, we find so much the opposite of an incessant advance towards the aim, that *standing still* and *doing nothing* is quite plainly the *normal condition* of an Army in the midst of War, *acting*, the *exception*. This must almost raise a doubt as to the correctness of our conception. But if military history leads to this conclusion when viewed in the mass the latest series of campaigns redeems our position. The War of the French Revolution shows too plainly its reality, and only proves too clearly its necessity. In these operations, and especially in the campaigns of Buonaparte, the conduct of War attained to that unlimited degree of energy which we have represented as the natural law of the element. This degree is therefore possible, and if it is possible then it is necessary.

How could any one in fact justify in the eyes of reason the expenditure of forces in War, if acting was not the object? The baker only heats his oven if he has bread to put into it; the horse is only yoked to the carriage if we mean to drive; why then make the enormous effort of a War if we look for nothing else by it but like efforts on the part of the enemy?

So much in justification of the general principle; now as to its modifications, as far as they lie in the nature of the thing and are independent of special cases.

There are three causes to be noticed here, which appear as innate counterpoises and prevent the over-rapid or uncontrollable movement of the wheel-work.

The first, which produces a constant tendency to delay, and is thereby a retarding principle, is the natural timidity and want of resolution in the human mind, a kind of inertia in the moral world, but which is produced not by attractive, but by repellent forces, that is to say, by dread of danger and responsibility.

In the burning element of War, ordinary natures appear to become heavier; the impulsion given must therefore be stronger and more frequently repeated if the motion is to be a continuous one. The mere idea of the object for which arms have been taken up is seldom sufficient to overcome this resistant force, and if a warlike enterprising spirit is not at the head, who feels himself in War in his natural element, as much as a fish in the ocean, or if there is not the pressure from above of some great responsibility, then standing still will be the order of the day, and progress will be the exception.

The second cause is the imperfection of human perception and judgment, which is greater in War than anywhere, because a person hardly knows exactly his own position from one moment to another, and can only conjecture on slight grounds that of the enemy, which is purposely concealed; this often gives rise to the case of both parties looking upon one and the same object as advantageous for them, while in reality the interest of one must preponderate; thus then each may think he acts wisely by waiting another moment, as we have already said in the fifth chapter of the second book.

The third cause which catches hold, like a ratchet wheel in machinery, from time to time producing a complete standstill, is the greater strength of the defensive form. A may feel too weak to attack B, from which it does not follow that B is strong enough for an attack on A. The addition of strength, which the defensive gives is not merely lost by assuming the offensive, but also passes to the enemy just as, figuratively expressed, the difference of a + b and a - b is equal to 2b. Therefore it may so happen that both parties, at one and the same time, not only feel themselves too weak to attack, but also are so in reality.

Thus even in the midst of the act of War itself, anxious sagacity and the apprehension of too great danger find vantage ground, by means of which they can exert their power, and tame the elementary impetuosity of War.

However, at the same time these causes without an exaggeration of their effect, would hardly explain the long states of inactivity which took place in military operations, in former times, in Wars undertaken about interests of no great importance, and in which inactivity consumed nine-tenths of the time that the troops remained under arms. This feature in these Wars, is to be traced principally to the influence which the demands of the one party, and the condition, and feeling of the other, exercised over the conduct of the operations, as has been already observed in the chapter on the essence and object of War.

These things may obtain such a preponderating influence as to make of War a half-and-half affair. A War is often nothing more than an armed neutrality, or a menacing attitude to support negotiations or an attempt to gain some small advantage by small exertions, and then to wait the tide of circumstances, or a disagreeable treaty obligation, which is fulfilled in the most niggardly way possible.

In all these cases in which the impulse given by interest is slight, and the principle of hostility feeble, in which there is no desire to do much, and also not much to dread from the enemy; in short, where no powerful motives press and drive, cabinets will not risk much in the game; hence this tame mode of carrying on War, in which the hostile spirit of real War is laid in irons.

The more War becomes in this manner devitalised so much the more its theory becomes destitute of the necessary firm pivots and buttresses for its reasoning; the necessary is constantly diminishing, the accidental constantly increasing.

Nevertheless in this kind of Warfare, there is also a certain shrewdness, indeed, its action is perhaps more diversified, and more extensive than in the other. Hazard played with realeaux of gold seems changed into a game of commerce with groschen. And on this field, where the conduct of War spins out the time with a number of small flourishes, with skirmishes at outposts, half in earnest half in jest, with long dispositions which end in nothing with positions and marches, which afterwards are designated as skilful only because their infinitesimally small causes are lost, and common sense can make nothing of them, here on this very field many theorists find the real Art of War at home: in these feints, parades, half and quarter thrusts of former Wars, they find the aim of all theory, the supremacy of mind over matter, and modern Wars appear to them mere savage fisticuffs, from which nothing is to be learnt, and which must be regarded as mere retrograde steps towards barbarism. This opinion is as frivolous as the objects to which it relates. Where great forces and great passions are wanting, it is certainly easier for a practised dexterity to show its game; but is then the command of great forces, not in itself a higher exercise of the intelligent faculties? Is then that kind of conventional sword-exercise not comprised in and belonging to the other mode of conducting War? Does it not bear the same relation to it as the motions upon a ship to the motion of the ship itself? Truly it can take place only under the tacit condition that the adversary does no better. And can we tell, how long he may choose to respect those conditions? Has not then the French Revolution fallen upon us in the midst of the fancied security of our old system of War, and driven us from Chalons to Moscow? And did not Frederick the Great in like manner surprise the Austrians reposing in their ancient habits of War, and make their monarchy tremble? Woe to the cabinet which, with a shilly-shally policy, and a routine-ridden military system, meets with an adversary who, like the rude element, knows no other law than that of his intrinsic force. Every deficiency in energy and exertion is then a weight in the scales in favour of the enemy; it is not so easy then to change from the fencing posture into that of an athlete, and a slight blow is often sufficient to knock down the whole.

The result of all the causes now adduced is, that the hostile action of a campaign does not progress by a continuous, but by an intermittent movement, and that, therefore, between the separate bloody acts, there is a period of watching, during which both parties fall into the defensive, and also that usually a higher object causes the principle of aggression to predominate on one side, and thus leaves it in general in an advancing position, by which then its proceedings become modified in some degree.

On the Character of Modern War

The attention which must be paid to the character of War as it is now made, has a great influence upon all plans, especially on strategic ones.

Since all methods formerly usual were upset by Buonaparte's luck and boldness, and first-rate Powers almost wiped out at a blow; since the Spaniards by their stubborn resistance have shown what the general arming of a nation and insurgent measures on a great scale can effect, in spite of weakness and porousness of individual parts; since Russia, by the campaign of 1812 has taught us, first, that an Empire of great dimensions is not to be conquered (which might have been easily known before), secondly, that the probability of final success does not in all cases diminish in the same measure as battles, capitals, and provinces are lost (which was formerly an incontrovertible principle with all diplomatists, and therefore made them always ready to enter at once into some bad temporary peace), but that a nation is often strongest in the heart of its country, if the enemy's offensive power has exhausted itself, and with what enormous force the defensive then springs over to the offensive; further, since Prussia (1813) has shown that sudden efforts may add to an Army sixfold by means of the militia, and that this militia is just as fit for service abroad as in its own country;—since all these events have shown what an enormous factor the heart and sentiments of a Nation may be in the product of its political and military strength, in fine, since governments have found out all these additional aids, it is not to be expected that they will let them lie idle in future Wars, whether it be that danger threatens their own existence, or that restless ambition drives them on.

That a War which is waged with the whole weight of the national power on each side must be organised differently in principle to those where everything is calculated according to the relations of standing Armies to each other, it is easy to perceive. Standing Armies once resembled fleets, the land force the sea force in their relations to the remainder of the State, and from that the Art of War on shore had in it something of naval tactics, which it has now quite lost.

Tension and Rest

The Dynamic Law of War

We have seen in the sixteenth chapter of this book, how, in most campaigns, much more time used to be spent in standing still and inaction than in activity.

Now, although, as observed in the preceding chapter we see quite a different character in the present form of War, still it is certain that real action will always be interrupted more or less by long pauses; and this leads to the necessity of our examining more closely the nature of these two phases of War.

If there is a suspension of action in War, that is, if neither party wills something positive, there is rest, and consequently equilibrium, but certainly an equilibrium in the largest signification, in which not only the moral and physical war-forces, but all relations and interests, come into calculation. As soon as ever one of the two parties proposes to himself a new positive object, and commences active steps towards it, even if it is only by preparations, and as soon as the adversary opposes this, there is a tension of powers; this lasts until the decision takes place—that is, until one party either gives up his object or the other has conceded it to him.

This decision—the foundation of which lies always in the combat—combinations which are made on each side—is followed by a movement in one or other direction.

When this movement has exhausted itself, either in the difficulties which had to be mastered, in overcoming its own internal friction, or through new resistant forces prepared by the acts of the enemy, then either a state of rest takes place or a new tension with a decision, and then a new movement, in most cases in the opposite direction.

This speculative distinction between equilibrium, tension, and motion is more essential for practical action than may at first sight appear.

In a state of rest and of equilibrium a varied kind of activity may prevail on one side that results from opportunity, and does not aim at a great alteration. Such an activity may contain important combats—even pitched battles—but yet it is still of quite a different nature, and on that account generally different in its effects.

If a state of tension exists, the effects of the decision are always greater partly because a greater force of will and a greater pressure of circumstances manifest themselves therein; partly because everything has been prepared and arranged for a great movement. The decision in such cases resembles the effect of a mine well closed and tamped, whilst an event in itself perhaps just as great, in a state of rest, is more or less like a mass of powder puffed away in the open air.

At the same time, as a matter of course, the state of tension must be imagined in different degrees of intensity, and it may therefore approach gradually by many steps towards the state of rest, so that at the last there is a very slight difference between them.

Now the real use which we derive from these reflections is the conclusion that every measure which is taken during a state of tension is more important and more prolific in results than the same measure could be in a state of equilibrium, and that this importance increases immensely in the highest degrees of tension.

The cannonade of Valmy, September 20, 1792, decided more than the battle of Hochkirch, October 14, 1758.

In a tract of country which the enemy abandons to us because he cannot defend it, we can settle ourselves differently from what we should do if the retreat of the enemy was only made with the view to a decision under more favourable circumstances. Again, a strategic attack in course of execution, a faulty position, a single false march, may be decisive in its consequence; whilst in a state of equilibrium such errors must be of a very glaring kind, even to excite the activity of the enemy in a general way.

Most bygone Wars, as we have already said, consisted, so far as regards the greater part of the time, in this state of equilibrium, or at least in such short tensions with long intervals between them, and weak in their effects, that the events to which they gave rise were seldom great successes, often they were theatrical exhibitions, got up in honour of a royal birthday (Hochkirch), often a mere satisfying of the honour of the arms (Kunersdorf), or the personal vanity of the commander (Freiberg).

That a Commander should thoroughly understand these states, that he should have the tact to act in the spirit of them, we hold to be a great requisite, and we have had experience in the campaign of 1806 how far it is sometimes wanting. In that tremendous tension, when everything pressed on towards a supreme decision, and that alone with all its consequences should have occupied the whole soul of the Commander, measures were proposed and even partly carried out (such as the reconnaissance towards Franconia), which at the most might have given a kind of gentle play of oscillation within a state of equilibrium. Over these blundering schemes and views, absorbing the activity of the Army, the really necessary means, which could alone save, were lost sight of.

But this speculative distinction which we have made is also necessary for our further progress in the construction of our theory, because all that we have to say on the relation of attack and defence, and on the completion of this double-sided act, concerns the state of the crisis in which the forces are placed during the tension and motion, and because all the activity which can take place during the condition of equilibrium can only be regarded and treated as a corollary; for that crisis is the real War and this state of equilibrium only its reflection.

Introductory

Having in the foregoing book examined the subjects which may be regarded as the efficient elements of War, we shall now turn our attention to the combat as the real activity in Warfare, which, by its physical and moral effects, embraces sometimes more simply, sometimes in a more complex manner, the object of the whole campaign. In this activity and in its effects these elements must therefore, reappear.

The formation of the combat is tactical in its nature; we only glance at it here in a general way in order to get acquainted with it in its aspect as a whole. In practice the minor or more immediate objects give every combat a characteristic form; these minor objects we shall not discuss until hereafter. But these peculiarities are in comparison to the general characteristics of a combat mostly only insignificant, so that most combats are very like one another, and, therefore, in order to avoid repeating that which is general at every stage, we are compelled to look into it here, before taking up the subject of its more special application.

In the first place, therefore, we shall give in the next chapter, in a few words, the characteristics of the modern battle in its tactical course, because that lies at the foundation of our conceptions of what the battle really is.

Character of the Modern Battle

According to the notion we have formed of tactics and strategy, it follows, as a matter of course, that if the nature of the former is changed, that change must have an influence on the latter. If tactical facts in one case are entirely different from those in another, then the strategic, must be so also, if they are to continue consistent and reasonable. It is therefore important to characterise a general action in its modern form before we advance with the study of its employment in strategy.

What do we do now usually in a great battle? We place ourselves quietly in great masses arranged contiguous to and behind one another. We deploy relatively only a small portion of the whole, and let it wring itself out in a fire-combat which lasts for several hours, only interrupted now and again, and removed hither and thither by separate small shocks from charges with the bayonet and cavalry attacks. When this line has gradually exhausted part of its warlike ardour in this manner and there remains nothing more than the cinders, it is withdrawn [The relief of the fighting line played a great part in the battles of the Smooth-Bore era; it was necessitated by the fouling of the muskets, physical fatigue of the men and consumption of ammunition, and was recognised as both necessary and advisable by Napoleon himself.—Editor.] and replaced by another.

In this manner the battle on a modified principle burns slowly away like wet powder, and if the veil of night commands it to stop, because neither party can any longer see, and neither chooses to run the risk of blind chance, then an account is taken by each side respectively of the masses remaining, which can be called still effective, that is, which have not yet quite collapsed like extinct volcanoes; account is taken of the ground gained or lost, and of how stands the security of the rear; these results with the special impressions as to bravery and cowardice, ability and stupidity, which are thought to have been observed in ourselves and in the enemy are collected into one single total impression, out of which there springs the resolution to quit the field or to renew the combat on the morrow.

This description, which is not intended as a finished picture of a modern battle, but only to give its general tone, suits for the offensive and defensive, and the special traits which are given, by the object proposed, the country, etc. etc., may be introduced into it, without materially altering the conception.

But modern battles are not so by accident; they are so because the parties find themselves nearly on a level as regards military organisation and the knowledge of the Art of War, and because the warlike element inflamed by great national interests has broken through artificial limits and now flows in its natural channel. Under these two conditions, battles will always preserve this character.

This general idea of the modern battle will be useful to us in the sequel in more places than one, if we want to estimate the value of the particular co-efficients of strength, country, etc. etc. It is only for general, great, and decisive combats, and such as come near to them that this description stands good; inferior ones have changed their character also in the same direction but less than great ones. The proof of this belongs to tactics; we shall, however, have an opportunity hereafter of making this subject plainer by giving a few particulars.

The Combat in General

The Combat is the real warlike activity, everything else is only its auxiliary; let us therefore take an attentive look at its nature.

Combat means fighting, and in this the destruction or conquest of the enemy is the object, and the enemy, in the particular combat, is the armed force which stands opposed to us.

This is the simple idea; we shall return to it, but before we can do that we must insert a series of others.

If we suppose the State and its military force as a unit, then the most natural idea is to imagine the War also as one great combat, and in the simple relations of savage nations it is also not much otherwise. But our Wars are made up of a number of great and small simultaneous or consecutive combats, and this severance of the activity into so many separate actions is owing to the great multiplicity of the relations out of which War arises with us.

In point of fact, the ultimate object of our Wars, the political one, is not always quite a simple one; and even were it so, still the action is bound up with such a number of conditions and considerations to be taken into account, that the object can no longer be attained by one single great act but only through a number of greater or smaller acts which are bound up into a whole; each of these separate acts is therefore a part of a whole, and has consequently a special object by which it is bound to this whole.

We have already said that every strategic act can be referred to the idea of a combat, because it is an employment of the military force, and at the root of that there always lies the idea of fighting. We may therefore reduce every military activity in the province of Strategy to the unit of single combats, and occupy ourselves with the object of these only; we shall get acquainted with these special objects by degrees as we come to speak of the causes which produce them; here we content ourselves with saying that every combat, great or small, has its own peculiar object in subordination to the main object. If this is the case then, the destruction and conquest of the enemy is only to be regarded as the means of gaining this object; as it unquestionably is.

But this result is true only in its form, and important only on account of the connection which the ideas have between themselves, and we have only sought it out to get rid of it at once.

What is overcoming the enemy? Invariably the destruction of his military force, whether it be by death, or wounds, or any means; whether it be completely or only to such a degree that he can no longer continue the contest; therefore as long as we set aside all special objects of combats, we may look upon the complete or partial destruction of the enemy as the only object of all combats.

Now we maintain that in the majority of cases, and especially in great battles, the special object by which the battle is individualised and bound up with the great whole is only a weak modification of that general object, or an ancillary object bound up with it, important enough to individualise the battle, but always insignificant in comparison with that general object; so that if that ancillary object alone should be obtained, only an unimportant part of the purpose of the combat is fulfilled. If this assertion is correct, then we see that the idea, according to which the destruction of the enemy's force is only the means, and something else always the object, can only be true in form, but, that it would lead to false conclusions if we did not recollect that this destruction of the enemy's force is comprised in that object, and that this object is only a weak modification of it. Forgetfulness of this led to completely false views before the Wars of the last period, and created tendencies as well as fragments of systems, in which theory thought it raised itself so much the more above handicraft, the less it supposed itself to stand in need of the use of the real instrument, that is the destruction of the enemy's force.

Certainly such a system could not have arisen unless supported by other false suppositions, and unless in place of the destruction of the enemy, other things had been substituted to which an efficacy was ascribed which did not rightly belong to them. We shall attack these falsehoods whenever occasion requires, but we could not treat of the combat without claiming for it the real importance and value which belong to it, and giving warning against the errors to which merely formal truth might lead.

But now how shall we manage to show that in most cases, and in those of most importance, the destruction of the enemy's Army is the chief thing? How shall we manage to combat that extremely subtle idea, which supposes it possible, through the use of a special artificial form, to effect by a small direct destruction of the enemy's forces a much greater destruction indirectly, or by means of small but extremely well-directed blows to produce such paralysation of the enemy's forces, such a command over the enemy's will, that this mode of proceeding is to be viewed as a great shortening of the road? Undoubtedly a victory at one point may be of more value than at another. Undoubtedly there is a scientific arrangement of battles amongst themselves, even in Strategy, which is in fact nothing but the Art of thus arranging them. To deny that is not our intention, but we assert that the direct destruction of the enemy's forces is everywhere predominant; we contend here for the overruling importance of this destructive principle and nothing else.

We must, however, call to mind that we are now engaged with Strategy, not with tactics, therefore we do not speak of the means which the former may have of destroying at a small expense a large body of the enemy's forces, but under direct destruction we understand the tactical results, and that, therefore, our assertion is that only great tactical results can lead to great strategical ones, or, as we have already once before more distinctly expressed it, *the tactical successes* are of paramount importance in the conduct of War.

The proof of this assertion seems to us simple enough, it lies in the time which every complicated (artificial) combination requires. The question whether a simple attack, or one more carefully prepared, i.e., more artificial, will produce greater effects, may undoubtedly be decided in favour of the latter as long as the enemy is assumed to remain quite passive. But every carefully combined attack requires time for its preparation, and if a counter-stroke by the enemy intervenes, our whole design may be upset. Now if the enemy should decide upon some simple attack, which can be executed in a shorter time, then he gains the initiative, and destroys the effect of the great plan. Therefore, together with the expediency of a complicated attack we must consider all the dangers which we run during its preparation, and should only adopt it if there is no reason to fear that the enemy will disconcert our scheme. Whenever this is the case we must ourselves choose the simpler, i.e., quicker way, and lower our views in this sense as far as the character, the relations of the enemy, and other circumstances may render necessary. If we quit the weak impressions of abstract ideas and descend to the region of practical life, then it is evident that a bold, courageous, resolute enemy will not let us have time for wide-reaching skilful combinations, and it is just against such a one we should require skill the most. By this it appears to us that the advantage of simple and direct results over those that are complicated is conclusively shown.

Our opinion is not on that account that the simple blow is the best, but that we must not lift the arm too far for the time given to strike, and that this condition will always lead more to direct conflict the more warlike our opponent is. Therefore, far from making it our aim to gain upon the enemy by complicated plans, we must rather seek to be beforehand with him by greater simplicity in our designs.

If we seek for the lowest foundation-stones of these converse propositions we find that in the one it is ability, in the other, courage. Now, there is something very attractive in the notion that a moderate degree of courage joined to great ability will produce greater effects than moderate ability with great courage. But unless we suppose these elements in a disproportionate relation, not logical, we have no right to assign to ability this advantage over courage in a field which is called danger, and which must be regarded as the true domain of courage.

After this abstract view we shall only add that experience, very far from leading to a different conclusion, is rather the sole cause which has impelled us in this direction, and given rise to such reflections.

Whoever reads history with a mind free from prejudice cannot fail to arrive at a conviction that of all military virtues, energy in the conduct of operations has always contributed the most to the glory and success of arms.

How we make good our principle of regarding the destruction of the enemy's force as the principal object, not only in the War as a whole but also in each separate combat, and how that principle suits all the forms and conditions necessarily demanded by the relations out of which War springs, the sequel will show. For the present all that we desire is to uphold its general importance, and with this result we return again to the combat.

The Combat in General (Continuation)

In the last chapter we showed the destruction of the enemy as the true object of the combat, and we have sought to prove by a special consideration of the point, that this is true in the majority of cases, and in respect to the most important battles, because the destruction of the enemy's Army is always the preponderating object in War. The other objects which may be mixed up with this destruction of the enemy's force, and may have more or less influence, we shall describe generally in the next chapter, and become better acquainted with by degrees afterwards; here we divest the combat of them entirely, and look upon the destruction of the enemy as the complete and sufficient object of any combat.

What are we now to understand by destruction of the enemy's Army? A diminution of it relatively greater than that on our own side. If we have a great superiority in numbers over the enemy, then naturally the same absolute amount of loss on both sides is for us a smaller one than for him, and consequently may be regarded in itself as an advantage. As we are here considering the combat as divested of all (other) objects, we must also exclude from our consideration the case in which the combat is used only indirectly for a greater destruction of the enemy's force; consequently also, only that direct gain which has been made in the mutual process of destruction, is to be regarded as the object, for this is an absolute gain, which runs through the whole campaign, and at the end of it will always appear as pure profit. But every other kind of victory over our opponent will either have its motive in other objects, which we have completely excluded here, or it will only yield a temporary relative advantage. An example will make this plain.

If by a skilful disposition we have reduced our opponent to such a dilemma, that he cannot continue the combat without danger, and after some resistance he retires, then we may say, that we have conquered him at that point; but if in this victory we have expended just as many forces as the enemy, then in closing the account of the campaign, there is no gain remaining from this victory, if such a result can be called a victory. Therefore the overcoming the enemy, that is, placing him in such a position that he must give up the fight, counts for nothing in itself, and for that reason cannot come under the definition of object. There remains, therefore, as we have said, nothing over except the direct gain which we have made in the process of destruction; but to this belong not only the losses which have taken place in the course of the combat, but also those which, after the withdrawal of the conquered part, take place as direct consequences of the same.

Now it is known by experience, that the losses in physical forces in the course of a battle seldom present a great difference between victor and vanquished respectively, often none at all, sometimes even one bearing an inverse relation to the result, and that the most decisive losses on the side of the vanquished only commence with the retreat, that is, those which the conqueror does not share with him. The weak remains of battalions already in disorder are cut down by cavalry, exhausted men strew the ground, disabled guns and broken caissons are abandoned, others in the bad state of the roads cannot be removed quickly enough, and are captured by the enemy's troops, during the night numbers lose their way, and fall defenceless into the enemy's hands, and thus the victory mostly gains bodily substance after it is already decided. Here would be a paradox, if it did not solve itself in the following manner.

The loss in physical force is not the only one which the two sides suffer in the course of the combat; the moral forces also are shaken, broken, and go to ruin. It is not only the loss in men, horses and guns, but in order, courage, confidence, cohesion and plan, which come into consideration when it is a question whether the fight can be still continued or not. It is principally the moral forces which decide here, and in all cases in which the conqueror has lost as heavily as the conquered, it is these alone.

The comparative relation of the physical losses is difficult to estimate in a battle, but not so the relation of the moral ones. Two things principally make it known. The one is the loss of the ground on which the fight has taken place, the other the superiority of the enemy's. The more our reserves have diminished as compared with those of the enemy, the more force we have used to maintain the equilibrium; in this at once, an evident proof of the moral superiority of the enemy is given which seldom fails to stir up in the soul of the Commander a certain bitterness of feeling, and a sort of contempt for his own troops. But the principal thing is, that men who have been engaged for a long continuance of time are more or less like burnt-out cinders; their ammunition is consumed; they have melted away to a certain extent; physical and moral energies are exhausted, perhaps their courage is broken as well. Such a force, irrespective of the diminution in its number, if viewed as an organic whole, is very different from what it was before the combat; and thus it is that the loss of moral force may be measured by the reserves that have been used as if it were on a foot-rule.

Lost ground and want of fresh reserves, are, therefore, usually the principal causes which determine a retreat; but at the same time we by no means exclude or desire to throw in the shade other reasons, which may lie in the interdependence of parts of the Army, in the general plan, etc.

Every combat is therefore the bloody and destructive measuring of the strength of forces, physical and moral; whoever at the close has the greatest amount of both left is the conqueror.

In the combat the loss of moral force is the chief cause of the decision; after that is given, this loss continues to increase until it reaches its culminating-point at the close of the whole act. This then is the opportunity the victor should seize to reap his harvest by the utmost possible restrictions of his enemy's forces, the real object of engaging in the combat. On the beaten side, the loss of all order and control often makes the prolongation of resistance by individual units, by the further punishment they are certain to suffer, more injurious than useful to the whole. The spirit of the mass is broken; the original excitement about losing or winning, through which danger was forgotten, is spent, and to the majority danger now appears no longer an appeal to their courage, but rather the endurance of a cruel punishment. Thus the instrument in the first moment of the enemy's victory is weakened and blunted, and therefore no longer fit to repay danger by danger.

This period, however, passes; the moral forces of the conquered will recover by degrees, order will be restored, courage will revive, and in the majority of cases there remains only a small part of the superiority obtained, often none at all. In some cases, even, although rarely, the spirit of revenge and intensified hostility may bring about an opposite result. On the other hand, whatever is gained in killed, wounded, prisoners, and guns captured can never disappear from the account.

The losses in a battle consist more in killed and wounded; those after the battle, more in artillery taken and prisoners. The first the conqueror shares with the conquered, more or less, but the second not; and for that reason they usually only take place on one side of the conflict, at least, they are considerably in excess on one side.

Artillery and prisoners are therefore at all times regarded as the true trophies of victory, as well as its measure, because through these things its extent is declared beyond a doubt. Even the degree of moral superiority may be better judged of by them than by any other relation, especially if the number of killed and wounded is compared therewith; and here arises a new power increasing the moral effects.

We have said that the moral forces, beaten to the ground in the battle and in the immediately succeeding movements, recover themselves gradually, and often bear no traces of injury; this is the case with small divisions of the whole, less frequently with large divisions; it may, however, also be the case with the main Army, but seldom or never in the State or Government to which the Army belongs. These estimate the situation more impartially, and from a more elevated point of view, and recognise in the number of trophies taken by the enemy, and their relation to the number of killed and wounded, only too easily and well, the measure of their own weakness and inefficiency.

In point of fact, the lost balance of moral power must not be treated lightly because it has no absolute value, and because it does not of necessity appear in all cases in the amount of the results at the final close; it may become of such excessive weight as to bring down everything with an irresistible force. On that account it may often become a great aim of the operations of which we shall speak elsewhere. Here we have still to examine some of its fundamental relations.

The moral effect of a victory increases, not merely in proportion to the extent of the forces engaged, but in a progressive ratio—that is to say, not only in extent, but also in its intensity. In a beaten detachment order is easily restored. As a single frozen limb is easily revived by the rest of the body, so the courage of a defeated detachment is easily raised again by the courage of the rest of the Army as soon as it rejoins it. If, therefore, the effects of a small victory are not completely done away with, still they are partly lost to the enemy. This is not the case if the Army itself sustains a great defeat; then one with the other fall together. A great fire attains quite a different heat from several small ones.

Another relation which determines the moral value of a victory is the numerical relation of the forces which have been in conflict with each other. To beat many with few is not only a double success, but shows also a greater, especially a more general superiority, which the conquered must always be fearful of encountering again. At the same time this influence is in reality hardly observable in such a case. In the moment of real action, the notions of the actual strength of the enemy are generally so uncertain, the estimate of our own commonly so incorrect, that the party superior in numbers either does not admit the disproportion, or is very far from admitting the full truth, owing to which, he evades almost entirely the moral disadvantages which would spring from it. It is only hereafter in history that the truth, long suppressed through ignorance, vanity, or a wise discretion, makes its appearance, and then it certainly casts a lustre on the Army and its Leader, but it can then do nothing more by its moral influence for events long past.

If prisoners and captured guns are those things by which the victory principally gains substance, its true crystallisations, then the plan of the battle should have those things specially in view; the destruction of the enemy by death and wounds appears here merely as a means to an end.

How far this may influence the dispositions in the battle is not an affair of Strategy, but the decision to fight the battle is in intimate connection with it, as is shown by the direction given to our forces, and their general grouping, whether we threaten the enemy's flank or rear, or he threatens ours. On this point, the number of prisoners and captured guns depends very much, and it is a point which, in many cases, tactics alone cannot satisfy, particularly if the strategic relations are too much in opposition to it.

The risk of having to fight on two sides, and the still more dangerous position of having no line of retreat left open, paralyse the movements and the power of resistance; further, in case of defeat, they increase the loss, often raising it to its extreme point, that is, to destruction. Therefore, the rear being endangered makes defeat more probable, and, at the same time, more decisive.

From this arises, in the whole conduct of the War, especially in great and small combats, a perfect instinct to secure our own line of retreat and to seize that of the enemy; this follows from the conception of victory, which, as we have seen, is something beyond mere slaughter.

In this effort we see, therefore, the first immediate purpose in the combat, and one which is quite universal. No combat is imaginable in which this effort, either in its double or single form, does not go hand in hand with the plain and simple stroke of force. Even the smallest troop will not throw itself upon its enemy without thinking of its line of retreat, and, in most cases, it will have an eye upon that of the enemy also.

We should have to digress to show how often this instinct is prevented from going the direct road, how often it must yield to the difficulties arising from more important considerations: we shall, therefore, rest contented with affirming it to be a general natural law of the combat.

It is, therefore, active; presses everywhere with its natural weight, and so becomes the pivot on which almost all tactical and strategic manoeuvres turn.

If we now take a look at the conception of victory as a whole, we find in it three elements:—

1. The greater loss of the enemy in physical power.
2. In moral power.
3. His open avowal of this by the relinquishment of his intentions.

The returns made up on each side of losses in killed and wounded, are never exact, seldom truthful, and in most cases, full of intentional misrepresentations. Even the statement of the number of trophies is seldom to be quite depended on; consequently, when it is not considerable it may also cast a doubt even on the reality of the victory. Of the loss in moral forces there is no reliable measure, except in the trophies: therefore, in many cases, the giving up the contest is the only real evidence of the victory. It is, therefore, to be regarded as a confession of inferiority—as the lowering of the flag, by which, in this particular instance, right and superiority are conceded to the enemy, and this degree of humiliation and disgrace, which, however, must be distinguished from all the other moral consequences of the loss of equilibrium, is an essential part of the victory. It is this part alone which acts upon the public opinion outside the Army, upon the people and the Government in both belligerent States, and upon all others in any way concerned.

But renouncement of the general object is not quite identical with quitting the field of battle, even when the battle has been very obstinate and long kept up; no one says of advanced posts, when they retire after an obstinate combat, that they have given up their object; even in combats aimed at the destruction of the enemy's Army, the retreat from the battlefield is not always to be regarded as a relinquishment of this aim, as for instance, in retreats planned beforehand, in which the ground is disputed foot by foot; all this belongs to that part of our subject where we shall speak of the separate object of the combat; here we only wish to draw attention to the fact that in most cases the giving up of the object is very difficult to distinguish from the retirement from the battlefield, and that the impression produced by the latter, both in and out of the Army, is not to be treated lightly.

For Generals and Armies whose reputation is not made, this is in itself one of the difficulties in many operations, justified by circumstances when a succession of combats, each ending in retreat, may appear as a succession of defeats, without being so in reality, and when that appearance may exercise a very depressing influence. It is impossible for the retreating General by making known his real intentions to prevent the moral effect spreading to the public and his troops, for to do that with effect he must disclose his plans completely, which of course would run counter to his principal interests to too great a degree.

In order to draw attention to the special importance of this conception of victory we shall only refer to the battle of Soor, the trophies from which were not important (a few thousand prisoners and twenty guns), and where Frederick proclaimed his victory by remaining for five days after on the field of battle, although his retreat into Silesia had been previously determined on, and was a measure natural to his whole situation. According to his own account, he thought he would hasten a peace by the moral effect of his victory. Now although a couple of other successes were likewise required, namely, the battle at Katholisch Hennersdorf, in Lusatia, and the battle of Kesseldorf, before this peace took place, still we cannot say that the moral effect of the battle of Soor was nil.

If it is chiefly the moral force which is shaken by defeat, and if the number of trophies reaped by the enemy mounts up to an unusual height, then the lost combat becomes a rout, but this is not the necessary consequence of every victory. A rout only sets in when the moral force of the defeated is very severely shaken then there often ensues a complete incapability of further resistance, and the whole action consists of giving way, that is of flight.

Jena and Belle Alliance were routs, but not so Borodino.

Although without pedantry we can here give no single line of separation, because the difference between the things is one of degrees, yet still the retention of the conception is essential as a central point to give clearness to our theoretical ideas and it is a want in our terminology that for a victory over the enemy tantamount to a rout, and a conquest of the enemy only tantamount to a simple victory, there is only one and the same word to use.

On the Signification of the Combat

Having in the preceding chapter examined the combat in its absolute form, as the miniature picture of the whole War, we now turn to the relations which it bears to the other parts of the great whole. First we inquire what is more precisely the signification of a combat.

As War is nothing else but a mutual process of destruction, then the most natural answer in conception, and perhaps also in reality, appears to be that all the powers of each party unite in one great volume and all results in one great shock of these masses. There is certainly much truth in this idea, and it seems to be very advisable that we should adhere to it and should on that account look upon small combats at first only as necessary loss, like the shavings from a carpenter's plane. Still, however, the thing cannot be settled so easily.

That a multiplication of combats should arise from a fractioning of forces is a matter of course, and the more immediate objects of separate combats will therefore come before us in the subject of a fractioning of forces; but these objects, and together with them, the whole mass of combats may in a general way be brought under certain classes, and the knowledge of these classes will contribute to make our observations more intelligible.

Destruction of the enemy's military forces is in reality the object of all combats; but other objects may be joined thereto, and these other objects may be at the same time predominant; we must therefore draw a distinction between those in which the destruction of the enemy's force is the principal object, and those in which it is more the means. The destruction of the enemy's force, the possession of a place or the possession of some object may be the general motive for a combat, and it may be either one of these alone or several together, in which case however usually one is the principal motive. Now the two principal forms of War, the offensive and defensive, of which we shall shortly speak, do not modify the first of these motives, but they certainly do modify the other two, and therefore if we arrange them in a scheme they would appear thus:— offensive. Defensive. 1. Destruction of enemy's force 1. Destruction of enemy's force. 2. Conquest of a place. 2. Defence of a place. 3. Conquest of some object. 3. Defence of some object.

These motives, however, do not seem to embrace completely the whole of the subject, if we recollect that there are reconnaissances and demonstrations, in which plainly none of these three points is the object of the combat. In reality we must, therefore, on this account be allowed a fourth class. Strictly speaking, in reconnaissances in which we wish the enemy to show himself, in alarms by which we wish to wear him out, in demonstrations by which we wish to prevent his leaving some point or to draw him off to another, the objects are all such as can only be attained indirectly and *under the pretext of one of the three objects specified in the table*, usually of the second; for the enemy whose aim is to reconnoitre must draw up his force as if he really intended to attack and defeat us, or drive us off, etc. etc. But this pretended object is not the real one, and our present question is only as to the latter; therefore, we must to the above three objects of the offensive further add a fourth, which is to lead the enemy to make a false conclusion. That offensive means are conceivable in connection with this object, lies in the nature of the thing.

On the other hand we must observe that the defence of a place may be of two kinds, either absolute, if as a general question the point is not to be given up, or relative if it is only required for a certain time. The latter happens perpetually in the combats of advanced posts and rear guards.

That the nature of these different intentions of a combat must have an essential influence on the dispositions which are its preliminaries, is a thing clear in itself. We act differently if our object is merely to drive an enemy's post out of its place from what we should if our object was to beat him completely; differently, if we mean to defend a place to the last extremity from what we should do if our design is only to detain the enemy for a certain time. In the first case we trouble ourselves little about the line of retreat, in the latter it is the principal point, etc.

But these reflections belong properly to tactics, and are only introduced here by way of example for the sake of greater clearness. What Strategy has to say on the different objects of the combat will appear in the chapters which touch upon these objects. Here we have only a few general observations to make, first, that the importance of the object decreases nearly in the order as they stand above, therefore, that the first of these objects must always predominate in the great battle; lastly, that the two last in a defensive battle are in reality such as yield no fruit, they are, that is to say, purely negative, and can, therefore, only be serviceable, indirectly, by facilitating something else which is positive. *It is, therefore, a bad sign of the strategic situation if battles of this kind become too frequent.*

Duration of the Combat

If we consider the combat no longer in itself but in relation to the other forces of War, then its duration acquires a special importance.

This duration is to be regarded to a certain extent as a second subordinate success. For the conqueror the combat can never be finished too quickly, for the vanquished it can never last too long. A speedy victory indicates a higher power of victory, a tardy decision is, on the side of the defeated, some compensation for the loss.

This is in general true, but it acquires a practical importance in its application to those combats, the object of which is a relative defence.

Here the whole success often lies in the mere duration. This is the reason why we have included it amongst the strategic elements.

The duration of a combat is necessarily bound up with its essential relations. These relations are, absolute magnitude of force, relation of force and of the different arms mutually, and nature of the country. Twenty thousand men do not wear themselves out upon one another as quickly as two thousand: we cannot resist an enemy double or three times our strength as long as one of the same strength; a cavalry combat is decided sooner than an infantry combat; and a combat between infantry only, quicker than if there is artillery [The increase in the relative range of artillery and the introduction of shrapnel has altogether modified this conclusion.] as well; in hills and forests we cannot advance as quickly as on a level country; all this is clear enough.

From this it follows, therefore, that strength, relation of the three arms, and position, must be considered if the combat is to fulfil an object by its duration; but to set up this rule was of less importance to us in our present considerations than to connect with it at once the chief results which experience gives us on the subject.

Even the resistance of an ordinary Division of 8000 to 10,000 men of all arms even opposed to an enemy considerably superior in numbers, will last several hours, if the advantages of country are not too preponderating, and if the enemy is only a little, or not at all, superior in numbers, the combat will last half a day. A Corps of three or four Divisions will prolong it to double the time; an Army of 80,000 or 100,000 to three or four times. Therefore the masses may be left to themselves for that length of time, and no separate combat takes place if within that time other forces can be brought up, whose co-operation mingles then at once into one stream with the results of the combat which has taken place.

These calculations are the result of experience; but it is important to us at the same time to characterise more particularly the moment of the decision, and consequently the termination.

Decision of the Combat

No battle is decided in a single moment, although in every battle there arise moments of crisis, on which the result depends. The loss of a battle is, therefore, a gradual falling of the scale. But there is in every combat a point of time [Under the then existing conditions of armament understood. This point is of supreme importance, as practically the whole conduct of a great battle depends on a correct solution of this question—viz., How long can a given command prolong its resistance? If this is incorrectly answered in practice—the whole manoeuvre depending on it may collapse—e.g., Kouroupatkin at Liao-Yang, September 1904.]

When it may be regarded as decided, in such a way that the renewal of the fight would be a new battle, not a continuation of the old one. To have a clear notion on this point of time, is very important, in order to be able to decide whether, with the prompt assistance of reinforcements, the combat can again be resumed with advantage.

Often in combats which are beyond restoration new forces are sacrificed in vain; often through neglect the decision has not been seized when it might easily have been secured. Here are two examples, which could not be more to the point:

When the Prince of Hohenlohe, in 1806, at Jena, [October 14, 1806.] with 35,000 men opposed to from 60,000 to 70,000, under Buonaparte, had accepted battle, and lost it—but lost it in such a way that the 35,000 might be regarded as dissolved—General Ruchel undertook to renew the fight with about 12,000; the consequence was that in a moment his force was scattered in like manner.

On the other hand, on the same day at Auerstadt, the Prussians maintained a combat with 25,000, against Davoust, who had 28,000, until mid-day, without success, it is true, but still without the force being reduced to a state of dissolution without even greater loss than the enemy, who was very deficient in cavalry;—but they neglected to use the reserve of 18,000, under General Kalkreuth, to restore the battle which, under these circumstances, it would have been impossible to lose.

Each combat is a whole in which the partial combats combine themselves into one total result. In this total result lies the decision of the combat. This success need not be exactly a victory such as we have denoted in the sixth chapter, for often the preparations for that have not been made, often there is no opportunity if the enemy gives way too soon, and in most cases the decision, even when the resistance has been obstinate, takes place before such a degree of success is attained as would completely satisfy the idea of a victory.

We therefore ask, Which is commonly the moment of the decision, that is to say, that moment when a fresh, effective, of course not disproportionate, force, can no longer turn a disadvantageous battle?

If we pass over false attacks, which in accordance with their nature are properly without decision, then,

1. If the possession of a movable object was the object of the combat, the loss of the same is always the decision.

2. If the possession of ground was the object of the combat, then the decision generally lies in its loss. Still not always, only if this ground is of peculiar strength, ground which is easy to pass over, however important it may be in other respects, can be re-taken without much danger.

3. But in all other cases, when these two circumstances have not already decided the combat, therefore, particularly in case the destruction of the enemy's force is the principal object, the decision is reached at that moment when the conqueror ceases to feel himself in a state of disintegration, that is, of unserviceableness to a certain extent, when therefore, there is no further advantage in using the successive efforts spoken of in the twelfth chapter of the third book. On this ground we have given the strategic unity of the battle its place here.

A battle, therefore, in which the assailant has not lost his condition of order and perfect efficiency at all, or, at least, only in a small part of his force, whilst the opposing forces are, more or less, disorganised throughout, is also not to be retrieved; and just as little if the enemy has recovered his efficiency.

The smaller, therefore, that part of a force is which has really been engaged, the greater that portion which as reserve has contributed to the result only by its presence. So much the less will any new force of the enemy wrest again the victory from our hands, and that Commander who carries out to the furthest with his Army the principle of conducting the combat with the greatest economy of forces, and making the most of the moral effect of strong reserves, goes the surest way to victory. We must allow that the French, in modern times, especially when led by Buonaparte, have shown a thorough mastery in this.

Further, the moment when the crisis-stage of the combat ceases with the conqueror, and his original state of order is restored, takes place sooner the smaller the unit he controls. A picket of cavalry pursuing an enemy at full gallop will in a few minutes resume its proper order, and the crisis ceases. A whole regiment of cavalry requires a longer time. It lasts still longer with infantry, if extended in single lines of skirmishers, and longer again with Divisions of all arms, when it happens by chance that one part has taken one direction and another part another direction, and the combat has therefore caused a loss of the order of formation, which usually becomes still worse from no part knowing exactly where the other is. Thus, therefore, the point of time when the conqueror has collected the instruments he has been using, and which are mixed up and partly out of order, the moment when he has in some measure rearranged them and put them in their proper places, and thus brought the battle-workshop into a little order, this moment, we say, is always later, the greater the total force.

Again, this moment comes later if night overtakes the conqueror in the crisis, and, lastly, it comes later still if the country is broken and thickly wooded. But with regard to these two points, we must observe that night is also a great means of protection, and it is only seldom that circumstances favour the expectation of a successful result from a night attack, as on March 10, 1814, at Laon, [The celebrated charge at night upon Marmont's Corps.] where York against Marmont gives us an example completely in place here. In the same way a wooded and broken country will afford protection against a reaction to those who are engaged in the long crisis of victory. Both, therefore, the night as well as the wooded and broken country are obstacles which make the renewal of the same battle more difficult instead of facilitating it.

Hitherto, we have considered assistance arriving for the losing side as a mere increase of force, therefore, as a reinforcement coming up directly from the rear, which is the most usual case. But the case is quite different if these fresh forces come upon the enemy in flank or rear.

On the effect of flank or rear attacks so far as they belong to Strategy, we shall speak in another place: such a one as we have here in view, intended for the restoration of the combat, belongs chiefly to tactics, and is only mentioned because we are here speaking of tactical results, our ideas, therefore, must trench upon the province of tactics.

By directing a force against the enemy's flank and rear its efficacy may be much intensified; but this is so far from being a necessary result always that the efficacy may, on the other hand, be just as much weakened. The circumstances under which the combat has taken place decide upon this part of the plan as well as upon every other, without our being able to enter thereupon here. But, at the same time, there are in it two things of importance for our subject: first, *flank and rear attacks have, as a rule, a more favourable effect on the consequences of the decision than upon the decision itself.*

Now as concerns the retrieving a battle, the first thing to be arrived at above all is a favourable decision and not magnitude of success. In this view one would therefore think that a force which comes to re-establish our combat is of less assistance if it falls upon the enemy in flank and rear, therefore separated from us, than if it joins itself to us directly; certainly, cases are not wanting where it is so, but we must say that the majority are on the other side, and they are so on account of the second point which is here important to us.

This second point *is the moral effect of the surprise, which, as a rule, a reinforcement coming up to re-establish a combat has generally in its favour.* Now the effect of a surprise is always heightened if it takes place in the flank or rear, and an enemy completely engaged in the crisis of victory in his extended and scattered order, is less in a state to counteract it. Who does not feel that an attack in flank or rear, which at the commencement of the battle, when the forces are concentrated and prepared for such an event would be of little importance, gains quite another weight in the last moment of the combat.

We must, therefore, at once admit that in most cases a reinforcement coming up on the flank or rear of the enemy will be more efficacious, will be like the same weight at the end of a longer lever, and therefore that under these circumstances, we may undertake to restore the battle with the same force which employed in a direct attack would be quite insufficient. Here results almost defy calculation, because the moral forces gain completely the ascendency. This is therefore the right field for boldness and daring.

The eye must, therefore, be directed on all these objects, all these moments of co-operating forces must be taken into consideration, when we have to decide in doubtful cases whether or not it is still possible to restore a combat which has taken an unfavourable turn.

If the combat is to be regarded as not yet ended, then the new contest which is opened by the arrival of assistance fuses into the former; therefore they flow together into one common result, and the first disadvantage vanishes completely out of the calculation. But this is not the case if the combat was already decided; then there are two results separate from each other. Now if the assistance which arrives is only of a relative strength, that is, if it is not in itself alone a match for the enemy, then a favourable result is hardly to be expected from this second combat: but if it is so strong that it can undertake the second combat without regard to the first, then it may be able by a favourable issue to compensate or even overbalance the first combat, but never to make it disappear altogether from the account.

At the battle of Kunersdorf, [August 12, 1759.] Frederick the Great at the first onset carried the left of the Russian position, and took seventy pieces of artillery; at the end of the battle both were lost again, and the whole result of the first combat was wiped out of the account. Had it been possible to stop at the first success, and to put off the second part of the battle to the coming day, then, even if the King had lost it, the advantages of the first would always have been a set off to the second.

But when a battle proceeding disadvantageously is arrested and turned before its conclusion, its minus result on our side not only disappears from the account, but also becomes the foundation of a greater victory. If, for instance, we picture to ourselves exactly the tactical course of the battle, we may easily see that until it is finally concluded all successes in partial combats are only decisions in suspense, which by the capital decision may not only be destroyed, but changed into the opposite. The more our forces have suffered, the more the enemy will have expended on his side; the greater, therefore, will be the crisis for the enemy, and the more the superiority of our fresh troops will tell. If now the total result turns in our favour, if we wrest from the enemy the field of battle and recover all the trophies again, then all the forces which he has sacrificed in obtaining them become sheer gain for us, and our former defeat becomes a stepping-stone to a greater triumph. The most brilliant feats which with victory the enemy would have so highly prized that the loss of forces which they cost would have been disregarded, leave nothing now behind but regret at the sacrifice entailed. Such is the alteration which the magic of victory and the curse of defeat produces in the specific weight of the same elements.

Therefore, even if we are decidedly superior in strength, and are able to repay the enemy his victory by a greater still, it is always better to forestall the conclusion of a disadvantageous combat, if it is of proportionate importance, so as to turn its course rather than to deliver a second battle.

Field-Marshal Daun attempted in the year 1760 to come to the assistance of General Laudon at Leignitz, whilst the battle lasted; but when he failed, he did not attack the King next day, although he did not want for means to do so.

For these reasons serious combats of advance guards which precede a battle are to be looked upon only as necessary evils, and when not necessary they are to be avoided. [This, however, was not Napoleon's view. A vigorous attack of his advance guard he held to be necessary always, to fix the enemy's attention and "paralyse his independent will-power." It was the failure to make this point which, in August 1870, led von Moltke repeatedly into the very jaws of defeat, from which only the lethargy of Bazaine on the one hand and the initiative of his subordinates, notably of von Alvensleben, rescued him. This is the essence of the new Strategic Doctrine of the French General Staff. See the works of Bonnal, Foch, etc.—Editor]

We have still another conclusion to examine.

If on a regular pitched battle, the decision has gone against one, this does not constitute a motive for determining on a new one. The determination for this new one must proceed from other relations. This conclusion, however, is opposed by a moral force, which we must take into account: it is the feeling of rage and revenge. From the oldest Field-Marshal to the youngest drummer-boy this feeling is general, and, therefore, troops are never in better spirits for fighting than when they have to wipe out a stain. This is, however, only on the supposition that the beaten portion is not too great in proportion to the whole, because otherwise the above feeling is lost in that of powerlessness.

There is therefore a very natural tendency to use this moral force to repair the disaster on the spot, and on that account chiefly to seek another battle if other circumstances permit. It then lies in the nature of the case that this second battle must be an offensive one.

In the catalogue of battles of second-rate importance there are many examples to be found of such retaliatory battles; but great battles have generally too many other determining causes to be brought on by this weaker motive.

Such a feeling must undoubtedly have led the noble Bluecher with his third Corps to the field of battle on February 14, 1814, when the other two had been beaten three days before at Montmirail. Had he known that he would have come upon Buonaparte in person, then, naturally, preponderating reasons would have determined him to put off his revenge to another day: but he hoped to revenge himself on Marmont, and instead of gaining the reward of his desire for honourable satisfaction, he suffered the penalty of his erroneous calculation.

On the duration of the combat and the moment of its decision depend the distances from each other at which those masses should be placed which are intended to fight *in conjunction with* each other. This disposition would be a tactical arrangement in so far as it relates to one and the same battle; it can, however, only be regarded as such, provided the position of the troops is so compact that two separate combats cannot be imagined, and consequently that the space which the whole occupies can be regarded strategically as a mere point. But in War, cases frequently occur where even those forces intended to fight *In unison* must be so far separated from each other that while their union for one common combat certainly remains the principal object, still the occurrence of separate combats remains possible. Such a disposition is therefore strategic.

Dispositions of this kind are: marches in separate masses and columns, the formation of advance guards, and flanking columns, also the grouping of reserves intended to serve as supports for more than one strategic point; the concentration of several Corps from widely extended cantonments, etc. etc. We can see that the necessity for these arrangements may constantly arise, and may consider them something like the small change in the strategic economy, whilst the capital battles, and all that rank with them are the gold and silver pieces.

Mutual Understanding as to a Battle

No battle can take place unless by mutual consent; and in this idea, which constitutes the whole basis of a duel, is the root of a certain phraseology used by historical writers, which leads to many indefinite and false conceptions.

According to the view of the writers to whom we refer, it has frequently happened that one Commander has offered battle to the other, and the latter has not accepted it.

But the battle is a very modified duel, and its foundation is not merely in the mutual wish to fight, that is in consent, but in the objects which are bound up with the battle: these belong always to a greater whole, and that so much the more, as even the whole war considered as a "combat-unit" has political objects and conditions which belong to a higher standpoint. The mere desire to conquer each other therefore falls into quite a subordinate relation, or rather it ceases completely to be anything of itself, and only becomes the nerve which conveys the impulse of action from the higher will.

Amongst the ancients, and then again during the early period of standing Armies, the expression that we had offered battle to the enemy in vain, had more sense in it than it has now. By the ancients everything was constituted with a view to measuring each other's strength in the open field free from anything in the nature of a hindrance, [Note the custom of sending formal challenges, fix time and place for action, and "enhazelug" the battlefield in Anglo-Saxon times.—Ed.] and the whole Art of War consisted in the organisation, and formation of the Army, that is in the order of battle.

Now as their Armies regularly entrenched themselves in their camps, therefore the position in a camp was regarded as something unassailable, and a battle did not become possible until the enemy left his camp, and placed himself in a practicable country, as it were entered the lists.

If therefore we hear about Hannibal having offered battle to Fabius in vain, that tells us nothing more as regards the latter than that a battle was not part of his plan, and in itself neither proves the physical nor moral superiority of Hannibal; but with respect to him the expression is still correct enough in the sense that Hannibal really wished a battle.

In the early period of modern Armies, the relations were similar in great combats and battles. That is to say, great masses were brought into action, and managed throughout it by means of an order of battle, which like a great helpless whole required a more or less level plain and was neither suited to attack, nor yet to defence in a broken, close or even mountainous country. The defender therefore had here also to some extent the means of avoiding battle. These relations although gradually becoming modified, continued until the first Silesian War, and it was not until the Seven Years' War that attacks on an enemy posted in a difficult country gradually became feasible, and of ordinary occurrence: ground did not certainly cease to be a principle of strength to those making use of its aid, but it was no longer a charmed circle, which shut out the natural forces of War.

During the past thirty years War has perfected itself much more in this respect, and there is no longer anything which stands in the way of a General who is in earnest about a decision by means of battle; he can seek out his enemy, and attack him: if he does not do so he cannot take credit for having wished to fight, and the expression he offered a battle which his opponent did not accept, therefore now means nothing more than that he did not find circumstances advantageous enough for a battle, an admission which the above expression does not suit, but which it only strives to throw a veil over.

It is true the defensive side can no longer refuse a battle, yet he may still avoid it by giving up his position, and the role with which that position was connected: this is however half a victory for the offensive side, and an acknowledgment of his superiority for the present.

This idea in connection with the cartel of defiance can therefore no longer be made use of in order by such rhodomontade to qualify the inaction of him whose part it is to advance, that is, the offensive. The defender who as long as he does not give way, must have the credit of willing the battle, may certainly say, he has offered it if he is not attacked, if that is not understood of itself.

But on the other hand, he who now wishes to, and can retreat cannot easily be forced to give battle. Now as the advantages to the aggressor from this retreat are often not sufficient, and a substantial victory is a matter of urgent necessity for him, in that way the few means which there are to compel such an opponent also to give battle are often sought for and applied with particular skill.

The principal means for this are—first *surrounding* the enemy so as to make his retreat impossible, or at least so difficult that it is better for him to accept battle; and, secondly, *surprising* him. This last way, for which there was a motive formerly in the extreme difficulty of all movements, has become in modern times very inefficacious.

From the pliability and manoeuvring capabilities of troops in the present day, one does not hesitate to commence a retreat even in sight of the enemy, and only some special obstacles in the nature of the country can cause serious difficulties in the operation.

As an example of this kind the battle of Neresheim may be given, fought by the Archduke Charles with Moreau in the Rauhe Alp, August 11, 1796, merely with a view to facilitate his retreat, although we freely confess we have never been able quite to understand the argument of the renowned general and author himself in this case.

The battle of Rosbach [November 5, 1757.] is another example, if we suppose the commander of the allied army had not really the intention of attacking Frederick the Great.

Of the battle of Soor, [Or Sohr, September 30, 1745.] the King himself says that it was only fought because a retreat in the presence of the enemy appeared to him a critical operation; at the same time the King has also given other reasons for the battle.

On the whole, regular night surprises excepted, such cases will always be of rare occurrence, and those in which an enemy is compelled to fight by being practically surrounded, will happen mostly to single corps only, like Mortier's at Durrenstein 1809, and Vandamme at Kulm, 1813.

The Battle

[Clausewitz still uses the word "die Hauptschlacht" but modern usage employs only the word "die Schlacht" to designate the decisive act of a whole campaign—encounters arising from the collision or troops marching towards the strategic culmination of each portion or the campaign are spoken of either as "Treffen," i.e., "engagements" or "Gefecht," i.e., "combat" or "action." Thus technically, Gravelotte was a "Schlacht," i.e., "battle," but Spicheren, Woerth, Borny, even Vionville were only "Treffen."]

Its Decision

What is a battle? A conflict of the main body, but not an unimportant one about a secondary object, not a mere attempt which is given up when we see betimes that our object is hardly within our reach: it is a conflict waged with all our forces for the attainment of a decisive victory.

Minor objects may also be mixed up with the principal object, and it will take many different tones of colour from the circumstances out of which it originates, for a battle belongs also to a greater whole of which it is only a part, but because the essence of War is conflict, and the battle is the conflict of the main Armies, it is always to be regarded as the real centre of gravity of the War, and therefore its distinguishing character is, that unlike all other encounters, it is arranged for, and undertaken with the sole purpose of obtaining a decisive victory.

This has an influence on the *manner of its decision*, on the *Effect of the victory contained in it*, and determines *the value which theory is to assign to it as a means to an end.*

On that account we make it the subject of our special consideration, and at this stage before we enter upon the special ends which may be bound up with it, but which do not essentially alter its character if it really deserves to be termed a battle.

If a battle takes place principally on its own account, the elements of its decision must be contained in itself; in other words, victory must be striven for as long as a possibility or hope remains. It must not, therefore, be given up on account of secondary circumstances, but only and alone in the event of the forces appearing completely insufficient.

Now how is that precise moment to be described?

If a certain artificial formation and cohesion of an Army is the principal condition under which the bravery of the troops can gain a victory, as was the case during a great part of the period of the modern Art of War, *then the breaking up of this formation* is the decision. A beaten wing which is put out of joint decides the fate of all that was connected with it. If as was the case at another time the essence of the defence consists in an intimate alliance of the Army with the ground on which it fights and its obstacles, so that Army and position are only one, then the *conquest of an essential point* in this position is the decision. It is said the key of the position is lost, it cannot therefore be defended any further; the battle cannot be continued. In both cases the beaten Armies are very much like the broken strings of an instrument which cannot do their work.

That geometrical as well as this geographical principle which had a tendency to place an Army in a state of crystallising tension which did not allow of the available powers being made use of up to the last man, have at least so far lost their influence that they no longer predominate. Armies are still led into battle in a certain order, but that order is no longer of decisive importance; obstacles of ground are also still turned to account to strengthen a position, but they are no longer the only support.

We attempted in the second chapter of this book to take a general view of the nature of the modern battle. According to our conception of it, the order of battle is only a disposition of the forces suitable to the convenient use of them, and the course of the battle a mutual slow wearing away of these forces upon one another, to see which will have soonest exhausted his adversary.

The resolution therefore to give up the fight arises, in a battle more than in any other combat, from the relation of the fresh reserves remaining available; for only these still retain all their moral vigour, and the cinders of the battered, knocked-about battalions, already burnt out in the destroying element, must not be placed on a level with them; also lost ground as we have elsewhere said, is a standard of lost moral force; it therefore comes also into account, but more as a sign of loss suffered than for the loss itself, and the number of fresh reserves is always the chief point to be looked at by both Commanders.

In general, an action inclines in one direction from the very commencement, but in a manner little observable. This direction is also frequently given in a very decided manner by the arrangements which have been made previously, and then it shows a want of discernment in that General who commences battle under these unfavourable circumstances without being aware of them. Even when this does not occur it lies in the nature of things that the course of a battle resembles rather a slow disturbance of equilibrium which commences soon, but as we have said almost imperceptibly at first, and then with each moment of time becomes stronger and more visible, than an oscillating to and fro, as those who are misled by mendacious descriptions usually suppose.

But whether it happens that the balance is for a long time little disturbed, or that even after it has been lost on one side it rights itself again, and is then lost on the other side, it is certain at all events that in most instances the defeated General foresees his fate long before he retreats, and that cases in which some critical event acts with unexpected force upon the course of the whole have their existence mostly in the colouring with which every one depicts his lost battle.

We can only here appeal to the decision of unprejudiced men of experience, who will, we are sure, assent to what we have said, and answer for us to such of our readers as do not know War from their own experience. To develop the necessity of this course from the nature of the thing would lead us too far into the province of tactics, to which this branch of the subject belongs; we are here only concerned with its results.

If we say that the defeated General foresees the unfavourable result usually some time before he makes up his mind to give up the battle, we admit that there are also instances to the contrary, because otherwise we should maintain a proposition contradictory in itself. If at the moment of each decisive tendency of a battle it should be considered as lost, then also no further forces should be used to give it a turn, and consequently this decisive tendency could not precede the retreat by any length of time. Certainly there are instances of battles which after having taken a decided turn to one side have still ended in favour of the other; but they are rare, not usual; these exceptional cases, however, are reckoned upon by every General against whom fortune declares itself, and he must reckon upon them as long as there remains a possibility of a turn of fortune. He hopes by stronger efforts, by raising the remaining moral forces, by surpassing himself, or also by some fortunate chance that the next moment will bring a change, and pursues this as far as his courage and his judgment can agree. We shall have something more to say on this subject, but before that we must show what are the signs of the scales turning.

The result of the whole combat consists in the sum total of the results of all partial combats; but these results of separate combats are settled by different considerations.

First by the pure moral power in the mind of the leading officers. If a General of Division has seen his battalions forced to succumb, it will have an influence on his demeanour and his reports, and these again will have an influence on the measures of the Commander-in-Chief; therefore even those unsuccessful partial combats which to all appearance are retrieved, are not lost in their results, and the impressions from them sum themselves up in the mind of the Commander without much trouble, and even against his will.

Secondly, by the quicker melting away of our troops, which can be easily estimated in the slow and relatively [Relatively, that is say to the shock of former days.] little tumultuary course of our battles.

Thirdly, by lost ground.

All these things serve for the eye of the General as a compass to tell the course of the battle in which he is embarked. If whole batteries have been lost and none of the enemy's taken; if battalions have been overthrown by the enemy's cavalry, whilst those of the enemy everywhere present impenetrable masses; if the line of fire from his order of battle wavers involuntarily from one point to another; if fruitless efforts have been made to gain certain points, and the assaulting battalions each, time been scattered by well-directed volleys of grape and case;—if our artillery begins to reply feebly to that of the enemy—if the battalions under fire diminish unusually, fast, because with the wounded crowds of unwounded men go to the rear;—if single Divisions have been cut off and made prisoners through the disruption of the plan of the battle;—if the line of retreat begins to be endangered: the Commander may tell very well in which direction he is going with his battle. The longer this direction continues, the more decided it becomes, so much the more difficult will be the turning, so much the nearer the moment when he must give up the battle. We shall now make some observations on this moment.

We have already said more than once that the final decision is ruled mostly by the relative number of the fresh reserves remaining at the last; that Commander who sees his adversary is decidedly superior to him in this respect makes up his mind to retreat. It is the characteristic of modern battles that all mischances and losses which take place in the course of the same can be retrieved by fresh forces, because the arrangement of the modern order of battle, and the way in which troops are brought into action, allow of their use almost generally, and in each position. So long, therefore, as that Commander against whom the issue seems to declare itself still retains a superiority in reserve force, he will not give up the day. But from the moment that his reserves begin to become weaker than his enemy's, the decision may be regarded as settled, and what he now does depends partly on special circumstances, partly on the degree of courage and perseverance which he personally possesses, and which may degenerate into foolish obstinacy. How a Commander can attain to the power of estimating correctly the still remaining reserves on both sides is an affair of skilful practical genius, which does not in any way belong to this place; we keep ourselves to the result as it forms itself in his mind. But this conclusion is still not the moment of decision properly, for a motive which only arises gradually does not answer to that, but is only a general motive towards resolution, and the resolution itself requires still some special immediate causes. Of these there are two chief ones which constantly recur, that is, the danger of retreat, and the arrival of night.

If the retreat with every new step which the battle takes in its course becomes constantly in greater danger, and if the reserves are so much diminished that they are no longer adequate to get breathing room, then there is nothing left but to submit to fate, and by a well-conducted retreat to save what, by a longer delay ending in flight and disaster, would be lost.

But night as a rule puts an end to all battles, because a night combat holds out no hope of advantage except under particular circumstances; and as night is better suited for a retreat than the day, so, therefore, the Commander who must look at the retreat as a thing inevitable, or as most probable, will prefer to make use of the night for his purpose.

That there are, besides the above two usual and chief causes, yet many others also, which are less or more individual and not to be overlooked, is a matter of course; for the more a battle tends towards a complete upset of equilibrium the more sensible is the influence of each partial result in hastening the turn. Thus the loss of a battery, a successful charge of a couple of regiments of cavalry, may call into life the resolution to retreat already ripening.

As a conclusion to this subject, we must dwell for a moment on the point at which the courage of the Commander engages in a sort of conflict with his reason.

If, on the one hand the overbearing pride of a victorious conqueror, if the inflexible will of a naturally obstinate spirit, if the strenuous resistance of noble feelings will not yield the battlefield, where they must leave their honour, yet on the other hand, reason counsels not to give up everything, not to risk the last upon the game, but to retain as much over as is necessary for an orderly retreat. However highly we must esteem courage and firmness in War, and however little prospect there is of victory to him who cannot resolve to seek it by the exertion of all his power, still there is a point beyond which perseverance can only be termed desperate folly, and therefore can meet with no approbation from any critic. In the most celebrated of all battles, that of Belle-Alliance, Buonaparte used his last reserve in an effort to retrieve a battle which was past being retrieved. He spent his last farthing, and then, as a beggar, abandoned both the battle-field and his crown.

Effects of Victory

According to the point from which our view is taken, we may feel as much astonished at the extraordinary results of some great battles as at the want of results in others. We shall dwell for a moment on the nature of the effect of a great victory.

Three things may easily be distinguished here: the effect upon the instrument itself, that is, upon the Generals and their Armies; the effect upon the States interested in the War; and the particular result of these effects as manifested in the subsequent course of the campaign.

If we only think of the trifling difference which there usually is between victor and vanquished in killed, wounded, prisoners, and artillery lost on the field of battle itself, the consequences which are developed out of this insignificant point seem often quite incomprehensible, and yet, usually, everything only happens quite naturally.

We have already said in the seventh chapter that the magnitude of a victory increases not merely in the same measure as the vanquished forces increase in number, but in a higher ratio. The moral effects resulting from the issue of a great battle are greater on the side of the conquered than on that of the conqueror: they lead to greater losses in physical force, which then in turn react on the moral element, and so they go on mutually supporting and intensifying each other. On this moral effect we must therefore lay special weight. It takes an opposite direction on the one side from that on the other; as it undermines the energies of the conquered so it elevates the powers and energy of the conqueror. But its chief effect is upon the vanquished, because here it is the direct cause of fresh losses, and besides it is homogeneous in nature with danger, with the fatigues, the hardships, and generally with all those embarrassing circumstances by which War is surrounded, therefore enters into league with them and increases by their help, whilst with the conqueror all these things are like weights which give a higher swing to his courage. It is therefore found, that the vanquished sinks much further below the original line of equilibrium than the conqueror raises himself above it; on this account, if we speak of the effects of victory we allude more particularly to those which manifest themselves in the army. If this effect is more powerful in an important combat than in a smaller one, so again it is much more powerful in a great battle than in a minor one. The great battle takes place for the sake of itself, for the sake of the victory which it is to give, and which is sought for with the utmost effort. Here on this spot, in this very hour, to conquer the enemy is the purpose in which the plan of the War with all its threads converges, in which all distant hopes, all dim glimmerings of the future meet, fate steps in before us to give an answer to the bold question.—This is the state of mental tension not only of the Commander but of his whole Army down to the lowest waggon-driver, no doubt in decreasing strength but also in decreasing importance.

According to the nature of the thing, a great battle has never at any time been an unprepared, unexpected, blind routine service, but a grand act, which, partly of itself and partly from the aim of the Commander, stands out from amongst the mass of ordinary efforts, sufficiently to raise the tension of all minds to a higher degree. But the higher this tension with respect to the issue, the more powerful must be the effect of that issue.

Again, the moral effect of victory in our battles is greater than it was in the earlier ones of modern military history. If the former are as we have depicted them, a real struggle of forces to the utmost, then the sum total of all these forces, of the physical as well as the moral, must decide more than certain special dispositions or mere chance.

A single fault committed may be repaired next time; from good fortune and chance we can hope for more favour on another occasion; but the sum total of moral and physical powers cannot be so quickly altered, and, therefore, what the award of a victory has decided appears of much greater importance for all futurity. Very probably, of all concerned in battles, whether in or out of the Army, very few have given a thought to this difference, but the course of the battle itself impresses on the minds of all present in it such a conviction, and the relation of this course in public documents, however much it may be coloured by twisting particular circumstances, shows also, more or less, to the world at large that the causes were more of a general than of a particular nature.

He who has not been present at the loss of a great battle will have difficulty in forming for himself a living or quite true idea of it, and the abstract notions of this or that small untoward affair will never come up to the perfect conception of a lost battle. Let us stop a moment at the picture.

The first thing which overpowers the imagination—and we may indeed say, also the understanding—is the diminution of the masses; then the loss of ground, which takes place always, more or less, and, therefore, on the side of the assailant also, if he is not fortunate; then the rupture of the original formation, the jumbling together of troops, the risks of retreat, which, with few exceptions may always be seen sometimes in a less sometimes in a greater degree; next the retreat, the most part of which commences at night, or, at least, goes on throughout the night. On this first march we must at once leave behind, a number of men completely worn out and scattered about, often just the bravest, who have been foremost in the fight who held out the longest: the feeling of being conquered, which only seized the superior officers on the battlefield, now spreads through all ranks, even down to the common soldiers, aggravated by the horrible idea of being obliged to leave in the enemy's hands so many brave comrades, who but a moment since were of such value to us in the battle, and aggravated by a rising distrust of the chief, to whom, more or less, every subordinate attributes as a fault the fruitless efforts he has made; and this feeling of being conquered is no ideal picture over which one might become master; it is an evident truth that the enemy is superior to us; a truth of which the causes might have been so latent before that they were not to be discovered, but which, in the issue, comes out clear and palpable, or which was also, perhaps, before suspected, but which in the want of any certainty, we had to oppose by the hope of chance, reliance on good fortune, Providence or a bold attitude. Now, all this has proved insufficient, and the bitter truth meets us harsh and imperious.

All these feelings are widely different from a panic, which in an army fortified by military virtue never, and in any other, only exceptionally, follows the loss of a battle. They must arise even in the best of Armies, and although long habituation to War and victory together with great confidence in a Commander may modify them a little here and there, they are never entirely wanting in the first moment. They are not the pure consequences of lost trophies; these are usually lost at a later period, and the loss of them does not become generally known so quickly; they will therefore not fail to appear even when the scale turns in the slowest and most gradual manner, and they constitute that effect of a victory upon which we can always count in every case.

We have already said that the number of trophies intensifies this effect.

It is evident that an Army in this condition, looked at as an instrument, is weakened! How can we expect that when reduced to such a degree that, as we said before, it finds new enemies in all the ordinary difficulties of making War, it will be able to recover by fresh efforts what has been lost! Before the battle there was a real or assumed equilibrium between the two sides; this is lost, and, therefore, some external assistance is requisite to restore it; every new effort without such external support can only lead to fresh losses.

Thus, therefore, the most moderate victory of the chief Army must tend to cause a constant sinking of the scale on the opponent's side, until new external circumstances bring about a change. If these are not near, if the conqueror is an eager opponent, who, thirsting for glory, pursues great aims, then a first-rate Commander, and in the beaten Army a true military spirit, hardened by many campaigns are required, in order to stop the swollen stream of prosperity from bursting all bounds, and to moderate its course by small but reiterated acts of resistance, until the force of victory has spent itself at the goal of its career.

And now as to the effect of defeat beyond the Army, upon the Nation and Government! It is the sudden collapse of hopes stretched to the utmost, the downfall of all self-reliance. In place of these extinct forces, fear, with its destructive properties of expansion, rushes into the vacuum left, and completes the prostration. It is a real shock upon the nerves, which one of the two athletes receives from the electric spark of victory. And that effect, however different in its degrees, is never completely wanting. Instead of every one hastening with a spirit of determination to aid in repairing the disaster, every one fears that his efforts will only be in vain, and stops, hesitating with himself, when he should rush forward; or in despondency he lets his arm drop, leaving everything to fate.

The consequence which this effect of victory brings forth in the course of the War itself depend in part on the character and talent of the victorious General, but more on the circumstances from which the victory proceeds, and to which it leads. Without boldness and an enterprising spirit on the part of the leader, the most brilliant victory will lead to no great success, and its force exhausts itself all the sooner on circumstances, if these offer a strong and stubborn opposition to it. How very differently from Daun, Frederick the Great would have used the victory at Kollin; and what different consequences France, in place of Prussia, might have given a battle of Leuthen!

The conditions which allow us to expect great results from a great victory we shall learn when we come to the subjects with which they are connected; then it will be possible to explain the disproportion which appears at first sight between the magnitude of a victory and its results, and which is only too readily attributed to a want of energy on the part of the conqueror. Here, where we have to do with the great battle in itself, we shall merely say that the effects now depicted never fail to attend a victory, that they mount up with the intensive strength of the victory—mount up more the more the whole strength of the Army has been concentrated in it, the more the whole military power of the Nation is contained in that Army, and the State in that military power.

But then the question may be asked, Can theory accept this effect of victory as absolutely necessary?—must it not rather endeavour to find out counteracting means capable of neutralising these effects? It seems quite natural to answer this question in the affirmative; but heaven defend us from taking that wrong course of most theories, out of which is begotten a mutually devouring Pro et Contra.

Certainly that effect is perfectly necessary, for it has its foundation in the nature of things, and it exists, even if we find means to struggle against it; just as the motion of a cannon ball is always in the direction of the terrestrial, although when fired from east to west part of the general velocity is destroyed by this opposite motion.

All War supposes human weakness, and against that it is directed.

Therefore, if hereafter in another place we examine what is to be done after the loss of a great battle, if we bring under review the resources which still remain, even in the most desperate cases, if we should express a belief in the possibility of retrieving all, even in such a case; it must not be supposed we mean thereby that the effects of such a defeat can by degrees be completely wiped out, for the forces and means used to repair the disaster might have been applied to the realisation of some positive object; and this applies both to the moral and physical forces.

Another question is, whether, through the loss of a great battle, forces are not perhaps roused into existence, which otherwise would never have come to life. This case is certainly conceivable, and it is what has actually occurred with many Nations. But to produce this intensified reaction is beyond the province of military art, which can only take account of it where it might be assumed as a possibility.

If there are cases in which the fruits of a victory appear rather of a destructive nature in consequence of the reaction of the forces which it had the effect of rousing into activity—cases which certainly are very exceptional—then it must the more surely be granted, that there is a difference in the effects which one and the same victory may produce according to the character of the people or state, which has been conquered.

The Use of the Battle

Whatever form the conduct of War may take in particular cases, and whatever we may have to admit in the sequel as necessary respecting it: we have only to refer to the conception of War to be convinced of what follows:

1. The destruction of the enemy's military force, is the leading principle of War, and for the whole chapter of positive action the direct way to the object.

2. This destruction of the enemy's force, must be principally effected by means of battle.

3. Only great and general battles can produce great results.

4. The results will be greatest when combats unite themselves in one great battle.

5. It is only in a great battle that the General-in-Chief commands in person, and it is in the nature of things, that he should place more confidence in himself than in his subordinates.

From these truths a double law follows, the parts of which mutually support each other; namely, that the destruction of the enemy's military force is to be sought for principally by great battles, and their results; and that the chief object of great battles must be the destruction of the enemy's military force.

No doubt the annihilation-principle is to be found more or less in other means—granted there are instances in which through favourable circumstances in a minor combat, the destruction of the enemy's forces has been disproportionately great (Maxen), and on the other hand in a battle, the taking or holding a single post may be predominant in importance as an object—but as a general rule it remains a paramount truth, that battles are only fought with a view to the destruction of the enemy's Army, and that this destruction can only be effected by their means.

The battle may therefore be regarded as War concentrated, as the centre of effort of the whole War or campaign. As the sun's rays unite in the focus of the concave mirror in a perfect image, and in the fulness of their heat; to the forces and circumstances of War, unite in a focus in the great battle for one concentrated utmost effort.

The very assemblage of forces in one great whole, which takes place more or less in all Wars, indicates an intention to strike a decisive blow with this whole, either voluntarily as assailant, or constrained by the opposite party as defender. When this great blow does not follow, then some modifying, and retarding motives have attached themselves to the original motive of hostility, and have weakened, altered or completely checked the movement. But also, even in this condition of mutual inaction which has been the key-note in so many Wars, the idea of a possible battle serves always for both parties as a point of direction, a distant focus in the construction of their plans. The more War is War in earnest, the more it is a venting of animosity and hostility, a mutual struggle to overpower, so much the more will all activities join deadly contest, and also the more prominent in importance becomes the battle.

In general, when the object aimed at is of a great and positive nature, one therefore in which the interests of the enemy are deeply concerned, the battle offers itself as the most natural means; it is, therefore, also the best as we shall show more plainly hereafter: and, as a rule, when it is evaded from aversion to the great decision, punishment follows.

The positive object belong to the offensive, and therefore the battle is also more particularly his means. But without examining the conception of offensive and defensive more minutely here, we must still observe that, even for the defender in most cases, there is no other effectual means with which to meet the exigencies of his situation, to solve the problem presented to him.

The battle is the bloodiest way of solution. True, it is not merely reciprocal slaughter, and its effect is more a killing of the enemy's courage than of the enemy's soldiers, as we shall see more plainly in the next chapter—but still blood is always its price, and slaughter its character as well as name; ["Schlacht", from schlachten = to slaughter.] from this the humanity in the General's mind recoils with horror.

But the soul of the man trembles still more at the thought of the decision to be given with one single blow. *In one point* of space and time all action is here pressed together, and at such a moment there is stirred up within us a dim feeling as if in this narrow space all our forces could not develop themselves and come into activity, as if we had already gained much by mere time, although this time owes us nothing at all. This is all mere illusion, but even as illusion it is something, and the same weakness which seizes upon the man in every other momentous decision may well be felt more powerfully by the General, when he must stake interests of such enormous weight upon one venture.

Thus, then, Statesmen and Generals have at all times endeavoured to avoid the decisive battle, seeking either to attain their aim without it, or dropping that aim unperceived. Writers on history and theory have then busied themselves to discover in some other feature in these campaigns not only an equivalent for the decision by battle which has been avoided, but even a higher art. In this way, in the present age, it came very near to this, that a battle in the economy of War was looked upon as an evil, rendered necessary through some error committed, a morbid paroxysm to which a regular prudent system of War would never lead: only those Generals were to deserve laurels who knew how to carry on War without spilling blood, and the theory of War—a real business for Brahmins—was to be specially directed to teaching this.

Contemporary history has destroyed this illusion, but no one can guarantee that it will not sooner or later reproduce itself, and lead those at the head of affairs to perversities which please man's weakness, and therefore have the greater affinity for his nature. Perhaps, by-and-by, Buonaparte's campaigns and battles will be looked upon as mere acts of barbarism and stupidity, and we shall once more turn with satisfaction and confidence to the dress-sword of obsolete and musty institutions and forms. If theory gives a caution against this, then it renders a real service to those who listen to its warning voice. *May we succeed in lending a hand to those who in our dear native land are called upon to speak with authority on these matters, that we may be their guide into this field of inquiry, and excite them to make a candid examination of the subject.* [On the Continent only, it still preserves full vitality in the minds of British politicians and pressmen.—Editor.] [This prayer was abundantly granted—vide the German victories of 1870.—Editor.]

Not only the conception of War but experience also leads us to look for a great decision only in a great battle. From time immemorial, only great victories have led to great successes on the offensive side in the absolute form, on the defensive side in a manner more or less satisfactory. Even Buonaparte would not have seen the day of Ulm, unique in its kind, if he had shrunk from shedding blood; it is rather to be regarded as only a second crop from the victorious events in his preceding campaigns. It is not only bold, rash, and presumptuous Generals who have sought to complete their work by the great venture of a decisive battle, but also fortunate ones as well; and we may rest satisfied with the answer which they have thus given to this vast question.

Let us not hear of Generals who conquer without bloodshed. If a bloody slaughter is a horrible sight, then that is a ground for paying more respect to War, but not for making the sword we wear blunter and blunter by degrees from feelings of humanity, until some one steps in with one that is sharp and lops off the arm from our body.

We look upon a great battle as a principal decision, but certainly not as the only one necessary for a War or a campaign. Instances of a great battle deciding a whole campaign, have been frequent only in modern times, those which have decided a whole War, belong to the class of rare exceptions.

A decision which is brought about by a great battle depends naturally not on the battle itself, that is on the mass of combatants engaged in it, and on the intensity of the victory, but also on a number of other relations between the military forces opposed to each other, and between the States to which these forces belong. But at the same time that the principal mass of the force available is brought to the great duel, a great decision is also brought on, the extent of which may perhaps be foreseen in many respects, though not in all, and which although not the only one, still is the *first* decision, and as such, has an influence on those which succeed. Therefore a deliberately planned great battle, according to its relations, is more or less, but always in some degree, to be regarded as the leading means and central point of the whole system. The more a General takes the field in the true spirit of War as well as of every contest, with the feeling and the idea, that is the conviction, that he must and will conquer, the more he will strive to throw every weight into the scale in the first battle, hope and strive to win everything by it. Buonaparte hardly ever entered upon a War without thinking of conquering his enemy at once in the first battle, [This was Moltke's essential idea in his preparations for the War of 1870. See his secret memorandum issued to G.O.C.s on May 7. 1870, pointing to a battle on the Upper Saar as his primary Purpose.—Editor.] and Frederick the Great, although in a more limited sphere, and with interests of less magnitude at stake, thought the same when, at the head of a small Army, he sought to disengage his rear from the Russians or the Federal Imperial Army.

The decision which is given by the great battle, depends, we have said, partly on the battle itself, that is on the number of troops engaged, and partly on the magnitude of the success.

How the General may increase its importance in respect to the first point is evident in itself and we shall merely observe that according to the importance of the great battle, the number of cases which are decided along with it increases, and that therefore Generals who, confident in themselves have been lovers of great decisions, have always managed to make use of the greater part of their troops in it without neglecting on that account essential points elsewhere.

As regards the consequences or speaking more correctly the effectiveness of a victory, that depends chiefly on four points:

1. On the tactical form adopted as the order of battle.
2. On the nature of the country.
3. On the relative proportions of the three arms.
4. On the relative strength of the two Armies.

A battle with parallel fronts and without any action against a flank will seldom yield as great success as one in which the defeated Army has been turned, or compelled to change front more or less. In a broken or hilly country the successes are likewise smaller, because the power of the blow is everywhere less.

If the cavalry of the vanquished is equal or superior to that of the victor, then the effects of the pursuit are diminished, and by that great part of the results of victory are lost.

Finally it is easy to understand that if superior numbers are on the side of the conqueror, and he uses his advantage in that respect to turn the flank of his adversary, or compel him to change front, greater results will follow than if the conqueror had been weaker in numbers than the vanquished. The battle of Leuthen may certainly be quoted as a practical refutation of this principle, but we beg permission for once to say what we otherwise do not like, *no rule without an exception.*

In all these ways, therefore, the Commander has the means of giving his battle a decisive character; certainly he thus exposes himself to an increased amount of danger, but his whole line of action is subject to that dynamic law of the moral world.

There is then nothing in War which can be put in comparison with the great battle in point of importance, *and the acme of strategic ability is displayed in the provision of means for this great event, in the skilful determination of place and time, and direction of troops, and its the good use made of success.*

But it does not follow from the importance of these things that they must be of a very complicated and recondite nature; all is here rather simple, the art of combination by no means great; but there is great need of quickness in judging of circumstances, need of energy, steady resolution, a youthful spirit of enterprise—heroic qualities, to which we shall often have to refer. There is, therefore, but little wanted here of that which can be taught by books and there is much that, if it can be taught at all, must come to the General through some other medium than printer's type.

The impulse towards a great battle, the voluntary, sure progress to it, must proceed from a feeling of innate power and a clear sense of the necessity; in other words, it must proceed from inborn courage and from perceptions sharpened by contact with the higher interests of life.

Great examples are the best teachers, but it is certainly a misfortune if a cloud of theoretical prejudices comes between, for even the sunbeam is refracted and tinted by the clouds. To destroy such prejudices, which many a time rise and spread themselves like a miasma, is an imperative duty of theory, for the misbegotten offspring of human reason can also be in turn destroyed by pure reason.

Strategic Means of Utilising Victory

The more difficult part, viz., that of perfectly preparing the victory, is a silent service of which the merit belongs to Strategy and yet for which it is hardly sufficiently commended. It appears brilliant and full of renown by turning to good account a victory gained.

What may be the special object of a battle, how it is connected with the whole system of a War, whither the career of victory may lead according to the nature of circumstances, where its culminating-point lies—all these are things which we shall not enter upon until hereafter. But under any conceivable circumstances the fact holds good, that without a pursuit no victory can have a great effect, and that, however short the career of victory may be, it must always lead beyond the first steps in pursuit; and in order to avoid the frequent repetition of this, we shall now dwell for a moment on this necessary supplement of victory in general.

The pursuit of a beaten Army commences at the moment that Army, giving up the combat, leaves its position; all previous movements in one direction and another belong not to that but to the progress of the battle itself. Usually victory at the moment here described, even if it is certain, is still as yet small and weak in its proportions, and would not rank as an event of any great positive advantage if not completed by a pursuit on the first day. Then it is mostly, as we have before said, that the trophies which give substance to the victory begin to be gathered up. Of this pursuit we shall speak in the next place.

Usually both sides come into action with their physical powers considerably deteriorated, for the movements immediately preceding have generally the character of very urgent circumstances. The efforts which the forging out of a great combat costs, complete the exhaustion; from this it follows that the victorious party is very little less disorganised and out of his original formation than the vanquished, and therefore requires time to reform, to collect stragglers, and issue fresh ammunition to those who are without. All these things place the conqueror himself in the state of crisis of which we have already spoken. If now the defeated force is only a detached portion of the enemy's Army, or if it has otherwise to expect a considerable reinforcement, then the conqueror may easily run into the obvious danger of having to pay dear for his victory, and this consideration, in such a case, very soon puts an end to pursuit, or at least restricts it materially. Even when a strong accession of force by the enemy is not to be feared, the conqueror finds in the above circumstances a powerful check to the vivacity of his pursuit. There is no reason to fear that the victory will be snatched away, but adverse combats are still possible, and may diminish the advantages which up to the present have been gained. Moreover, at this moment the whole weight of all that is sensuous in an Army, its wants and weaknesses, are dependent on the will of the Commander. All the thousands under his command require rest and refreshment, and long to see a stop put to toil and danger for the present; only a few, forming an exception, can see and feel beyond the present moment, it is only amongst this little number that there is sufficient mental vigour to think, after what is absolutely necessary at the moment has been done, upon those results which at such a moment only appear to the rest as mere embellishments of victory—as a luxury of triumph. But all these thousands have a voice in the council of the General, for through the various steps of the military hierarchy these interests of the sensuous creature have their sure conductor into the heart of the Commander. He himself, through mental and bodily fatigue, is more or less weakened in his natural activity, and thus it happens then that, mostly from these causes, purely incidental to human nature, less is done than might have been done, and that generally what is done is to be ascribed entirely to the *thirst for glory*, the energy, indeed also the *hard-heartedness* of the General-in-Chief. It is only thus we can explain the hesitating manner in which many Generals follow up a victory which superior numbers have given them. The first pursuit of the enemy we limit in general to the extent of the first day, including the night following the victory. At the end of that period the necessity of rest ourselves prescribes a halt in any case.

This first pursuit has different natural degrees.

The first is, if cavalry alone are employed; in that case it amounts usually more to alarming and watching than to pressing the enemy in reality, because the smallest obstacle of ground is generally sufficient to check the pursuit. Useful as cavalry may be against single bodies of broken demoralised troops, still when opposed to the bulk of the beaten Army it becomes again only the auxiliary arm, because the troops in retreat can employ fresh reserves to cover the movement, and, therefore, at the next trifling obstacle of ground, by combining all arms they can make a stand with success. The only exception to this is in the case of an army in actual flight in a complete state of dissolution.

The second degree is, if the pursuit is made by a strong advance-guard composed of all arms, the greater part consisting naturally of cavalry. Such a pursuit generally drives the enemy as far as the nearest strong position for his rear-guard, or the next position affording space for his Army. Neither can usually be found at once, and, therefore, the pursuit can be carried further; generally, however, it does not extend beyond the distance of one or at most a couple of leagues, because otherwise the advance-guard would not feel itself sufficiently supported. The third and most vigorous degree is when the victorious Army itself continues to advance as far as its physical powers can endure. In this case the beaten Army will generally quit such ordinary positions as a country usually offers on the mere show of an attack, or of an intention to turn its flank; and the rear-guard will be still less likely to engage in an obstinate resistance.

In all three cases the night, if it sets in before the conclusion of the whole act, usually puts an end to it, and the few instances in which this has not taken place, and the pursuit has been continued throughout the night, must be regarded as pursuits in an exceptionally vigorous form.

If we reflect that in fighting by night everything must be, more or less, abandoned to chance, and that at the conclusion of a battle the regular cohesion and order of things in an army must inevitably be disturbed, we may easily conceive the reluctance of both Generals to carrying on their business under such disadvantageous conditions. If a complete dissolution of the vanquished Army, or a rare superiority of the victorious Army in military virtue does not ensure success, everything would in a manner be given up to fate, which can never be for the interest of any one, even of the most fool-hardy General. As a rule, therefore, night puts an end to pursuit, even when the battle has only been decided shortly before darkness sets in. This allows the conquered either time for rest and to rally immediately, or, if he retreats during the night it gives him a march in advance. After this break the conquered is decidedly in a better condition; much of that which had been thrown into confusion has been brought again into order, ammunition has been renewed, the whole has been put into a fresh formation. Whatever further encounter now takes place with the enemy is a new battle not a continuation of the old, and although it may be far from promising absolute success, still it is a fresh combat, and not merely a gathering up of the debris by the victor.

When, therefore, the conqueror can continue the pursuit itself throughout the night, if only with a strong advance-guard composed of all arms of the service, the effect of the victory is immensely increased, of this the battles of Leuthen and La Belle Alliance [Waterloo.] are examples.

The whole action of this pursuit is mainly tactical, and we only dwell upon it here in order to make plain the difference which through it may be produced in the effect of a victory.

This first pursuit, as far as the nearest stopping-point, belongs as a right to every conqueror, and is hardly in any way connected with his further plans and combinations. These may considerably diminish the positive results of a victory gained with the main body of the Army, but they cannot make this first use of it impossible; at least cases of that kind, if conceivable at all, must be so uncommon that they should have no appreciable influence on theory. And here certainly we must say that the example afforded by modern Wars opens up quite a new field for energy. In preceding Wars, resting on a narrower basis, and altogether more circumscribed in their scope, there were many unnecessary conventional restrictions in various ways, but particularly in this point. *The conception, honour of victory* seemed to Generals so much by far the chief thing that they thought the less of the complete destruction of the enemy's military force, as in point of fact that destruction of force appeared to them only as one of the many means in War, not by any means as the principal, much less as the only means; so that they the more readily put the sword in its sheath the moment the enemy had lowered his. Nothing seemed more natural to them than to stop the combat as soon as the decision was obtained, and to regard all further carnage as unnecessary cruelty. Even if this false philosophy did not determine their resolutions entirely, still it was a point of view by which representations of the exhaustion of all powers, and physical impossibility of continuing the struggle, obtained readier evidence and greater weight. Certainly the sparing one's own instrument of victory is a vital question if we only possess this one, and foresee that soon the time may arrive when it will not be sufficient for all that remains to be done, for every continuation of the offensive must lead ultimately to complete exhaustion. But this calculation was still so far false, as the further loss of forces by a continuance of the pursuit could bear no proportion to that which the enemy must suffer. That view, therefore, again could only exist because the military forces were not considered the vital factor. And so we find that in former Wars real heroes only—such as Charles XII., Marlborough, Eugene, Frederick the Great—added a vigorous pursuit to their victories when they were decisive enough, and that other Generals usually contented themselves with the possession of the field of battle. In modern times the greater energy infused into the conduct of Wars through the greater importance of the circumstances from which they have proceeded has thrown down these conventional barriers; the pursuit has become an all-important business for the conqueror; trophies have on that account multiplied in extent, and if there are cases also in modern Warfare in which this has not been the case, still they belong to the list of exceptions, and are to be accounted for by peculiar circumstances.

At Gorschen [Gorschen or Lutzen, May 2, 1813; Gross Beeren and Dennewitz, August 22, 1813; Bautzen. May 22, 1913; Laon, March 10 1813.] and Bautzen nothing but the superiority of the allied cavalry prevented a complete rout, at Gross Beeren and Dennewitz the ill-will of Bernadotte, the Crown Prince of Sweden; at Laon the enfeebled personal condition of Bluecher, who was then seventy years old and at the moment confined to a dark room owing to an injury to his eyes.

But Borodino is also an illustration to the point here, and we cannot resist saying a few more words about it, partly because we do not consider the circumstances are explained simply by attaching blame to Buonaparte, partly because it might appear as if this, and with it a great number of similar cases, belonged to that class which we have designated as so extremely rare, cases in which the general relations seize and fetter the General at the very beginning of the battle. French authors in particular, and great admirers of Buonaparte (Vaudancourt, Chambray, Se'gur), have blamed him decidedly because he did not drive the Russian Army completely off the field, and use his last reserves to scatter it, because then what was only a lost battle would have been a complete rout. We should be obliged to diverge too far to describe circumstantially the mutual situation of the two Armies; but this much is evident, that when Buonaparte passed the Niemen with his Army the same corps which afterwards fought at Borodino numbered 300,000 men, of whom now only 120,000 remained, he might therefore well be apprehensive that he would not have enough left to march upon Moscow, the point on which everything seemed to depend. The victory which he had just gained gave him nearly a certainty of taking that capital, for that the Russians would be in a condition to fight a second battle within eight days seemed in the highest degree improbable; and in Moscow he hoped to find peace. No doubt the complete dispersion of the Russian Army would have made this peace much more certain; but still the first consideration was to get to Moscow, that is, to get there with a force with which he should appear dictator over the capital, and through that over the Empire and the Government. The force which he brought with him to Moscow was no longer sufficient for that, as shown in the sequel, but it would have been still less so if, in scattering the Russian Army, he had scattered his own at the same time. Buonaparte was thoroughly alive to all this, and in our eyes he stands completely justified. But on that account this case is still not to be reckoned amongst those in which, through the general relations, the General is interdicted from following up his victory, for there never was in his case any question of mere pursuit. The victory was decided at four o'clock in the afternoon, but the Russians still occupied the greater part of the field of battle; they were not yet disposed to give up the ground, and if the attack had been renewed, they would still have offered a most determined resistance, which would have undoubtedly ended in their complete defeat, but would have cost the conqueror much further bloodshed. We must therefore reckon the Battle of Borodino as amongst battles, like Bautzen, left unfinished. At Bautzen the vanquished preferred to quit the field sooner; at Borodino the conqueror preferred to content himself with a half victory, not because the decision appeared doubtful, but because he was not rich enough to pay for the whole.

Returning now to our subject, the deduction from our reflections in relation to the first stage of pursuit is, that the energy thrown into it chiefly determines the value of the victory; that this pursuit is a second act of the victory, in many cases more important also than the first, and that strategy, whilst here approaching tactics to receive from it the harvest of success, exercises the first act of her authority by demanding this completion of the victory.

But further, the effects of victory are very seldom found to stop with this first pursuit; now first begins the real career to which victory lent velocity. This course is conditioned as we have already said, by other relations of which it is not yet time to speak. But we must here mention, what there is of a general character in the pursuit in order to avoid repetition when the subject occurs again.

In the further stages of pursuit, again, we can distinguish three degrees: the simple pursuit, a hard pursuit, and a parallel march to intercept.

The simple *following* or *pursuing* causes the enemy to continue his retreat, until he thinks he can risk another battle. It will therefore in its effect suffice to exhaust the advantages gained, and besides that, all that the enemy cannot carry with him, sick, wounded, and disabled from fatigue, quantities of baggage, and carriages of all kinds, will fall into our hands, but this mere following does not tend to heighten the disorder in the enemy's Army, an effect which is produced by the two following causes.

If, for instance, instead of contenting ourselves with taking up every day the camp the enemy has just vacated, occupying just as much of the country as he chooses to abandon, we make our arrangements so as every day to encroach further, and accordingly with our advance-guard organised for the purpose, attack his rear-guard every time it attempts to halt, then such a course will hasten his retreat, and consequently tend to increase his disorganisation.—This it will principally effect by the character of continuous flight, which his retreat will thus assume. Nothing has such a depressing influence on the soldier, as the sound of the enemy's cannon afresh at the moment when, after a forced march he seeks some rest; if this excitement is continued from day to day for some time, it may lead to a complete rout.

There lies in it a constant admission of being obliged to obey the law of the enemy, and of being unfit for any resistance, and the consciousness of this cannot do otherwise than weaken the moral of an Army in a high degree. The effect of pressing the enemy in this way attains a maximum when it drives the enemy to make night marches. If the conqueror scares away the discomfited opponent at sunset from a camp which has just been taken up either for the main body of the Army, or for the rear-guard, the conquered must either make a night march, or alter his position in the night, retiring further away, which is much the same thing; the victorious party can on the other hand pass the night in quiet.

The arrangement of marches, and the choice of positions depend in this case also upon so many other things, especially on the supply of the Army, on strong natural obstacles in the country, on large towns, etc. etc., that it would be ridiculous pedantry to attempt to show by a geometrical analysis how the pursuer, being able to impose his laws on the retreating enemy, can compel him to march at night while he takes his rest. But nevertheless it is true and practicable that marches in pursuit may be so planned as to have this tendency, and that the efficacy of the pursuit is very much enchanced thereby. If this is seldom attended to in the execution, it is because such a procedure is more difficult for the pursuing Army, than a regular adherence to ordinary marches in the daytime. To start in good time in the morning, to encamp at mid-day, to occupy the rest of the day in providing for the ordinary wants of the Army, and to use the night for repose, is a much more convenient method than to regulate one's movements exactly according to those of the enemy, therefore to determine nothing till the last moment, to start on the march, sometimes in the morning, sometimes in the evening, to be always for several hours in the presence of the enemy, and exchanging cannon shots with him, and keeping up skirmishing fire, to plan manoeuvres to turn him, in short, to make the whole outlay of tactical means which such a course renders necessary. All that naturally bears with a heavy weight on the pursuing Army, and in War, where there are so many burdens to be borne, men are always inclined to strip off those which do not seem absolutely necessary. These observations are true, whether applied to a whole Army or as in the more usual case, to a strong advance-guard. For the reasons just mentioned, this second method of pursuit, this continued pressing of the enemy pursued is rather a rare occurrence; even Buonaparte in his Russian campaign, 1812, practised it but little, for the reasons here apparent, that the difficulties and hardships of this campaign, already threatened his Army with destruction before it could reach its object; on the other hand, the French in their other campaigns have distinguished themselves by their energy in this point also.

Lastly, the third and most effectual form of pursuit is, the parallel march to the immediate object of the retreat.

Every defeated Army will naturally have behind it, at a greater or less distance, some point, the attainment of which is the first purpose in view, whether it be that failing in this its further retreat might be compromised, as in the case of a defile, or that it is important for the point itself to reach it before the enemy, as in the case of a great city, magazines, etc., or, lastly, that the Army at this point will gain new powers of defence, such as a strong position, or junction with other corps.

Now if the conqueror directs his march on this point by a lateral road, it is evident how that may quicken the retreat of the beaten Army in a destructive manner, convert it into hurry, perhaps into flight. [This point is exceptionally well treated by von Bernhardi in his "Cavalry in Future Wars." London: Murray, 1906.] The conquered has only three ways to counteract this: the first is to throw himself in front of the enemy, in order by an unexpected attack to gain that probability of success which is lost to him in general from his position; this plainly supposes an enterprising bold General, and an excellent Army, beaten but not utterly defeated; therefore, it can only be employed by a beaten Army in very few cases.

The second way is hastening the retreat; but this is just what the conqueror wants, and it easily leads to immoderate efforts on the part of the troops, by which enormous losses are sustained, in stragglers, broken guns, and carriages of all kinds.

The third way is to make a detour, and get round the nearest point of interception, to march with more ease at a greater distance from the enemy, and thus to render the haste required less damaging. This last way is the worst of all, it generally turns out like a new debt contracted by an insolvent debtor, and leads to greater embarrassment. There are cases in which this course is advisable; others where there is nothing else left; also instances in which it has been successful; but upon the whole it is certainly true that its adoption is usually influenced less by a clear persuasion of its being the surest way of attaining the aim than by another inadmissible motive—this motive is the dread of encountering the enemy. Woe to the Commander who gives in to this! However much the moral of his Army may have deteriorated, and however well founded may be his apprehensions of being at a disadvantage in any conflict with the enemy, the evil will only be made worse by too anxiously avoiding every possible risk of collision. Buonaparte in 1813 would never have brought over the Rhine with him the 30,000 or 40,000 men who remained after the battle of Hanau, [At Hanau (October 30, 1813), the Bavarians some 50,000 strong threw themselves across the line of Napoleon's retreat from Leipsic. By a masterly use of its artillery the French tore the

Bavarians asunder and marched on over their bodies.—Editor.] if he had avoided that battle and tried to pass the Rhine at Mannheim or Coblenz. It is just by means of small combats carefully prepared and executed, and in which the defeated army being on the defensive, has always the assistance of the ground—it is just by these that the moral strength of the Army can first be resuscitated.

The beneficial effect of the smallest successes is incredible; but with most Generals the adoption of this plan implies great self-command. The other way, that of evading all encounter, appears at first so much easier, that there is a natural preference for its adoption. It is therefore usually just this system of evasion which best, promotes the view of the pursuer, and often ends with the complete downfall of the pursued; we must, however, recollect here that we are speaking of a whole Army, not of a single Division, which, having been cut off, is seeking to join the main Army by making a de'tour; in such a case circumstances are different, and success is not uncommon. But there is one condition requisite to the success of this race of two Corps for an object, which is that a Division of the pursuing army should follow by the same road which the pursued has taken, in order to pick up stragglers, and keep up the impression which the presence of the enemy never fails to make. Bluecher neglected this in his, in other respects unexceptionable, pursuit after La Belle Alliance.

Such marches tell upon the pursuer as well as the pursued, and they are not advisable if the enemy's Army rallies itself upon another considerable one; if it has a distinguished General at its head, and if its destruction is not already well prepared. But when this means can be adopted, it acts also like a great mechanical power. The losses of the beaten Army from sickness and fatigue are on such a disproportionate scale, the spirit of the Army is so weakened and lowered by the constant solicitude about impending ruin, that at last anything like a well organised stand is out of the question; every day thousands of prisoners fall into the enemy's hands without striking a blow. In such a season of complete good fortune, the conqueror need not hesitate about dividing his forces in order to draw into the vortex of destruction everything within reach of his Army, to cut off detachments, to take fortresses unprepared for defence, to occupy large towns, etc. etc. He may do anything until a new state of things arises, and the more he ventures in this way the longer will it be before that change will take place. There is no want of examples of brilliant results from grand decisive victories, and of great and vigorous pursuits in the wars of Buonaparte. We need only quote Jena 1806, Ratisbonne 1809, Leipsic 1813, and Belle-Alliance 1815.

Retreat after a Lost Battle

In a lost battle the power of an Army is broken, the moral to a greater degree than the physical. A second battle unless fresh favourable circumstances come into play, would lead to a complete defeat, perhaps, to destruction. This is a military axiom. According to the usual course the retreat is continued up to that point where the equilibrium of forces is restored, either by reinforcements, or by the protection of strong fortresses, or by great defensive positions afforded by the country, or by a separation of the enemy's force. The magnitude of the losses sustained, the extent of the defeat, but still more the character of the enemy, will bring nearer or put off the instant of this equilibrium. How many instances may be found of a beaten Army rallied again at a short distance, without its circumstances having altered in any way since the battle. The cause of this may be traced to the moral weakness of the adversary, or to the preponderance gained in the battle not having been sufficient to make lasting impression.

To profit by this weakness or mistake of the enemy, not to yield one inch breadth more than the pressure of circumstances demands, but above all things, in order to keep up the moral forces to as advantageous a point as possible, a slow retreat, offering incessant resistance, and bold courageous counterstrokes, whenever the enemy seeks to gain any excessive advantages, are absolutely necessary. Retreats of great Generals and of Armies inured to War have always resembled the retreat of a wounded lion, such is, undoubtedly, also the best theory.

It is true that at the moment of quitting a dangerous position we have often seen trifling formalities observed which caused a waste of time, and were, therefore, attended with danger, whilst in such cases everything depends on getting out of the place speedily. Practised Generals reckon this maxim a very important one. But such cases must not be confounded with a general retreat after a lost battle. Whoever then thinks by a few rapid marches to gain a start, and more easily to recover a firm standing, commits a great error. The first movements should be as small as possible, and it is a maxim in general not to suffer ourselves to be dictated to by the enemy. This maxim cannot be followed without bloody fighting with the enemy at our heels, but the gain is worth the sacrifice; without it we get into an accelerated pace which soon turns into a headlong rush, and costs merely in stragglers more men than rear-guard combats, and besides that extinguishes the last remnants of the spirit of resistance.

A strong rear-guard composed of picked troops, commanded by the bravest General, and supported by the whole Army at critical moments, a careful utilisation of ground, strong ambuscades wherever the boldness of the enemy's advance-guard, and the ground, afford opportunity; in short, the preparation and the system of regular small battles,—these are the means of following this principle.

The difficulties of a retreat are naturally greater or less according as the battle has been fought under more or less favourable circumstances, and according as it has been more or less obstinately contested. The battle of Jena and La Belle-Alliance show how impossible anything like a regular retreat may become, if the last man is used up against a powerful enemy.

Now and again it has been suggested [Allusion is here made to the works of Lloyd Bullow and others.] to divide for the purpose of retreating, therefore to retreat in separate divisions or even eccentrically. Such a separation as is made merely for convenience, and along with which concentrated action continues possible and is kept in view, is not what we now refer to; any other kind is extremely dangerous, contrary to the nature of the thing, and therefore a great error. Every lost battle is a principle of weakness and disorganisation; and the first and immediate desideratum is to concentrate, and in concentration to recover order, courage, and confidence. The idea of harassing the enemy by separate corps on both flanks at the moment when he is following up his victory, is a perfect anomaly; a faint-hearted pedant might be overawed by his enemy in that manner, and for such a case it may answer; but where we are not sure of this failing in our opponent it is better let alone. If the strategic relations after a battle require that we should cover ourselves right and left by detachments, so much must be done, as from circumstances is unavoidable, but this fractioning must always be regarded as an evil, and we are seldom in a state to commence it the day after the battle itself.

If Frederick the Great after the battle of Kollin, [June 19, 1757.] and the raising of the siege of Prague retreated in three columns that was done not out of choice, but because the position of his forces, and the necessity of covering Saxony, left him no alternative, Buonaparte after the battle of Brienne, [January 30, 1814.] sent Marmont back to the Aube, whilst he himself passed the Seine, and turned towards Troyes; but that this did not end in disaster, was solely owing to the circumstance that the Allies, instead of pursuing divided their forces in like manner, turning with the one part

(Bluecher) towards the Marne, while with the other (Schwartzenberg), from fear of being too weak, they advanced with exaggerated caution.

Night Fighting

The manner of conducting a combat at night, and what concerns the details of its course, is a tactical subject; we only examine it here so far as in its totality it appears as a special strategic means.

Fundamentally every night attack is only a more vehement form of surprise. Now at the first look of the thing such an attack appears quite pre-eminently advantageous, for we suppose the enemy to be taken by surprise, the assailant naturally to be prepared for everything which can happen. What an inequality! Imagination paints to itself a picture of the most complete confusion on the one side, and on the other side the assailant only occupied in reaping the fruits of his advantage. Hence the constant creation of schemes for night attacks by those who have not to lead them, and have no responsibility, whilst these attacks seldom take place in reality.

These ideal schemes are all based on the hypothesis that the assailant knows the arrangements of the defender because they have been made and announced beforehand, and could not escape notice in his reconnaissances, and inquiries; that on the other hand, the measures of the assailant, being only taken at the moment of execution, cannot be known to the enemy. But the last of these is not always quite the case, and still less is the first. If we are not so near the enemy as to have him completely under our eye, as the Austrians had Frederick the Great before the battle of Hochkirch (1758), then all that we know of his position must always be imperfect, as it is obtained by reconnaissances, patrols, information from prisoners, and spies, sources on which no firm reliance can be placed because intelligence thus obtained is always more or less of an old date, and the position of the enemy may have been altered in the meantime. Moreover, with the tactics and mode of encampment of former times it was much easier than it is now to examine the position of the enemy. A line of tents is much easier to distinguish than a line of huts or a bivouac; and an encampment on a line of front, fully and regularly drawn out, also easier than one of Divisions formed in columns, the mode often used at present. We may have the ground on which a Division bivouacs in that manner completely under our eye, and yet not be able to arrive at any accurate idea.

But the position again is not all that we want to know the measures which the defender may take in the course of the combat are just as important, and do not by any means consist in mere random shots. These measures also make night attacks more difficult in modern Wars than formerly, because they have in these campaigns an advantage over those already taken. In our combats the position of the defender is more temporary than definitive, and on that account the defender is better able to surprise his adversary with unexpected blows, than he could formerly. [All these difficulties obviously become increased as the power of the weapons in use tends to keep the combatants further Apart.—Editor.]

Therefore what the assailant knows of the defensive previous to a night attack, is seldom or never sufficient to supply the want of direct observation.

But the defender has on his side another small advantage as well, which is that he is more at home than the assailant, on the ground which forms his position, and therefore, like the inhabitant of a room, will find his way about it in the dark with more ease than a stranger. He knows better where to find each part of his force, and therefore can more readily get at it than is the case with his adversary.

From this it follows, that the assailant in a combat at night feels the want of his eyes just as much as the defender, and that therefore, only particular reasons can make a night attack advisable.

Now these reasons arise mostly in connection with subordinate parts of an Army, rarely with the Army itself; it follows that a night attack also as a rule can only take place with secondary combats, and seldom with great battles.

We may attack a portion of the enemy's Army with a very superior force, consequently enveloping it with a view either to take the whole, or to inflict very severe loss on it by an unequal combat, provided that other circumstances are in our favour. But such a scheme can never succeed except by a great surprise, because no fractional part of the enemy's Army would engage in such an unequal combat, but would retire instead. But a surprise on an important scale except in rare instances in a very close country, can only be effected at night. If therefore we wish to gain such an advantage as this from the faulty disposition of a portion of the enemy's Army, then we must make use of the night, at all events, to finish the preliminary part even if the combat itself should not open till towards daybreak. This is therefore what takes place in all the little enterprises by night against outposts, and other small bodies, the main point being invariably through superior numbers, and getting round his position, to entangle him unexpectedly in such a disadvantageous combat, that he cannot disengage himself without great loss.

The larger the body attacked the more difficult the undertaking, because a strong force has greater resources within itself to maintain the fight long enough for help to arrive.

On that account the whole of the enemy's Army can never in ordinary cases be the object of such an attack for although it has no assistance to expect from any quarter outside itself, still, it contains within itself sufficient means of repelling attacks from several sides particularly in our day, when every one from the commencement is prepared for this very usual form of attack. Whether the enemy can attack us on several sides with success depends generally on conditions quite different from that of its being done unexpectedly; without entering here into the nature of these conditions, we confine ourselves to observing, that with turning an enemy, great results, as well as great dangers are connected; that therefore, if we set aside special circumstances, nothing justifies it but a great superiority, just such as we should use against a fractional part of the enemy's Army.

But the turning and surrounding a small fraction of the enemy, and particularly in the darkness of night, is also more practicable for this reason, that whatever we stake upon it, and however superior the force used may be, still probably it constitutes only a limited portion of our Army, and we can sooner stake that than the whole on the risk of a great venture. Besides, the greater part or perhaps the whole serves as a support and rallying-point for the portion risked, which again very much diminishes the danger of the enterprise.

Not only the risk, but the difficulty of execution as well confines night enterprises to small bodies. As surprise is the real essence of them so also stealthy approach is the chief condition of execution: but this is more easily done with small bodies than with large, and for the columns of a whole Army is seldom practicable. For this reason such enterprises are in general only directed against single outposts, and can only be feasible against greater bodies if they are without sufficient outposts, like Frederick the Great at Hochkirch.[October 14, 1758.] This will happen seldomer in future to Armies themselves than to minor divisions.

In recent times, when War has been carried on with so much more rapidity and vigour, it has in consequence often happened that Armies have encamped very close to each other, without having a very strong system of outposts, because those circumstances have generally occurred just at the crisis which precedes a great decision.

But then at such times the readiness for battle on both sides is also more perfect; on the other hand, in former Wars it was a frequent practice for armies to take up camps in sight of each other, when they had no other object but that of mutually holding each other in check, consequently for a longer period. How often Frederick the Great stood for weeks so near to the Austrians, that the two might have exchanged cannon shots with each other.

But these practices, certainly more favourable to night attacks, have been discontinued in later days; and armies being now no longer in regard to subsistence and requirements for encampment, such independent bodies complete in themselves, find it necessary to keep usually a day's march between themselves and the enemy. If we now keep in view especially the night attack of an army, it follows that sufficient motives for it can seldom occur, and that they fall under one or other of the following classes.

1. An unusual degree of carelessness or audacity which very rarely occurs, and when it does is compensated for by a great superiority in moral force.

2. A panic in the enemy's army, or generally such a degree of superiority in moral force on our side, that this is sufficient to supply the place of guidance in action.

3. Cutting through an enemy's army of superior force, which keeps us enveloped, because in this all depends on surprise, and the object of merely making a passage by force, allows a much greater concentration of forces.

4. Finally, in desperate cases, when our forces have such a disproportion to the enemy's, that we see no possibility of success, except through extraordinary daring.

But in all these cases there is still the condition that the enemy's army is under our eyes, and protected by no advance-guard.

As for the rest, most night combats are so conducted as to end with daylight, so that only the approach and the first attack are made under cover of darkness, because the assailant in that manner can better profit by the consequences of the state of confusion into which he throws his adversary; and combats of this description which do not commence until daybreak, in which the night therefore is only made use of to approach, are not to be counted as night combats.

The Art of War
by Niccolò Machiavelli

Table of Contents

Many, Lorenzo, have held and still hold the opinion, that there is nothing which has less in common with another, and that is so dissimilar, as civilian life is from the military. Whence it is often observed, if anyone designs to avail himself of an enlistment in the army, that he soon changes, not only his clothes, but also his customs, his habits, his voice, and in the presence of any civilian custom, he goes to pieces; for I do not believe that any man can dress in civilian clothes who wants to be quick and ready for any violence; nor can that man have civilian customs and habits, who judges those customs to be effeminate and those habits not conducive to his actions; nor does it seem right to him to maintain his ordinary appearance and voice who, with his beard and cursing, wants to make other men afraid: which makes such an opinion in these times to be very true. But if they should consider the ancient institutions, they would not find matter more united, more in conformity, and which, of necessity, should be like to each other as much as these (civilian and military); for in all the arts that are established in a society for the sake of the common good of men, all those institutions created to (make people) live in fear of the laws and of God would be in vain, if their defense had not been provided for and which, if well arranged, will maintain not only these, but also those that are not well established. And so (on the contrary), good institutions without the help of the military are not much differently disordered than the habitation of a superb and regal palace, which, even though adorned with jewels and gold, if it is not roofed over will not have anything to protect it from the rain. And, if in any other institutions of a City and of a Republic every diligence is employed in keeping men loyal, peaceful, and full of the fear of God, it is doubled in the military; for in what man ought the country look for greater loyalty than in that man who has to promise to die for her? In whom ought there to be a greater love of peace, than in him who can only be injured by war? In whom ought there to be a greater fear of God than in him who, undergoing infinite dangers every day, has more need for His aid? If these necessities in forming the life of the soldier are well considered, they are found to be praised by those who gave the laws to the Commanders and by those who were put in charge of military training, and followed and imitated with all diligence by others.

But because military institutions have become completely corrupt and far removed from the ancient ways, these sinister opinions have arisen which make the military hated and intercourse with those who train them avoided. And I, judging, by what I have seen and read, that it is not impossible to restore its ancient ways and return some form of past virtue to it, have decided not to let this leisure time of mine pass without doing something, to write what I know of the art of war, to the satisfaction of those who are lovers of the ancient deeds. And although it requires courage to treat of those matters of which others have made a profession, none the less, I do not believe that it is a mistake to occupy a position with words, which may, with greater presumption, have been occupied with deeds; for the errors which I should make in writing can be corrected without injury to anyone, but those which are made with deeds cannot be found out except by the ruin of the Commanders.

You, Lorenzo, will therefore consider the quality of these efforts of mine, and will give in your judgement of them that censure or praise which will appear to you to be merited. I send you these, as much as to show myself grateful for all the benefits I have received from you, although I will not include in them the (review) of this work of mine, as well as also, because being accustomed to honor similar works of those who shine because of their nobility, wealth, genius, and liberality, I know you do not have many equals in wealth and nobility, few in ingenuity, and no one in liberality.

First Book

As I believe that it is possible for one to praise, without concern, any man after he is dead since every reason and supervision for adulation is lacking, I am not apprehensive in praising our own Cosimo Ruccelai, whose name is never remembered by me without tears, as I have recognized in him those parts which can be desired in a good friend among friends and in a citizen of his country. For I do not know what pertained to him more than to spend himself willingly, not excepting that courage of his, for his friends, and I do not know of any enterprise that dismayed him when he knew it was for the good of his country. And I confess freely not to have met among so many men whom I have known and worked with, a man in whom there was a mind more fired with great and magnificent things. Nor does one grieve with the friends of another of his death, except for his having been born to die young unhonored within his own home, without having been able to benefit anyone with that mind of his, for one would know that no one could speak of him, except (to say) that a good friend had died. It does not remain for us, however, or for anyone else who, like us, knew him, to be able because of this to keep the faith (since deeds do not seem to) to his laudable qualities. It is true however, that fortune was not so unfriendly to him that it did not leave some brief memory of the dexterity of his genius, as was demonstrated by some of his writings and compositions of amorous verses, in which (as he was not in love) he (employed as an) exercise in order not to use his time uselessly in his juvenile years, in order that fortune might lead him to higher thoughts. Here, it can be clearly comprehended, that if his objective was exercise, how very happily he described his ideas, and how much he was honored in his poetry. Fortune, however, having deprived us of the use of so great a friend, it appears to me it is not possible to find any other better remedy than for us to seek to benefit from his memory, and recover from it any matter that was either keenly observed or wisely discussed. And as there is nothing of his more recent than the discussions which the Lord Fabrizio Colonna had with him in his gardens, where matters pertaining to war were discussed at length by that Lord, with (questions) keenly and prudently asked by Cosimo, it seemed proper to me having been present with other friends of ours, to recall him to memory, so that reading it, the friends of Cosimo who met there will renew in their minds the memory of his virtue, and another part grieving for not having been there, will learn in part of many things discussed wisely by a most sagacious man useful not only to the military way of life, but to the civilian as well. I will relate, therefore, how Fabrizio Colonna, when he returned from Lombardy where he had fought a long time gloriously for the Catholic King, decided to pass through Florence to rest several days in that City in order to visit His Excellency the Duke, and see again several gentlemen with whom he had been familiar in the past. Whence it appeared proper to Cosimo to invite him to a banquet in his gardens, not so much to show his generosity as to have reason to talk to him at length, and to learn and understand several things from him, according as one can hope to from such a man, for it appeared to him to give him an opportunity to spend a day discussing such matters as would satisfy his mind.

Fabrizio, therefore, came as planned, and was received by Cosimo together with several other loyal friends of his, among whom were Zanobi Buondelmonti, Battista Della Palla, and Luigi Alamanni, young men most ardent in the same studies and loved by him, whose good qualities, because they were also praised daily by himself, we will omit. Fabrizio, therefore, was honored according to the times and the place, with all the highest honors they could give him. As soon as the convivial pleasures were past and the table cleared and every arrangement of feasting finished, which, in the presence of great men and those who have their minds turned to honorable thoughts is soon accomplished, and because the day was long and the heat intense, Cosimo, in order to satisfy their desire better, judged it would be well to take the opportunity to escape the heat by leading them to the more secret and shadowy part of his garden: when they arrived there and chairs brought out, some sat on the grass which was most fresh in the place, some sat on chairs placed in those parts under the shadow of very high trees; Fabrizio praised the place as most delightful, and looking especially at the trees, he did not recognize one of them, and looked puzzled. Cosimo, becoming aware of this said: Perhaps you have no knowledge of some of these trees, but do not wonder about them, because here are some which were more widely known by the ancients than are those commonly seen today. And giving him the name of some and telling him that Bernardo, his grandfather, had worked hard in their culture, Fabrizio replied: I was thinking that it was what you said I was, and this place and this study make me remember several Princes of the Kingdom, who delighted in their ancient culture and the shadow they cast. And stopping speaking of this, and somewhat upon himself as though in suspense, he added: If I did not think I would offend you, I would give you my opinion: but I do not believe in talking and discussing things with friends in this manner that I insult them. How much better would they have done (it is said with peace to everyone) to seek to imitate the ancients in the strong and rugged things, not in the soft and delicate, and in the things they did under the sun, not in the shadows, to adopt the honest and perfect ways of antiquity, not the false and corrupt; for while these practices were pleasing to my Romans, my country (without them) was ruined. To which Cosimo replied (but to avoid the necessity of having to repeat so many times who is speaking, and what the other adds, only the names of those speaking will be noted, without repeating the others). Cosimo, therefore, said: You have opened the way for a discussion which I desired, and I pray you to speak without regard, for I will question you without regard; and if, in questioning or in replying, I accuse or excuse anyone, it will not be for accusing or excusing, but to understand the truth from you.

FABRIZIO: And I will be much content to tell you what I know of all that you ask me; whether it be true or not, I will leave to your judgement. And I will be grateful if you ask me, for I am about to learn as much from what you ask me, as you will from me replying to you, because many times a wise questioner causes one to consider many things and understand many others which, without having been asked, would never have been understood.

COSIMO: I want to return to what you first were saying, that my grandfather and those of yours had more wisely imitated the ancients in rugged things than in delicate ones, and I want to excuse my side because I will let you excuse the other (your side). I do not believe that in your time there was a man who disliked living as softly as he, and that he was so much a lover of that rugged life which you praise: none the less he recognized he could not practice it in his personal life, nor in that of his sons, having been born in so corrupted an age, where anyone who wanted to depart from the common usage would be deformed and despised by everyone. For if anyone in a naked state should thrash upon the sand under the highest sun, or upon the snow in the most icy months of winter, as did Diogenes, he would be considered mad. If anyone (like the Spartan) should raise his children on a farm, make them sleep in the open, go with head and feet bare, bathe in cold water in order to harden them to endure vicissitudes, so that they then might love life less and fear death less, he would be praised by few and followed by none. So that dismayed at these ways of living, he presently leaves the ways of the ancients, and in imitating antiquity, does only that which he can with little wonderment.

FABRIZIO: You have excused him strongly in this part, and certainly you speak the truth: but I did not speak so much of these rugged ways of living, as of those other more human ways which have a greater conformity to the ways of living today, which I do not believe should have been difficult to introduce by one who is numbered among the Princes of a City. I will never forego my examples of my Romans. If their way of living should be examined, and the institutions in their Republic, there will be observed in her many things not impossible to introduce in a Society where there yet might be something of good.

COSIMO: What are those things similar to the ancients that you would introduce?

FABRIZIO: To honor and reward virtu, not to have contempt for poverty, to esteem the modes and orders of military discipline, to constrain citizens to love one another, to live without factions, to esteem less the private than the public good, and other such things which could easily be added in these times. It is not difficult to persuade (people) to these ways, when one considers these at length and approaches them in the usual manner, for the truth will appear in such (examinations) that every common talent is capable of undertaking them. Anyone can arrange these things; (for example), one plants trees under the shadow of which he lives more happily and merrily than if he had not (planted them).

COSIMO: I do not want to reply to anything of what you have spoken, but I do want leave to give a judgment on these, which can be easily judged, and I shall address myself to you who accuse those who in serious and important actions are not imitators of the ancients, thinking that in this way I can more easily carry out my intentions. I should want, therefore, to know from you whence it arises that, on the one hand you condemn those who do not imitate the ancients in their actions, on the other hand, in matters of war which is your profession and in which you are judged to be excellent, it is not observed that you have employed any of the ancient methods, or those which have some similarity.

FABRIZIO: You have come to the point where I expected you to, for what I said did not merit any other question, nor did I wish for any other. And although I am able to save myself with a simple excuse, none the less I want, for your greater satisfaction and mine, since the season (weather) allows it, to enter into a much longer discussion. Men who want to do something, ought first to prepare themselves with all industry, in order ((when the opportunity is seen)) to be prepared to achieve that which they have proposed. And whenever the preparations are undertaken cautiously, unknown to anyone, no none can be accused of negligence unless he is first discovered by the occasion; in which if it is not then successful, it is seen that either he has not sufficiently prepared himself, or that he has not in some part given thought to it. And as the opportunity has not come to me to be able to show the preparations I would make to bring the military to your ancient organization, and it I have not done so, I cannot be blamed either by you or by others. I believe this excuse is enough to respond to your accusation.

COSIMO: It would be enough if I was certain that the opportunity did not present itself.

FABRIZIO: But because I know you could doubt whether this opportunity had come about or not, I want to discuss at length ((if you will listen to me with patience)) which preparations are necessary to be made first, what occasion needs to arise, what difficulty impedes the preparations from becoming beneficial and the occasion from arriving, and that this is ((which appears a paradox)) most difficult and most easy to do.

COSIMO: You cannot do anything more pleasing for me and for the others than this. But if it is not painful for you to speak, it will never be painful for us to listen. But at this discussion may be long, I want help from these, my friends, and with your permission, and they and I pray you one thing, that you do not become annoyed if we sometimes interrupt you with some opportune question.

FABRIZIO: I am most content that you, Cosimo, with these other young people here, should question me, for I believe that young men will become more familiar with military matters, and will more easily understand what I have to say. The others, whose hair (head) is white and whose blood is icy, in part are enemies of war and in part incorrigible, as those who believe that the times and not the evil ways constrain men to live in such a fashion. So ask anything of me, with assurance and without regard; I desire this, as much because it will afford me a little rest, as because it will give me pleasure not to leave any doubts in your minds. I want to begin from your words, where you said to me that in war ((which is my profession)) I have not employed any of the ancient methods. Upon this I say, that this being a profession by which men of every time were not able to live honestly, it cannot be employed as a profession except by a Republic or a Kingdom; and both of these, if well established, will never allow any of their citizens or subjects to employ it as a profession: for he who practices it will never be judged to be good, as to gain some usefulness from it at any time he must be rapacious, deceitful, violent, and have many qualities, which of necessity, do not make him good: nor can men who employ this as a profession, the great as well as the least, be made otherwise, for this profession does not provide for them in peace. Whence they are obliged, either to hope that there will be no peace or to gain so much for themselves in times of war, that they can provide for themselves in times of peace. And wherever one of these two thoughts exists, it does not occur in a good man; for, from the desire to provide for oneself in every circumstance, robberies, violence and assassinations result, which such soldiers do to friends as well as to enemies: and from not desiring peace, there arises those deceptions which Captains perpetrate upon those whom they lead, because war hardens them: and even if peace occurs frequently, it happens that the leaders, being deprived of their stipends and of their licentious mode of living, raise a flag of piracy, and without any mercy sack a province.

Do you not have within the memory of events of your time, many soldiers in Italy, finding themselves without employment because of the termination of wars, gathered themselves into very troublesome gangs, calling themselves companies, and went about levying tribute on the towns and sacking the country, without there being any remedy able to be applied? Have you not read how the Carthaginian soldiers, when the first war they engaged in with the Romans under Matus and Spendius was ended, tumultuously chose two leaders, and waged a more dangerous war against the Carthaginians than that which they had just concluded with the Romans? And in the time of our fathers, Francesco Sforza, in order to be able to live honorably (comfortably) in times of peace, not only deceived the Milanese, in whose pay he was, but took away their liberty and became their Prince. All the other soldiers of Italy, who have employed the military as their particular profession, have been like this man; and if, through their malignity, they have not become Dukes of Milan, so much more do they merit to be censured; for without such a return ((if their lives were to be examined)), they all have the same cares. Sforza, father of Francesco, constrained Queen Giovanna to throw herself into the arms of the King of Aragon, having abandoned her suddenly, and left her disarmed amid her enemies, only in order to satisfy his ambition of either levying tribute or taking the Kingdom. Braccio, with the same industry, sought to occupy the Kingdom of Naples, and would have succeeded, had he not been routed and killed at Aquilla. Such evils do not result from anything else other than the existence of men who employ the practice of soldiering as their own profession. Do you not have a proverb which strengthens my argument, which says: War makes robbers, and peace hangs them? For those who do not know how to live by another practice, and not finding any one who will support them in that, and not having so much virtu that they know how to come and live together honorably, are forced by necessity to roam the streets, and justice is forced to extinguish them.

COSIMO: You have made me turn this profession (art) of soldiering back almost to nothing, and I had supposed it to be the most excellent and most honorable of any: so that if you do not clarify this better, I will not be satisfied; for if it is as you say, I do not know whence arises the glory of Caesar, Pompey, Scipio, Marcellus, and of so many Roman Captains who are celebrated for their fame as the Gods.

FABRIZIO: I have not yet finished discussing all that I proposed, which included two things: the one, that a good man was not able to undertake this practice because of his profession: the other, that a well established Republic or Kingdom would never permit its subjects or citizens to employ it for their profession. Concerning the first, I have spoken as much as has occurred to me: it remains for me to talk of the second, where I shall reply to this last question of yours, and I say that Pompey and Caesar, and almost all those Captains who were in Rome after the last Carthaginian war, acquired fame as valiant men, not as good men: but those who had lived before them acquired glory as valiant and good men: which results from the fact that these latter did not take up the practice of war as their profession; and those whom I named first as those who employed it as their profession. And while the Republic lived immaculately, no great citizen ever presumed by means of such a practice to enrich himself during (periods of) peace by breaking laws, despoiling the provinces, usurping and tyrannizing the country, and imposing himself in every way; nor did anyone of the lowest fortune think of violating the sacred agreement, adhere himself to any private individual, not fearing the Senate, or to perform any disgraceful act of tyranny in order to live at all times by the profession of war. But those who were Captains, being content with the triumph, returned with a desire for the private life; and those who were members (of the army) returned with a desire to lay down the arms they had taken up; and everyone returned to the art (trade or profession) by which they ordinarily lived; nor was there ever anyone who hoped to provide for himself by plunder and by means of these arts. A clear and evident example of this as it applies to great citizens can be found in the Regent Attilio, who, when he was captain of the Roman armies in Africa, and having almost defeated the Carthaginians, asked the Senate for permission to return to his house to look after his farms which were being spoiled by his laborers. Whence it is clearer than the sun, that if that man had practiced war as his profession, and by means of it thought to obtain some advantage for himself, having so many provinces which (he could) plunder, he would not have asked permission to return to take care of his fields, as each day he could have obtained more than the value of all his possessions. But as these good men, who do not practice war as their profession, do not expect to gain anything from it except hard work, danger, and glory, as soon as they are sufficiently glorious, desire to return to their homes and live from the practice of their own profession. As to men of lower status and gregarious soldiers, it is also true that every one voluntarily withdrew from such a practice, for when he was not fighting would have desired to fight, but when he was fighting wanted to be dismissed. Which illustrates the many ways, and especially in seeing that it was among the first privileges, that the Roman people gave to one of its Citizens, that he should not be constrained unwillingly to fight. Rome, therefore, while she was well organized ((which it was up to the time of the Gracchi)) did not have one soldier who had to take up this practice as a profession, and therefore had few bad ones, and these were severely punished. A well ordered City, therefore, ought to desire that this training for war ought to be employed in times of peace as an exercise, and in times of war as a necessity and for glory, and allow the public only to use it as a profession, as Rome did. And any citizen who has other aims in (using) such exercises is not good, and any City which governs itself otherwise, is not well ordered.

COSIMO: I am very much content and satisfied with what you have said up to now, and this conclusion which you have made pleases me greatly: and I believe it will be true when expected from a Republic, but as to Kings, I do not yet know why I should believe that a King would not want particularly to have around him those who take up such a practice as their profession.

FABRIZIO: A well ordered Kingdom ought so much the more avoid such artifices, for these only are the things which corrupt the King and all the Ministers in a Tyranny. And do not, on the other side, tell me of some present Kingdom, for I will not admit them to be all well ordered Kingdoms; for Kingdoms that are well ordered do not give absolute (power to) Rule to their Kings, except in the armies, for only there is a quick decision necessary, and, therefore, he who (rules) there must have this unique power: in other matters, he cannot do anything without counsel, and those who counsel him have to fear those whom he may have near him who, in times of peace, desire war because they are unable to live without it. But I want to dwell a little longer on this subject, and look for a Kingdom totally good, but similar to those that exist today, where those who take up the profession of war for themselves still ought to be feared by the King, for the sinews of armies without any doubt are the infantry. So that if a King does not organize himself in such a way that his infantry in time of peace are content to return to their homes and live from the practice of their own professions, it must happen of necessity that he will be ruined; for there is not to be found a more dangerous infantry than that which is composed of those who make the waging of war their profession; for you are forced to make war always, or pay them always, or to risk the danger that they take away the Kingdom from you. To make war always is not possible: (and) one cannot pay always; and, hence, that danger is run of losing the State. My Romans ((as I have said)), as long as they were wise and good, never permitted that their citizens should take up this practice as their profession, notwithstanding that they were able to raise them at all times, for they made war at all times: but in order to avoid the harm which this continuous practice of theirs could do to them, since the times did not change, they changed the men, and kept turning men over in their legions so that every fifteen years they always completely re-manned them: and thus they desired men in the flower of their age, which is from eighteen to thirty five years, during which time their legs, their hands, and their eyes, worked together, nor did they expect that their strength should decrease in them, or that malice should grow in them, as they did in corrupt times.

Ottavianus first, and then Tiberius, thinking more of their own power than the public usefulness, in order to rule over the Roman people more easily, begun to disarm them and to keep the same armies continually at the frontiers of the Empire. And because they did not think it sufficient to hold the Roman People and the Senate in check, they instituted an army called the Praetorian (Guard), which was kept near the walls of Rome in a fort adjacent to that City. And as they now begun freely to permit men assigned to the army to practice military matters as their profession, there soon resulted that these men became insolent, and they became formidable to the Senate and damaging to the Emperor. Whence there resulted that many men were killed because of their insolence, for they gave the Empire and took it away from anyone they wished, and it often occurred that at one time there were many Emperors created by the several armies. From which state of affairs proceeded first the division of the Empire and finally its ruin. Kings ought, therefore, if they want to live securely, have their infantry composed of men, who, when it is necessary for him to wage war, will willingly go forth to it for love of him, and afterwards when peace comes, more willingly return to their homes; which will always happen if he selects men who know how to live by a profession other than this. And thus he ought to desire, with the coming of peace, that his Princes return to governing their people, gentlemen to the cultivation of their possessions, and the infantry to their particular arts (trades or professions); and everyone of these will willingly make war in order to have peace, and will not seek to disturb the peace to have war.

COSIMO: Truly, this reasoning of yours appears to me well considered: none the less, as it is almost contrary to what I have thought up to now, my mind is not yet purged of every doubt. For I see many Lords and Gentlemen who provide for themselves in times of peace through the training for war, as do your equals who obtain provisions from Princes and the Community. I also see almost all the men at arms remaining in the garrisons of the city and of the fortresses. So that it appears to me that there is a long time of peace for everyone.

FABRIZIO: I do not believe that you believe this, that everyone has a place in time of peace; for other reasons can be cited for their being stationed there, and the small number of people who remain in the places mentioned by you will answer your question. What is the proportion of infantry needed to be employed in time of war to that in peace? for while the fortresses and the city are garrisoned in times of peace, they are much more garrisoned in times of war; to this should be added the soldiers kept in the field who are a great number, but all of whom are released in time of peace. And concerning the garrisons of States, who are a small number, Pope Julius and you have shown how much they are to be feared who do not know any other profession than war, as you have taken them out of your garrisons because of their insolence, and placed the Swiss there, who are born and raised under the laws and are chosen by the community in an honest election; so do not say further that in peace there is a place for every man. As to the men at arms continued in their enlistment in peace time, the answer appears more difficult. None the less, whoever considers everything well, will easily find the answer, for this thing of keeping on the men at arms is a corrupt thing and not good. The reason is this; as there are men who do not have any art (trade or profession), a thousand evils will arise every day in those States where they exist, and especially so if they were to be joined by a great number of companions: but as they are few, and unable by themselves to constitute an army, they therefore, cannot do any serious damage. None the less, they have done so many times, as I said of Francesco and of Sforza, his father, and of Braccio of Perugia. So I do not approve of this custom of keeping men at arms, both because it is corrupt and because it can cause great evils.

COSIMO: Would you do without them?, or if you keep them, how would you do so?

FABRIZIO: By means of an ordinance, not like those of the King of France, because they are as dangerous and insolent as ours, but like those of the ancients, who created horsemen (cavalry) from their subjects, and in times of peace sent them back to their homes to live from the practice of their own profession, as I shall discuss at length before I finish this discussion. So, if this part of the army can now live by such a practice even when there is peace, it stems from a corrupt order. As to the provisions that are reserved for me and the other leaders, I say to you that this likewise is a most corrupt order, for a wise Republic ought not to give them to anyone, rather it ought to employ its citizens as leaders in war, and in time of peace desire that they return to their professions. Thus also, a wise King ought not to give (provisions) to them, or if he does give them, the reasons ought to be either as a reward for some excellent act, or in order to avail himself of such a man in peace as well as in war. And because you have mentioned me, I want the example to include me, and I say I have never practiced war as a profession, for my profession is to govern my subjects, and defend them, and in order to defend them, I must love peace but know how to make war; and my King does not reward and esteem me so much for what I know of war, as because I know also how to counsel him in peace. Any King ought not, therefore, to want to have next to him anyone who is not thusly constituted, if he is wise and wants to govern prudently; for if he has around him either too many lovers of peace or too many lovers of war, they will cause him to err. I cannot, in this first discussion of mine and according to my suggestion, say otherwise, and if this is not enough for you, you must seek one which satisfies you better. You can begin to recognize how much difficulty there is in bringing the ancient methods into modem wars, and what preparations a wise man must make, and what opportunities he can hope for to put them into execution. But little by little you will know these things better if the discussion on bringing any part of the ancient institutions to the present order of things does not weary you.

COSIMO: If we first desired to hear your discussion of these matters, truly what you have said up to now redoubles that desire. We thank you, therefore, for what we have had and ask you for the rest.

FABRIZIO: Since this is your pleasure, I want to begin to treat of this matter from the beginning being able in that way to demonstrate it more fully, so that it may be better understood. The aim of those who want to make war is to be able to combat in the field with every (kind) of enemy, and to be able to win the engagement. To want to do this, they must raise an army. In raising an army, it is necessary to find men, arm them, organize them, train them in small and large (battle) orders, lodge them, and expose them to the enemy afterwards, either at a standstill or while marching. All the industry of war in the field is placed in these things, which are the more necessary and honored (in the waging of war). And if one does well in offering battle to the enemy, all the other errors he may make in the conduct of the war are supportable: but if he lacks this organization, even though he be valiant in other particulars, he will never carry on a war to victory (and honor). For, as one engagement that you win cancels out every other bad action of yours, so likewise, when you lose one, all the things you have done well before become useless. Since it is necessary, therefore, first to find men, you must come to the Deletto (Draft) of them, as thus the ancients called it, and which we call Scelta (Selection): but in order to call it by a more honored name, I want us to preserve the name of Deletto. Those who have drawn up regulations for war want men to be chosen from temperate countries as they have spirit and are prudent; for warm countries give rise to men who are prudent but not spirited, and cold (countries) to men who are spirited but not prudent. This regulation is drawn up well for one who is the Prince of all the world, and is therefore permitted to draw men from those places that appear best to him: but wanting to draw up a regulation that anyone can use, one must say that every Republic and every Kingdom ought to take soldiers from their own country, whether it is hot, cold, or temperate. For, from ancient examples, it is seen that in every country, good soldiers are made by training; because where nature is lacking, industry supplies it, which, in this case, is worth more than nature: And selecting them from another place cannot be called Deletto, because Deletto means to say to take the best of a province, and to have the power to select as well those who do not want to fight as those who do want to. This Deletto therefore, cannot be made unless the places are subject to you; for you cannot take whoever you want in the countries that are not yours, but you need to take those who want to come.

COSIMO: And of those who want to come, it can even be said, that they turn and leave you, and because of this, it can then be called a Deletto.

FABRIZIO: In a certain way, you say what is true: but consider the defects that such as Deletto has in itself, for often it happens that it is not a Deletto. The first thing (to consider), is that those who are not your subjects and do not willingly want to fight, are not of the best, rather they are of the worst of a province; for if nay are troublesome, idle, without restraint, without religion, subject to the rule of the father, blasphemous, gamblers, and in every way badly brought up, they are those who want to fight, (and) these habits cannot be more contrary to a true and good military life. When there are so many of such men offered to you that they exceed the number you had designated, you can select them; but if the material is bad, it is impossible for the Deletto to be good: but many times it happens that they are not so many as (are needed) to fill the number you require: so that being forced to take them all, it results that it can no longer be called the making of a Deletto, but in enlisting of infantry. The armies of Italy and other places are raised today with these evils, except in Germany, where no one is enlisted by command of the Prince, but according to the wishes of those who want to fight. Think, therefore, what methods of those ancients can now be introduced in an army of men put together by similar means.

COSIMO: What means should be taken therefore?

FABRIZIO: What I have just said: select them from your own subjects, and with the authority of the Prince.

COSIMO: Would you introduce any ancient form in those thus selected?

FABRIZIO: You know well it would be so; if it is a Principality, he who should command should be their Prince or an ordinary Lord; or if it is a Republic, a citizen who for the time should be Captain: otherwise it is difficult to do the thing well.

COSIMO: Why?

FABRIZIO: I will tell you in time: for now, I want this to suffice for you, that it cannot be done well in any other way.

COSIMO: If you have, therefore, to make ibis Deletto in your country, whence do you judge it better to draw them, from the City or the Countryside?

FABRIZIO: Those who have written of this all agree that it is better to select them from the Countryside, as they are men accustomed to discomfort, brought up on hard work, accustomed to be in the sun and avoid the shade, know how to handle the sword, dig a ditch, carry a load, and are without cunning or malice. But on this subject, my opinion would be, that as soldiers are of two kinds, afoot and on horseback, that those afoot be selected from the Countryside, and those on horseback from the City.

COSIMO: Of what age would you draw them?

FABRIZIO: If I had to raise an (entirely) new army, I would draw them from seventeen to forty years of age; if the army already exists and I had to replenish it, at seventeen years of age always.

COSIMO: I do not understand this distinction well.

FABRIZIO: I will tell you: if I should have to organize an army where there is none, it would be necessary to select all those men who were more capable, as long as they were of military age, in order to instruct them as I would tell them: but if I should have to make the Deletto in places where the army was (already) organized, in order to supplement it, I would take those of seventeen years of age, because the others having been taken for some time would have been selected and instructed.

COSIMO: Therefore you would want to make an ordinance similar to that which exists in our countries.

FABRIZIO: You say well: it is true that I would arm them, captain them, train them, and organize them, in a way which I do not know whether or not you have organized them similarly.

COSIMO: Therefore you praise the ordinance?

FABRIZIO: Why would you want me to condemn it?

COSIMO: Because many wise men have censured it.

FABRIZIO: You say something contrary, when you say a wise man censured the ordinance: for he can be held a wise man and to have censured them wrongly.

COSIMO: The wrong conclusion that he has made will always cause us to have such a opinion.

FABRIZIO: Watch out that the defect is not yours, but his: as that which you recognized before this discussion furnishes proof.

COSIMO: You do a most gracious thing. But I want to tell you that you should be able to justify yourself better in that of which those men are accused. These men say thusly: either that it is useless and our trusting in it will cause us to lose the State: or it is of virtue, and he who governs through it can easily deprive her of it. They cite the Romans, who by their own arms lost their liberty: They cite the Venetians and the King of France, of whom they say that the former, in order not to obey one of its Citizens employed the arms of others, and the King disarmed his People so as to be able to command them more easily. But they fear the uselessness of this much more; for which uselessness they cite two principal reasons: the one, because they are inexpert; the other, for having to fight by force: because they say that they never learn anything from great men, and nothing good is ever done by force.

FABRIZIO: All the reasons that you mention are from men who are not far sighted, as I shall clearly show. And first, as to the uselessness, I say to you that no army is of more use than your own, nor can an army of your own be organized except in this way. And as there is no debating over this, which all the examples of ancient history does for us, I do not want to lose time over it. And because they cite inexperience and force, I say ((as it is true)) that inept experience gives rise to little spirit (enthusiasm) and force makes for discontent: but experience and enthusiasm gains for themselves the means for arming, training, and organizing them, as you will see in the first part of this discussion. But as to force, you must understand that as men are brought to the army by commandment of the Prince, they have to come, whether it is entirely by force or entirely voluntarily: for if it were entirely from desire, there would not be a Deletto as only a few of them would go; so also, the (going) entirely by force would produce bad results; therefore, a middle way ought to be taken where neither the entirely forced or entirely voluntarily (means are used), but they should come, drawn by the regard they have for the Prince, where they are more afraid of of his anger then the immediate punishment: and it will always happen that there will be a compulsion mixed with willingness, from which that discontent cannot arise which causes bad effects. Yet I do not claim that an army thus constituted cannot be defeated; for many times the Roman armies were overcome, and the army of Hannibal was defeated: so that it can be seen that no army can be so organized that a promise can be given that it cannot be routed. These wise men of yours, therefore, ought not measure this uselessness from having lost one time, but to believe that just as they can lose, so too they can win and remedy the cause of the defeat. And if they should look into this, they will find that it would not have happened because of a defect in the means, but of the organization which was not sufficiently perfect. And, as I have said, they ought to provide for you, not by censuring the organization, but by correcting it: as to how this ought to be done, you will come to know little by little.

As to being apprehensive that such organization will not deprive you of the State by one who makes himself a leader, I reply, that the arms carried by his citizens or subjects, given to them by laws and ordinances, never do him harm, but rather are always of some usefulness, and preserve the City uncorrupted for a longer time by means of these (arms), than without (them). Rome remained free four hundred years while armed: Sparta eight hundred:

Many other Cities have been dis-armed, and have been free less than forty years; for Cities have need of arms, and if they do not have arms of their own, they hire them from foreigners, and the arms of foreigners more readily do harm to the public good than their own; for they are easier to corrupt, and a citizen who becomes powerful can more readily avail himself, and can also manage the people more readily as he has to oppress men who are disarmed. In addition to this, a City ought to fear two enemies more than one. One which avails itself of foreigners immediately has to fear not only its citizens, but the foreigners that it enlists; and, remembering what I told you a short while ago of Francesco Sforza, (you will see that) that fear ought to exist. One which employs its own arms, has not other fear except of its own Citizens. But of all the reasons which can be given, I want this one to serve me, that no one ever established any Republic or Kingdom who did not think that it should be defended by those who lived there with arms: and if the Venetians had been as wise in this as in their other institutions, they would have created a new world Kingdom; but who so much more merit censure, because they had been the first who were armed by their founders. And not having dominion on land, they armed themselves on the sea, where they waged war with virtu, and with arms in hand enlarged their country. But when the time came when they had to wage war on land to defend Venice and where they ought to have sent their own citizens to fight (on land), they enlisted as their captain (a foreigner), the Marquis of Mantua. This was the sinister course which prevented them from rising to the skies and expanding. And they did this in the belief that, as they knew how to wage war at sea, they should not trust themselves in waging it on land; which was an unwise belief (distrust), because a Sea captain, who is accustomed to combat with winds, water, and men, could more easily become a Captain on land where the combat is with men only, than a land Captain become a sea one. And my Romans, knowing how to combat on land and not on the sea, when the war broke out with the Carthaginians who were powerful on the sea, did not enlist Greeks or Spaniards experienced at sea, but imposed that change on those citizens they sent (to fight) on land, and they won. If they did this in order that one of their citizens should not become Tyrant, it was a fear that was given little consideration; for, in addition to the other reasons mentioned a short while ago concerning such a proposal, if a citizen (skilled) in (the use of) arms at sea had never been made a Tyrant in a City situated in the sea, so much less would he be able to do this if he were (skilled) in (the use of arms) on land. And, because of this, they ought to have seen that arms in the hands of their own citizens could not create Tyrants, but the evil institutions of a Government are those which cause a City to be tyrannized; and, as they had a good Government, did not have to fear arms of their own citizens. They took an imprudent course, therefore, which was the cause of their being deprived of much glory and happiness. As to the error which the King of France makes in not having his people disciplined to war, from what has been cited from examples previously mentioned, there is no one ((devoid of some particular passion of theirs)) who does not judge this defect to be in the Republic, and that this negligence alone is what makes it weak. But I have made too great a digression and have gotten away from my subject: yet I have done this to answer you and to show you, that no reliance can be had on arms other than ones own, and ones own arms cannot be established otherwise than by way of an ordinance, nor can forms of armies be introduced in any place, nor military discipline instituted. If you have read the arrangements which the first Kings made in Rome, and most especially of Servius Tullus, you will find that the institution of classes is none other than an arrangement to be able quickly to put together an army for the defense of that City. But turning to our Deletto, I say again, that having to replenish an established (old) organization, I would take the seventeen year olds, but having to create a new one, I would take them of every age between seventeen and forty in order to avail myself of them quickly.

COSIMO: Would you make a difference of what profession (art) you would choose them from?

FABRIZIO: These writers do so, for they do not want that bird hunters, fishermen, cooks, procurers, and anyone who makes amusement his calling should be taken, but they want that, in addition to tillers of the soil, smiths and blacksmiths, carpenters, butchers, hunters, and such like, should be taken. But I would make little difference in conjecturing from his calling how good the man may be, but how much I can use him with the greatest usefulness. And for this reason, the peasants, who are accustomed to working the land, are more useful than anyone else, for of all the professions (arts), this one is used more than any other in the army: After this, are the forgers (smiths), carpenters, blacksmiths, shoemakers; of whom it is useful to have many, for their skills succeed in many things, as they are a very good thing for a soldier to have, from whom you draw double service.

COSIMO: How are those who are or are not suitable to fight chosen?

FABRIZIO: I want to talk of the manner of selecting a new organization in order to make it after wards into an army; which yet also apply in the discussion of the selection that should be made in re-manning an old (established) organization. I say, therefore, that how good the man is that you have to select as a soldier is recognized either from his experience, shown by some excellent deeds of his, or by conjecture. The proof of virtu cannot be found in men who are newly selected, and who never before have been selected; and of the former, few or none are found in an organization which is newly established. It is necessary, therefore, lacking experience to have recourse to conjecture, which is derived from their age, profession, and physical appearance. The first two have been discussed: it remains to talk of the third. And yet I say that some have wanted that the soldier be big, among whom was Pyrrhus: Some others have chosen them only from the strength of the body, as Caesar did: which strength of body is conjectured from the composition of the members and the gracefulness of aspect. And yet some of those who write say that he should have lively and merry eyes, a nervy neck, a large breast, muscular arms, long fingers, a small stomach, round hips, sleek legs and feet: which parts usually render a man strong and agile, which are the two things sought above everything else in a soldier. He ought, above all, to have regard for his habits and that there should be in him a (sense of) honesty and shame, otherwise there will be selected only an instrument of trouble and a beginning of corruption; for there is no one who believes that in a dishonest education and in a brutish mind, there can exist some virtu which in some part may be praiseworthy. Nor does it appear to me superfluous, rather I believe it necessary, in order for you to understand better the importance of this selection, to tell you the method that the Roman Consuls at the start of their Magistracy observed in selecting the Roman legions. In which Deletto, because those who had to be selected were to be a mixture of new and veteran men ((because of the continuing wars)), they proceeded from experience with regard to the old (veteran) men, and from conjecture with regard to the new. And this ought to be noted, that these Deletti are made, either for immediate training and use, or for future employment.

I have talked, and will talk, of those that are made for future employment, because my intention is to show you how an army can be organized in countries where there is no military (organization), in which countries I cannot have Deletti in order to make use of them. But in countries where it is the custom to call out armies, and by means of the Prince, these (Deletti) exist, as was observed at Rome and is today observed among the Swiss. For in these Deletti, if they are for the (selection of) new men, there are so many others accustomed to being under military orders, that the old (veteran) and new, being mixed together, make a good and united body. Notwithstanding this, the Emperors, when they began to hold fixed the (term of service of the) soldiers, placed new men in charge over the soldiers, whom they called Tironi, as teachers to train them, as is seen in the life of the Emperor Maximus: which thing, while Rome was free, was instituted, not in the army, but within the City: and as the military exercises where the young men were trained were in the City, there resulted that those then chosen to go to war, being accustomed in the method of mock warfare, could easily adapt themselves to real war. But afterwards, when these Emperors discontinued these exercises, it was necessary to employ the methods I have described to you. Arriving, therefore, at the methods of the Roman Selection, I say that, as soon as the Roman Consuls, on whom was imposed the carrying on of the war, had assumed the Magistracy, in wanting to organize their armies ((as it was the custom that each of them had two legions of Roman men, who were the nerve (center) of their armies)), created twenty four military Tribunes, proposing six for each legion, who filled that office which today is done by those whom we call Constables. After they had assembled all the Roman men adept at carrying arms, and placed the Tribunes of each legion apart from each of the others. Afterwards, by lot they drew the Tribes, from which the first Selection was to be made, and of that Tribe they selected four of their best men, from whom one was selected by the Tribunes of the first legion, and of the other three, one was selected by the Tribunes of the second legion; of the other two, one was selected by the Tribunes of the third, and that last belonged to the fourth legion. After these four, four others were selected, of whom the first man was selected by the Tribunes of the second legion, the second by those of the third, the third by those of the fourth, the fourth remained to the first. After, another four were chosen: the first man was selected by the (Tribunes of the) third (legion), the second by the fourth, the third by the first, the fourth remained to the second. And thus this method of selection changed successively, so that the selection came to be equal, and the legions equalized. And as we said above, this was done where the men were to be used immediately: and as it was formed of men of whom a good part were experienced in real warfare, and everyone in mock battles, this Deletto was able to be based on conjecture and experience. But when a new army was to be organized and the selection made for future employment, this Deletto cannot be based except on conjecture, which is done by age and physical appearance.

COSIMO: I believe what you have said is entirely true: but before you pass on to other discussion, I want to ask about one thing which you have made me remember, when you said that the Deletto which should be made where these men are not accustomed to fighting should be done by conjecture: for I have heard our organization censured in many of its parts, and especially as to number; for many say that a lesser number ought to be taken, of whom those that are drawn would be better and the selection better, as there would not be as much hardship imposed on the men, and some reward given them, by means of which they would be more content and could be better commanded. Whence I would like to know your opinion on this part, and if you preferred a greater rather than a smaller number, and what methods you would use in selecting both numbers.

FABRIZIO: Without doubt the greater number is more desirable and more necessary than the smaller: rather, to say better, where a great number are not available, a perfect organization cannot be made, and I will easily refute all the reasons cited in favor of this. I say, therefore, first, that where there are many people, as there are for example in Tuscany, does not cause you to have better ones, or that the Deletto is more selective; for desiring in the selection of men to judge them on the basis of experience, only a very few would probably be found in that country who would have had this experience, as much because few have been in a war, as because of those few who have been, very few have ever been put to the test, so that because of this they merit to be chosen before the others: so that whoever is in a similar situation should select them, must leave experience to one side and take them by conjecture: and if I were brought to such a necessity, I would want to see, if twenty young men of good physical appearance should come before me, with what rule rule I ought to take some or reject some: so that without doubt I believe that every man will confess that it is a much smaller error to take them all in arming and training them, being unable to know (beforehand) which of them are better, and to reserve to oneself afterwards to make a more certain Deletto where, during the exercises with the army, those of greater courage and vitality may be observed. So that, considering everything, the selection in this case of a few in order to have them better, is entirely false. As to causing less hardship to the country and to the men, I say that the ordinance, whether it is bad or insufficient, does not cause any hardship: for this order does not take men away from their business, and does not bind them so that they cannot go to carry out their business, because it only obliges them to come together for training on their free days, which proposition does not do any harm either to the country or the men; rather, to the young, it ought to be delightful, for where, on holidays they remain basely indolent in their hangouts, they would now attend these exercises with pleasure, for the drawing of arms, as it is a beautiful spectacle, is thus delightful to the young men. As to being able to pay (more to) the lesser number, and thereby keeping them more content and obedient, I reply, that no organizion of so few can be made, who are paid so continually, that their pay satisfies them. For instance, if an army of five thousand infantry should be organized, in wanting to pay them so that it should be believed they would be contented, they must be given at least ten thousand ducats a month. To begin with, this number of infantry is not enough to make an army, and the payment is unendurable to a State; and on the other hand, it is not sufficient to keep the men content and obligated to respect your position. So that in doing this although much would be spent, it would provide little strength, and would not be sufficient to defend you, or enable you to undertake any enterprise. If you should give them more, or take on more, so much more impossible would it be for you to pay them: if you should give them less, or take on fewer, so much less would be content and so much less useful would they be to you. Therefore, those who consider things which are either useless or impossible. But it is indeed necessary to pay them when they are levied to send to war.

But even if such an arrangement should give some hardship to those enrolled in it in times of peace, which I do not see, they are still recompensed by all those benefits which an army established in a City bring; for without them, nothing is secure. I conclude that whoever desires a small number in order to be able to pay them, or for any other reason cited by you, does not know (what he is doing); for it will also happen, in my opinion, that any number will always diminish in your hands, because of the infinite impediments that men have; so that the small number will succeed at nothing. However, when you have a large organization, you can at your election avail yourself of few or of many. In addition to this, it serves you in fact and reputation, for the large number will always give you reputation. Moreover, in creating the organization, in order to keep men trained, if you enroll a small number of men in many countries, and the armies are very distant from each other, you cannot without the gravest injury to them assemble them for (joint) exercises, and without this training the organization is useless, as will be shown in its proper place.

COSIMO: What you have said is enough on my question: but I now desire that you resolve another doubt for me. There are those who say that such a multitude of armed men would cause confusion, trouble, and disorder in the country.

FABRIZIO: This is another vain opinion for the reason I will tell you. These organized under arms can cause disorders in two ways: either among themselves, or against others; both of these can be obviated where discipline by itself should not do so: for as to troubles among themselves, the organization removes them, not brings them up, because in the organization you give them arms and leaders. If the country where you organize them is so unwarlike that there are not arms among its men, and so united that there are no leaders, such an organization will make them more ferocious against the foreigner, but in no way will make it more disunited, because men well organized, whether armed or unarmed, fear the laws, and can never change, unless the leaders you give them cause a change; and I will later tell you the manner of doing this. But if the country where you have organized an army is warlike and disunited, this organization alone is reason enough to unite them, for these men have arms and leaders for themselves: but the arms are useless for war, and the leaders causes of troubles; but this organization gives them arms useful for war, and leaders who will extinguish troubles; for as soon as some one is injured in that country, he has recourse to his (leader) of the party, who, to maintain his reputation, advises him to avenge himself, (and) not to remain in peace. The public leader does the contrary. So that by this means, the causes for trouble are removed, and replaced by those for union; and provinces which are united but effeminate (unwarlike) lose their usefulness but maintain the union, while those that are disunited and troublesome remain united; and that disordinate ferocity which they usually employ, is turned to public usefulness.

As to desiring that they do us injury against others, it should be kept in mind that they cannot do this except by the leaders who govern them. In desiring that the leaders do not cause disorders, it is necessary to have care that they do not acquire too much authority over them. And you have to keep in mind that this authority is acquired either naturally or by accident: And as to nature, it must be provided that whoever is born in one place is not put in charge of men enrolled in another place, but is made a leader in those places where he does not have any natural connections. As to accidents, the organization should be such that each year the leaders are exchanged from command to command; for continuous authority over the same men generates so much unity among them, which can easily be converted into prejudice against the Prince. As to these exchanges being useful to those who have employed them, and injurious to those who have not observed them, is known from the example of the Kingdom of Assyria and from the Empire of the Romans, in which it is seen that the former Kingdom endured a thousand years without tumult and without civil war; which did not result from anything else than the exchanges of those Captains, who were placed in charge of the care of the armies, from place to place every year. Nor, for other reasons, (did it result) in the Roman Empire; once the blood (race) of Caesar was extinguished, so many civil wars arose among the Captains of the armies, and so many conspiracies of the above mentioned Captains against the Emperors, resulting from the continuing of those Captains in their same Commands. And if any of those Emperors, and any who later held the Empire by reputation, such as Hadrian, Marcus, Severus, and others like them, would have observed such happenings, and would have introduced this custom of exchanging Captains in that Empire, without doubt they would have made it more tranquil and lasting; for the Captains would have had fewer opportunities for creating tumults, and the Emperors fewer causes to fear them, and the Senate, when there was a lack in the succession, would have had more authority in the election of Emperors, and consequently, better conditions would have resulted. But the bad customs of men, whether from ignorance or little diligence, or from examples of good or bad, are never put aside.

COSIMO: I do not know if, with my question, I have gone outside the limits you set; for from the Deletto we have entered into another discussion, and if I should not be excused a little, I shall believe I merit some reproach.

FABRIZIO: This did us no harm; for all this discussion was necessary in wanting to discuss the Organization (of an Army), which, being censured by many, it was necessary to explain it, if it is desired that this should take place before the Deletto. And before I discuss the other parts, I want to discuss the Deletto for men on horseback. This (selection) was done by the ancients from among the more wealthy, having regard both for the age and quality of the men, selecting three hundred for each legion: so that the Roman cavalry in every Consular army did not exceed six hundred.

COSIMO: Did you organize the cavalry in order to train them at home and avail yourself of them in the future?

FABRIZIO: Actually it is a necessity and cannot be done otherwise, if you want to have them take up arms for you, and not to want to take them away from those who make a profession of them.

COSIMO: How would you select them?

FABRIZIO: I would imitate the Romans: I would take the more wealthy, and give them leaders in the same manner as they are given to others today, and I would arm them, and train them.

COSIMO: Would it be well to give these men some provision?

FABRIZIO: Yes, indeed: but only as much as is necessary to take care of the horse; for, as it brings an expense to your subjects, they could complain of you. It would be necessary, therefore, to pay them for the horse and its upkeep.

COSIMO: How many would you make? How would you arm them?

FABRIZIO: You pass into another discussion. I will tell you in its place, which will be when I have said how the infantry ought to be armed, and how they should prepare for an engagement.

Second Book

I believe that it is necessary, once the men are found, to arm them; and in wanting to do this, I believe it is necessary to examine what arms the ancients used, and from them select the best. The Romans divided their infantry into the heavily and lightly armed. The light armed they gave the name Veliti. Under this name they included all those who operated with the sling, cross-bow, and darts: and the greater part of them carried a helmet (head covering) and a shield on the arm for their defense. These men fought outside the regular ranks, and apart from the heavy armor, which was a Casque that came up to the shoulders, they also carried a Cuirass which, with the skirt, came down to the knees, and their arms and legs were covered by shin-guards and bracelets; they also carried a shield on the arm, two arms in length and one in width, which had an iron hoop on it to be able to sustain a blow, and another underneath, so that in rubbing on the ground, it should not be worn out. For attacking, they had cinched on their left side a sword of an arm and a half length, and a dagger on the right side. They carried a spear, which they called Pilus, and which they hurled at the enemy at the start of a battle. These were the important Roman arms, with which they conquered the world. And although some of the ancient writers also gave them, in addition to the aforementioned arms, a shaft in the hand in the manner of a spit, I do not know how a staff can be used by one who holds a shield, for in managing it with two hands it is impeded by the shield, and he cannot do anything worthwhile with one hand because of its heaviness. In addition to this, to combat in the ranks with the staff (as arms) is useless, except in the front rank where there is ample space to deploy the entire staff, which cannot be done in the inner ranks, because the nature of the battalions ((as I will tell you in their organization)) is to press its ranks continually closer together, as this is feared less, even though inconvenient, than for the ranks to spread further apart, where the danger is most apparent. So that all the arms which exceed two arms in length are useless in tight places; for if you have a staff and want to use it with both hands, and handled so that the shield should not annoy you, you cannot attack an enemy with it who is next to you. If you take it in one hand in order to serve yourself of the shield, you cannot pick it up except in the middle, and there remains so much of the staff in the back part, that those who are behind impede you in using it. And that this is true, that the Romans did not have the staff, or, having it, they valued it little, you will read in all the engagements noted by Titus Livius in his history, where you will see that only very rarely is mention made of the shaft, rather he always says that, after hurling the spears, they put their hands on the sword. Therefore I want to leave this staff, and relate how much the Romans used the sword for offense, and for defense, the shield together with the other arms mentioned above.

The Greeks did not arm so heavily for defense as did the Romans, but in the offense relied more on this staff than on the sword, and especially the Phalanxes of Macedonia, who carried staffs which they called Sarisse, a good ten arms in length, with which they opened the ranks of the enemy and maintained order in the Phalanxes. And although other writers say they also had a shield, I do not know ((for the reasons given above)) how the Sarisse and the shield could exist together. In addition to this, in the engagement that Paulus Emilius had with Perseus, King of Macedonia, I do not remember mention being made of shields, but only of the Sarisse and the difficulty the Romans had in overcoming them. So that I conjecture that a Macedonian Phalanx was nothing else than a battalion of Swiss is today, who have all their strength and power in their pikes. The Romans ((in addition to the arms)) ornamented the infantry with plumes; which things make the sight of an army beautiful to friends, and terrible to the enemy. The arms for men on horseback in the original ancient Roman (army) was a round shield, and they had the head covered, but the rest (of the body) without armor. They had a sword and a staff with an iron point, long and thin; whence they were unable to hold the shield firm, and only make weak movements with the staff, and because they had no armor, they were exposed to wounds. Afterwards, with time, they were armed like the infantry, but the shield was much smaller and square, and the staff more solid and with two iron tips, so that if the one side was encumbered, they could avail themselves of the other. With these arms, both for the infantry and the cavalry, my Romans occupied all the world, and it must be believed, from the fruits that are observed, that they were the best armed armies that ever existed.

And Titus Livius, in his histories, gives many proofs, where, in coming to the comparison with enemy armies, he says, "but the Romans were superior in virtu, kinds of arms, and discipline". And, therefore, I have discussed more in particular the arms of the victors than those of the losers. It appears proper to me to discuss only the present methods of arming. The infantry have for their defense a breast plate of iron, and for offense a lance nine armlengths long, which they call a pike, and a sword at their side, rather round in the point than sharp. This is the ordinary armament of the infantry today, for few have their arms and shins (protected by) armor, no one the head; and those few carry a halberd in place of a pike, the shaft of which ((as you know)) is three armlengths long, and has the iron attached as an axe. Among them they have three Scoppettieri (Exploders, i.e., Gunners), who, with a burst of fire fill that office which anciently was done by slingers and bow-men. This method of arming was established by the Germans, and especially by the Swiss, who, being poor and wanting to live in freedom, were, and are, obliged to combat with the ambitions of the Princes of Germany, who were rich and could raise horses, which that people could not do because of poverty: whence it happened that being on foot and wanting to defend themselves from enemies who were on horseback, it behooved them to search the ancient orders and find arms which should defend them from the fury of horses. This necessity has caused them to maintain or rediscover the ancient orders, without which, as every prudent man affirms, the infantry is entirely useless. They therefore take up pikes as arms, which are most useful not only in sustaining (the attacks of) horses, but to overcome them. And because of the virtu of these arms and ancient orders, the Germans have assumed so much audacity, that fifteen or twenty thousand of them would assault any great number of horse, and there have been many examples of this seen in the last twenty five years. And this example of their virtu founded on these arms and these orders have been so powerful, that after King Charles passed into Italy, every nation has imitated them: so that the Spanish armies have come into a very great reputation.

COSIMO: What method of arms do you praise more, this German one or the ancient Roman?

FABRIZIO: The Roman without any doubt, and I will tell you the good and the bad of one and the other. The German infantry can sustain and overcome the cavalry. They are more expeditious in marching and in organizing themselves, because they are not burdened with arms. On the other hand, they are exposed to blows from near and far because of being unarmed. They are useless in land battles and in every fight where there is stalwart resistance. But the Romans sustained and overcame the cavalry, as these (Germans) do. They were safe from blows near and far because they were covered with armor. They were better able to attack and sustain attacks having the shields. They could more actively in tight places avail themselves of the sword than these (Germans) with the pike; and even if the latter had the sword, being without a shield, they become, in such a case, (equally) useless. They (the Romans) could safely assault towns, having the body covered, and being able to cover it even better with the shield. So that they had no other inconvenience than the heaviness of the arms (armor) and the annoyance of having to carry them; which inconveniences they overcame by accustoming the body to hardships and inducing it to endure hard work. And you know we do not suffer from things to which we are accustomed. And you must understand this, that the infantry must be able to fight with infantry and cavalry, and those are always useless who cannot sustain the (attacks of the) cavalry, or if they are able to sustain them, none the less have fear of infantry who are better armed and organized than they. Now if you will consider the German and the Roman infantry, you will find in the German ((as we have said)) the aptitude of overcoming cavalry, but great disadvantages when fighting with an infantry organized as they are, and armed as the Roman. So that there will be this advantage of the one over the other, that the Romans could overcome both the infantry and the cavalry, and the Germans only the cavalry.

COSIMO: I would desire that you give some more particular example, so that we might understand it better.

FABRIZIO: I say thusly, that in many places in our histories you will find the Roman infantry to have defeated numberless cavalry, but you will never find them to have been defeated by men on foot because of some defect they may have had in their arms or because of some advantage the enemy had in his. For if their manner of arming had been defective, it was necessary for them to follow one of two courses: either when they found one who was better armed than they, not to go on further with the conquest, or that they take up the manner of the foreigner, and leave off theirs: and since neither ensued, there follows, what can be easily conjectured, that this method of arming was better than that of anyone else. This has not yet occurred with the German infantry; for it has been seen that anytime they have had to combat with men on foot organized and as obstinate as they, they have made a bad showing; which results from the disadvantage they have in trying themselves against the arms of the enemy. When Filippo Visconti, Duke of Milan, was assaulted by eighteen thousand Swiss, he sent against them Count Carmingnuola, who was his Captain at that time. This man with six thousand cavalry and a few infantry went to encounter them, and, coming hand to hand with them, was repulsed with very great damage. Whence Carmingnuola as a prudent man quickly recognized the power of the enemy arms, and how much they prevailed against cavalry, and the weakness of cavalry against those on foot so organized; and regrouping his forces, again went to meet the Swiss, and as they came near he made his men-at-arms descend from their horses, and in that manner fought with them, and killed all but three thousand, who, seeing themselves consumed without having any remedy, threw their arms on the ground and surrendered.

COSIMO: Whence arises such a disadvantage?

FABRIZIO: I have told you a little while ago, but since you have not understood it, I will repeat it to you. The German infantry ((as was said a little while ago)) has almost no armor in defending itself, and use pikes and swords for offense. They come with these arms and order of battle to meet the enemy, who ((if he is well equipped with armor to defend himself, as were the men-at-arms of Carmingnuola who made them descend to their feet)) comes with his sword and order of battle to meet him, and he has no other difficulty than to come near the Swiss until he makes contact with them with the sword; for as soon as he makes contact with them, he combats them safely, for the German cannot use the pike against the enemy who is next to him because of the length of the staff, so he must use the sword, which is useless to him, as he has no armor and has to meet an enemy that is (protected) fully by armor. Whence, whoever considers the advantages and disadvantages of one and the other, will see that the one without armor has no remedy, but the one well armored will have no difficulty in overcoming the first blow and the first passes of the pike: for in battles, as you will understand better when I have demonstrated how they are put together, the men go so that of necessity they accost each other in a way that they are attacked on the breast, and if one is killed or thrown to the ground by the pike, those on foot who remain are so numerous that they are sufficient for victory. From this there resulted that Carmingnuola won with such a massacre of the Swiss, and with little loss to himself.

COSIMO: I see that those with Carmingnuola were men-at-arms, who, although they were on foot, were all covered with iron (armor), and, therefore, could make the attempt that they made; so that I think it would be necessary to arm the infantry in the same way if they want to make a similar attempt.

FABRIZIO: If you had remembered how I said the Romans were armed, you would not think this way. For an infantryman who has his head covered with iron, his breast protected by a cuirass and a shield, his arms and legs with armor, is much more apt to defend himself from pikes, and enter among them, than is a man-at-arms (cavalryman) on foot. I want to give you a small modem example. The Spanish infantry had descended from Sicily into the Kingdom of Naples in order to go and meet Consalvo who was besieged in Barletta by the French. They came to an encounter against Monsignor D'Obigni with his men-at-arms, and with about four thousand German infantry. The Germans, coming hand to hand with their pikes low, penetrated the (ranks of the) Spanish infantry; but the latter, aided by their spurs and the agility of their bodies, intermingled themselves with the Germans, so that they (the Germans) could not get near them with their swords; whence resulted the death of almost all of them, and the victory of the Spaniards.

Everyone knows how many German infantry were killed in the engagement at Ravenna, which resulted from the same causes, for the Spanish infantry got as close as the reach of their swords to the German infantry, and would have destroyed all of them, if the German infantry had not been succored by the French Cavalry: none the less, the Spaniards pressing together made themselves secure in that place. I conclude, therefore, that a good infantry not only is able to sustain the (attack) of cavalry, but does not have fear of infantry, which ((as I have said many times)) proceeds from its arms (armor) and organization (discipline).

COSIMO: Tell us, therefore, how you would arm them.

FABRIZIO: I would take both the Roman arms and the German, and would want half to be armed as the Romans, and the other half as the Germans. For, if in six thousand infantry ((as I shall explain a little later)) I should have three thousand infantry with shields like the Romans, and two thousand pikes and a thousand gunners like the Germans, they would be enough for me; for I would place the pikes either in the front lines of the battle, or where I should fear the cavalry most; and of those with the shield and the sword, I would serve myself to back up the pikes and to win the engagement, as I will show you. So that I believe that an infantry so organized should surpass any other infantry today.

COSIMO: What you have said to us is enough as regards infantry, but as to cavalry, we desire to learn which seems the more strongly armed to you, ours or that of the ancients?

FABRIZIO: I believe in these times, with respect to saddles and stirrups not used by the ancients, one stays more securely on the horse than at that time. I believe we arm more securely: so that today one squadron of very heavily (armed) men-at-arms comes to be sustained with much more difficulty than was the ancient cavalry. With all of this, I judge, none the less, that no more account ought to be taken of the cavalry than was taken anciently; for ((as has been said above)) they have often in our times been subjected to disgrace by the infantry armed (armored) and organized as (described) above. Tigranus, King of Armenia, came against the Roman army of which Lucullus was Captain, with (an army) of one hundred fifty thousand cavalry, among whom were many armed as our men-at-arms, whom they called Catafratti, while on the other side the Romans did not total more than six thousand (cavalry) and fifteen thousand infantry; so that Tigranus, when he saw the army of the enemy, said: "These are just about enough horsemen for an embassy". None the less, when they came to battle, he was routed; and he who writes of that battle blames those Catafratti, showing them to be useless, because, he says, that having their faces covered, their vision was impaired and they were little adept at seeing and attacking the enemy, and as they were heavily burdened by the armor, they could not regain their feet when they fell, nor in any way make use of their persons. I say, therefore, that those People or Kingdoms which esteem the cavalry more than the infantry, are always weaker and more exposed to complete ruin, as has been observed in Italy in our times, which has been plundered, ruined, and overrun by foreigners, not for any other fault than because they had paid little attention to the foot soldiers and had mounted all their soldiers on horses. Cavalry ought to be used, but as a second and not the first reliance of an army; for they are necessary and most useful in undertaking reconnaissance, in overrunning and despoiling the enemy country, and to keep harassing and troubling the enemy army so as to keep it continually under arms, and to impede its provisions; but as to engagements and battles in the field, which are the important things in war and the object for which armies are organized, they are more useful in pursuing than in routing the enemy, and are much more inferior to the foot soldier in accomplishing the things necessary in accomplishing such (defeats).

COSIMO: But two doubts occur to me: the one, that I know that the Parthians did not engage in war except with cavalry, yet they divided the world with the Romans: the other, that I would like you to tell me how the (attack of) the cavalry can be sustained by the infantry, and whence arises the virtu of the latter and the weakness of the former?

FABRIZIO: Either I have told you, or I meant to tell you, that my discussion on matters of war is not going beyond the limits of Europe. Since this is so, I am not obliged to give reasons for that which is the custom in Asia. Yet, I have this to say, that the army of Parthia was completely opposite to that of the Romans, as the Parthians fought entirely on horseback, and in the fighting was about confused and disrupted, and was a way of fighting unstable and full of uncertainties. The Romans, it may be recalled, were almost all on foot, and fought pressed closely together, and at various times one won over the other, according as the site (of the battle) was open or tight; for in the latter the Romans were superior, but in the former the Parthians, who were able to make a great trial with that army with respect to the region they had to defend, which was very open with a seacoast a thousand miles distant, rivers two or three days (journey) apart from each other, towns likewise, and inhabitants rare: so that a Roman army, heavy and slow because of its arms and organization, could not pursue him without suffering great harm, because those who defended the country were on horses and very speedy, so that he would be in one place today, and tomorrow fifty miles distant. Because of this, the Parthians were able to prevail with cavalry alone, and thus resulted the ruin of the army of Crassus, and the dangers to those of Marcantonio. But ((as I have said)) I did not intend in this discussion of mine to speak of armies outside of Europe; and, therefore, I want to continue on those which the Romans and Greeks had organized in their time, and that the Germans do today.

But let us come to the other question of yours, in which you desire to know what organization or what natural virtu causes the infantry to be superior to the cavalry. And I tell you, first, that the horses cannot go in all the places that the infantry do, because it is necessary for them either to turn back after they have come forward, or turning back to go forward, or to move from a stand-still, or to stand still after moving, so that, without doubt, the cavalry cannot do precisely thus as the infantry. Horses cannot, after being put into disorder from some attack, return to the order (of the ranks) except with difficulty, and even if the attack does not occur; the infantry rarely do this. In addition to this, it often occurs that a courageous man is mounted on a base horse, and a base man on a courageous horse, whence it must happen that this difference in courage causes disorders. Nor should anyone wonder that a Knot (group) of infantry sustains every attack of the cavalry, for the horse is a sensible animal and knows the dangers, and goes in unwillingly. And if you would think about what forces make him (the horse) go forward and what keep him back, without doubt you will see that those which hold him back are greater than those which push him; for spurs make him go forward, and, on the other hand, the sword and the pike retain him. So that from both ancient and modem experiences, it has been seen that a small group of infantry can be very secure from, and even actually insuperable to, the cavalry. And if you should argue on this that the Elan with which he comes makes it more furious in hurling himself against whoever wants to sustain his attack, and he responds less to the pike than the spur, I say that, as soon as the horse so disposed begins to see himself at the point of being struck by the points of the pikes, either he will by himself check his gait, so that he will stop as soon as he sees himself about to be pricked by them, or, being pricked by them, he will turn to the right or left. If you want to make a test of this, try to run a horse against a wall, and rarely will you find one that will run into it, no matter with what Elan you attempt it. Caesar, when he had to combat the Swiss in Gaul, dismounted and made everyone dismount to their feet, and had the horses removed from the ranks, as they were more adept at fleeing than fighting.

But, notwithstanding these natural impediments that horses have, the Captain who leads the infantry ought to select roads that have as many obstacles for horses as possible, and rarely will it happen that the men will not be able to provide for their safety from the kind of country. If one marches among hills, the location of the march should be such that you may be free from those attacks of which you may be apprehensive; and if you go on the plains, rarely will you find one that does not have crops or woods which will provide some safety for you, for every bush and embankment, even though small, breaks up that dash, and every cultivated area where there are vines and other trees impedes the horses. And if you come to an engagement, the same will happen to you as when marching, because every little impediment which the horse meets cause him to lose his fury. None the less, I do not want to forget to tell you one thing, that although the Romans esteemed much their own discipline and trusted very much on their arms (and armor), that if they had to select a place, either so rough to protect themselves from horses and where they could not be able to deploy their forces, or one where they had more to fear from the horses but where they were able to spread out, they would always take the latter and leave the former.

But, as it is time to pass on to the training (of the men), having armed this infantry according to the ancient and modem usage, we shall see what training they gave to the Romans before the infantry were led to battle. Although they were well selected and better armed, they were trained with the greatest attention, because without this training a soldier was never any good. This training consisted of three parts. The first, to harden the body and accustom it to endure hardships, to act faster, and more dexterously. Next, to teach the use of arms: The third, to teach the trainees the observance of orders in marching as well as fighting and encamping. These are the three principal actions which make an army: for if any army marches, encamps, and fights, in a regular and practical manner, the Captain retains his honor even though the engagement should not have a good ending. All the ancient Republics, therefore, provided such training, and both by custom and law, no part was left out. They therefore trained their youth so as to make them speedy in running, dextrous in jumping, strong in driving stakes and wrestling. And these three qualities are almost necessary in a soldier; for speed makes him adept at occupying places before the enemy, to come upon him unexpectedly, and to pursue him when he is routed. Dexterity makes him adept at avoiding blows, jumping a ditch and climbing over an embankment. Strength makes him better to carry arms, hurl himself against an enemy, and sustain an attack. And above all, to make the body more inured to hardships, they accustom it to carry great weights. This accustoming is necessary, for in difficult expeditions it often happens that the soldier, in addition to his arms, must carry provisions for many days, and if he had not been accustomed to this hard work, he would not be able to do it, and, hence, he could neither flee from a danger nor acquire a victory with fame.

As to the teaching of the use of arms, they were trained in this way. They had the young men put on arms (armor) which weighed more than twice that of the real (regular) ones, and, as a sword, they gave them a leaded club which in comparison was very heavy. They made each one of them drive a pole into the ground so that three arm-lengths remained (above ground), and so firmly fixed that blows would not drive it to one side or have it fall to the ground; against this pole, the young men were trained with the shield and the club as against an enemy, and sometime they went against it as if they wanted to wound the head or the face, another time as if they wanted to puncture the flank, sometimes the legs, sometime they drew back, another time they went forward. And in this training, they had in mind making themselves adept at covering (protecting) themselves and wounding the enemy; and since the feigned arms were very heavy, the real ones afterwards seemed light. The Romans wanted their soldiers to wound (the enemy) by the driving of a point against him, rather than by cutting (slashing), as much because such a blow was more fatal and had less defense against it, as also because it left less uncovered (unprotected) those who were wounding, making him more adept at repeating his attack, than by slashing. Do you not wonder that those ancients should think of these minute details, for they reasoned that where men had to come hand to hand (in battle), every little advantage is of the greatest importance; and I will remind you of that, because the writers say of this that I have taught it to you. Nor did the ancients esteem it a more fortunate thing in a Republic than to have many of its men trained in arms; for it is not the splendor of jewels and gold that makes the enemy submit themselves to you, but only the fear of arms. Moreover, errors made in other things can sometimes be corrected afterwards, but those that are made in war, as the punishment happens immediately, cannot be corrected. In addition to this, knowing how to fight makes men more audacious, as no one fears to do the things which appear to him he has been taught to do. The ancients, therefore, wanted their citizens to train in every warlike activity; and even had them throw darts against the pole heavier than the actual ones: which exercise, in addition to making men expert in throwing, also makes the arm more limber and stronger. They also taught them how to draw the bow and the sling, and placed teachers in charge of doing all these things: so that when (men) were selected to go to war, they were already soldiers in spirit and disposition. Nor did these remain to teach them anything else than to go by the orders and maintain themselves in them whether marching or combatting: which they easily taught by mixing themselves with them, so that by knowing how to keep (obey) the orders, they could exist longer in the army.

COSIMO: Would you have them train this way now?

FABRIZIO: Many of those which have been mentioned, like running wrestling, making them jump, making them work hard under arms heavier than the ordinary, making them draw the crossbow and the sling; to which

I would add the light gun, a new instrument ((as you know)), and a necessary one. And I would accustom all the youth of my State to this training: but that part of them whom I have enrolled to fight, I would (especially) train with greater industry and more solicitude, and I would train them always on their free days. I would also desire that they be taught to swim, which is a very useful thing, because there are not always bridges at rivers, nor ships ready: so that if your army does not know how to swim, it may be deprived of many advantages, and many opportunities, to act well are taken away. The Romans, therefore, arranged that the young men be trained on the field of Mars, so that having the river Tiber nearby, they would be able after working hard in exercises on land to refresh themselves in the water, and also exercise them in their swimming.

I would also do as the ancients and train those who fight on horseback: which is very necessary, for in addition to knowing how to ride, they would know how to avail themselves of the horse (in maneuvering him). And, therefore, they arranged horses of wood on which they straddled, and jumped over them armed and unarmed without any help and without using their hands: which made possible that in a moment, and at a sign from the Captain, the cavalry to become as foot soldiers, and also at another sign, for them to be remounted. And as such exercises, both on foot and horseback, were easy at that time, so now it should not be difficult for that Republic or that Prince to put them in practice on their youth, as is seen from the experience of Western Cities, where these methods similar to these institutions are yet kept alive.

They divide all their inhabitants into several parts, and assign one kind of arms of those they use in war to each part. And as they used pikes, halberds, bows, and light guns, they called them pikemen, halberdiers, archers, and gunners. It therefore behooved all the inhabitants to declare in what order they wanted to be enrolled. And as all, whether because of age or other impediment, are not fit for war (combat), they make a selection from each order and they call them the Giurati (Sworn Ones), who, on their free days, are obliged to exercise themselves in those arms in which they are enrolled: and each one is assigned his place by the public where such exercises are to be carried on, and those who are of that order but are not sworn, participate by (contributing) money for those expenses which are necessary for such exercises. That which they do, therefore, we can do, but our little prudence does not allow us to take up any good proceeding.

From these exercises, it resulted that the ancients had good infantry, and that now those of the West have better infantry than ours, for the ancients exercised either at home as did those Republics, or in the armies as did those Emperors, for the reasons mentioned above. But we do not want to exercise at home, and we cannot do so in the field because they are not our subjects and we cannot obligate them to other exercises than they themselves want. This reason has caused the armies to die out first, and then the institutions, so that the Kingdoms and the Republics, especially the Italian, exist in such a weak condition today.

But let us return to our subject, and pursuing this matter of training, I say, that it is not enough in undertaking good training to have hardened the men, made them strong, fast and dextrous, but it is also necessary to teach them to keep discipline, obey the signs, the sounds (of the bugle), and the voice of the Captain; to know when to stand, to retire, to go forward, and when to combat, to march, to maintain ranks; for without this discipline, despite every careful diligence observed and practiced, an army is never good. And without doubt, bold but undisciplined men are more weak than the timid but disciplined ones; for discipline drives away fear from men, lack of discipline makes the bold act foolishly. And so that you may better understand what will be mentioned below, you have to know that every nation has made its men train in the discipline of war, or rather its army as the principal part, which, if they have varied in name, they have varied little in the numbers of men involved, as all have comprised six to eight thousand men. This number was called a Legion by the Romans, a Phalanx by the Greeks, a Caterna by the Gauls. This same number, by the Swiss, who alone retain any of that ancient military umbrage, in our times is called in their language what in ours signifies a Battalion. It is true that each one is further subdivided into small Battaglia (Companies), and organized according to its purpose. It appears to me, therefore, more suitable to base our talk on this more notable name, and then according to the ancient and modern systems, arrange them as best as is possible. And as the Roman Legions were composed of five or six thousand men, in ten Cohorts, I want to divide our Battalion into ten Companies, and compose it of six thousand men on foot; and assign four hundred fifty men to each Company, of whom four hundred are heavily armed and fifty lightly armed: the heavily armed include three hundred with shields and swords, and will be called Scudati (shield bearers), and a hundred with pikes, and will be called pikemen: the lightly armed are fifty infantry armed with light guns, cross-bows, halberds, and bucklers, and these, from an ancient name, are called regular (ordinary) Veliti: the whole ten Companies, therefore, come to three thousand shield bearers; a thousand ordinary pikemen, and one hundred fifty ordinary Veliti, all of whom comprise (a number of) four thousand five hundred infantry. And we said we wanted to make a Battalion of six thousand men; therefore it is necessary to add another one thousand five hundred infantry, of whom I would make a thousand with pikes, whom I will call extraordinary pikemen, (and five hundred light armed, whom I will call extraordinary Veliti): and thus my infantry would come ((according as was said a little while ago)) to be composed half of shield bearers and half among pikemen and other arms (carriers). In every Company, I would put in charge a Constable, four Centurions, and forty Heads of Ten, and in addition, a Head of the ordinary Veliti with five Heads of Ten. To the thousand extraordinary pikemen, I would assign three Constables, ten Centurions, and a hundred Heads of Ten: to the extraordinary Veliti, two Constables, five Centurions, and fifty Heads of Ten. I would also assign a general Head for the whole Battalion. I would want each Constable to have a distinct flag and (bugle) sound.

Summarizing, therefore, a Battalion would be composed of ten Companies, of three thousand shield bearers, a thousand ordinary pikemen, a thousand extraordinary pikemen, five hundred ordinary Veliti, and five hundred extraordinary Veliti: thus they would come to be six thousand infantry, among whom there would be one thousand five hundred Heads of Ten, and in addition fifteen Constables, with fifteen Buglers and fifteen flags, fifty five Centurions, ten Captains of ordinary Veliti, and one Captain for the whole Battalion with its flag and Bugler. And I have knowingly repeated this arrangement many times, so that then, when I show you the methods for organizing the Companies and the armies, you will not be confounded.

I say, therefore, that any King or Republic which would want to organize its subjects in arms, would provide them with these parties and these arms, and create as many battalions in the country as it is capable of doing: and if it had organized it according to the division mentioned above, and wanting to train it according to the orders, they need only to be trained Company by Company. And although the number of men in each of them could not be themselves provide a reasonably (sized) army, none the less, each man can learn to do what applies to him in particular, for two orders are observed in the armies: the one, what men ought to do in each Company: the other, what the Company ought to do afterwards when it is with others in an army: and those men who carry out the first, will easily observe the second: but without the first, one can never arrive at the discipline of the second. Each of these Companies, therefore, can by themselves learn to maintain (discipline in) their ranks in every kind and place of action, and then to know how to assemble, to know its (particular bugle) call, through which it is commanded in battle; to know how to recognize by it ((as galleys do from the whistle)) as to what they have to do, whether to stay put, or go forward, or turn back, or the time and place to use their arms. So that knowing how to maintain ranks well, so that neither the action nor the place disorganizes them, they understand well the commands of the leader by means of the (bugle) calls, and knowing how to reassemble quickly, these Companies then can easily ((as I have said)), when many have come together, learn to do what each body of them is obligated to do together with other Companies in operating as a reasonably (sized) army. And as such a general practice also is not to be esteemed little, all the Battalions can be brought together once or twice in the years of peace, and give them a form of a complete army, training it for several days as if it should engage in battle, placing the front lines, the flanks, and auxiliaries in their (proper) places.

And as a Captain arranges his army for the engagement either taking into account the enemy he sees, or for that which he does not see but is apprehensive of, the army ought to be trained for both contingencies, and instructed so that it can march and fight when the need arises; showing your soldiers how they should conduct themselves if they should be assaulted by this band or that. And when you instruct them to fight against an enemy they can see, show them how the battle is enkindled, where they have to retire without being repulsed, who has to take their places, what signs, what (bugle) calls, and what voice they should obey, and to practice them so with Companies and by mock attacks, that they have the desire for real battle. For a courageous army is not so because the men in it are courageous, but because the ranks are well disciplined; for if I am of the first line fighters, and being overcome, I know where I have to retire, and who is to take my place, I will always fight with courage seeing my succor nearby: If I am of the second line fighters, I would not be dismayed at the first line being pushed back and repulsed, for I would have presupposed it could happen, and I would have desired it in order to be he who, as it was not them, would give the victory to my patron. Such training is most necessary where a new army is created; and where the army is old (veteran), it is also necessary for, as the Romans show, although they knew the organization of their army from childhood, none the less, those Captains, before they came to an encounter with the enemy, continually exercised them in those disciplines. And Joseph in his history says, that the continual training of the Roman armies resulted in all the disturbance which usually goes on for gain in a camp, was of no effect in an engagement, because everyone knew how to obey orders and to fight by observing them. But in the armies of new men which you have to put together to combat at the time, or that you caused to be organized to combat in time, nothing is done without this training, as the Companies are different as in a complete army; for as much discipline is necessary, it must be taught with double the industry and effort to those who do not have it, and be maintained in those who have it, as is seen from the fact that many excellent Captains have tired themselves without any regard to themselves.

COSIMO: And it appears to me that this discussion has somewhat carried you away, for while you have not yet mentioned the means with which

Companies are trained, you have discussed engagements and the complete army.

FABRIZIO: You say the truth, and truly the reason is the affection I have for these orders, and the sorrow that I feel seeing that they are not put into action: none the less, have no fear, but I shall return to the subject. As I have told you, of first importance in the training of the Company is to know how to maintain ranks. To do this, it is necessary to exercise them in those orders, which they called Chiocciole (Spiralling). And as I told you that one of these Companies ought to consist of four hundred heavily armed infantry, I will stand on this number. They should, therefore, be arranged into eighty ranks (files), with five per file. Then continuing on either strongly or slowly, grouping them and dispersing them; which, when it is done, can be demonstrated better by deeds than by words: afterwards, it becomes less necessary, for anyone who is practiced in these exercises knows how this order proceeds, which is good for nothing else but to accustom the soldiers to maintain ranks. But let us come and put together one of those Companies.

I say that these can be formed in three ways: the first and most useful is to make it completely massive and give it the form of two squares: the second is to make the square with a homed front: the third is to make it with a space in the center, which they call Piazza (plaza). The method of putting together the first form can be in two steps. The first is to have the files doubled, that is, that the second file enters the first, the fourth into the third, and sixth into the fifth, and so on in succession; so that where there were eighty files and five (men) per file, they become forty files and ten per file. Then make them double another time in the same manner, placing one file within the other, and thus they become twenty files of twenty men per file. This makes almost a square, for although there are so many men on one side (of the square) as the other, none the less, on the side of the front, they come together so that (the side of) one man touches the next; but on the other side (of the square) the men are distant at least two arm lengths from each other, so that the square is longer from the front to the back (shoulders), then from one side (flank) to the other. (So that the rectangle thus formed is called two squares).

And as we have to talk often today of the parts in front, in the rear, and on the side of this Company, and of the complete army, you will understand that when I will say either head or front, I mean to say the part in front; when I say shoulder, the part behind (rear); when I say flanks, the parts on the side.

The fifty ordinary Veliti of the company are not mixed in with the other files, but when the company is formed, they extend along its flanks.

The other method of putting together (forming) the company is this; and because it is better than the first, I want to place in front of your eyes in detail how it ought to be organized. I believe you remember the number of men and the heads which compose it, and with what arms it is armed. The form, therefore, that this company ought to have is ((as I have said)) of twenty files, twenty men per file, five files of pikemen in front, and fifteen files of shield bearers on the shoulders (behind); two centurions are in front and two behind in the shoulders who have the office of those whom the ancients called Tergiduttori (Rear-leaders): The Constable, with the flag and bugler, is in that space which is between the five files of pikemen and the fifteen of shield-bearers: there is one of the Captains of the Ten on every flank, so that each one is alongside his men, those who are on the left side of his right hand, those on the right side on his left hand. The fifty Veliti are on the flanks and shoulders (rear) of the company. If it is desired, now, that regular infantry be employed, this company is put together in this form, and it must organize itself thusly: Have the infantry be brought to eighty files, five per file, as we said a little while ago; leaving the Veliti at the head and on the tail (rear), even though they are outside this arrangement; and it ought to be so arranged that each Centurion has twenty files behind him on the shoulders, and those immediately behind every Centurion are five files of pikemen, and the remaining shield-bearers: the Constable, with his flag and bugler, is in that space that is between the pikemen and the shield-bearers of the second Centurion, and occupies the places of three shield-bearers: twenty of the Heads of Ten are on the Flanks of the first Centurion on the left hand, and twenty are on the flanks of the last Centurion on the right hand. And you have to understand, that the Head of Ten who has to guide (lead) the pikemen ought to have a pike, and those who guide the shield-bearers ought to have similar arms.

The files, therefore, being brought to this arrangement, and if it is desired, by marching, to bring them into the company to form the head (front), you have to cause the first Centurion to stop with the first file of twenty, and the second to continue to march; and turning to the right (hand) he goes along the flanks of the twenty stopped files, so that he comes head-to-head with the other Centurion, where he too stops; and the third Centurion continues to march, also turning to the right (hand), and marches along the flanks of the stopped file so that he comes head-to-head with the other two Centurions; and when he also stops, the other Centurion follows with his file, also going to the right along the flanks of the stopped file, so that he arrives at the head (front) with the others, and then he stops; and the two Centurions who are alone quickly depart from the front and go to the rear of the company, which becomes formed in that manner and with those orders to the point which we showed a little while ago. The Veliti extend themselves along its flanks, according as they were disposed in the first method; which method is called

Doubling by the straight line, and this last (method) is called Doubling by the flanks.

The first method is easier, while this latter is better organized, and is more adaptable, and can be better controlled by you, for it must be carried out by the numbers, that from five you make ten, ten twenty, twenty forty: so that by doubling at your direction, you cannot make a front of fifteen, or twenty five or thirty or thirty five, but you must proceed to where the number is less. And yet, every day, it happens in particular situations, that you must make a front with six or eight hundred infantry, so that the doubling by the straight line will disarrange you: yet this (latter) method pleases me more, and what difficulty may exist, can be more easily overcome by the proper exercise and practice of it.

I say to you, therefore, that it is more important than anything to have soldiers who know how to form themselves quickly, and it is necessary in holding them in these Companies, to train them thoroughly, and have them proceed bravely forward or backward, to pass through difficult places without disturbing the order; for the soldiers who know how to do this well, are experienced soldiers, and although they may have never met the enemy face to face, they can be called seasoned soldiers; and, on the contrary, those who do not know how to maintain this order, even if they may have been in a thousand wars, ought always to be considered as new soldiers. This applies in forming them when they are marching in small files: but if they are formed, and then become broken because of some accident that results either from the location or from the enemy, to reorganize themselves immediately is the important and difficult thing, in which much training and practice is needed, and in which the ancients placed much emphasis. It is necessary, therefore, to do two things: first, to have many countersigns in the Company: the other, always to keep this arrangement, that the same infantry always remain in the same file. For instance, if one is commanded to be in the second (file), he will afterwards always stay there, and not only in this same file, but in the same position (in the file); it is to be observed ((as I have said)) how necessary are the great number of countersigns, so that, coming together with other companies, it may be recognized by its own men. Secondly, that the Constable and Centurion have tufts of feathers on their head-dress different and recognizable, and what is more important, to arrange that the Heads of Ten be recognized. To which the ancients paid very much attention, that nothing else would do, but that they wrote numbers on their bucklers, calling then the first, second, third, fourth, etc. And they were not above content with this, but each soldier had to write on his shield the number of his file, and the number of his place assigned him in that file. The men, therefore, being thus countersigned (assigned), and accustomed to stay within these limits, if they should be disorganized, it is easy to reorganize them all quickly, for the flag staying fixed, the Centurions and Heads of Ten can judge their place by eye, and bring the left from the right, or the right from the left, with the usual distances between; the infantry guided by their rules and by the difference in countersigns, can quickly take their proper places, just as, if you were the staves of a barrel which you had first countersigned, I would wager you would put it (the barrel) back together with great ease, but if you had not so countersigned them (the staves), it is impossible to reassemble (the barrel). This system, with diligence and practice, can be taught quickly, and can be quickly learned, and once learned are forgotten with difficulty; for new men are guided by the old, and in time, a province which has such training, would become entirely expert in war. It is also necessary to teach them to turn in step, and do so when he should turn from the flanks and by the soldiers in the front, or from the front to the flanks or shoulders (rear). This is very easy, for it is sufficient only that each man turns his body toward the side he is commanded to, and the direction in which they turned becomes the front. It is true that when they turn by the flank, the ranks which turn go outside their usual area, because there is a small space between the breast to the shoulder, while from one flank to the other there is much space, which is all contrary to the regular formation of the company. Hence, care should be used in employing it. But this is more important and where more practice is needed, is when a company wants to turn entirely, as if it was a solid body. Here, great care and practice must be employed, for if it is desired to turn to the left, for instance, it is necessary that the left wing be halted, and those who are closer to the halted one, march much slower then those who are in the right wing and have to run; otherwise everything would be in confusion.

But as it always happens when an army marches from place to place, that the companies not situated in front, not having to combat at the front, or at the flanks or shoulders (rear), have to move from the flank or shoulder quickly to the front, and when such companies in such cases have the space necessary as we indicated above, it is necessary that the pikemen they have on that flank become the front, and the Heads of the Ten, Centurions, and Constables belonging to it relocate to their proper places. Therefore, in wanting to do this, when forming them it is necessary to arrange the eighty files of five per file, placing all the pikemen in the first twenty files, and placing five of the Heads of Ten (of it) in the front of them and five in the rear: the other sixty files situated behind are all shield-bearers, who total to three hundred. It should therefore be so arranged, that the first and last file of every hundred of Heads of Ten; the Constable with his flag and bugler be in the middle of the first hundred (century) of shield-bearers; and the Centurions at the head of every century. Thus arranged, when you want the pikemen to be on the left flank, you have to double them, century by century, from the right flank: if you want them to be on the right flank, you have to double them from the left. And thus this company turns with the pikemen on the flank, with the Heads of Ten on the front and rear, with the Centurions at the front of them, and the Constable in the middle. Which formation holds when going forward; but when the enemy comes and the time for the (companies) to move from the flanks to the front, it cannot be done unless all the soldiers face toward the flank where the pikemen are, and then the company is turned with its files and heads in that manner that was described above; for the Centurions being on the outside, and all the men in their places, the Centurions quickly enter them (the ranks) without difficulty. But when they are marching frontwards, and have to combat in the rear, they must arrange the files so that, in forming the company, the pikes are situated in the rear; and to do this, no other order has to be maintained except that where, in the formation of the company ordinarily every Century has five files of pikemen in front, it now has them behind, but in all the other parts, observe the order that I have mentioned.

COSIMO: You have said ((if I remember well)) that this method of training is to enable them to form these companies into an army, and that this training serves to enable them to be arranged within it. But if it should occur that these four hundred fifty infantry have to operate as a separate party, how would you arrange them?

FABRIZIO: I will now guide you in judging where he wants to place the pikes, and who should carry them, which is not in any way contrary to the arrangement mentioned above, for although it may be the method that is observed when, together with other companies, it comes to an engagement, none the less, it is a rule that serves for all those methods, in which it should happen that you have to manage it. But in showing you the other two methods for arranging the companies, proposed by me, I will also better satisfy your question; for either they are never used, or they are used when the company is above, and not in the company of others.

And to come to the method of forming it with two horns (wings), I say, that you ought to arrange the eighty files at five per file in this way: place a Centurion in the middle, and behind him twenty five files that have two pikemen (each) on the left side, and three shield-bearers on the right: and after the first five, in the next twenty, twenty Heads of Ten be placed, all between the pikemen and shield-bearers, except that those (Heads) who carry pikes stay with the pikemen. Behind these twenty five files thusly arranged, another Centurion is placed who has fifteen files of shield-bearers behind him. After these, the Constable between the flag and the bugler, who also has behind him another fifteen files of shield-bearers. The third

Centurion is placed behind these, and he has twenty five files behind him, in each of which are three shield-bearers on the left left side and two pikemen on the right: and after the first five files are twenty Heads of Ten placed between the pikemen and the shield-bearers. After these files, there is the fourth Centurion. If it is desired, therefore, to arrange these files to form a company with two horns (wings), the first Centurion has to be halted with the twenty five files which are behind him. The second Centurion then has to be moved with the fifteen shield-bearers who are on his rear, and turning to the right, and on the right flank of the twenty five files to proceed so far that he comes to the fifteen files, and here he halts. After, the Constable has to be moved with the fifteen files of shield bearers who are behind, and turning around toward the right, over by the right flank of the fifteen files which were moved first, marches so that he comes to their front, and here he halts. After, move the third Centurion with the twenty five files and with the fourth Centurion who is behind them, and turning to the right, march by the left flank of the last fifteen files of shield-bearers, and he does not halt until he is at the head of them, but continues marching up until the last files of twenty five are in line with the files behind. And, having done this, the Centurion who was Head of the first fifteen files of shield-bearers leaves the place where he was, and goes to the rear of the left angle. And thus he will turn a company of twenty five solid files, of twenty infantry per file, with two wings, on each side of his front, and there will remain a space between then, as much as would (be occupied by) by ten men side by side. The Captain will be between the two wings, and a Centurion in each corner of the wing. There will be two files of pikemen and twenty Heads of Ten on each flank. These two wings (serve to) hold between them that artillery, whenever the company has any with it, and the carriages. The Veliti have to stay along the flanks beneath the pikemen. But, in wanting to bring this winged (formed) company into the form of the piazza (plaza), nothing else need be done than to take eight of the fifteen files of twenty per file and place them between the points of the two horns (wings), which then from wings become the rear (shoulder) of the piazza (plaza). The carriages are kept in this plaza, and the Captain and the flag there, but not the artillery, which is put either in the front or along the flanks. These are the methods which can be used by a company when it has to pass by suspicious places by itself. None the less, the solid company, without wings and without the plaza, is best. But in wanting to make safe the disarmed ones, that winged one is necessary.

The Swiss also have many forms of companies, among which they form one in the manner of a cross, as in the spaces between the arms, they keep their gunners safe from the attacks of the enemy. But since such companies are good in fighting by themselves, and my intention is to show how several companies united together combat with the enemy, I do not belabor myself further in describing it.

COSIMO: And it appears to me I have very well comprehended the method that ought to be employed in training the men in these companies, but ((if I remember well)) you said that in addition to the ten companies in a Battalion, you add a thousand extraordinary pikemen and four hundred extraordinary Veliti. Would you not describe how to train these?

FABRIZIO: I would, and with the greatest diligence: and I would train the pikemen, group by group, at least in the formations of the companies, as the others; for I would serve myself of these more than of the ordinary companies, in all the particular actions, how to escort, to raid, and such things. But the Veliti I would train at home without bringing them together with the others, for as it is their office to combat brokenly (in the open, separately), it is not as necessary that they come together with the others or to train in common exercises, than to train them well in particular exercises. They ought, therefore, ((as was said in the beginning, and now it appears to me laborious to repeat it)) to train their own men in these companies so that they know how to maintain their ranks, know their places, return there quickly when either the evening or the location disrupts them; for when this is caused to be done, they can easily be taught the place the company has to hold and what its office should be in the armies. And if a Prince or a Republic works hard and puts diligence in these formations and in this training, it will always happen that there will be good soldiers in that country, and they will be superior to their neighbors, and will be those who give, and not receive, laws from other men. But ((as I have told you)) the disorder in which one exists, causes them to disregard and not to esteem these things, and, therefore, our training is not good: and even if there should be some heads or members naturally of virtue, they are unable to demonstrate it.

COSIMO: What carriages would you want each of these companies to have?

FABRIZIO: The first thing I would want is that the Centurions or the Heads of Ten should not go on horseback: and if the Constables want to ride mounted, I would want them to have a mule and not a horse. I would permit them two carriages, and one to each Centurion, and two to every three Heads of Ten, for they would quarter so many in each encampment, as we will narrate in its proper place. So that each company would have thirty six carriages, which I would have (them) to carry the necessary tents, cooking utensils, hatchets, digging bars, sufficient to make the encampment, and after that anything else of convenience.

COSIMO: I believe that Heads assigned by you in each of the companies are necessary: none the less, I would be apprehensive that so many commanders would be confusing.

FABRIZIO: They would be so if I would refer to one, but as I refer to many, they make for order; actually, without those (orders), it would be impossible to control them, for a wall which inclines on every side would need many and frequent supports, even if they are not so strong, but if few, they must be strong, for the virtu of only one, despite its spacing, can remedy any ruin. And so it must be that in the armies and among every ten men there is one of more life, of more heart, or at least of more authority, who with his courage, with words and by example keeps the others firm and disposed to fight. And these things mentioned by me, as the heads, the flags, the buglers, are necessary in an army, and it is seen that we have all these in our (present day) armies, but no one does his duty. First, the Heads of Ten, in desiring that those things be done because they are ordered, it is necessary ((as I have said)) for each of them to have his men separate, lodge with them, go into action with them, stay in the ranks with them, for when they are in their places, they are all of mind and temperament to maintain their ranks straight and firm, and it is impossible for them to become disrupted, or if they become disrupted, do not quickly reform their ranks. But today, they do not serve us for anything other than to give them more pay than the others, and to have them do some particular thing. The same happens with the flags, for they are kept rather to make a beautiful show, than for any military use. But the ancients served themselves of it as a guide and to reorganize themselves, for everyone, when the flag was standing firm, knew the place that he had to be near his flag, and always returned there. He also knew that if it were moving or standing still, he had to move or halt. It is necessary in an army, therefore, that there be many bodies, and that each body have its own flag and its own guide; for if they have this, it needs must be they have much courage and consequently, are livelier. The infantry, therefore, ought to march according to the flag, and the flag move according to the bugle (call), which call, if given well, commands the army, which proceeding in step with those, comes to serve the orders easily. Whence the ancients having whistles (pipes), fifes, and bugles, controlled (modulated) them perfectly; for, as he who dances proceeds in time with the music, and keeping with it does not make a miss-step, so an army obedient in its movement to that call (sound), will not become disorganized. And, therefore, they varied the calls according as they wanted to enkindle or quiet, or firm the spirits of men. And as the sounds were various, so they named them variously. The Doric call (sound) brought on constancy, Frigio, fury (boldness): whence they tell, that Alexander being at table, and someone sounding the Frigio call, it so excited his spirit that he took up arms. It would be necessary to rediscover all these methods, and if this is difficult, it ought not at least to be (totally) put aside by those who teach the soldier to obey; which each one can vary and arrange in his own way, so long as with practice he accustoms the ears of his soldiers to recognize them. But today, no benefit is gotten from these sounds in great part, other than to make noise.

COSIMO: I would desire to learn from you, if you have ever pondered this with yourself, whence such baseness and disorganization arises, and such negligence of this training in our times?

FABRIZIO: I will tell you willingly what I think. You know of the men excellent in war there have been many famed in Europe, few in Africa, and less in Asia. This results from (the fact that) these last two parts of the world have had a Principality or two, and few Republics; but Europe alone has had some Kingdoms and an infinite number of Republics. And men become excellent, and show their virtu, according as they are employed and recognized by their Prince, Republic, or King, whichever it may be. It happens, therefore, that where there is much power, many valiant men spring up, where there is little, few. In Asia, there are found Ninus, Cyrus, Artafersus, Mithradates, and very few others to accompany these. In Africa, there are noted ((omitting those of ancient Egypt)) Maximinius, Jugurtha, and those Captains who were raised by the Carthaginian Republic, and these are very few compared to those of Europe; for in Europe there are excellent men without number, and there would be many more, if there should be named together with them those others who have been forgotten by the malignity of the time, since the world has been more virtuous when there have been many States which have favored virtu, either from necessity or from other human passion. Few men, therefore, spring up in Asia, because, as that province was entirely subject to one Kingdom, in which because of its greatness there was indolence for the most part, it could not give rise to excellent men in business (activity). The same happened in Africa: yet several, with respect to the Carthaginian Republic, did arise. More excellent men come out of Republics than from Kingdoms, because in the former virtu is honored much of the time, in the Kingdom it is feared; whence it results that in the former, men of virtu are raised, in the latter they are extinguished. Whoever, therefore, considers the part of Europe, will find it to have been full of Republics and Principalities, which from the fear one had of the other, were constrained to keep alive their military organizations, and honor those who greatly prevailed in them. For in Greece, in addition to the Kingdom of the Macedonians, there were many Republics, and many most excellent men arose in each of them. In Italy, there were the Romans, the Samnites, the Tuscans, the Cisalpine Gauls. France and Germany were full of Republics and

Princes. Spain, the very same. And although in comparison with the Romans, very few others were noted, it resulted from the malignity of the writers, who pursued fortune and to whom it was often enough to honor the victors. For it is not reasonable that among the Samnites and Tuscans, who fought fifty years with the Roman People before they were defeated, many excellent men should not have sprung up. And so likewise in France and Spain. But that virtu which the writers do not commemorate in particular men, they commemorate generally in the peoples, in which they exalt to the stars (skies) the obstinacy which existed in them in defending their liberty. It is true, therefore, that where there are many Empires, more valiant men spring up, and it follows, of necessity, that those being extinguished, little by little, virtu is extinguished, as there is less reason which causes men to become virtuous. And as the Roman Empire afterwards kept growing, and having extinguished all the Republics and Principalities of Europe and Africa, and in greater part those of Asis, no other path to virtu was left, except Rome. Whence it resulted that men of virtu began to be few in Europe as in Asia, which virtu ultimately came to decline; for all the virtu being brought to Rome, and as it was corrupted, so almost the whole world came to be corrupted, and the Scythian people were able to come to plunder that Empire, which had extinguished the virtu of others, but did not know how to maintain its own. And although afterwards that Empire, because of the inundation of those barbarians, became divided into several parts, this virtu was not renewed: first, because a price is paid to recover institutions when they are spoiled; another, because the mode of living today, with regard to the Christian religion, does not impose that necessity to defend it that anciently existed, in which at the time men, defeated in war, were either put to death or remained slaves in perpetuity, where they led lives of misery: the conquered lands were either desolated or the inhabitants driven out, their goods taken away, and they were sent dispersed throughout the world, so that those overcome in war suffered every last misery. Men were terrified from the fear of this, and they kept their military exercises alive, and honored those who were excellent in them. But today, this fear in large part is lost, and few of the defeated are put to death, and no one is kept prisoner long, for they are easily liberated. The Citizens, although they should rebel a thousand times, are not destroyed, goods are left to their people, so that the greatest evil that is feared is a ransom; so that men do not want to subject themselves to dangers which they little fear. Afterwards, these provinces of Europe exist under very few Heads as compared to the past, for all of France obeys a King, all of Spain another, and Italy exists in a few parts; so that weak Cities defend themselves by allying themselves with the victors, and strong States, for the reasons mentioned, do not fear an ultimate ruin.

COSIMO: And in the last twenty five years, many towns have been seen to be pillaged, and lost their Kingdoms; which examples ought to teach others to live and reassume some of the ancient orders.

FABRIZIO: That is what you say, but if you would note which towns are pillaged, you would not find them to be the Heads (Chief ones) of the States, but only members: as is seen in the sacking of Tortona and not Milan, Capua and not Naples, Brescia and not Venice, Ravenna and not Rome. Which examples do not cause the present thinking which governs to change, rather it causes them to remain in that opinion of being able to recover themselves by ransom: and because of this, they do not want to subject themselves to the bother of military training, as it appears to them partly unnecessary, partly a tangle they do not understand. Those others who are slave, to whom such examples ought to cause fear, do not have the power of remedying (their situation), and those Princes who have lost the State, are no longer in time, and those who have (the State) do not have (military training) and those Princes who have lost the State, are no longer in time, and those who have (the State) do not have (military training) or want it; for they want without any hardship to remain (in power) through fortune, not through their own virtu, and who see that, because there is so little virtu, fortune governs everything, and they want it to master them, not they master it. And that that which I have discussed is true, consider Germany, in which, because there are many Principalities and Republics, there is much virtu, and all that is good in our present army, depends on the example of those people, who, being completely jealous of their State ((as they fear servitude, which elsewhere is not feared)) maintain and honor themselves all us Lords. I want this to suffice to have said in showing the reasons for the present business according to my opinion. I do not know if it appears the same to you, or if some other apprehension should have risen from this discussion.

COSIMO: None, rather I am most satisfied with everything. I desire above, returning to our principal subject, to learn from you how you would arrange the cavalry with these companies, and how many, how captained, and how armed.

FABRIZIO: And it, perhaps, appears to you that I have omitted these, at which do not be surprised, for I speak little of them for two reasons: one, because this part of the army is less corrupt than that of the infantry, for it is not stronger than the ancient, it is on a par with it. However, a short while before, the method of training them has been mentioned. And as to arming them, I would arm them as is presently done, both as to the light cavalry as to the men-at-arms. But I would want the light cavalry to be all archers, with some light gunners among them, who, although of little use in other actions of war, are most useful in terrifying the peasants, and place them above a pass that is to be guarded by them, for one gunner causes more fear to them (the enemy) than twenty other armed men. And as to numbers, I say that departing from imitating the Roman army, I would have not less than three hundred effective cavalry for each battalion, of which I would want one hundred fifty to be men-at-arms, and a hundred fifty light cavalry; and I would give a leader to each of these parts, creating among them fifteen Heads of Ten per hand, and give each one a flag and a bugler. I would want that every ten men-at-arms have five carriages and every ten light cavalrymen two, which, like those of the infantry, should carry the tents, (cooking) utensils, hitches, poles, and in addition over the others, their tools. And do not think this is out of place seeing that men-at-arms have four horses at their service, and that such a practice is a corrupting one; for in Germany, it is seen that those men-at-arms are alone with their horses, and only every twenty have a cart which carries the necessary things behind them. The horsemen of the Romans were likewise alone: it is true that the Triari encamped near the cavalry and were obliged to render aid to it in the handling of the horses: this can easily be imitated by us, as will be shown in the distribution of quarters. That, therefore, which the Romans did, and that which the Germans do, we also can do; and in not doing it, we make a mistake. These cavalrymen, enrolled and organized together with a battalion, can often be assembled when the companies are assembled, and caused to make some semblance of attack among them, which should be done more so that they may be recognized among them than for any necessity. But I have said enough on this subject for now, and let us descend to forming an army which is able to offer battle to the enemy, and hope to win it; which is the end for which an army is organized, and so much study put into it.

Third Book

COSIMO: Since we are changing the discussion, I would like the questioner to be changed, so that I may not be held to be presumptuous, which I have always censured in others. I, therefore, resign the speakership, and I surrender it to any of these friends of mine who want it.

ZANOBI: It would be most gracious of you to continue: but since you do not want to, you ought at least to tell us which of us should succeed in your place.

COSIMO: I would like to pass this burden on the Lord Fabrizio.

FABRIZIO: I am content to accept it, and would like to follow the Venetian custom, that the youngest talks first; for this being an exercise for young men, I am persuaded that young men are more adept at reasoning, than they are quick to follow.

COSIMO: It therefore falls to you Luigi: and I am pleased with such a successor, as long as you are satisfied with such a questioner.

FABRIZIO: I am certain that, in wanting to show how an army is well organized for undertaking an engagement, it would be necessary to narrate how the Greeks and the Romans arranged the ranks in their armies. None the less, as you yourselves are able to read and consider these things, through the medium of ancient writers, I shall omit many particulars, and will cite only those things that appear necessary for me to imitate, in the desire in our times to give some (part of) perfection to our army. This will be done, and, in time, I will show how an army is arranged for an engagement, how it faces a real battle, and how it can be trained in mock ones. The greatest mistake that those men make who arrange an army for an engagement, is to give it only one front, and commit it to only one onrush and one attempt (fortune). This results from having lost the method the ancients employed of receiving one rank into the other; for without this method, one cannot help the rank in front, or defend them, or change them by rotation in battle, which was practiced best by the Romans. In explaining this method, therefore, I want to tell how the Romans divided each Legion into three parts, namely, the Astati, the Princeps, and the Triari; of whom the Astati were placed in the first line of the army in solid and deep ranks, (and) behind them were the Princeps, but placed with their ranks more open: and behind these they placed the

Triari, and with ranks so sparse, as to be able, if necessary, to receive the Princeps and the Astati between them. In addition to these, they had slingers, bow-men (archers), and other lightly armed, who were not in these ranks, but were situated at the head of the army between the cavalry and the infantry. These light armed men, therefore, enkindled the battle, and if they won ((which rarely happened)), they pursued the victory: if they were repulsed, they retired by way of the flanks of the army, or into the intervals (gaps) provided for such a result, and were led back among those who were not armed: after this proceeding, the Astati came hand to hand with the enemy, and who, if they saw themselves being overcome, retired little by little through the open spaces in the ranks of the Princeps, and, together with them, renewed the fight. If these also were forced back, they all retired into the thin lines of the Triari, and all together, en masse, recommenced the battle; and if these were defeated, there was no other remedy, as there was no way left to reform themselves. The cavalry were on the flanks of the army, placed like two wings on a body, and they some times fought on horseback, and sometimes helped the infantry, according as the need required. This method of reforming themselves three times is almost impossible to surpass, as it is necessary that fortune abandon you three times, and that the enemy has so much virtu that he overcomes you three times. The Greeks, with their Phalanxes, did not have this method of reforming themselves, and although these had many ranks and Leaders within them, none the less, they constituted one body, or rather, one front. So that in order to help one another, they did not retire from one rank into the other, as the Romans, but one man took the place of another, which they did in this way. Their Phalanxes were (made up) of ranks, and supposing they had placed fifty men per rank, when their front came against the enemy, only the first six ranks of all of them were able to fight, because their lances, which they called Sarisse, were so long, that the points of the lances of those in the sixth rank reached past the front rank. When they fought, therefore, if any of the first rank fell, either killed or wounded, whoever was behind him in the second rank immediately entered into his place, and whoever was behind him in the third rank immediately entered into the place in the second rank which had become vacant, and thus successively all at once the ranks behind restored the deficiencies of those in front, so that the ranks were always remained complete, and no position of the combatants was vacant except in the last rank, which became depleted because there was no one in its rear to restore it. So that the injuries which the first rank suffered, depleted the last, and the first rank always remained complete; and thus the Phalanxes, because of their arrangement, were able rather to become depleted than broken, since the large (size of its) body made it more immobile. The Romans, in the beginning, also employed Phalanxes, and instructed their Legions in a way similar to theirs. Afterwards, they were not satisfied with this arrangement, and divided the Legion into several bodies; that is, into Cohorts and Maniples; for they judged ((as was said a little while ago)) that that body should have more life in it (be more active) which should have more spirit, and that it should be composed of several parts, and each regulate itself. The Battalions of the Swiss, in these times, employed all the methods of the Phalanxes, as much in the size and entirety of their organization, as in the method of helping one another, and when coming to an engagement they place the Battalions one on the flank of the other, or they place them one behind the other. They have no way in which the first rank, if it should retire, to be received by the second, but with this arrangement, in order to help one another, they place one Battalion in front and another behind it to the right, so that if the first has need of aid, the latter can go forward and succor it. They put a third Battalion behind these, but distant a gun shot. This they do, because if the other two are repulsed, this (third) one can make its way forward, and the others have room in which to retire, and avoid the onrush of the one which is going forward; for a large multitude cannot be received (in the same way) as a small body, and, therefore, the small and separate bodies that existed in a Roman Legion could be so placed together as to be able to receive one another among themselves, and help each other easily. And that this arrangement of the Swiss is not as good as that of the ancient Romans is demonstrated by the many examples of the Roman Legions when they engaged in battle with the Greek Phalanxes, and the latter were always destroyed by the former, because the kinds of arms ((as I mentioned before)) and this method of reforming themselves, was not able to maintain the solidity of the Phalanx. With these examples, therefore, if I had to organize an army, I would prefer to retain the arms and the methods, partly of the Greek Phalanxes, partly of the Roman Legions; and therefore I have mentioned wanting in a Battalion two thousand pikes, which are the arms of the Macedonian Phalanxes, and three thousand swords and shield, which are the arms of the Romans. I have divided the Battalion into ten Companies, as the Romans (divided) the Legion into ten Cohorts. I have organized the Veliti, that is the light armed, to enkindle the battle, as they (the Romans did). And thus, as the arms are mixed, being shared by both nations and as also the organizations are shared, I have arranged that each company have five ranks of pikes (pikemen) in front, and the remainder shields (swordsmen with shields), in order to be able with this front to resist the cavalry, and easily penetrate the enemy companies on foot, and the enemy at the first encounter would meet the pikes, which I would hope would suffice to resist him, and then the shields (swordsmen) would defeat him. And if you would note the virtu of this arrangement, you will see all these arms will execute their office completely. First, because pikes are useful against cavalry, and when they come against infantry, they do their duty well before the battle closes in, for when they are pressed, they become useless. Whence the Swiss, to avoid this disadvantage, after every three ranks of pikemen place one of halberds, which, while it is not enough, gives the pikemen room (to maneuver). Placing, therefore, our pikes in the front and the shields (swordsmen) behind, they manage to resist the cavalry, and in enkindling the battle, lay open and attack the infantry: but when the battle closes in, and they become useless, the shields and swords take their place, who are able to take care of themselves in every strait.

LUIGI: We now await with desire to learn how you would arrange the army for battle with these arms and with these organizations.

FABRIZIO: I do not now want to show you anything else other than this. You have to understand that in a regular Roman army, which they called a Consular Army, there were not more than two Legions of Roman Citizens, which consist of six hundred cavalry and about eleven thousand infantry. They also had as many more infantry and cavalry which were sent to them by their friends and confederates, which they divided into two parts, and they called one the right wing, and the other the left wing, and they never permitted this (latter) infantry to exceed the number of the infantry of the Legion. They were well content that the cavalry should be greater in number. With this army which consisted of twenty two thousand infantry and about two thousand cavalry effectives, a Consul undertook every action and went on every enterprise. And when it was necessary to face a large force, they brought together two Consuls with two armies. You ought also to note that ordinarily in all three of the principal activities in which armies engage, that is, marching, camping, and fighting, they place the Legion in the middle, because they wanted that virtu in which they should trust most should be greater unity, as the discussion of all these three activities will show you. Those auxiliary infantry, because of the training they had with the infantry of the Legion, were as effective as the latter, as they were disciplined as they were, and therefore they arranged them in a similar way when organizing (for) and engagement. Whoever, therefore, knows how they deployed the entire (army). Therefore, having told you how they divided a Legion into three lines, and how one line would receive the other, I have come to tell you how the entire army was organized for an engagement.

If I would want, therefore, to arrange (an army for) an engagement in imitation of the Romans, just as they had two Legions, I would take two Battalions, and these having been deployed, the disposition of an entire Army would be known: for by adding more people, nothing else is accomplished than to enlarge the organization. I do not believe it is necessary that I remind you how many infantry there are in a Battalion, and that it has ten companies, and what Leaders there are per company, and what arms they have, and who are the ordinary (regular) pikemen and Veliti, and who the extraordinary, because a little while I distinctly told you, and I reminded you to commit it to memory as something necessary if you should want to understand all the other arrangements: and, therefore, I will come to the demonstration of the arrangement, without repeating these again. And it appears to me that ten Companies of a Battalion should be placed on the left flank, and the ten others of the other on the right. Those on the left should be arranged in this way. The five companies should be placed one alongside the other on the front, so that between one and the next there would be a space of four arm lengths which come to occupy an area of one hundred forty one arm lengths long, and forty wide. Behind these five Companies I would place three others, distant in a straight line from the first ones by forty arm lengths, two of which should come behind in a straight line at the ends of the five, and the other should occupy the space in the middle. Thus these three would come to occupy in length and width the same space as the five: but where the five would have a distance of four arm lengths between one another, this one would have thirty three. Behind these I would place the last two companies, also in a straight line behind the three, and distant from those three forty arm lengths, and I would place each of them behind the ends of the three, so that the space between them would be ninety one arm lengths. All of these companies arranged thusly would therefore cover (an area of) one hundred forty one arm lengths long and two hundred wide. The extraordinary pikemen I would extend along the flanks of these companies on the left side, distant twenty arm lengths from it, creating a hundred forty three files of seven per file, so that they should cover the entire length of the ten companies arranged as I have previously described; and there would remain forty files for protecting the wagons and the unarmed people in the tail of the army, (and) assigning the Heads of Ten and the Centurions in their (proper) places: and, of the three Constables, I would put one at the head, another in the middle, and the third in the last file, who should fill the office of Tergiduttore, as the ancients called the one placed in charge of the rear of the Army. But returning to the head (van) of the Army I say, that I would place the extraordinary Veliti alongside the extraordinary pikemen, which, as you know, are five hundred, and would place them at a distance of forty arm lengths. On the side of these, also on the left hand; I would place the men-at-arms, and would assign them a distance of a hundred fifty arm lengths away. Behind these, the light cavalry, to whom I would assign the same space as the men-at-arms. The ordinary Veliti I would leave around their companies, who would occupy those spaces which I placed between one company and another, who would act to minister to those (companies) unless I had already placed them under the extraordinary pikemen; which I would do or not do according as it should benefit my plans. The general Head of all the Battalions I would place in that space that exists between the first and second order of companies, or rather at the head, and in that space with exists between the last of the first five companies and the extraordinary pikemen, according as it should benefit my plans, surrounded by thirty or sixty picked men, (and) who should know how to execute a commission prudently, and stalwartly resist an attack, and should also be in the middle of the buglers and flag carriers. This is the order in which I would deploy a Battalion on the left side, which would be the deployment of half the Army, and would cover an area five hundred and eleven arm lengths long and as much as mentioned above in width, not including the space which that part of the extraordinary pikemen should occupy who act as a shield for the unarmed men, which would be about one hundred arm lengths. The other Battalions I would deploy on the right side exactly in the same way as I deployed those on the left, having a space of thirty arm lengths between our battalions and the other, in the head of which space I would place some artillery pieces, behind which would be the Captain general of the entire Army, who should have around him in addition to the buglers and flag carriers at least two hundred picked men, the greater portion on foot, among whom should be ten or more adept at executing every command, and should be so provided with arms and a horse as to be able to go on horseback or afoot as the needs requires. Ten cannon of the artillery of the Army suffice for the reduction of towns, which should not exceed fifty pounds per charge, of which in the field I would employ more in the defense of the encampment than in waging a battle, and the other artillery should all be rather often than fifteen pounds per charge. This I would place in front of the entire army, unless the country should be such that I could situate it on the flank in a safe place, where it should not be able to be attacked by the enemy.

This formation of the Army thusly arranged, in combat, can maintain the order both of the Phalanxes and of the Roman Legions, because the pikemen are in front and all the infantry so arranged in ranks, that coming to battle with the enemy, and resisting him, they should be able to reform the first ranks from those behind according to the usage of the Phalanxes. On the other hand, if they are attacked so that they are compelled to break ranks and retire, they can enter into the spaces of the second company behind them, and uniting with them, (and) en masse be able to resist and combat the enemy again: and if this should not be enough, they can in the same way retire a second time, and combat a third time, so that in this arrangement, as to combatting, they can reform according to both the Greek method, and the Roman. As to the strength of the Army, it cannot be arranged any stronger, for both wings are amply provided with both leaders and arms, and no part is left weak except that part behind which is unarmed, and even that part has its flanks protected by the extraordinary pikemen. Nor can the enemy assault it in any part where he will not find them organized, and the part in the back cannot be assaulted, because there cannot be an enemy who has so much power that he can assail every side equally, for it there is one, you don't have to take the field with him. But if he should be a third greater than you, and as well organized as you, if he weakens himself by assaulting you in several places, as soon as you defeat one part, all will go badly for him. If his cavalry should be greater than yours, be most assured, for the ranks of pikemen that gird you will defend you from every onrush of theirs, even if your cavalry should be repulsed. In addition to this, the Heads are placed on the side so that they are able easily to command and obey. And the spaces that exist between one company and the next one, and between one rank and the next, not only serve to enable one to receive the other, but also to provide a place for the messengers who go and come by order of the Captain. And as I told you before, as the Romans had about twenty thousand men in an Army, so too ought this one have: and as other soldiers borrowed their mode of fighting and the formation of their Army from the Legions, so too those soldiers that you assembled into your two Battalions would have to borrow their formation and organization. Having given an example of these things, it is an easy matter to initiate it: for if the army is increased either by two Battalions, or by as many men as are contained in them, nothing else has to be done than to double the arrangements, and where ten companies are placed on the left side, twenty are now placed, either by increasing or extending the ranks, according as the place or the enemy should command you.

LUIGI: Truly, (my) Lord, I have so imagined this army, that I see it now, and have a desire to see it facing us, and not for anything in the world would I desire you to become Fabius Maximus, having thoughts of holding the enemy at bay and delaying the engagement, for I would say worse of you, than the Roman people said of him.

FABRIZIO: Do not be apprehensive. Do you not hear the artillery? Ours has already fired, but harmed the enemy little; and the extraordinary Veliti come forth from their places together with the light cavalry, and spread out, and with as much fury and the loudest shouts of which they are capable, assault the enemy, whose artillery has fired one time, and has passed over the heads of our infantry without doing them an injury. And as it is not able to fire a second time, our Veliti and cavalry have already seized it, and to defend it, the enemy has moved forward, so that neither that of friend or enemy can perform its office. You see with what virtu our men fight, and with what discipline they have become accustomed because of the training they have had, and from the confidence they have in the Army, which you see with their stride, and with the men-at-arms alongside, in marching order, going to rekindle the battle with the adversary. Your see our artillery, which to make place for them, and to leave the space free, has retired to the place from which the Veliti went forth. You see the Captain who encourages them and points out to them certain victory. You see the Veliti and light cavalry have spread out and returned to the flanks of the Army, in order to see if they can cause any injury to the enemy from the flanks. Look, the armies are facing each other: watch with what virtu they have withstood the onrush of the enemy, and with what silence, and how the Captain commands the men-at-arms that they should resist and not attack, and do not detach themselves from the ranks of the infantry. You see how our light cavalry are gone to attack a band of enemy gunners who wanted to attach by the flank, and how the enemy cavalry have succored them, so that, caught between the cavalry of the one and the other, they cannot fire, and retire behind their companies. You see with what fury our pikemen attack them, and how the infantry is already so near each other that they can no longer manage their pikes: so that, according to the discipline taught by us, our pikemen retire little by little among the shields (swordsmen). Watch how in this (encounter), so great an enemy band of men-at-arms has pushed back our men-at-arms on the left side and how ours, according to discipline, have retired under the extraordinary pikemen, and having reformed the front with their aid, have repulsed the adversary, and killed a good part of them. In fact all the ordinary pikemen of the first company have hidden themselves among the ranks of the shields (swordsmen), and having left the battle to the swordsmen, who, look with what virtu, security, and leisure, kill the enemy. Do you not see that, when fighting, the ranks are so straitened, that they can handle the swords only with much effort? Look with what hurry the enemy moves; for, armed with the pike and their swords useless ((the one because it is too long, the other because of finding the enemy too greatly armed)), in part they fall dead or wounded, in part they flee. See them flee on the right side. They also flee on the left. Look, the victory is ours. Have we not won an engagement very happily? But it would have been won with greater felicity if I should have been allowed to put them in action. And see that it was not necessary to avail ourselves of either the second or third ranks, that our first line was sufficient to overcome them. In this part, I have nothing else to tell you, except to dissolve any doubts that should arise in you.

LUIGI: You have won this engagement with so much fury, that I am astonished, and in fact so stupefied, that I do not believe I can well explain if there is any doubt left in my mind. Yet, trusting in your prudence, I will take courage to say that I intend. Tell me first, why did you not let your artillery fire more than one time? and why did you have them quickly retire within the army, nor afterward make any other mention of them? It seems to me also that you pointed the enemy artillery high, and arranged it so that it should be of much benefit to you. Yet, if it should occur ((and I believe it happens often)) that the lines are pierced, what remedy do you provide? And since I have commenced on artillery, I want to bring up all these questions so as not to have to discuss it any more. I have heard many disparage the arms and the organization of the ancient Armies, arguing that today they could do little, or rather how useless they would be against the fury of artillery, for these are superior to their arms and break the ranks, so that it appears to them to be madness to create an arrangement that cannot be held, and to endure hardship in carrying a weapon that cannot defend you.

FABRIZIO: This question of yours has need ((because it has so many items)) of a long answer. It is true that I did not have the artillery fire more than one time, and because of it one remains in doubt. The reason is, that it is more important to one to guard against being shot than shooting the enemy. You must understand that, if you do not want the artillery to injure you, it is necessary to stay where it cannot reach you, or to put yourself behind a wall or embankment. Nothing else will stop it; but it is necessary for them to be very strong. Those Captains who must make an engagement cannot remain behind walls or embankments, nor can they remain where it may reach them. They must, therefore, since they do not have a way of protecting themselves, find one by which they are injured less; nor can they do anything other than to undertake it quickly. The way of doing this is to go find it quickly and directly, not slowly or en masse; for, speed does not allow them to shoot again, and because the men are scattered, they can injure only a few of them. A band of organized men cannot do this, because if they march in a straight line, they become disorganized, and if they scatter, they do not give the enemy the hard work to rout them, for they have routed themselves. And therefore I would organize the Army so that it should be able to do both; for having placed a thousand Veliti in its wings, I would arrange, that after our artillery had fired, they should issue forth together with the light cavalry to seize the enemy artillery. And therefore I did not have my artillery fire again so as not to give the enemy time, for you cannot give me time and take it from others. And for that, the reason I did not have it fired a second time, was not to allow it to be fired first; because, to render the enemy artillery useless, there is no other remedy than to assault it; which, if the enemy abandons it, you seize it; if they want to defend it, it is necessary that they leave it behind, so that in the hands of the enemy or of friends, it cannot be fired. I believe that, even without examples, this discussion should be enough for you, yet, being able to give you some from the ancients, I will do so. Ventidius, coming to battle with the Parthians, the virtu of whom (the latter) in great part consisted in their bows and darts, he allowed them to come almost under his encampments before he led the Army out, which he only did in order to be able to seize them quickly and not give them time to fire. Caesar in Gaul tells, that in coming to battle with the enemy, he was assaulted by them with such fury, that his men did not have time to draw their darts according to the Roman custom. It is seen, therefore, that, being in the field, if you do not want something fired from a distance to injure you, there is no other remedy than to be able to seize it as quickly as possible. Another reason also caused me to do without firing the artillery, at which you may perhaps laugh, yet I do not judge it is to be disparaged. And there is nothing that causes greater confusion in an Army than to obstruct its vision, whence most stalwart Armies have been routed for having their vision obstructed either by dust or by the sun. There is also nothing that impedes the vision than the smoke which the artillery makes when fired: I would think, therefore, that it would be more prudent to let the enemy blind himself, than for you to go blindly to find him. I would, therefore, not fire, or ((as this would not be approved because of the reputation the artillery has)) I would put it in the wings of the Army, so that firing it, its smoke should not blind the front of what is most important of our forces. And that obstructing the vision of the enemy is something useful, can be adduced from the example of Epaminondas, who, to blind the enemy Army which was coming to engage him, had his light cavalry run in front of the enemy so that they raised the dust high, and which obstructed their vision, and gave him the victory in the engagement. As to it appearing to you that I aimed the shots of artillery in my own manner, making it pass over the heads of the infantry, I reply that there are more times, and without comparison, that the heavy artillery does not penetrate the infantry than it does, because the infantry lies so low, and they (the artillery) are so difficult to fire, that any little that you raise them, (causes) them to pass over the heads of the infantry, and if you lower them, they damage the ground, and the shot does not reach them (the infantry). Also, the unevenness of the ground saves them, for every little mound or height which exists between the infantry and it (the artillery), impedes it. And as to cavalry, and especially men-at-arms, because they are taller and can more easily be hit, they can be kept in the rear (tail) of the Army until the time the artillery has fired. It is true that often they injure the smaller artillery and the gunners more that the latter (cavalry), to which the best remedy is to come quickly to grips (hand to hand): and if in the first assault some are killed ((as some always do die)) a good Captain and a good

Army do not have to fear an injury that is confined, but a general one; and to imitate the Swiss, who never shun an engagement even if terrified by artillery, but rather they punish with the capital penalty those who because of fear of it either break ranks or by their person give the sign of fear. I made them ((once it had been fired)) to retire into the Army because it left the passage free to the companies. No other mention of it was made, as something useless, once the battle was started.

You have also said in regard to the fury of this instrument that many judge the arms and the systems of the ancients to be useless, and it appears from your talk that the modems have found arms and systems which are useful against the artillery. If you know this, I would be pleased for you to show it to me, for up to now I do not know of any that have been observed, nor do I believe any can be found. So that I would like to learn from those men for what reasons the soldiers on foot of our times wear the breastplate or the corselet of iron, and those on horseback go completely covered with armor, since, condemning the ancient armor as useless with respect to artillery, they ought also to shun these. I would also like to learn for what reason the Swiss, in imitation of the ancient systems, for a close (pressed) company of six or eight thousand infantry, and for what reason all the others have imitated them, bringing the same dangers to this system because of the artillery as the others brought which had been imitated from antiquity. I believe that they would not know what to answer; but if you asked the soldiers who should have some experience, they would answer, first that they go armed because, even if that armor does not protect them from the artillery, it does every other injury inflicted by an enemy, and they would also answer that they go closely together as the Swiss in order to be better able to attack the infantry, resist the cavalry, and give the enemy more difficulty in routing them. So that it is observed that soldiers have to fear many other things besides the artillery, from which they defend themselves with armor and organization. From which it follows that as much as an Army is better armed, and as much as its ranks are more serrated and more powerful, so much more is it secure. So that whoever is of the opinion you mentioned must be either of little prudence, or has thought very little on this matter; for if we see the least part of the ancient way of arming in use today, which is the pike, and the least part of those systems, which are the battalions of the Swiss, which do us so much good, and lend so much power to our Armies, why shouldn't we believe that the other arms and other systems that they left us are also useful? Moreover, if we do not have any regard for the artillery when we place ourselves close together, like the Swiss, what other system than that can make us afraid? inasmuch as there is no other arrangement that can make us afraid than that of being pressed together. In addition to this, if the enemy artillery does not frighten me when I lay siege to a town, where he may injure me with great safety to himself, and where I am unable to capture it as it is defended from the walls, but can stop him only with time with my artillery, so that he is able to redouble his shots as he wishes, why do I have to be afraid of him in the field where I am able to seize him quickly? So that I conclude this, that the artillery, according to my opinion, does not impede anyone who is able to use the methods of the ancients, and demonstrate the ancient virtu. And if I had not talked another time with you concerning this instrument, I would extend myself further, but I want to return to what I have now said.

LUIGI: We are able to have a very good understanding since you have so much discoursed about artillery, and in sum, it seems to me you have shown that the best remedy that one has against it when he is in the field and having an Army in an encounter, is to capture it quickly. Upon which, a doubt rises in me, for it seems to me the enemy can so locate it on a side of his army from which he can injure you, and would be so protected by the other sides, that it cannot be captured. You have ((if you will remember)) in your army's order for battle, created intervals of four arm lengths between one company and the next, and placed twenty of the extraordinary pikemen of the company there. If the enemy should organize his army similarly to yours, and place his artillery well within those intervals, I believe that from here he would be able to injure you with the greatest safety to himself, for it would not be possible to enter among the enemy forces to capture it.

FABRIZIO: You doubt very prudently, and I will endeavor either to resolve the doubt, or to give you a remedy. I have told you that these companies either when going out or when fighting are continually in motion, and by nature always end up close together, so that if you make the intervals small, in which you would place the artillery, in a short time, they would be so closed up that the artillery can no longer perform its function: if you make them large to avoid this danger, you incur a greater, so that, because of those intervals, you not only give the enemy the opportunity to capture your artillery, but to rout you. But you have to know that it is impossible to keep the artillery between the ranks, especially those that are mounted on carriages, for the artillery travel in one direction, and are fired in the other, so that if they are desired to be fired while travelling, it is necessary before they are fired that they be turned, and when they are being turned they need so much space, that fifty carriages of artillery would disrupt every Army. It is necessary, therefore, to keep them outside the ranks where they can be operated in the manner which we showed you a short time ago. But let us suppose they can be kept there, and that a middle way can be found, and of a kind which, when closed together, should not impede the artillery, yet not be so open as to provide a path for the enemy, I say that this is easily remedied at the time of the encounter by creating intervals in your army which give a free path for its shots, and thus its fury will be useless. Which can be easily done, because the enemy, if it wants its artillery to be safe, must place it in the end portions of the intervals, so that its shots, if they should not harm its own men, must pass in a straight line, and always in the same line, and, therefore, by giving them room, can be easily avoided. Because this is a general rule, that you must give way to those things which cannot be resisted, as the ancients did to the elephants and chariots with sickles. I believe, rather I am more than certain, that it must appear to you that I prepared and won an engagement in my own manner; none the less, I will repeat this, if what I have said up to now is now enough, that it would be impossible for an Army thus organized and armed not to overcome, at the first encounter, every other Army organized as modem Armies are organized, which often, unless they have shields (swordsmen), do not form a front, and are of an unarmed kind, which cannot defend themselves from a near-by enemy; and so organized that, that if they place their companies on the flanks next to each other, not having a way of receiving one another, they cause it to be confused, and apt to be easily disturbed. And although they give their Armies three names, and divide them into three ranks, the Vanguard, the Company (main body) and the Rearguard, none the less, they do not serve for anything else than to distinguish them in marching and in their quarters: but in an engagement, they are all pledged to the first attack and fortune.

LUIGI: I have also noted that in making your engagement, your cavalry was repulsed by the enemy cavalry, and that it retired among the extraordinary pikemen, whence it happened that with their aid, they withstood and repulsed the enemy in the rear. I believe the pikemen can withstand the cavalry, as you said, but not a large and strong Battalion, as the Swiss do, which, in your Army, have five ranks of pikemen at the head, and seven on the flank, so that I do not know how they are able to withstand them.

FABRIZIO: Although I have told you that six ranks were employed in the Phalanxes of Macedonia at one time, none the less, you have to know that a Swiss Battalion, if it were composed of ten thousand tanks could not employ but four, or at most five, because the pikes are nine arm lengths long and an arm length and a half is occupied by the hands; whence only seven and a half arm lengths of the pike remain to the first rank. The second rank, in addition to what the hand occupies, uses up an arm's length of the space that exists between one rank and the next; so that not even six arm lengths of pike remain of use. For the same reasons, these remain four and one half arm lengths to the third rank, three to the fourth, and one and a half to the fifth. The other ranks are useless to inflict injury; but they serve to replace the first ranks, as we have said, and serve as reinforcements for those (first) five ranks. If, therefore, five of their ranks can control cavalry, why cannot five of ours control them, to whom five ranks behind them are also not lacking to sustain them, and give the same support, even though they do not have pikes as the others do? And if the ranks of extraordinary pikemen which are placed along the flanks seem thin to you, they can be formed into a square and placed by the flank of the two companies which I place in the last ranks of the army, from which place they would all together be able easily to help the van and the rear of the army, and lend aid to the cavalry according as their need may require.

LUIGI: Would you always use this form of organization, when you would want to engage in battle?

FABRIZIO: Not in every case, for you have to vary the formation of the army according to the fitness of the site, the kind and numbers of the enemy, which will be shown before this discussion is furnished with an example. But this formation that is given here, not so much because it is stronger than others, which is in truth very strong, as much because from it is obtained a rule and a system, to know how to recognize the manner of organization of the others; for every science has its generations, upon which, in good part, it is based. One thing only, I would remind you, that you never organize an army so that whoever fights in the van cannot be helped by those situated behind, because whoever makes this error renders useless the great part of the army, and if any virtu is eliminated, he cannot win.

LUIGI: And on this part, some doubt has arisen in me. I have seen that in the disposition of the companies you form the front with five on each side the center with three, and the rear with two; and I would believe that it should be better to arrange them oppositely, because I think that an army can be routed with more difficulty, for whoever should attack it, the more he should penetrate into it, so much harder would he find it: but the arrangement made by you appears to me results, that the more one enters into it, the more he finds it weak.

FABRIZIO: If you would remember that the Triari, who were the third rank of the Roman Legions, were not assigned more than six hundred men, you would have less doubt, when you leave that they were placed in the last ranks, because you will see that I (motivated by this example) have placed two companies in the last ranks, which comprise nine-hundred infantry; so that I come to err rather with the Roman people in having taken away too many, than few. And although this example should suffice, I want to tell you the reasons, which is this. The first front (line) of the army is made solid and dense because it has to withstand the attack of the enemy, and does not have to receive any friends into it, and because of this, it must abound in men, for few men would make it weak both from their sparseness and their numbers. But the second line, because it has to relieve the friends from the first who have withstood the enemy, must have large intervals, and therefore must have a smaller number than the first; for if it should be of a greater or equal number, it would result in not leaving any intervals, which would cause disorder, or if some should be left, it would extend beyond the ends of those in front, which would make the formation of the army incomplete (imperfect). And what you say is not true, that the more the enemy enters into the Battalions, the weaker he will find them; for the enemy can never fight with the second line, if the first one is not joined up with it: so that he will come to find the center of the Battalion stronger and not weaker, having to fight with the first and second (lines) together. The same thing happens if the enemy should reach the third line, because here, he will not only have to fight with two fresh companies, but with the entire Battalion. And as this last part has to receive more men, its spaces must be larger, and those who receive them lesser in number.

LUIGI: And I like what you have said; but also answer me this. If the five companies retire among the second three, and afterwards, the eight among the third two, does it not seem possible that the eight come together then the ten together, are able to crowd together, whether they are eight or ten, into the same space which the five occupied.

FABRIZIO: The first thing that I answer is, that it is not the same space; for the five have four spaces between them, which they occupy when retiring between one Battalion and the next, and that which exists between the three or the two: there also remains that space which exists between the companies and the extraordinary pikemen, which spaces are all made large. There is added to this whatever other space the companies have when they are in the lines without being changed, for, when they are changed, the ranks are either compressed or enlarged. They become enlarged when they are so very much afraid, that they put themselves in flight: they become compressed when they become so afraid, that they seek to save themselves, not by flight, but by defense; so that in this case, they would compress themselves, and not spread out. There is added to this, that the five ranks of pikemen who are in front, once they have started the battle, have to retire among their companies in the rear (tail) of the army to make place for the shield-bearers (swordsmen) who are able to fight: and when they go into the tail of the army they can serve whoever the captain should judge should employ them well, whereas in the front, once the fight becomes mixed, they would be completely useless. And therefore, the arranged spaces come to be very capacious for the remaining forces. But even if these spaces should not suffice, the flanks on the side consist of men and not walls, who, when they give way and spread out, are able to create a space of such capacity, which should be sufficient to receive them.

LUIGI: The ranks of the extraordinary pikemen, which you place on the flank of the army when the first company retires into the second, do you want them to remain firm, and become as two wings of the army or do you also want them to retire with the company. Which, if they have to do this, I do not see how they can, as they do not have companies behind them with wide intervals which would receive them.

FABRIZIO: If the enemy does not fight them when he faces the companies to retire, they are able to remain firm in their ranks, and inflict injury on the enemy on the flank since the first companies had retired: but if they should also fight them, as seems reasonable, being so powerful as to be able to force the others to retire, they should cause them also to retire. Which they are very well able to do, even though they have no one behind who should receive them, for from the middle forward they are able to double on the right, one file entering into the other in the manner we discussed when we talked of the arrangement for doubling themselves. It is true, that when doubling, they should want to retire behind, other means must be found than that which I have shown you, since I told you that the second rank had to enter among the first, the fourth among the third, and so on little by little, and in this case, it would not be begun from the front, but from the rear, so that doubling the ranks, they should come to retire to the rear, and not to turn in front. But to reply to all of that, which (you have asked) concerning this engagement as shown by me, it should be repeated, (and) I again say that I have so organized this army, and will (again) explain this engagement to you for two reasons: one, to show you how it (the army) is organized: the other, to show you how it is trained. As to the systems, I believe you all most knowledgeable. As to the army, I tell you that it may often be put together in this form, for the Heads are taught to keep their companies in this order: and because it is the duty of each individual soldier to keep (well) the arrangement of each company, and it is the duty of each Head to keep (well) those in each part of the Army, and to know well how to obey the commands of the general Captain. They must know, therefore, how to join one company with another, and how to take their places instantly: and therefore, the banner of each company must have its number displayed openly, so that they may be commanded, and the Captain and the soldiers will more readily recognize that number. The Battalions ought also to be numbered, and have their number on their principal banner. One must know, therefore, what the number is of the Battalion placed on the left or right wing, the number of those placed in the front and the center, and so on for the others. I would want also that these numbers reflect the grades of positions in the Army. For instance, the first grade is the Head of Ten, the second is the head of fifty ordinary Veliti, the third the Centurion, the fourth the head of the first company, the fifth that of the second (company), the sixth of the third, and so on up to the tenth Company, which should be in the second place next to the general Captain of the Battalion; nor should anyone arrive to that Leadership, unless he had (first) risen through all these grades. And, as in addition to these Heads, there are the three Constables (in command) of the extraordinary pikemen, and the two of the extraordinary Veliti, I would want them to be of the grade of Constable of the first company, nor would I care if they were men of equal grade, as long as each of them should vie to be promoted to the second company. Each one of these Captains, therefore, knowing where his Company should be located, of necessity it will follow that, at the sound of the trumpet, once the Captain's flag was raised, all of the Army would be in its proper places. And this is the first exercise to which an Army ought to become accustomed, that is, to assemble itself quickly: and to do this, you must frequently each day arrange them and disarrange them.

LUIGI: What signs would you want the flags of the Army to have, in addition to the number?

FABRIZIO: I would want the one of the general Captain to have the emblem of the Army: all the others should also have the same emblem, but varying with the fields, or with the sign, as it should seem best to the Lord of the Army, but this matters little, so long as their effect results in their recognizing one another.

But let us pass on to another exercise in which an army ought to be trained, which is, to set it in motion, to march with a convenient step, and to see that, while in motion, it maintains order. The third exercise is, that they be taught to conduct themselves as they would afterwards in an engagement; to fire the artillery, and retire it; to have the extraordinary Veliti issue forth, and after a mock assault, have them retire; have the first company, as if they were being pressed, retire within the intervals of the second (company), and then both into the third, and from here each one return to its place; and so to accustom them in this exercise, that it become understood and familiar to everyone, which with practice and familiarity, will readily be learned. The fourth exercise is that they be taught to recognize commands of the Captain by virtue of his (bugle) calls and flags, as they will understand, without other command, the pronouncements made by voice. And as the importance of the commands depends on the (bugle) calls, I will tell you what sounds (calls) the ancients used. According as Thucydides affirms, whistles were used in the army of the Lacedemonians, for they judged that its pitch was more apt to make their Army proceed with seriousness and not with fury. Motivated by the same reason, the Carthaginians, in their first assault, used the zither.

Alliatus, King of the Lydians, used the zither and whistles in war; but Alexander the Great and the Romans used horns and trumpets, like those who thought the courage of the soldiers could be increased by virtue of such instruments, and cause them to combat more bravely. But just as we have borrowed from the Greek and Roman methods in equipping our Army, so also in choosing sounds should we serve ourselves of the customs of both those nations. I would, therefore, place the trumpets next to the general Captain, as their sound is apt not only to inflame the Army, but to be heard over every noise more than any other sound. I would want that the other sounds existing around the Constables and Heads of companies to be (made by) small drums and whistles, sounded not as they are presently, but as they are customarily sounded at banquets. I would want, therefore, for the Captain to use the trumpets in indicating when they should stop or go forward or turn back, when they should fire the artillery, when to move the extraordinary Veliti, and by changes in these sounds (calls) point out to the Army all those moves that generally are pointed out; and those trumpets afterwards followed by drums. And, as training in these matters are of great importance, I would follow them very much in training your Army. As to the cavalry, I would want to use the same trumpets, but of lower volume and different pitch of sounds from those of the Captain. This is all that occurs to me concerning the organization and training of the Army.

LUIGI: I beg you not to be so serious in clearing up another matter for me: why did you have the light cavalry and the extraordinary Veliti move with shouts and noise and fury when they attacked, but they in rejoining the Army you indicated the matter was accomplished with great silence: and as I do not understand the reason for this fact, I would desire you to clarify it for me.

FABRIZIO: When coming to battle, there have been various opinions held by the ancient Captains, whether they ought either to accelerate the step (of the soldiers) by sounds, or have them go slowly in silence. This last manner serves to keep the ranks firmer and have them understand the commands of the Captain better: the first serves to encourage the men more. And, as I believe consideration ought to be given to both these methods, I made the former move with sound, and the latter in silence. And it does not seem to me that in any case the sounds are planned to be continuous, for they would impede the commands, which is a pernicious thing. Nor is it reasonable that the Romans, after the first assault, should follow with such sounds, for it is frequently seen in their histories that soldiers who were fleeing were stopped by the words and advice of the Captains, and changed the orders in various ways by his command: which would not have occurred if the sounds had overcome his voice.

Fourth Book

LUIGI: Since an engagement has been won so honorably under my Rule, I think it is well if I do not tempt fortune further, knowing how changeable and unstable it is. And, therefore, I desire to resign my speakership, and that, wanting to follow the order that belongs to the youngest, Zanobi now assume this office of questioning. And I know he will not refuse this honor, or we would rather say, this hard work, as much in order to (give) pleasure, as also because he is naturally more courageous than I: nor should he be afraid to enter into these labors, where he can thus be overcome, as he can overcome.

ZANOBI: I intend to stay where you put me, even though I would more willingly stay to listen, because up to now I am more satisfied with your questions than those which occurred to me in listening to your discussions pleased me. But I believe it is well, Lords, that since you have time left, and have patience, we do not annoy you with these ceremonies of ours.

FABRIZIO: Rather you give me pleasure, because this change of questioners makes me know the various geniuses, and your various desires. Is there anything remaining of the matter discussed which you think should be added?

ZANOBI: There are two things I desire before we pass on to another part: the one is, that you would show me if there is another form of organizing the Army which may occur to you: the other, what considerations ought a Captain have before going to battle, and if some accident should arise concerning it, what remedies can be made.

FABRIZIO: I will make an effort to satisfy you, I will not reply to your questions in detail; for, when I answer one, often it will also answer another. I have told you that I proposed a form for the Army which should fill all the requirements according to the (nature of) the enemy and the site, because in this case, one proceeds according to the site and the enemy. But note this, that there is no greater peril than to over extend the front of your army, unless you have a very large and very brave Army: otherwise you have to make it rather wide and of short length, than of long length and very narrow. For when you have a small force compared to the enemy, you ought to seek other remedies; for example, arrange your army so that you are girded on a side by rivers or swamps, so that you cannot be surrounded or gird yourself on the flanks with ditches, as Caesar did in Gaul. In this case, you have to take the flexibility of being able to enlarge or compress your front, according to the numbers of the enemy: and if the enemy is of a lesser number, you ought to seek wide places, especially if you have your forces so disciplined, that you are able not only to surround the enemy, but extend your ranks, because in rough and difficult places, you do not have the advantage of being able to avail yourself of (all) your ranks. Hence it happened that the Romans almost always sought open fields, and avoided the difficult ones. On the other hand ((as I have said)) you ought to, if you have either a small force or a poorly disciplined one, for you have to seek places where a small number can defend you, or where inexperience may not cause you injury. Also, higher places ought to be sought so as to be able more easily to attack (the enemy). None the less, one ought to be aware not to arrange your Army on a beach and in a place near the adjoining hills, where the enemy Army can come; because in this case, with respect to the artillery, the higher place would be disadvantageous to you, because you could continuously and conveniently be harmed by the enemy artillery, without being able to undertake any remedy, and similarly, impeded by your own men, you cannot conveniently injure him. Whoever organizes an Army for battle, ought also to have regard for both the sun and the wind, that the one and the other do not strike the front, because both impede your vision, the one with its rays, the other with dust. And in addition, the wind does not aid the arms that are thrown at the enemy, and makes their blows more feeble. And as to the sun, it is not enough that you take care that it is not in your face at the time, but you must think about it not harming you when it comes up. And because of this, in arranging the army, I would have it (the sun) behind them, so that much time should pass before it should come in front of you. This method was observed by Hannibal at Cannae and by Marius against the Cimbrians. If you should be greatly inferior in cavalry, arrange your army between vines and trees, and such impediments, as the Spaniards did in our times when they routed the French in the Kingdom (of Naples) on the Cirignuola. And it has been frequently seen that the same soldiers, when they changed only their arrangement and the location, from being overcome became victorious, as happened to the Carthaginians, who, after having been often defeated by Marius Regulus, were afterwards victorious, through the counsel of Xantippe, the Lacedemonian, who had them descend to the plain, where, by the virtu of their cavalry and Elephants, they were able to overcome the Romans. And it appears to me, according to the examples of the ancients, that almost all the excellent Captains, when they learned that the enemy had strengthened one side of the company, did not attack the stronger side, but the weaker, and the other stronger side they oppose to the weaker: then, when starting a battle, they cornered the stronger part that it only resist the enemy, and not push it back, and the weaker part that it allow itself to be overcome, and retire into the rear ranks of the Army. This causes two great disorders to the enemy: the first, that he finds his strongest part surrounded: the second is, that as it appears to them they will obtain the victory quickly, it rarely happens that he will not become disorganized, whence his defeat quickly results. Cornelius Scipio, when he was in Spain, (fighting) against Hasdrubal, the Carthaginian, and knowing that Hasdrubal was noted, that in arranging the Army, placed his legions in the center, which constituted the strongest part of his Army, and therefore, when Hasdrubal was to proceed in this manner, afterwards, when he came to the engagement, changed the arrangement, and put his Legions in the wings of the Army, and placed his weakest forces in the center. Then when they came hand to hand, he quickly had those forces in the center to walk slowly, and the wings to move forward swiftly: so that only the wings of both armies fought, and the ranks in the center, being distant from each other, did not join (in battle), and thus the strongest part of (the army of) Scipio came to fight the weakest part of (that of) Hasdrubal, and defeated it. This method at that time was useful, but today, because of the artillery, could not be employed, because that space that existed between one and the other army, gives them time to fire, which is most pernicious, as we said above. This method, therefore, must be set aside, and be used, as was said a short time ago, when all the Army is engaged, and the weaker part made to yield. When a Captain finds himself to have an army larger than that of the enemy, and not wanting to be prevented from surrounding him, arranges his Army with fronts equal to those of the enemy: then when the battle is started, has his front retire and the flanks extend little by little, and it will always happen that the enemy will find himself surrounded without being aware of it. When a Captain wants to fight almost secure in not being routed, he arranges his army in a place where he has a safe refuge nearby, either amid swamps or mountains or in a powerful city; for, in this manner, he cannot be pursued by the enemy, but the enemy cannot be pursued by him. This means was employed by Hannibal when fortune began to become adverse for him, and he was apprehensive of the valor of Marcus Marcellus. Several, in order to disorganize the ranks of the enemy, have commanded those who are lightly armed, that they begin the fight, and having begun it, retire among the ranks; and when the Armies afterwards have joined fronts together, and each front is occupied in fighting, they have allowed them to issue forth from the flanks of the companies, and disorganized and routed them. If anyone finds himself inferior in cavalry, he can, in addition to the methods mentioned, place a company of pikemen behind his cavalry, and in the fighting, arrange for them to give way for the pikemen, and he will always remain superior. Many have accustomed some of the lightly armed infantry to get used to combat amidst the cavalry, and this has been a very great help to the cavalry. Of all those who have organized Armies for battle, the most praiseworthy have been Hannibal and Scipio when they were fighting in Africa: and as Hannibal had his Army composed of Carthaginians and auxiliaries of various kinds, he placed eighty Elephants in the first van, then placed the auxiliaries, after these he placed his Carthaginians, and in the rear, he placed the Italians, whom he trusted little. He arranged matters thusly, because the auxiliaries, having the enemy in front and their rear closed by his men, they could not flee: so that being compelled to fight, they should overcome or tire out the Romans, thinking afterwards with his forces of virtu, fresh, he could easily overcome the already tired Romans. In the encounter with this arrangement, Scipio placed the Astati, the Principi, and the Triari, in the accustomed fashion for one to be able to receive the other, and one to help the other. He made the vans of the army full of intervals; and so that they should not be seen through, but rather appear united, he filled them with Veliti, whom he commanded that, as soon as the Elephants arrived, they should give way, and enter through the regular spaces among the legions, and leave the way open to the Elephants: and thus come to render their attack vain, so that coming hand to hand with them, he was superior.

ZANOBI: You have made me remember in telling me of this engagement, that Scipio, during the fight, did not have the Astati retire into the ranks of the Principi, but divided them and had them retire into the wings of the army, so as to make room for the Principi, if he wanted to push them forward. I would desire, therefore, that you tell me what reason motivated him not to observe the accustomed arrangement.

FABRIZIO: I will tell you. Hannibal had placed all the virtu of his army in the second line; whence Scipio, in order to oppose a similar virtu to it, assembled the Principi and the Triari; so that the intervals of the Principi being occupied by the Triari, there was no place to receive the Astati, and therefore, he caused the Astati to be divided and enter the wings of the army, and did not bring them among the Principi. But take note that this method of opening up the first lines to make a place for the second, cannot be employed except when the other are superior, because then the convenience exists to be able to do it, as Scipio was able to. But being inferior and repulsed, it cannot be done except with your manifest ruin: and, therefore, you must have ranks in the rear which will receive you. But let us return to our discussion. The ancient Asiatics ((among other things thought up by them to injure the enemy)) used chariots which had scythes on their sides, so that they not only served to open up the lines with their attack, but also kill the adversary with the scythes. Provisions against these attacks were made in three ways. It was resisted by the density of the ranks, or they were received within the lines as were the Elephants, or a stalwart resistance was made with some stratagems, as did Sulla, the Roman, against Archelaus, who had many of those chariots which they called Falcati; he (Sulla), in order to resist them, fixed many poles in the ground behind the first ranks, by which the chariots, being resisted, lost their impetus. And note is to be taken of the new method which Sulla used against this man in arranging the army, since he put the Veliti and the cavalry in the rear, and all the heavily armed in front, leaving many intervals in order to be able to send those in the rear forward if necessity should require it; whence when the battle was started, with the aid of the cavalry, to whom he gave the way, he obtained the victory. To want to worry the enemy during the battle, something must be made to happen which dismays him, either by announcing new help which is arriving, or by showing things which look like it, so that the enemy, being deceived by that sight, becomes frightened; and when he is frightened, can be easily overcome. These methods were used by the Roman Consuls Minucius Rufus and Accilius Glabrius, Caius Sulpicius also placed many soldier-packs on mules and other animals useless in war, but in a manner that they looked like men-at-arms, and commanded that they appear on a hill while they were (in) hand to hand (combat) with the Gauls: whence his victory resulted. Marius did the same when he was fighting against the Germans. Feigned assaults, therefore, being of great value while the battle lasts, it happens that many are benefited by the real (assaults), especially if, improvised in the middle of the battle, it is able to attack the enemy from behind or on the sides. Which can be done only with difficulty, unless the (nature of the) country helps you; for if it is open, part of your forces cannot be speeded, as must be done in such enterprises: but in wooded or mountainous places, and hence capable of ambush, part of your forces can be well hidden, so that the enemy may be assaulted, suddenly and without his expecting it, which will always be the cause of giving you the victory. And sometimes it has been very important, while the battle goes on, to plant voices which announce the killed of the enemy Captain, or to have defeated some other part of the army; and this often has given the victory to whoever used it. The enemy cavalry may be easily disturbed by unusual forms (sights) or noises; as did Croesus, who opposed camels to the cavalry of his adversaries, and Pyrrhus who opposed elephants to the Roman cavalry, the sight of which disturbed and disorganized it. In our times, the Turk routed the Shah in Persia and the Soldan in Syria with nothing else than the noise of guns, which so affected their cavalry by their unaccustomed noises, that the Turk was able easily to defeat it. The Spaniards, to overcome the army of

Hamilcar, placed in their first lines chariots full of tow drawn by oxen, and when they had come to battle, set fire to them, whence the oxen, wanting to flee the fire, hurled themselves on the army of Hamilcar and dispersed it. As we mentioned, where the country is suitable, it is usual to deceive the enemy when in combat by drawing him into ambushes: but when it is open and spacious, many have employed the making (digging) of ditches, and then covering them lightly with earth and branches, but leaving several places (spaces) solid in order to be able to retire between them; then when the battle is started, retire through them, and the enemy pursuing, comes to ruin in them. If, during the battle, some accident befalls you which dismays your soldiers, it is a most prudent thing to know how to dissimulate and divert them to (something) good, as did Lucius Sulla, who, while the fighting was going on, seeing that a great part of his forces had gone over to the side of the enemy, and that this had dismayed his men, quickly caused it to be understood throughout the entire army that everything was happening by his order, and this not only did not disturb the army, but so increased its courage that it was victorious. It also happened to Sulla, that having sent certain soldiers to undertake certain business, and they having been killed, in order that his army would not be dismayed said, that because he had found them unfaithful, he had cunningly sent them into the hands of the enemy. Sertorious, when undertaking an engagement in Spain, killed one who had pointed out to him the slaying of one of his Heads, for fear that by telling the same to the others, he should dismay them. It is a difficult matter to stop an army already in flight, and return it to battle. And you have to make this distinction: either they are entirely in flight (motion), and here it is impossible to return them: or only a part are in flight, and here there is some remedy. Many Roman Captains, by getting in front of those fleeing, have stopped them, by making them ashamed of their flight, as did Lucius Sulla, who, when a part of his Legions had already turned, driven by the forces of Mithradates, with his sword in hand he got in front of them and shouted, "if anyone asks you where you have left your Captain, tell them, we have left him in Boetia fighting." The Consul Attilius opposed those who fled with those who did not flee, and made them understand that if they did not turn about, they would be killed by both friends and enemies. Phillip of Macedonia, when he learned that his men were afraid of the Scythian soldiers, put some of his most trusted cavalry behind his army, and commissioned them to kill anyone who fled; whence his men, preferring to die fighting rather than in flight, won. Many Romans, not so much in order to stop a flight, as to give his men an occasion to exhibit greater prowess, while they were fighting, have taken a banner out of their hands, and tossing it amid the enemy, offered rewards to whoever would recover it.

I do not believe it is out of order to add to this discussion those things that happen after a battle, especially as they are brief, and not to be omitted, and conform greatly to this discussion. I will tell you, therefore, how engagements are lost, or are won. When one wins, he ought to follow up the victory with all speed, and imitate Caesar in this case, and not Hannibal, who, because he had stopped after he had defeated the Romans at Cannae, lost the Empire of Rome. The other (Caesar) never rested after a victory, but pursued the routed enemy with great impetus and fury, until he had completely assaulted it. But when one loses, a Captain ought to see if something useful to him can result from this loss, especially if some residue of the army remains to him. An opportunity can arise from the unawareness of the enemy, which frequently becomes obscured after a victory, and gives you the occasion to attack him; as Martius, the Roman, attacked the Carthaginian army, which, having killed the two Scipios and defeated their armies, thought little of that remnant of the forces who, with Martius, remained alive; and was (in turn) attacked and routed by him. It is seen, therefore, that there is nothing so capable of success as that which the enemy believes you cannot attempt, because men are often injured more when they are less apprehensive. A Captain ought, therefore, when he cannot do this, at least endeavor with industry to restrict the injury caused by the defeat. And to do this, it is necessary for you to take steps that the enemy is not able to follow you easily, or give him cause for delay. In the first case some, after they realize they are losing, order their Leaders to flee in several parts by different paths, having (first) given an order where they should afterward reassemble, so that the enemy, fearing to divide his forces, would leave all or a greater part of them safe. In the second case, many have thrown down their most precious possessions in front of the enemy, so that being retarded by plundering, he gave them more time for flight. Titus Dimius used not a little astuteness in hiding the injury received in battle; for, after he had fought until nightfall with a loss of many of his men, caused a good many of them to be buried during the night; whence in the morning, the enemy seeing so many of their dead and so few Romans, believing they had had the disadvantage, fled. I believe I have thus confused you, as I said, (but) satisfied your question in good part: it is true, that concerning the shape of the army, there remains for me to tell you how sometimes it is customary for some Captains to make the front in the form of a wedge, judging in that way to be able more readily to open (penetrate) the Army of the enemy. In opposition to this shape they customarily would use a form of a scissor, so as to be able to receive that wedge into that space, and surround and fight it from every side. On this, I would like you to have this general rule, that the greatest remedy used against the design of the enemy, is to do that willingly which he designs for you to do by force, because doing it willingly you do it with order and to your advantage, but to his disadvantage: if you should do it by force, it would be to your ruin. As to the fortifying of this, I would not care to repeat anything already said. Does the adversary make a wedge in order to open your ranks? if you proceed with yours open, you disorganize him, and he does not disorganize you. Hannibal placed Elephants in front of his Army to open that of the Army of Scipio; Scipio went with his open and was the cause of his own victory and the ruin of the former (Hannibal). Hasdrubal placed his most stalwart forces in the center of the van of his Army to push back the forces of Scipio: Scipio commanded in like fashion that they should retire, and defeated him. So that such plans, when they are put forward, are the cause for the victory of him against whom they were organized. It remains for me yet, if I remember well, to tell you what considerations a Captain ought to take into account before going into battle: upon which I have to tell you first that a Captain never has to make an engagement, if he does not have the advantage, or if he is not compelled to. Advantages arise from the location, from the organization, and from having either greater or better forces. Necessity, (compulsion) arises when you see that, by not fighting, you must lose in an event; for example, when you see you are about to lack money, and therefore your Army has to be dissolved in any case; when hunger is about to assail you, or when you expect the enemy to be reinforced again by new forces. In these cases, one ought always to fight, even at your disadvantage; for it is much better to try your fortune when it can favor you, than by not trying, see your ruin sure: and in such a case, it is as serious an error for a Captain not to fight, as it is to pass up an opportunity to win, either from ignorance, or from cowardice. The enemy sometimes gives you the advantage, and sometimes (it derives from) your prudence. Many have been routed while crossing a river by an alert enemy of theirs, who waited until they were in the middle of the stream, and then assaulted them on every side; as Caesar did to the Swiss, where he destroyed a fourth part of them, after they had been split by the river. Some time you may find your enemy tired from having pursued you too inconsiderately, so that, finding yourself fresh, and rested, you ought not to lose such an opportunity. In addition to this, if an enemy offers you battle at a good hour of the morning, you can delay going out of your encampment for many hours: and if he has been under arms for a long time, and has lost that first ardor with which he started, you can then fight with him. Scipio and Metellus employed this method in Spain, the first against Hasdrubal, and the other against Sertorius. If the enemy has diminished in strength, either from having divided the Armies, as the Scipios (did) in Spain, or from some other cause, you ought to try (your) fortune. The greater part of prudent Captains would rather receive the onrush of the enemy, who impetuously go to assault them, for their fury is easily withstood by firm and resolute men, and that fury which was withstood, easily converts itself into cowardice. Fabius acted thusly against the Samnites and against the Gauls, and was victorious, but his colleague, Decius was killed. Some who feared the virtu of their enemy, have begun the battle at an hour near nightfall, so that if their men were defeated, they might be able to be protected by its darkness and save themselves. Some, having known that the enemy Army, because of certain superstitions, does not want to undertake fighting at such a time, selected that time for battle, and won: which Caesar did in Gaul against Ariovistus, and Vespatianus in Syria against the Jews. The greater and more important awareness that a Captain ought to have, is (to see) that he has about him, men loyal and most expert in war, and prudent, with whom he counsels continually, and discusses his forces and those of the enemy with them: which are the greater in number, which are better armed or better trained, which are more apt to suffer deprivation, which to confide in more, the infantry or the cavalry. Also, they consider the location in which they are, and if it is more suitable for the enemy than for themselves; which of them has the better convenience of supply; whether it is better to delay the engagement or undertake it, and what benefit the weather might give you or take away from them; for often when the soldiers see the war becoming long, they become irritable, and weary from hard work and tedium, will abandon you. Above all, it is important for the Captain to know the enemy, and who he has around him: if he is foolhardy or cautious: if timid or audacious. See whether you can trust the auxiliary soldiers. And above all, you ought to guard against leading an army into battle which is afraid, or distrustful in any way of victory, for the best indication of defeat is when one believes he cannot win. And, therefore, in this case, you ought to avoid an engagement, either by doing as Fabius Maximus did, who, by encamping in strong places, did not give Hannibal courage to go and meet him, or by believing that the enemy, also in strong places, should come to meet you, you should depart from the field, and divide your forces among your towns, so that the tedium of capturing them will tire him.

ZANOBI: Can he not avoid the engagement in other ways than by dividing it (the army) into several parts, and putting them in towns?

FABRIZIO: I believe at another time I have discussed with some of you that whoever is in the field, cannot avoid an engagement if he has an enemy who wants to fight in any case; and he has but one remedy, and that is to place himself with his Army at least fifty miles distant from his adversary, so as to be in time to get out of his way if he should come to meet him. And Fabius Maximus never avoided an engagement with Hannibal, but wanted it at his advantage; and Hannibal did not presume to be able to overcome him by going to meet him in the places where he was encamped. But if he supposed he could defeat him, it was necessary for Fabius to undertake an engagement with him in any case, or to flee. Phillip, King of Macedonia, he who was the father of Perseus, coming to war with the Romans, placed his encampment on a very high mountain so as not to have an engagement with them; but the Romans went to meet him on that mountain, and routed him. Vercingetorix, a Captain of the Gauls, in order to avoid an engagement with Caesar, who unexpectedly had crossed the river, placed himself miles distant with his forces. The Venetians in our times, if they did not want to come to an engagement with the King of France, ought not to have waited until the French Army had crossed the Adda, but should have placed themselves distant from him, as did Vercingetorix: whence, having waited for him, they did not know how to take the opportunity of undertaking an engagement during the crossing, nor how to avoid it; for the French being near to them, as the Venetians decamped, assaulted and routed them. And so it is, that an engagement cannot be avoided if the enemy at all events wants to undertake it. Nor does anyone cite Fabius, for he avoided an engagement in cases like that, just as much as did Hannibal. It often happens that your soldiers are not willing to fight, and you know that because of their number or the location, or from some other cause, you have a disadvantage, and would like them to change their minds. It also happens that necessity or opportunity constrains you to (come to) an engagement, and that your soldiers are discontent and little disposed to fight, whence it is necessary for you in one case to frighten them, and in the other to excite them. In the first instance, if persuasion is not enough, there is no better way to have both those who fight and those who would not believe you, than to give some of them over to the enemy as plunder. It may also be well to do with cunning that which happened to Fabius Maximus at home. The Army of Fabius desired ((as you know)) to fight with the Army of Hannibal: his Master of cavalry had the same desire. It did not seem proper to Fabius to attempt the battle, so that in order to dispel such (desires), he had to divide the Army. Fabius kept his men in the encampments: and the other (the Master of cavalry) going forth, and coming into great danger, would have been routed, if Fabius had not succored him. By this example, the Master of the cavalry, together with the entire army, realized it was a wise course to obey Fabius. As to exciting them to fight, it is well to make them angry at the enemy, by pointing out that (the enemy) say slanderous things of them, and showing them to have with their intelligence (in the enemy camp) and having corrupted some part, to encamp on the side where they see they enemy, and undertake some light skirmishes with them; because things that are seen daily are more easily disparaged. By showing yourself indignant, and by making an oration in which you reproach them for their laziness, you make them so ashamed by saying you want to fight only if they do not accompany you. And above every thing, to have this awareness, if you want to make the soldiers obstinate in battle, not to permit them to send home any of their possessions, or settle in any place, until the war ends, so that they understand that if flight saves them their lives, it will not save them their possessions, the love of the latter, not less than the former, renders men obstinate in defense.

ZANOBI: You have told how soldiers can be made to turn and fight, by talking to them. Do you mean by this that he has to talk to the entire Army, or to its Heads?

FABRIZIO: To persuade or dissuade a few from something, is very easy; for if words are not enough, you can use authority and force: but the difficulty is to take away a sinister idea from a multitude, whether it may be in agreement or contrary to your own opinion, where only words can be used, which, if you want to persuade everyone, must be heard by everyone. Captains, therefore, must be excellent Orators, for without knowing how to talk to the entire Army, good things can only be done with difficulty. Which, in these times of ours, is completely done away with. Read the life (biography) of Alexander the Great, and see how many times it was necessary to harangue and speak publicly to the Army; otherwise he could never have them led them ((having become rich and full of plunder)) through the deserts of Arabia and into India with so much hardship and trouble; for infinite numbers of things arose by which an Army is ruined if a Captain does not know how or is not accustomed to talking to it; for this speaking takes away fear, incites courage, increases obstinacy, and sweeps away deceptions, promises rewards, points out dangers and the ways to avoid them, reprimands, begs, threatens, fills with hope, praises, slanders, and does all those things by which human passion are extinguished or enkindled. Whence that Prince or Republic planning to raise a new army, and to give this army reputation, ought to accustom the soldiers to listen to the talk of the Captain, and the Captain to know how to talk to them. Religion was (also) of much value in keeping the ancient soldiers well disposed and an oath was given to (taken by) them when they came into the army; for whenever they made a mistake, they were threatened not only by those evils that can be feared by men, but also by those that can be expected from the Deity. This practice, mixed with other religious means, often made an entire enterprise easy for the ancient Captains, and would always be so whenever religion was feared and observed. Sertorius availed himself of this when he told of talking with a Hind (female stag), which promised him victory on the part of the Deity. Sulla was said to talk with a Statue which he had taken from the Temple of Apollo. Many have told of God appearing to them in their sleep, and admonishing them to fight. In the times of our fathers, Charles the seventh, King of France, in the war he waged against the English, was said to counsel with a young girl sent by God, who is called the Maid of France, and who was the cause for victory. You can also take means to make your (soldiers) value the enemy little, as Agesilaus the Spartan did, who showed his soldiers some Persians in the nude, so that seeing their delicate members, they should have no cause for being afraid of them. Some have constrained them to fight from necessity, by removing from their paths all hope of saving themselves, except through victory. This is the strongest and the best provision that can be made when you want to make your soldiers obstinate. Which obstinacy is increased by the confidence and the love either of the Captain or of the Country. Confidence is instilled by arms organization, fresh victories, and the knowledge of the Captain. Love of Country springs from nature: that of the Captain from (his) virtu more than any other good event. Necessities can be many, but that is the strongest, which constrains you either to win or to die.

Fifth Book

FABRIZIO: I have shown you how to organize an army to battle another army which is seen posted against you, and I have told you how it is overcome, and also of the many circumstances which can occur because of the various incidents surrounding it, so that it appears to me now to be the time to show you how to organize an army against an enemy which is unseen, but which you are continually afraid will assault you. This happens when marching through country which is hostile, or suspected (of being so). And first you have to understand that a Roman Army ordinarily always sent ahead some groups of cavalry as observers for the march. Afterwards the right wing followed. After this came all the wagons which pertained to it. After those, another Legion, and next its wagons. After these come the left wing with its wagon in the rear, and the remainder of the cavalry followed in the last part. This was in effect the manner in which one ordinarily marched. And if it happened that the Army should be assaulted on the march in front or from the rear, they quickly caused all the wagons to be withdrawn either on the right, or on the left, according as it happened, or rather as best they could depending on the location, and all the forces together, free from their baggage, set up a front on that side from which the enemy was coming. If they were assaulted on the flank, they would withdraw the wagons to the side which was secure, and set up a front on the other. This method being good, and prudently conducted, appears to me ought to be imitated, sending cavalry ahead to observe the country, then having four battalions, having them march in line, and each with its wagons in the rear. And as the wagons are of two kinds, that is, those pertaining to individual soldiers, and the public ones for use by the whole camp, I would divide the public wagons into four parts, and assign a part to each Battalion, also dividing the artillery and all the unarmed men, so that each one of those armed should have its equal share of impedimenta. But as it sometimes happens that one marches in a country not only suspect, but hostile in fact, that you are afraid of being attacked hourly, in order to go on more securely, you are compelled to change the formation of the march, and go on in the regular way, so that in some unforeseen place, neither the inhabitants nor the Army can injure you. In such a case, the ancient Captains usually went on with the Army in squares, for such they called these formations, not because it was entirely square, but because it was capable of fighting on four sides, and they said that they were going prepared either for marching or for battle. I do not want to stray far from this method, and want to arrange my two Battalions, which I have taken as a rule for an Army, in this manner. If you want, therefore, to walk securely through the enemy country, and be able to respond from every side, if you had been assaulted by surprise, and wanting, in accordance with the ancients, to bring it into a square, I would plan to make a square whose hollow was two hundred arm lengths on every side in this manner. I would first place the flanks, each distant from the other by two hundred twelve arm lengths, and would place five companies in each flank in a file along its length, and distant from each other three arm lengths; these would occupy their own space, each company occupying (a space) forty arm lengths by two hundred twelve arm lengths. Between the front and rear of these two flanks, I would place another ten companies, five on each side, arranging them in such a way that four should be next to the front of the right flank, and five at the rear of the left flank, leaving between each one an interval (gap) of four arm lengths: one of which should be next to the front of the left flank, and one at the rear of the right flank. And as the space existing between the one flank and the other is two hundred twelve arm lengths, and these companies placed alongside each other by their width and not length, they would come to occupy, with the intervals, one hundred thirty four arm lengths, (and) there would be between the four companies placed on the front of the right flank, and one placed on the left, a remaining space of seventy eight arm lengths, and a similar space be left among the companies placed in the rear parts; and there would be no other difference, except that one space would be on the rear side toward the right wing, the other would be on the front side toward the left wing. In the space of seventy eight arm lengths in front, I would place all the ordinary Veliti, and in that in the rear the extraordinary Veliti, who would come to be a thousand per space. And if you want that the space taken up by the Army should be two hundred twelve arm lengths on every side, I would see that five companies are placed in front, and those that are placed in the rear, should not occupy any space already occupied by the flanks, and therefore I would see that the five companies in the rear should have their front touch the rear of their flanks, and those in front should have their rear touch the front (of their flanks), so that on every side of that army, space would remain to receive another company. And as there are four spaces, I would take four banners away from the extraordinary pikemen and would put one on every corner: and the two banners of the aforementioned pikemen left to me, I would place in the middle of the hollow of their army (formed) in a square of companies, at the heads of which the general Captain would remain with his men around him. And as these companies so arranged all march in one direction, but not all fight in one, in putting them together, one has to arrange which sides are not guarded by other companies during the battle. And, therefore, it ought to be considered that the five companies in front protect all the other sides, except the front; and therefore these have to be assembled in an orderly manner (and) with the pikemen in front. The five companies behind protect all the sides, except the side in the back; and therefore ought to be assembled so that the pikemen are in the rear, as we will demonstrate in its place. The five companies on the right flank protect all the sides, from the right flank outward. The five on the left, engird all the sides, from the left flank outward: and therefore in arranging the companies, the pikemen ought to be placed so that they turn by that flank which in uncovered. And as the Heads of Ten are placed in the front and rear, so that when they have to fight, all the army and its members are in their proper places, the manner of accomplishing this was told when we discussed the methods of arranging the companies. I would divide the artillery, and one part I would place outside the right flank, and the other at the left. I would send the light cavalry ahead to reconnoiter the country. Of the men-at-arms, I would place part in the rear on the right wing, and part on the left, distant forty arms lengths from the companies. And no matter how you arrange your Army, you have to take up ((as the cavalry)) this general (rule), that you have to place them always either in the rear or on the flanks. Whoever places them ahead in front of the Army must do one of two things: either he places them so far ahead, that if they are repulsed they have so much room to give them time to be able to obtain shelter for themselves from your infantry and not collide with them; or to arrange them (the infantry) with so many intervals, that by means of them the cavalry can enter among them without disorganizing them. Let not anyone think little of this instruction, because many, not being aware of this, have been ruined, and have been disorganized and routed by themselves. The wagons and the unarmed men are placed in the plaza that exists within the Army, and so compartmented, that they easily make way for whoever wants to go from one side to the other, or from one front of the Army to the other. These companies, without artillery and cavalry, occupy two hundred eighty two arm lengths of space on the outside in every direction. And as this square is composed of two Battalions, it must be devised as to which part one Battalion makes up, and which part the other. And since the Battalions are called by number, and each of them has ((as you know)) ten companies and a general Head, I would have the first Battalion place its first five companies in the front, the other five on the left flank, and the Head should be in the left angle of the front. The first five companies of the second Battalion then should be placed on the right flank, and the other five in the rear, and the Head should be in the right angle, who would undertake the office of the Tergiduttore.

The Army organized in this manner is ready to move, and in its movement should completely observe this arrangement: and without doubt it is secure from all the tumults of the inhabitants. Nor ought the Captain make other provisions against these tumultuous assaults, than sometime to give a commission to some cavalry or band of Veliti to put them in their place. Nor will it ever happen that these tumultuous people will come to meet you within the drawing of a sword or pike, because disorderly people are afraid of order; and it will always be seen that they make a great assault with shouts and noises without otherwise approaching you in the way of yelping dogs around a mastiff. Hannibal, when he came to harm from the Romans in Italy, passed through all of France, and always took little account of the tumults of the French. When you want to march, you must have levellers and men with pick axes ahead who clear the road for you, and who are well protected by that cavalry sent ahead to reconnoiter. An Army will march in this order ten miles a day, and enough Sun (light will remain for them to dine and camp, since ordinarily an Army marches twenty miles. If it happens that it is assaulted by an organized Army, this assault cannot arise suddenly, because an organized Army travels at its own rate (step), so that you are always in time to reorganize for the engagement, and quickly bring yourself to that formation, or similar to that formation of the Army, which I showed you above. For if you are assaulted on the front side, you do nothing except (to have) the artillery in the flanks and the cavalry behind come forward and take those places and with those distances mentioned above. The thousand Veliti who are forward, come forth from their positions, and dividing into groups of a hundred, enter into their places between the cavalry and the wings of the Army. Then, into the voids left by them, enter the two bands of extraordinary pikemen which I had placed in the plaza of the Army. The thousand Veliti that I had placed in the rear depart from there, and distribute themselves among the flanks of the companies to strengthen them: and from the open space they leave all the wagons and unarmed men issue forth and place themselves at the rear of the companies. The plaza, therefore, remains vacant as everyone has gone to their places, and the five companies that I placed in the rear of the Army come forward through the open void that exists between the one and the other flank, and march toward the company in the front, and the three approach them at forty arm lengths with equal intervals between one another, and two remain behind distant another forty arm lengths. This formation can be organized quickly, and comes to be almost the same as the first disposition of the Army which we described before: and if it becomes more straitened in the front, it becomes larger in the flanks, which does not weaken it. But as the five companies in the back have their pikemen in the rear for the reasons mentioned above, it is necessary to have them come from the forward part, if you want them to get behind the front of the Army; and, therefore, one must either make them turn company by company, as a solid body, or make them enter quickly between the ranks of the shield-bearers (swordsmen), and bring them forward; which method is more swift and less disorderly than to make them turn. And thus you ought to do with all those who are in the rear in every kind of assault, as I will show you. If it should happen that the enemy comes from the rear, the first thing that ought to be done is to have everyone turn to face the enemy, so that at once the front of the army becomes the rear, and the rear the front. Then all those methods of organizing the front should be followed, which I mentioned above. If the enemy attacks on the right flank, the entire army ought to be made to face in that direction, and then those things ought to be done to strengthen that (new) front which were mentioned above, so that the cavalry, the Veliti, and the artillery are in the position assigned in this front. There is only this difference, that in the changing of fronts, of those who move about, some have to go further, and some less. It is indeed true that when a front is made of the right flank, the Veliti would have to enter the intervals (gaps) that exist between the wings of the Army, and the cavalry would be those nearer to the left flank, in the position of those who would have to enter into the two bands of extraordinary pikemen placed in the center. But before they enter, the wagons and unarmed men stationed at the openings, should clear the plaza and retire behind the left flank, which then becomes the rear of the army. And the other Veliti who should be placed in the rear according to the original arrangement, in this case should not be changed, as that place should not remain open, which, from being the rear, would become a flank. All the other things ought to be done as was said concerning the first front.

What has been said concerning making a front from the right flank, is intended also in making one from the left flank, since the same arrangements ought to be observed. If the enemy should happen to be large and organized to assault you on two sides, the two sides on which he assaults you ought to be strengthened from the two that are not assaulted, doubling the ranks in each one, and distributing the artillery, Veliti, and cavalry among each side. If he comes from three or four sides, it needs must be either you or he lacks prudence, for if you were wise, you would never put yourself on the side where the enemy could assault you from three or four sides with large and organized forces, and if he wanted to attach you in safety he must be so large and assault you on each side with a force almost as large as you have in your entire Army. And if you are so little prudent that you put yourself in the midst of the territory and forces of an enemy, who has three times the organized forces that you have, you cannot complain if evil happens to you, except of yourself. If it happens, not by your fault, but by some misadventure, the injury will be without shame, and it will happen to you as it did to the Scipios in Spain, and the Hasdrubal in Italy. But if the enemy has a much larger force than you, and in order to disorganize you wants to assault you on several sides, it will be his foolishness and his gamble; for to do this, he must go (spread) himself thin, that you can always attack on one side and resist on another, and in a brief time ruin him. This method of organizing an Army which is not seen, but who is feared, is necessary, and it is a most useful thing to accustom your soldiers to assemble, and march in such order, and in marching arrange themselves to fight according to the first front (planned), and then return to marching formation, from that make a front from the rear, and then from the flank, and from that return to the original formation. These exercises and accustomization are necessary matters if you want a disciplined and trained Army. Captains and Princes have to work hard at these things: nor is military discipline anything else, than to know how to command and how to execute these things, nor is a disciplined Army anything else, than an army which is well trained in these arrangements; nor would it be possible for anyone in these times who should well employ such discipline ever to be routed. And if this square formation which I have described is somewhat difficult, such difficulty is necessary, if you take it up as exercise; since knowing how to organize and maintain oneself well in this, one would afterwards know how to manage more easily those which not be as difficult.

ZANOBI: I believe as you say, that these arrangements are very necessary, and by myself, I would not know what to add or leave out. It is true that I desire to know two things from you: the one, when you want to make a front from the rear or from a flank, and you want them to turn, whether the command is given by voice or by sound (bugle call): the other, whether those you sent ahead to clear the roads in order to make a path for the Army, ought to be soldiers of your companies, or other lowly people assigned to such practices.

FABRIZIO: Your first question is very important, for often the commands of the Captain are not very well understood or poorly interpreted, have disorganized their Army; hence the voices with which they command in (times of) danger, ought to be loud and clear. And if you command with sounds (bugle calls), it ought to be done so that they are so different from each other that one cannot be mistaken for another; and if you command by voice, you ought to be alert to avoid general words, and use particular ones, and of the particular ones avoid those which might be able to be interpreted in an incorrect manner. Many times saying "go back, go back", has caused an

Army to be ruined: therefore this expression ought to be avoided, and in its place use "Retreat". If you want them to turn so as to change the front, either from the rear or from the flank, never use "Turn around", but say, "To the left", "To the right", "To the rear", "To the front". So too, all the other words have to be simple and clear, as "Hurry", "Hold still", "Forward", "Return". And all those things which can be done by words are done, the others are done by sounds (calls). As to the (road) clearers, which is your second question, I would have this job done by my own soldiers, as much because the ancient military did so, as also because there would be fewer unarmed men and less impediments in the army: and I would draw the number needed from every company, and I would have them take up the tools suitable for clearing, and leave their arms in those ranks that are closest to them, which would carry them so that if the enemy should come, they would have nothing to do but take them up again and return to their ranks.

ZANOBI: Who would carry the clearing equipment?

FABRIZIO: The wagons assigned to carry such equipment.

ZANOBI: I'm afraid you have never led these soldiers of ours to dig.

FABRIZIO: Everything will be discussed in its place. For now I want to leave these parts alone, and discuss the manner of living of the Army, for it appears to me that having worked them so hard, it is time to refresh and restore it with food. You have to understand that a Prince ought to organize his army as expeditiously as possible, and take away from it all those things that add burdens to it and make the enterprise difficult. Among those that cause more difficulty, are to have to keep the army provided with wine and baked bread. The ancients did not think of wine, for lacking it, they drank water tinted with a little vinegar, and not wine. They did not cook bread in ovens, as is customary throughout the cities; but they provided flour, and every soldier satisfied himself of that in his own way, having lard and grease for condiment, which gave flavor to the bread they made, and which kept them strong. So that the provisions of living (eating) for the army were Flour, Vinegar, Lard (Bacon) and Grease (Lard), and Barley for the horses. Ordinarily, they had herds of large and small beasts that followed the Army, which ((as they did not need to be carried)) did not impede them much. This arrangement permitted an ancient Army to march, sometimes for many days, through solitary and difficult places without suffering hardship of (lack of) provisions, for it lived from things which could be drawn behind. The contrary happens in modern Armies, which, as they do not want to lack wine and eat baked bread in the manner that those at home do, and of which they cannot make provision for long, often are hungry; or even if they are provided, it is done with hardship and at very great expense. I would therefore return my Army to this form of living, and I would not have them eat other bread than that which they should cook for themselves. As to wine, I would not prohibit its drinking, or that it should come into the army, but I would not use either industry or any hard work to obtain it, and as to other provisions, I would govern myself entirely as the ancients. If you would consider this matter well, you will see how much difficulty is removed, and how many troubles and hardships an army and a Captain avoid, and what great advantage it will give any enterprise which you may want to undertake.

ZANOBI: We have overcome the enemy in the field, and then marched on his country: reason wants that there be no booty, ransoming of towns, prisoners taken. Yet I would like to know how the ancients governed themselves in these matters.

FABRIZIO: Here, I will satisfy you. I believe you have considered ((since I have at another time discussed this with some of you)) that modern wars impoverish as much those Lords who win, as those who lose; for if one loses the State, the other loses his money and (movable) possessions. Which anciently did not happen, as the winner of a war (then) was enriched. This arises from not keeping track in these times of the booty (acquired), as was done anciently, but everything is left to the direction of the soldiers. This method makes for two very great disorders: the one, that of which I have spoken: the other, that a soldier becomes more desirous of booty and less an observer of orders: and it has often been said that the cupidity for booty has made him lose who had been victorious. The Romans, however, who were Princes in this matter, provided for both these inconveniences, ordering that all the booty belong to the public, and that hence the public should dispense it as it pleased. And so they had Quaestors in the Army, who were, as we would say, chamberlains, to whom all the ransoms and booty was given to hold: from which the Consul served himself to give the soldiers their regular pay, to help the wounded and infirm, and to provide for the other needs of the army. The Consul could indeed, and often did, concede a booty to the soldiers, but this concession did not cause disorders; for when the (enemy) army was routed, all the booty was placed in the middle and was distributed to each person, according to the merits of each. This method made for the soldiers attending to winning and not robbing, and the Roman legions defeating the enemy but not pursuing him: for they never departed from their orders: only the cavalry and lightly armed men pursued him, unless there were other soldiers than legionnaires, which, if the booty would have been kept by whoever acquired it, it was neither possible nor reasonable to (expect to) hold the Legion firm, and would bring on many dangers. From this it resulted, therefore that the public was enriched, and every Consul brought, with his triumphs, much treasure into the Treasury, which (consisted) entirely of ransoms and booty. Another thing well considered by the ancients, was the pay they gave to each soldier: they wanted a third part to be placed next to him who carried the flag of the company, who never was given any except that furnished by the war. They did this for two reasons: The first so that the soldier would make capital (save) of his pay: for the greater part of them being young and irresponsible, the more they had, the more they spent without need to. The other part because, knowing that their movable possessions were next to the flag, they would be forced to have greater care, and defend it with greater obstinacy: and thus this method made them savers, and strong. All of these things are necessary to observe if you want to bring the military up to your standards.

ZANOBI: I believe it is not possible for an army while marching from place to place not to encounter dangerous incidents, (and) where the industry of the Captain and the virtu of the soldier is needed if they are to be avoided; therefore, if you should have something that occurs to you, I would take care to listen.

FABRIZIO: I will willingly content you, especially as it is necessary, if I want to give you complete knowledge of the practice. The Captains, while they march with the Army, ought, above everything else, to guard against ambushes, which may happen in two ways: either you enter into them while marching, or the enemy cunningly draws you into them without your being aware of it. In the first case, if you want to avoid them, it is necessary to send ahead double the guard, who reconnoiter the country. And the more the country is suitable for ambush, as are wooded and mountainous countries, the more diligence ought to be used, for the enemy always place themselves either in woods or behind a hill. And, just as by not foreseeing an ambush you will be ruined, so by foreseeing it you will not be harmed. Birds or dust have often discovered the enemy, for where the enemy comes to meet you, he will always raise a great dust which will point out his coming to you. Thus often a Captain when he sees in a place whence he ought to pass, pigeons taking off and other birds flying about freely, circling and not setting, has recognized this to be the place of any enemy ambush, and knowing this has sent his forces forward, saving himself and injuring the enemy. As to the second case, being drawn into it ((which our men call being drawn into a trap)) you ought to look out not to believe readily those things that appear to be less reasonable than they should be: as would be (the case) if an enemy places some booty before you, you would believe that it to be (an act of) love, but would conceal deceit inside it. If many enemies are driven out by few of your man: if only a few of the enemy assault you: if the enemy takes to sudden and unreasonable flight: in such cases, you ought always to be afraid of deceit; and you should never believe that the enemy does not know his business, rather, if you want to deceive yourself less and bring on less danger, the more he appears weak, the more enemy appears more cautious, so much the more ought you to esteem (be wary) of him. And in this you have to use two different means, since you have to fear him with your thoughts and arrangements, but by words and other external demonstrations show him how much you disparage him; for this latter method causes your soldiers to have more hope in obtaining the victory, the former makes you more cautious and less apt to be deceived. And you have to understand that when you march through enemy country, you face more and greater dangers than in undertaking an engagement. And therefore, when marching, a Captain ought to double his diligence, and the first thing he ought to do, is to have all the country through which he marches described and depicted, so that he will know the places, the numbers, the distances, the roads, the mountains, the rivers, the marshes, and all their characteristics. And in getting to know this, in diverse ways one must have around him different people who know the places, and question them with diligence, and contrast their information, and make notes according as it checks out. He ought to send cavalry ahead, and with them prudent Heads, not so much to discover the enemy as to reconnoiter the country, to see whether it checks with the places and with the information received from them. He ought also to send out guides, guarded (kept loyal) by hopes of reward and fear of punishment. And above all, he ought to see to it that the Army does not know to which sides he guides them, since there is nothing more useful in war, than to keep silent (about) the things that have to be done. And so that a sudden assault does not disturb your soldiers, you ought to advise them to be prepared with their arms, since things that are foreseen cause less harm. Many have ((in order to avoid the confusion of the march)) placed the wagons and the unarmed men under the banners, and commanded them to follow them, so that having to stop or retire during the march, they are able to do so more easily: which I approve very much as something useful. He ought also to have an awareness during the march, that one part of the Army does not detach itself from another, or that one (part) going faster and the other more slowly, the Army does not become compacted (jumbled), which things cause disorganization. It is necessary, therefore, to place the Heads along the sides, who should maintain the steps uniform, restraining those which are too fast, and hastening the slow; which step cannot be better regulated than by sound (music). The roads ought to be widened, so that at least one company can always move in order. The customs and characteristics of the enemy ought to be considered, and if he wants to assault you in the morning, noon, or night, and if he is more powerful in infantry or cavalry, from what you have learned, you may organize and prepare yourself. But let us come to some incident in particular. It sometimes happens that as you are taking yourself away from in front of the enemy because you judge yourself to be inferior (to him), and therefore do not want to come to an engagement with him, he comes upon your rear as you arrive at the banks of a river, which causes you to lose times in its crossing, so that the enemy is about to join up and combat with you. There have been some who have found themselves in such a peril, their army girded on the rear side by a ditch, and filling it with tow, have set it afire, then have passed on with the army without being able to be impeded by the enemy, he being stopped by that fire which was in between.

ZANOBI: And it is hard for me to believe that this fire can check him, especially as I remember to have heard that Hanno, the Carthaginian, when he was besieged by the enemy, girded himself on that side from which he wanted to make an eruption with wood, and set fire to it. Whence the enemy not being intent to guard that side, had his army pass over the flames, having each (soldier) protect his face from the fire and smoke with his shield.

FABRIZIO: You say well; but consider what I have said and what Hanno did: for I said that he dug a ditch and filled it with tow, so that whoever wanted to pass had to contend with the ditch and the fire. Hanno made the fire without a ditch, and as he wanted to pass through it did not make it very large (strong), since it would have impeded him even without the ditch. Do you not know that Nabidus, the Spartan, when he was besieged in Sparta by the Romans, set fire to part of his own town in order to stop the passage of the Romans, who had already entered inside? and by those flames not only stopped their passage, but pushed them out. But let us return to our subject. Quintus Luttatius, the Roman, having the Cimbri at his rear, and arriving at a river, so that the enemy should give him time to cross, made as if to give him time to combat him, and therefore feigned to make camp there, and had ditches dug, and some pavilions raised, and sent some horses to the camps to be shod: so that the Cimbri believing he was encamping, they also encamped, and divided themselves into several parts to provide themselves with food: of which Luttatius becoming aware, he crossed the river without being able to be impeded by them. Some, in order to cross a river, not having a bridge, have diverted it, and having drawn a part of it in their rear, the other then became so low that they crossed it easily. If the rivers are rapid, (and) desiring that the infantry should cross more safely, the more capable horses are placed on the side above which holds back the water, and another part below which succor the infantry if any, in crossing, should be overcome by the river. Rivers that are not forded, are crossed by bridges, boats, and rafts: and it is therefore well to have skills in your Armies capable of doing all these things. It sometimes happens that in crossing a river, the enemy on the opposite bank impedes you. If you want to overcome this difficulty there is no better example known than that of Caesar, who, having his army on the bank of a river in Gaul, and his crossing being impeded by Vercingetorix, the Gaul, who had his forces on the other side of the river, marched for several days along the river, and the enemy did the same. And Caesar having made an encampment in a woody place (and) suitable to conceal his forces, withdrew three cohorts from every Legion, and had them stop in that place, commanding then that as soon as he should depart, they should throw a bridge across and fortify it, and he with the rest of his forces continued the march: Whence Vercingetorix seeing the number of Legions, and believing that no part had remained behind, also continued the march: but Caesar, as soon as he thought the bridge had been completed, turned back, and finding everything in order, crossed the river without difficulty.

ZANOBI: Do you have any rule for recognizing the fords?

FABRIZIO: Yes, we have. The river, in that part between the stagnant water and the current, always looks like a line to whoever looks at it, is shallower, and is a place more suitable for fording than elsewhere, for the river always places more material, and in a pack, which it draws (with it) from the bottom. Which thing, as it has been experienced many times, is very true.

ZANOBI: If it happens that the river has washed away the bottom of the ford, so that horses sink, what remedy do you have?

FABRIZIO: Make grids of wood, and place them on the bottom of the river, and cross over those. But let us pursue our discussion. If it happens that a Captain with his army is led (caught) between two mountains, and has but two ways of saving himself, either that in front, or the one in the rear, and both being occupied by the enemy, has, as a remedy, to do what some have done in the past, which is to dig a large ditch, difficult to cross, and show the enemy that by it you want to be able to hold him with all his forces, without having to fear those forces in the rear for which the road in front remains open. The enemy believing this, fortifies himself on the side open, and abandons the (side) closed, and he then throws a wooden bridge, planned for such a result, over the ditch, and without any impediment, passes on that side and freed himself from the hands of the enemy. Lucius Minutius, the Roman Consul, was in Liguria with the Armies, and had been enclosed between certain mountains by the enemy, from which he could not go out. He therefore sent some soldiers of Numidia, whom he had in his army, who were badly armed, and mounted on small and scrawny horses, toward those places which were guarded by the enemy, and the first sight of whom caused the enemy to assemble to defend the pass: but then when they saw those forces poorly organized, and also poorly mounted, they esteemed them little and loosened their guard. As soon as the Numidians saw this, giving spurs to their horses and attacking them, they passed by without the enemy being able to take any remedy; and having passed, they wasted and plundered the country, constraining the enemy to leave the pass free to the army of Lucius. Some Captain, who has found himself assaulted by a great multitude of the enemy, has tightened his ranks, and given the enemy the faculty of completely surrounding him, and then has applied force to that part which he has recognized as being weaker, and has made a path in that way, and saved himself. Marcantonio, while retiring before the army of the Parthians, became aware that every day at daybreak as he moved, the enemy assaulted him, and infested him throughout the march: so that he took the course of not departing before midday. So that the Parthians, believing he should not want to decamp that day returned to their quarters, and Marcantonio was able then for the remainder of the day to march without being molested. This same man, to escape the darts of the Parthians, commanded that, when the Parthians came toward them, they should kneel, and the second rank of the company should place their shields on the heads of (those in the) first, the third on (those of the) second, the fourth on the third, and so on successively: so that the entire Army came to be as under a roof, and protected from the darts of the enemy. This is as much as occurs to me to tell you of what can happen to an army when marching: therefore, if nothing else occurs to you, I will pass on to another part.

Sixth Book

ZANOBI: I believe it is well, since the discussion ought to be changed, that Battista take up his office, and I resign mine; and in this case we would come to imitate the good Captains, according as I have already learned here from the Lord, who place the best soldiers in the front and in the rear of the Army, as it appears necessary to them to have those who bravely enkindle the battle, and those in the rear who bravely sustain it. Cosimo, therefore, begun this discussion prudently, and Battista will prudently finish it. Luigi and I have come in between these. And as each one of us has taken up his part willingly, so too I believe Battista is about to close it.

BATTISTA: I have allowed myself to be governed up to now, so too I will allow myself (to be governed) in the future. Be content, therefore, (my) Lords, to continue your discussions, and if we interrupt you with these questions (practices), you have to excuse us.

FABRIZIO: You do me, as I have already told you, a very great favor, since these interruptions of yours do not take away my imagination, rather they refresh it. But if we want to pursue our subject I say, that it is now time that we quarter this Army of ours, since you know that everything desires repose, and safety; since to repose oneself, and not to repose safely, is not complete (perfect) repose. I am afraid, indeed, that you should not desire that I should first quarter them, then had them march, and lastly to fight, and we have done the contrary. Necessity has led us to this, for in wanting to show when marching, how an army turns from a marching formation to that of battle, it was necessary first to show how they were organized for battle. But returning to our subject I say, that if you want the encampment to be safe, it must be Strong and Organized. The industry of the Captain makes it organized: Arts or the site make it Strong. The Greeks sought strong locations, and never took positions where there was neither grottoes (caves), or banks of rivers, or a multitude of trees, or other natural cover which should protect them. But the Romans did not encamp safely so much from the location as by arts, nor ever made an encampment in places where they should not have been able to spread out all their forces, according to their discipline. From this resulted that the Romans were always able to have one form of encampment, for they wanted the site to obey them, and not they the site. The Greeks were not able to observe this, for as they obeyed the site, and the sites changing the formation, it behooved them that they too should change the mode of encamping and the form of their encampment. The Romans, therefore, where the site lacked strength, supplied it with (their) art and industry. And since in this narration of mine, I have wanted that the Romans be imitated, I will not depart from their mode of encamping, not, however, observing all their arrangements: but taking (only) that part which at the present time seems appropriate to me. I have often told you that the Romans had two Legions of Roman men in their consular armies, which comprised some eleven thousand infantry of forces sent by friends (allies) to aid them; but they never had more foreign soldiers in their armies than Romans, except for cavalry, which they did not care if they exceeded the number in their Legions; and that in every action of theirs, they place the Legions in the center, and the Auxiliaries on the sides. Which method they observed even when they encamped, as you yourselves have been able to read in those who write of their affairs; and therefore I am not about to narrate in detail how they encamped, but will tell you only how I would at present arrange to encamp my army, and then you will know what part of the Roman methods I have treated. You know that at the encounter of two Roman Legions I have taken two Battalions of six thousand infantry and three hundred cavalry effective for each Battalion, and I have divided them by companies, by arms, and names. You know that in organizing the army for marching and fighting, I have not made mention of other forces, but have only shown that in doubling the forces, nothing else had to be done but to double the orders (arrangements).

Since at present I want to show you the manner of encamping, it appears proper to me not to stay only with two Battalions, but to assemble a fair army, and composed like the Roman of two Battalions and as many auxiliary forces. I know that the form of an encampment is more perfect, when a complete army is quartered: which matter did not appear necessary to me in the previous demonstration. If I want, therefore, to quarter a fair (sized) army of twenty four thousand infantry and two thousand cavalry effectives, being divided into four companies, two of your own forces and two of foreigners, I would employ this method. When I had found the site where I should want to encamp, I would raise the Captain's flag, and around it I would draw a square which would have each face distant from it fifty arm lengths, of which each should look out on one of the four regions of the sky, that is, east, west, south and north, in which space I would put the quarters of the Captain. And as I believe it prudent, and because thus the Romans did in good part, I would divide the armed men from the unarmed, and separate the men who carry burdens from the unburdened ones. I would quarter all or a greater part of the armed men on the east side, and the unarmed and burdened ones on the west side, making the east the front and the west the rear of the encampment, and the south and north would be the flanks. And to distinguish the quarters of the armed men, I would employ this method. I would run a line from the Captain's flag, and would lead it easterly for a distance of six hundred eighty (680) arm lengths. I would also run two other lines which I would place in the middle of it, and be of the same length as the former, but distant from each of them by fifteen arm lengths, at the extremity of which, I would want the east gate to be (placed): and the space which exists between the two extreme (end) lines, I would make a road that would go from the gate to the quarters of the Captain, which would be thirty arm lengths in width and six hundred thirty (630) long ((since the Captain's quarters would occupy fifty arm lengths)) and call this the Captain's Way. I would then make another road from the south gate up to the north gate, and cross by the head of the Captain's Way, and along the east side of the Captain's quarters which would be one thousand two hundred fifty (1250) arm lengths long ((since it would occupy the entire width of the encampment)) and also be thirty arm lengths wide and be called the Cross Way. The quarters of the Captain and these two roads having been designed, therefore the quarters of the two battalions of your own men should begin to be designed; and I would quarter one on the right hand (side) of the Captain's Way, and one on the left. And hence beyond the space which is occupied by the width of the Cross Way, I would place thirty two quarters on the left side of the Captain's Way, and thirty two on the right side, leaving a space of thirty arm lengths between the sixteenth and seventeenth quarters which should serve as a transverse road which should cross through all of the quarters of the battalions, as will be seen in their partitioning. Of these two arrangements of quarters, in the first tents that would be adjacent to the Cross Way, I would quarter the heads of men-at-arms, and since each company has one hundred and fifty men-at-arms, there would be assigned ten men-at-arms to each of the quarters. The area (space) of the quarters of the Heads should be forty arm lengths wide and ten arm lengths long. And it is to be noted that whenever I say width, I mean from south to north, and when I say length, that from west to east. Those of the men-at-arms should be fifteen arm lengths long and thirty wide. In the next fifteen quarters which in all cases are next ((which should have their beginning across the transverse road, and which would have the same space as those of the men-at-arms)) I would quarter the light cavalry, which, since they are one hundred fifty, ten cavalrymen would be assigned to each quarter, and in the sixteenth which would be left, I would quarter their Head, giving him the same space which is given to the Head of men-at-arms. And thus the quarters of the cavalry of the two battalions would come to place the Captain's Way in the center and give a rule for the quarters of the infantry, as I will narrate. You have noted that I have quartered the three hundred cavalry of each battalion with their heads in thirty two quarters situated on the Captain's Way, and beginning with the Cross Way, and that from the sixteenth to the seventeenth there is a space of thirty arm lengths to make a transverse road. If I want, therefore, to quarter the twenty companies which constitute the two regular Battalions, I would place the quarters of every two companies behind the quarters of the cavalry, each of which should be fifteen arm lengths long and thirty wide, as those of the cavalry, and should be joined on the rear where they touch one another. And in every first quarter of each band that fronts on the Cross Way, I would quarter the Constable of one company, which would come to correspond with the quartering of the Head of the men-at-arms: and their quarters alone would have a space twenty arm lengths in width and ten in length. And in the other fifteen quarters in each group which follow after this up the Transverse Way, I would quarter a company of infantry on each side, which, as they are four hundred fifty, thirty would be assigned to each quarter. I would place the other fifteen quarters contiguous in each group to those of the cavalry with the same space, in which I would quarter a company of infantry from each group. In the last quarter of each group I would place the Constable of the company, who would come to be adjacent to the Head of the light cavalry, with a space of ten arm lengths long and twenty wide. And thus these first two rows of quarters would be half of cavalry and half of infantry.

And as I want ((as I told you in its place)) these cavalry to be all effective, and hence without retainers who help taking care of the horses or other necessary things, I would want these infantry quartered behind the cavalry should be obligated to help the owners (of the horses) in providing and taking care of them, and because of this should be exempt from other activities of the camp, which was the manner observed by the Romans. I would also leave behind these quarters on all sides a space of thirty arm lengths to make a road, and I would call one of the First Road on the right hand (side) and the other the First Road on the left, and in each area I would place another row of thirty two double quarters which should face one another on the rear, with the same spaces as those which I have mentioned, and also divided at the sixteenth in the same manner to create a Transverse Road, in which I would quarter in each area four companies of infantry with the Constables in the front at the head and foot (of each row). I would also leave on each side another space of thirty arm lengths to create a road which should be called the Second Road on the right hand (side) and on the other side the Second Road to the left; I would place another row in each area of thirty two double quarters, with the same distances and divisions, in which I would quarter on every side four companies (of infantry) with their Constables. And thus there would come to be quartered in three rows of quarters per area the cavalry and the companies (of infantry) of the two regular battalions, in the center of which I would place the Captain's Way. The two battalions of auxiliaries ((since I had them composed of the same men)) I would quarter on each side of these two regular battalions with the same arrangement of double quarters, placing first a row of quarters in which I should quarter half with cavalry and half infantry, distant thirty arm lengths from each other, to create two roads which I should call, one the Third Road on the right hand (side), the other the Third on the left hand. And then I would place on each side two other rows of quarters, separate but arranged in the same way, which are those of the regular battalions, which would create two other roads, and all of these would be called by the number and the band (side) where they should be situated. So that all this part of the Army would come to be quartered in twelve rows of double quarters, and on thirteen roads, counting the Captain's Way and the Cross Way.

I would want a space of one hundred arm lengths all around left between the quarters and the ditch (moat). And if you count all those spaces, you will see, that from the middle of the quarters of the Captain to the east gate, there are seven hundred arm lengths. There remains to us now two spaces, of which one is from the quarters of the Captain to the south gate, the other from there to the north gate, each of which comes to be, measuring from the center point, six hundred thirty five (635) arm lengths. I then subtract from each of these spaces fifty arm lengths which the quarters of the Captain occupies, and forty five arm lengths of plaza which I want to give to each side, and thirty arm lengths of road, which divides each of the mentioned spaces in the middle, and a hundred arm lengths which are left on each side between the quarters and the ditch, and there remains in each area a space left for quarters four hundred arm lengths wide and a hundred long, measuring the length to include the space occupied by the Captain's quarters. Dividing the said length in the middle, therefore, there would be on each side of the Captain forty quarters fifty arm lengths long and twenty wide, which would total eighty quarters, in which would be quartered the general Heads of the battalions, the Chamberlains, the Masters of the camps, and all those who should have an office (duty) in the army, leaving some vacant for some foreigners who might arrive, and for those who should fight through the courtesy of the Captain. On the rear side of the Captain's quarters, I would create a road thirty arm lengths wide from north to south, and call it the Front Road, which would come to be located along the eighty quarters mentioned, since this road and the Cross Way would have between them the

Captain's quarters and the eighty quarters on their flanks. From this Front road and opposite to the Captain's quarters, I would create another road which should go from there to the west gate, also thirty arm lengths wide, and corresponding in location and length to the Captain's Way, and I should call it the Way of the Plaza. These two roads being located, I would arrange the plaza where the market should be made, which I would place at the head of the Way of the Plaza, opposite to the Captain's quarters, and next to the Front Road, and would want it to be square, and would allow it a hundred twenty one arm lengths per side. And from the right hand and left hand of the said plaza, I would make two rows of quarters, and each row have eight double quarters, which would take up twelve arm lengths in length and thirty in width so that they should be on each side of the plaza, in which there would be sixteen quarters, and total thirty two all together, in which I would quarter that cavalry left over from the auxiliary battalions, and if this should not be enough, I would assign them some of the quarters about the Captain, and especially those which face the ditch.

It remains for us now to quarter the extraordinary pikemen and Veliti, which every battalion has; which you know, according to our arrangement, in addition to the ten companies (of infantry), each has a thousand extraordinary pikemen, and five hundred Veliti; so that each of the two regular battalions have two thousand extraordinary pikemen, and a thousand extraordinary pikemen, and five hundred Veliti; so that each of the two regular battalions have two thousand extraordinary pikemen, and a thousand extraordinary Veliti, and the auxiliary as many as they; so that one also comes to have to quarter six thousand infantry, all of whom I would quarter on the west side along the ditches. From the point, therefore, of the Front Road, and northward, leaving the space of a hundred arm lengths from those (quarters) to the ditch, I would place a row of five double quarters which would be seventy five arm lengths long and sixty in width: so that with the width divided, each quarters would be allowed fifteen arm lengths for length and thirty for width. And as there would be ten quarters, I would quarter three hundred infantry, assigning thirty infantry to each quarters. Leaving then a space of thirty one arm lengths, I would place another row of five double quarters in a similar manner and with similar spaces, and then another, so that there would be five rows of five double quarters, which would come to be fifty quarters placed in a straight line on the north side, each distant one hundred arm lengths from the ditches, which would quarter one thousand five hundred infantry. Turning then on the left hand side toward the west gate, I would want in all that tract between them and the said gate, five other rows of double quarters, in a similar manner and with the same spaces, ((it is true that from one row to the other there would not be more than fifteen arm lengths of space)) in which there would also be quartered a thousand five hundred infantry: and thus from the north gate to that on the west, following the ditches, in a hundred quarters, divided into ten rows of five double quarters per row, the extraordinary pikemen and Veliti of the regular battalions would be quartered. And so, too, from the west gate to that on the south, following the ditches, in exactly the same manner, in another ten rows of ten quarters per row, the extraordinary pikemen and Veliti of the auxiliary battalions would be quartered. Their Heads, or rather their Constables, could take those quarters on the side toward the ditches which appeared most convenient for themselves.

I would dispose the artillery all along the embankments of the ditches: and in all the other space remaining toward the west, I would quarter all the unarmed men and all the baggage (impedimenta) of the Camp. And it has to be understood that under this name of impedimenta ((as you know)) the ancients intended all those carriages (wagons) and all those things which are necessary to an Army, except the soldiers; as are carpenters (wood workers), smiths, blacksmiths, shoe makers, engineers, and bombardiers, and others which should be placed among the number of the armed: herdsmen with their herds of castrated sheep and oxen, which are used for feeding the Army: and in addition, masters of every art (trade), together with public wagons for the public provisions of food and arms. And I would not particularly distinguish their quarters: I would only designate the roads that should not be occupied by them. Then the other spaces remaining between the roads, which would be four, I would assign in general to all the impedimenta mentioned, that is, one to the herdsmen, another to Artificers and workmen, another to the public wagons for provisions, and the fourth to the armorers. The roads which I would want left unoccupied would be the Way of the Plaza, the Front Road, and in addition, a road that should be called the Center Road, which should take off at the north and proceed toward the south, and pass through the center of the Way of the Plaza, which, on the west side, should have the same effect as has the Transverse Road on the east side. And in addition to this a Road that should go around the rear along the quarters of the extraordinary pikemen and Veliti. And all these roads should be thirty arm lengths wide. And I would dispose the artillery along the ditches on the rear of the camp.

BATTISTA: I confess I do not understand, and I also do not believe that to say so makes me ashamed, as this is not my profession. None the less, I like this organization very much: I would want only that you should resolve these doubts for me. The one, why you make the roads and the spaces around the quarters so wide. The other, which annoys me more, is this, how are these spaces that you designate for quarters to be used.

FABRIZIO: You know that I made all the roads thirty arm lengths wide, so that a company of infantry is able to go through them in order (formation): which, if you remember well, I told you that each of these (formations) were twenty five to thirty arm lengths wide. The space between the ditch and the quarters, which is a hundred arm lengths wide, is necessary, since the companies and the artillery can be handled here, through which booty is taken, (and) when space is needed into which to retire, new ditches and embankments are made. The quarters very distant from the ditches are better, for they are more distant from the fires and other things that might be able to draw the enemy to attack them. As to the second question, my intention is not that every space designated by me is covered by only one pavilion, but is to be used as an all-round convenience for those who are quartered, with several or few tents, so long as they do not go outside its limits. And in designing these quarters, the men must be most experienced and excellent architects, who, as soon as the Captain has selected the site, know how to give it form, and divide it, and distinguishing the roads, dividing the quarters with cords and hatchets in such a practical manner, that they might be divided and arranged quickly. And if confusion is not to arise, the camp must always face the same way, so that everyone will know on which Road and in which space he has to find his quarters. And this ought to be observed at all times, in every place, and in a manner that it appears to be a movable City, which, wherever it goes, brings with it the same roads, the same houses, and the same appearance: which cannot be observed by those men who, seeking strong locations, have to change the form according to the variations in the sites. But the Romans made the places strong with ditches, ramparts, and embankments, for they placed a space around the camp, and in front of it they dug a ditch and ordinarily six arm lengths wide and three deep, which spaces they increased according to the (length of) time they resided in the one place, and according as they feared the enemy. For myself, I would not at present erect a stockade (rampart), unless I should want to winter in a place. I would, however, dig the ditch and embankment, not less than that mentioned, but greater according to the necessity. With respect to the artillery, on every side of the encampment, I would have a half circle ditch, from which the artillery should be able to batter on the flanks whoever should come to attack the moats (ditches). The soldiers ought also to be trained in this practice of knowing how to arrange an encampment, and work with them so they may aid him in designing it, and the soldiers quick in knowing their places. And none of these is difficult, as will be told in its proper place. For now I want to pass on to the protection of the camp, which, without the distribution (assignment) of guards, all the other efforts would be useless.

BATTISTA: Before you pass on to the guards, I would want you to tell me, what methods are employed when others want to place the camp near the enemy, for I do not know whether there is time to be able to organize it without danger.

FABRIZIO: You have to know this, that no Captain encamps near the enemy, unless he is disposed to come to an engagement whenever the enemy wants; and if the others are so disposed, there is no danger except the ordinary, since two parts of the army are organized to make an engagement, while the other part makes the encampment. In cases like this, the Romans assigned this method of fortifying the quarters to the Triari, while the Principi and the Astati remained under arms. They did this, because the Triari, being the last to combat, were in time to leave the work if the enemy came, and take up their arms and take their places. If you want to imitate the Romans, you have to assign the making of the encampment to that company which you would want to put in the place of the Triari in the last part of the army.

But let us return to the discussion of the guards. I do not seem to find in connection with the ancients guarding the camp at night, that they had guards outside, distant from the ditches, as is the custom today, which they call the watch. I believe I should do this, when I think how the army could be easily deceived, because of the difficulty which exists in checking (reviewing) them, for they may be corrupted or attacked by the enemy, so that they judged it dangerous to trust them entirely or in part. And therefore all the power of their protection was within the ditches, which they dug with very great diligence and order, punishing capitally anyone who deviated from such an order. How this was arranged by them, I will not talk to you further in order not to tire you, since you are able to see it by yourselves, if you have not seen it up to now. I will say only briefly what would be done by me. I would regularly have a third of the army remain armed every night, and a fourth of them always on foot, who would be distributed throughout the embankments and all the places of the army, with double guards posted at each of its squares, where a part should remain, and a part continually go from one side of the encampment to the other. And this arrangement I describe, I would also observe by day if I had the enemy near. As to giving it a name, and renewing it every night, and doing the other things that are done in such guarding, since they are things (already) known, I will not talk further of them. I would only remind you of a most important matter, and by observing it do much good, by not observing it do much evil; which is, that great diligence be used as to who does not lodge within the camp at night, and who arrives there anew. And this is an easy matter, to review who is quartered there, with those arrangements we have designated, since every quarter having a predetermined number of men, it is an easy thing to see if there are any men missing or if any are left over; and when they are missing without permission, to punish them as fugitives, and if they are left over, to learn who they are, what they know, and what are their conditions. Such diligence results in the enemy not being able to have correspondence with your Heads, and not to have co-knowledge of your counsels. If this had not been observed with diligence by the Romans, Claudius Nero could not, when he had Hannibal near to him, have departed from the encampment he had in Lucania, and go and return from the Marches, without Hannibal having been aware of it. But it is not enough to make these good arrangements, unless they are made to be observed by great security, for there is nothing that wants so much observance as any required in the army. Therefore, the laws for their enforcement should be harsh and hard, and the executor very hard. The Roman punished with the capital penalty whoever was missing from the guard, whoever abandoned the place given him in combat, whoever brought anything concealed from outside the encampment; if anyone should tell of having performed some great act in battle, and should not have done it; if anyone should have fought except at the command of the Captain, if anyone from fear had thrown aside his arms. And if it occurred that an entire Cohort or an entire Legion had made a similar error, in order that they not all be put to death, they put their names in a purse, and drew the tenth part, and those they put to death. Which penalty was so carried out, that if everyone did not hear of it, they at least feared it. And because where there are severe punishments, there also ought to be rewards, so that men should fear and hope at the same time, they proposed rewards for every great deed; such as to him who, during the fighting, saved the life of one of its citizens, to whoever first climbed the walls of enemy towns, to whoever first entered the encampment of the enemy, to whoever in battle wounded or killed an enemy, to whoever had thrown him from his horse. And thus any act of virtu was recognized and rewarded by the Consuls, and publicly praised by everyone: and those who received gifts for any of these things, in addition to the glory and fame they acquired among the soldiers, when they returned to their country, exhibited them with solemn pomp and with great demonstrations among their friends and relatives. It is not to marvel therefore, if that people acquired so much empire, when they had so great an observance of punishment and reward toward them, which operated either for their good or evil, should merit either praise or censure; it behooves us to observe the greater part of these things. And it does not appear proper for me to be silent on a method of punishment observed by them, which was, that as the miscreant was convicted before the Tribune or the Consul, he was struck lightly by him with a rod: after which striking of the criminal, he was allowed to flee, and all the soldiers allowed to kill him, so that immediately each of them threw stones or darts, or hit him with other arms, of a kind from which he went little alive, and rarely returned to camp; and to such that did return to camp, he was not allowed to return home except with so much inconvenience and ignominy, that it was much better for him to die. You see this method almost observed by the Swiss, who have the condemned publicly put to death by the other soldiers. Which is well considered and done for the best, for if it is desired that one be not a defender of a criminal, the better remedy that is found, is to make him the punisher of him (the criminal); for in some respects he favors him while from other desires he longs for his punishment, if he himself is the executioner, than if the execution is carried out by another. If you want, therefore, that one is not to be favored in his mistakes by a people, a good remedy is to see to it that the public judged him. In support of this, the example of Manlius Capitol that can be cited, who, when he was accused by the Senate, was defended so much by the public up to the point where it no longer became the judge: but having become arbiter of his cause, condemned him to death. It is, therefore, a method of punishing this, of doing away with tumults, and of having justice observed. And since in restraining armed men, the fear of laws, or of men, is not enough, the ancients added the authority of God: and, therefore, with very great ceremony, they made their soldiers swear to observe the military discipline, so that if they did the contrary, they not only had to fear the laws and men, but God; and they used every industry to fill them with Religion.

BATTISTA: Did the Romans permit women to be in their armies, or that they indulge in indolent games that are used to day?

FABRIZIO: They prohibited both of them, and this prohibition was not very difficult, because the exercises which they gave each day to the soldiers were so many, sometimes being occupied all together, sometimes individually, that no time was left to them to think either of Venery, or of games, or of other things which make soldiers seditious and useless.

BATTISTA: I like that. But tell me, when the army had to take off, what arrangements did they have?

FABRIZIO: The captain's trumpet was sounded three times: at the first sound the tents were taken down and piled into heaps, at the second they loaded the burdens, and at the third they moved in the manner mentioned above, with the impedimenta behind, the armed men on every side, placing the Legions in the center. And, therefore, you would have to have a battalion of auxiliaries move, and behind it its particular impedimenta, and with those the fourth part of the public impedimenta, which would be all those who should be quartered in one of those (sections of the camp) which we showed a short while back. And, therefore, it would be well to have each one of them assigned to a battalion, so that when the army moved, everyone would know where his place was in marching. And every battalion ought to proceed on its way in this fashion with its own impedimenta, and with a quarter of the public (impedimenta) at its rear, as we showed the Roman army marched.

BATTISTA: In placing the encampment, did they have other considerations than those you mentioned?

FABRIZIO: I tell you again, that in their encampments, the Romans wanted to be able to employ the usual form of their method, in the observance of which, they took no other consideration. But as to other considerations, they had two principal ones: the one, to locate themselves in a healthy place: to locate themselves where the enemy should be unable to besiege them, and cut off their supply of water and provisions. To avoid this weakness, therefore, they avoided marshy places, or exposure to noxious winds. They recognized these, not so much from the characteristics of the site, but from the looks of the inhabitants: and if they saw them with poor color, or short winded, or full of other infections, they did not encamp there. As to the other part of not being besieged, the nature of the place must be considered, where the friends are, and where the enemy, and from these make a conjecture whether or not you can be besieged. And, therefore, the Captain must be very expert concerning sites of the countries, and have around him many others who have the same expertness. They also avoided sickness and hunger so as not to disorganize the army; for if you want to keep it healthy, you must see to it that the soldiers sleep under tents, that they are quartered, where there are trees to create shade, where there is wood to cook the food, and not to march in the heat. You need, therefore, to consider the encampment the day before you arrive there, and in winter guard against marching in the snow and through ice without the convenience of making a fire, and not lack necessary clothing, and not to drink bad water. Those who get sick in the house, have them taken care of by doctors; for a captain has no remedy when he has to fight both sickness and the enemy. But nothing is more useful in maintaining an army healthy than exercise: and therefore the ancients made them exercise every day. Whence it is seen how much exercise is of value, for in the quarters it keeps you healthy, and in battle it makes you victorious. As to hunger, not only is it necessary to see that the enemy does not impede your provisions, but to provide whence you are to obtain them, and to see that those you have are not lost. And, therefore, you must always have provisions (on hand) for the army for a month, and beyond that to tax the neighboring friends that they provide you daily, keep the provisions in a strong place, and, above all, dispense it with diligence, giving each one a reasonable measure each day, and so observe this part that they do not become disorganized; for every other thing in war can be overcome with time, this only with time overcomes you. Never make anyone your enemy, who, while seeking to overcome you with the sword (iron), can overcome you by hunger, because if such a victory is not as honorable, it is more secure and more certain. That army, therefore, cannot escape hunger which does not observe justice, and licentiously consume whatever it please, for one evil causes the provisions not to arrive, and the other that when they arrive, they are uselessly consumed: therefore the ancients arranged that what was given was eaten, and in the time they assigned, so that no soldier ate except when the Captain did. Which, as to being observed by the modern armies, everyone does (the contrary), and deservedly they cannot be called orderly and sober as the ancients, but licentious and drunkards.

BATTISTA: You have said in the beginning of arranging the encampment, that you did not want to stay only with two battalions, but took up four, to show how a fair (sized) army was quartered. Therefore I would want you to tell me two things: the one, if I have more or less men, how should I quarter them: the other, what number of soldiers would be enough to fight against any enemy?

FABRIZIO: To the first question, I reply, that if the army has four or six thousand soldiers more or less, rows of quarters are taken away or added as are needed, and in this way it is possible to accommodate more or fewer infinitely. None the less, when the Romans joined together two consular armies, they made two encampments and had the parts of the disarmed men face each other. As to the second question, I reply, that the regular Roman army had about twenty four thousand soldiers: but when a great force pressed them, the most they assembled were fifty thousand. With this number they opposed two hundred thousand Gauls whom they assaulted after the first war which they had with the Carthaginians. With the same number, they opposed Hannibal. And you have to note that the Romans and Greeks had made war with few (soldiers), strengthened by order and by art; the westerners and easterners had made it with a multitude: but one of these nations serves itself of natural fury, as are the westerners; the other of the great obedience which its men show to their King. But in Greece and Italy, as there is not this natural fury, nor the natural reverence toward their King, it has been necessary to turn to discipline; which is so powerful, that it made the few able to overcome the fury and natural obstinacy of the many. I tell you, therefore, if you want to imitate the Romans and Greeks, the number of fifty thousand soldiers ought not to be exceeded, rather they should actually be less; for the many cause confusion, and do not allow discipline to be observed nor the orders learned. And Pyrrhus used to say that with fifteen thousand men he would assail the world.

But let us pass on to another part. We have made our army win an engagement, and I showed the troubles that can occur in battle; we have made it march, and I have narrated with what impedimenta it can be surrounded while marching: and lastly we have quartered it: where not only a little repose from past hardship ought to be taken, but also to think about how the war ought to be concluded; for in the quarters, many things are discussed, especially if there remain enemies in the field, towns under suspicion, of which it is well to reassure oneself, and to capture those which are hostile. It is necessary, therefore, to come to these demonstrations, and to pass over this difficulty with that (same) glory with which we have fought up to the present. Coming down to particulars, therefore, that if it should happen to you that many men or many peoples should do something, which might be useful to you and very harmful to them, as would be the destruction of the walls of their City, or the sending of many of themselves into exile, it is necessary that you either deceive them in a way that everyone should believe he is affected, so that one not helping the other, all find themselves oppressed without a remedy, or rather, to command everyone what they ought to do on the same day, so that each one believing himself to be alone to whom the command is given, thinks of obeying it, and not of a remedy; and thus, without tumult, your command is executed by everyone. If you should have suspicion of the loyalty of any people, and should want to assure yourself and occupy them without notice, in order to disguise your design more easily, you cannot do better than to communicate to him some of your design, requesting his aid, and indicate to him you want to undertake another enterprise, and to have a mind alien to every thought of his: which will cause him not to think of his defense, as he does not believe you are thinking of attacking him, and he will give you the opportunity which will enable you to satisfy your desire easily. If you should have present in your army someone who keeps the enemy advised of your designs, you cannot do better if you want to avail yourself of his evil intentions, than to communicate to him those things you do not want to do, and keep silent those things you want to do, and tell him you are apprehensive of the things of which you are not apprehensive, and conceal those things of which you are apprehensive: which will cause the enemy to undertake some enterprise, in the belief that he knows your designs, in which you can deceive him and defeat him. If you should design ((as did Claudius Nero)) to decrease your army, sending aid to some friend, and they should not be aware of it, it is necessary that the encampment be not decreased, but to maintain entire all the signs and arrangements, making the same fires and posting the same guards as for the entire army. Likewise, if you should attach a new force to your army, and do not want the enemy to know you have enlarged it, it is necessary that the encampment be not increased, for it is always most useful to keep your designs secret. Whence Metellus, when he was with the armies in Spain, to one who asked him what he was going to do the next day, answered that if his shirt knew it, he would bum it. Marcus Crassus, to one who asked him when he was going to move his army, said: "do you believe you are alone in not hearing the trumpets?" If you should desire to learn the secrets of your enemy and know his arrangement, some used to send ambassadors, and with them men expert in war disguised in the clothing of the family, who, taking the opportunity to observe the enemy army, and consideration of his strengths and weaknesses, have given them the occasion to defeat him. Some have sent a close friend of theirs into exile, and through him have learned the designs of their adversary. You may also learn similar secrets from the enemy if you should take prisoners for this purpose. Marius, in the war he waged against Cimbri, in order to learn the loyalty of those Gauls who lived in Lombardy and were leagued with the Roman people, sent them letters, open and sealed: and in the open ones he wrote them that they should not open the sealed ones except at such a time: and before that time, he called for them to be returned, and finding them opened, knew their loyalty was not complete. Some Captains, when they were assaulted have not wanted to go to meet the enemy, but have gone to assail his country, and constrain him to return to defend his home. This often has turned out well, because your soldiers begin to win and fill themselves with booty and confidence, while those of the enemy become dismayed, it appearing to them that from being winners, they have become losers. So that to whoever has made this diversion, it has turned out well. But this can only be done by that man who has his country stronger than that of the enemy, for if it were otherwise, he would go on to lose. It has often been a useful thing for a Captain who finds himself besieged in the quarters of the enemy, to set in motion proceedings for an accord, and to make a truce with him for several days; which only any enemy negligent in every way will do, so that availing yourself of his negligence, you can easily obtain the opportunity to get out of his hands. Sulla twice freed himself from his enemies in this manner, and with this same deceit, Hannibal in Spain got away from the forces of Claudius Nero, who had besieged him.

It also helps one in freeing himself from the enemy to do something in addition to those mentioned, which keeps him at bay. This is done in two ways: either by assaulting him with part of your forces, so that intent on the battle, he gives the rest of your forces the opportunity to be able to save themselves, or to have some new incident spring up, which, by the novelty of the thing, makes him wonder, and for this reason to become apprehensive and stand still, as you know Hannibal did, who, being trapped by Fabius Maximus, at night placed some torches between the horns of many oxen, so that Fabius is suspense over this novelty, did not think further of impeding his passage. A Captain ought, among all the other actions of his, endeavor with every art to divide the forces of the enemy, either by making him suspicious of his men in whom he trusted, or by giving him cause that he has to separate his forces, and, because of this, become weaker. The first method is accomplished by watching the things of some of those whom he has next to him, as exists in war, to save his possessions, maintaining his children or other of his necessities without charge. You know how Hannibal, having burned all the fields around Rome, caused only those of Fabius Maximus to remain safe. You know how Coriolanus, when he came with the army to Rome, saved the possessions of the Nobles, and burned and sacked those of the Plebs. When Metellus led the army against Jugurtha, all me ambassadors, sent to him by Jugurtha, were requested by him to give up Jugurtha as a prisoner; afterwards, writing letters to these same people on the same subject, wrote in such a way that in a little while Jugurtha became suspicious of all his counsellors, and in different ways, dismissed them. Hannibal, having taken refuge with Antiochus, the Roman ambassadors frequented him so much at home, that Antiochus becoming suspicious of him, did not afterwards have any faith in his counsels. As to dividing the enemy forces, there is no more certain way than to have one country assaulted by part of them (your forces), so that being constrained to go to defend it, they (of that country) abandon the war. This is the method employed by Fabius when his Army had encountered the forces of the Gauls, the Tuscans, Umbrians, and Samnites. Titus Didius, having a small force in comparison with those of the enemy, and awaiting a Legion from Rome, the enemy wanted to go out to meet it; so that in order that it should not do so, he gave out by voice throughout his army that he wanted to undertake an engagement with the enemy on the next day; then he took steps that some of the prisoners he had were given the opportunity to escape, who carried back the order of the Consul to fight on the next day, (and) caused the enemy, in order not to diminish his forces, not to go out to meet that Legion: and in this way, kept himself safe. Which method did not serve to divide the forces of the enemy, but to double his own. Some, in order to divide his (the enemy) forces, have employed allowing him to enter their country, and (in proof) allowed him to take many towns so that by placing guards in them, he diminished his forces, and in this manner having made him weak, assaulted and defeated him. Some others, when they wanted to go into one province, feigned making an assault on another, and used so much industry, that as soon as they extended toward that one where there was no fear they would enter, have overcome it before the enemy had time to succor it. For the enemy, as he is not certain whether you are to return back to the place first threatened by you, is constrained not to abandon the one place and succor the other, and thus often he does not defend either. In addition to the matters mentioned, it is important to a

Captain when sedition or discord arises among the soldiers, to know how to extinguish it with art. The better way is to castigate the heads of this folly (error); but to do it in a way that you are able to punish them before they are able to become aware of it. The method is, if they are far from you, not to call only the guilty ones, but all the others together with them, so that as they do not believe there is any cause to punish them, they are not disobedient, but provide the opportunity for punishment. When they are present, one ought to strengthen himself with the guiltless, and by their aid, punish them. If there should be discord among them, the best way is to expose them to danger, which fear will always make them united. But, above all, what keeps the Army united, is the reputation of its Captain, which only results from his virtu, for neither blood (birth) or authority attain it without virtu. And the first thing a Captain is expected to do, is to see to it that the soldiers are paid and punished; for any time payment is missed, punishment must also be dispensed with, because you cannot castigate a soldier you rob, unless you pay him; and as he wants to live, he can abstain from being robbed. But if you pay him but do not punish him, he becomes insolent in every way, because you become of little esteem, and to whomever it happens, he cannot maintain the dignity of his position; and if he does not maintain it, of necessity, tumults and discords follow, which are the ruin of an Army. The Ancient Captains had a molestation from which the present ones are almost free, which was the interpretation of sinister omen to their undertakings; for if an arrow fell in an army, if the Sun or the Moon was obscured, if an earthquake occurred, if the Captain fell while either mounting or dismounting from his horse, it was interpreted in a sinister fashion by the soldiers, and instilled so much fear in them, that when they came to an engagement, they were easily defeated. And, therefore, as soon as such an incident occurred, the ancient Captains either demonstrated the cause of it or reduced it to its natural causes, or interpreted it to (favor) their own purposes. When Caesar went to Africa, and having fallen while he was putting out to sea, said, "Africa, I have taken you": and many have profited from an eclipse of the Moon and from earthquakes: these things cannot happen in our time, as much because our men are not as superstitious, as because our Religion, by itself, entirely takes away such ideas. Yet if it should occur, the orders of the ancients should be imitated. When, either from hunger, or other natural necessity, or human passion, your enemy is brought to extreme desperation, and, driven by it, comes to fight with you, you ought to remain within your quarters, and avoid battle as much as you can. Thus the Lacedemonians did against the Messinians: thus Caesar did against Afranius and Petreius. When Fulvius was Consul against the Cimbri, he had the cavalry assault the enemy continually for many days, and considered how they would issue forth from their quarters in order to pursue them; whence he placed an ambush behind the quarters of the Cimbri, and had them assaulted by the cavalry, and when the Cimbri came out of their quarters to pursue them, Fulvius seized them and plundered them. It has been very effective for a Captain, when his army is in the vicinity of the enemy army, to send his forces with the insignia of the enemy, to rob and burn his own country: whence the enemy, believing they were forces coming to their aid, also ran out to help them plunder, and, because of this, have become disorganized and given the adversary the faculty of overcoming them. Alexander of Epirus used these means fighting against the Illirici, and Leptenus the Syracusan against the Carthaginians, and the design succeeded happily for both. Many have overcome the enemy by giving him the faculty of eating and drinking beyond his means, feigning being afraid, and leaving his quarters full of wine and herds, and when the enemy had filled himself beyond every natural limit, they assaulted him and overcome him with injury to him. Thus Tamirus did against Cyrus, and Tiberius Gracchus against the Spaniards. Some have poisoned the wine and other things to eat in order to be able to overcome them more easily. A little while ago, I said I did not find the ancients had kept a night Watch outside, and I thought they did it to avoid the evils that could happen, for it has been found that sometimes, the sentries posted in the daytime to keep watch for the enemy, have been the ruin of him who posted them; for it has happened often that when they had been taken, and by force had been made to give the signal by which they called their own men, who, coming at the signal, have been either killed or taken. Sometimes it helps to deceive the enemy by changing one of your habits, relying on which, he is ruined: as a Captain had already done, who, when he wanted to have a signal made to his men indicating the coming of the enemy, at night with fire and in the daytime with smoke, commanded that both smoke and flame be made without any intermission; so that when the enemy came, he should remain in the belief that he came without being seen, as he did not see the signals (usually) made to indicate his discovery, made ((because of his going disorganized)) the victory of his adversary easier. Menno Rodius, when he wanted to draw the enemy from the strong places, sent one in the disguise of a fugitive, who affirmed that his army was full of discord, and that the greater part were deserting, and to give proof of the matter, had certain tumults started among the quarters: whence to the enemy, thinking he was able to break him, assaulted him and was routed.

In addition to the things mentioned, one ought to take care not to bring the enemy to extreme desperation; which Caesar did when he fought the Germans, who, having blocked the way to them, seeing that they were unable to flee, and necessity having made them brave, desired rather to undergo the hardship of pursuing them if they defended themselves. Lucullus, when he saw that some Macedonian cavalry who were with him, had gone over to the side of the enemy, quickly sounded the call to battle, and commanded the other forces to pursue it: whence the enemy, believing that Lucullus did not want to start the battle, went to attack the Macedonians with such fury, that they were constrained to defend themselves, and thus, against their will, they became fighters of the fugitives. Knowing how to make yourself secure of a town when you have doubts of its loyalty once you have conquered it, or before, is also important; which some examples of the ancients teach you. Pompey, when he had doubts of the Catanians, begged them to accept some infirm people he had in his army, and having sent some very robust men in the disguise of infirm ones, occupied the town. Publius Valerius, fearful of the loyalty of the Epidaurians, announced an amnesty to be held, as we will tell you, at a Church outside the town, and when all the public had gone there for the amnesty, he locked the doors, and then let no one out from inside except those whom he trusted. Alexander the Great, when he wanted to go into Asia and secure Thrace for himself, took with him all the chiefs of this province, giving them provisions, and placed lowborn men in charge of the common people of Thrace; and thus he kept the chiefs content by paying them, and the common people quiet by not having Heads who should disquiet them. But among all the things by which Captains gain the people over to themselves, are the examples of chastity and justice, as was that of Scipio in Spain when he returned that girl, beautiful in body, to her husband and father, which did more than arms in gaining over Spain. Caesar, when he paid for the lumber that he used to make the stockades around his army in Gaul, gained such a name for himself of being just, that he facilitated the acquisition of that province for himself. I do not know what else remains for me to talk about regarding such events, and there does not remain any part of this matter that has not been discussed by us. The only thing lacking is to tell of the methods of capturing and defending towns, which I am about to do willingly, if it is not painful for you now.

BATTISTA: Your humaneness is so great, that it makes us pursue our desires without being afraid of being held presumptuous, since you have offered it willingly, that we would be ashamed to ask you. Therefore we say only this to you, that you cannot do a greater or more thankful benefit to us than to furnish us this discussion. But before you pass on to that other matter, resolve a doubt for us: whether it is better to continue the war even in winter, as is done today, or wage it only in the summer, and go into quarters in the winter, as the ancients did.

FABRIZIO: Here, if there had not been the prudence of the questioner, some part that merits consideration would have been omitted. I tell you again that the ancients did everything better and with more prudence than we; and if some error is made in other things, all are made in matters of war. There is nothing more imprudent or more perilous to a Captain than to wage war in winter, and more dangerous to him who brings it, than to him who awaits it. The reason is this: all the industry used in military discipline, is used in order to be organized to undertake an engagement with your enemy, as this is the end toward which a Captain must aim, for the engagement makes you win or lose a war. Therefore, whoever know how to organize it better, and who has his army better disciplined, has the greater advantage in this, and can hope more to win it. On the other hand, there is nothing more inimical to organization than the rough sites, or cold and wet seasons; for the rough side does not allow you to use the plentitude (of your forces) according to discipline, and the cold and wet seasons do not allow you to keep your forces together, and you cannot have them face the enemy united, but of necessity, you must quarter them separately, and without order, having to take into account the castles, hamlets, and farm houses that receive you; so that all the hard work employed by you in disciplining your army is in vain. And do not marvel if they war in winter time today, for as the armies are without discipline, and do not know the harm that is done to them by not being quartered together, for their annoyance does not enable those arrangements to be made and to observe that discipline which they do not have. Yet, the injury caused by campaigning in the field in the winter ought to be observed, remembering that the French in the year one thousand five hundred three (1503) were routed on the Garigliano by the winter, and not by the Spaniards. For, as I have told you, whoever assaults has even greater disadvantage, because weather harms him more when he is in the territory of others, and wants to make war. Whence he is compelled either to withstand the inconveniences of water and cold in order to keep together, or to divide his forces to escape them. But whoever waits, can select the place to his liking, and await him (the enemy) with fresh forces, and can unite them in a moment, and go out to find the enemy forces who cannot withstand their fury. Thus were the French routed, and thus are those always routed who assault an enemy in winter time, who in itself has prudence. Whoever, therefore, does not want the forces, organization, discipline, and virtu, in some part, to be of value, makes war in the field in the winter time. And because the Romans wanted to avail themselves of all of these things, into which they put so much industry, avoided not only the winter time, but rough mountains and difficult places, and anything else which could impede their ability to demonstrate their skill and virtu. So this suffices to (answer) your question; and now let us come to treat of the attacking and defending of towns, and of the sites, and of their edifices.

Seventh Book

You ought to know that towns and fortresses can be strong either by nature or industry. Those are strong by nature which are surrounded by rivers or marshes, as is Mantua or Ferrara, or those situated on a rock or sloping mountain, as Monaco and San Leo; for those situated on mountains which are not difficult to climb, today are ((with respect to caves and artillery)) very weak. And, therefore, very often today a plain is sought on which to build (a city) to make it strong by industry. The first industry is, to make the walls twisted and full of turned recesses; which pattern results in the enemy not being able to approach them, as they will be able to be attacked easily not only from the front, but on the flanks. If the walls are made too high, they are excessively exposed to the blows of the artillery; if they are made too low, they are very easily scaled. If you dig ditches (moats) in front of them to make it difficult (to employ) ladders, if it should happen that the enemy fills them ((which a large army can do easily)) the wall becomes prey to the enemy. I believe, therefore, ((subject to a better judgement)) that if you want to make provision against both evils the wall ought to be made high, with the ditches inside and not outside. This is the strongest way to build that is possible, for it protects you from artillery and ladders, and does not give the enemy the faculty of filling the ditches. The wall, therefore, ought to be as high as occurs to you, and not less than three arm lengths wide, to make it more difficult to be ruined. It ought to have towers placed at intervals of two hundred arm lengths. The ditch inside ought to be at least thirty arm lengths wide and twelve deep, and all the earth that is excavated in making the ditch is thrown toward the city, and is sustained by a wall that is part of the base of the ditch, and extends again as much above the ground, as that a man may take cover behind it: which has the effect of making the depth of the ditch greater. In the base of the ditch, every two hundred arm lengths, there should be a matted enclosure, which with the artillery, causes injury to anyone who should descend into it. The heavy artillery which defends the city, are placed behind the wall enclosing the ditch; for to defend the wall from the front, as it is high, it is not possible to use conveniently anything else other than small or middle sized guns. If the enemy comes to scale your wall, the height of the first wall easily protects you. If he comes with artillery, he must first batter down the first wall: but once it is battered down, because the nature of all batterings is to cause the wall to fall toward the battered side, the ruin of the wall will result ((since it does not find a ditch which receives and hides it)) in doubling the depth of the ditch, so that it is not possible for you to pass on further as you will find a ruin that holds you back and a ditch which will impede you, and from the wall of the ditch, in safety, the enemy artillery kills you. The only remedy there exists for you, is to fill up the ditch: which is very difficult, as much because its capacity is large, as from the difficulty you have in approaching it, since the walls being winding and recessed, you can enter among them only with difficulty, for the reasons previously mentioned; and then, having to climb over the ruin with the material in hand, causes you a very great difficulty: so that I know a city so organized is completely indestructible.

BATTISTA: If, in addition to the ditch inside, there should be one also on the outside, wouldn't (the encampment) be stronger?

FABRIZIO: It would be, without doubt; but my reasoning is, that if you want to dig one ditch only, it is better inside than outside.

BATTISTA: Would you have water in the ditch, or would you leave them dry?

FABRIZIO: Opinions are different; for ditches full of water protect you from (subterranean) tunnels, the ditches without water make it more difficult for you to fill them in again. But, considering everything, I would have them without water; for they are more secure, and, as it has been observed that in winter time the ditches ice over, the capture of a city is made easy, as happened at Mirandola when Pope Julius besieged it. And to protect yourself from tunnels, I would dig them so deep, that whoever should want to go (tunnel) deeper, should find water. I would also build the fortresses in a way similar to the walls and ditches, so that similar difficulty would be encountered in destroying it I want to call to mind one good thing to anyone who defends a city. This is, that they do not erect bastions outside, and they be distant from its wall. And another to anyone who builds the fortresses: And this is, that he not build any redoubts in them, into which whoever is inside can retire when the wall is lost. What makes me give the first counsel is, that no one ought to do anything, through the medium of which, you begin to lose your reputation without any remedy, the loss of which makes others esteem you less, and dismay those who undertake your defense. And what I say will always happen to you if you erect bastions outside the town you have to defend, for you will always lose them, as you are unable to defend small things when they are placed under the fury of the artillery; so that in losing them, they become the beginning and the cause of your ruin. Genoa, when it rebelled from King Louis of France, erected some bastions on the hills outside the City, which, as soon as they were lost, and they were lost quickly, also caused the city to be lost. As to the second counsel, I affirm there is nothing more dangerous concerning a fortress, than to be able to retire into it, for the hope that men have (lose) when they abandon a place, cause it to be lost, and when it is lost, it then causes the entire fortress to be lost. For an example, there is the recent loss of the fortress of Forli when the Countess Catherine defended it against Caesare Borgia, son of Pope Alexander the Sixth, who had led the army of the King of France. That entire fortress was full of places by both of them: For it was originally a citadel. There was a moat before coming to the fortress, so that it was entered by means of a draw bridge. The fortress was divided into three parts, and each part separated by a ditch, and with water between them; and one passed from one place to another by means of bridges: whence the Duke battered one of those parts of the fortress with artillery, and opened up part of a wall; whence Messer Giovanni Da Casale, who was in charge of the garrison, did not think of defending that opening, but abandoned to retire into the other places; so that the forces of the Duke, having entered that part without opposition, immediately seized all of it, for they became masters of the bridges that connected the members (parts) with each other. He lost the fort which was held to be indestructible because of two mistakes: one, because it had so many redoubts: the other, because no one was made master of his bridges (they were unprotected). The poorly built fortress and the little prudence of the defender, therefore, brought disgrace to the magnanimous enterprise of the Countess, who had the courage to face an army which neither the King of Naples, nor the Duke of Milan, had faced. And although his (the Duke) efforts did not have a good ending, none the less, he became noted for those honors which his virtu merited. Which was testified to by the many epigrams made in those times praising him. If I should therefore have to build a fortress, I would make its walls strong, and ditches in the manner we have discussed, nor would I build anything else to live in but houses, and they would be weak and low, so that they would not impede the sight of the walls to anyone who might be in the plaza, so that the Captain should be able to see with (his own) eyes where he could be of help, and that everyone should understand that if the walls and the ditch were lost, the entire fortress would be lost. And even if I should build some redoubts, I would have the bridges so separated, that each part should be master of (protect) the bridge in its own area, arranging that it be buttressed on its pilasters in the middle of the ditch.

BATTISTA: You have said that, today, the little things can not be defended, and it seems to me I have understood the opposite, that the smaller the thing was, the better it was defended.

FABRIZIO: You have not understood well, for today that place can not be called strong, where he who defends it does not have room to retire among new ditches and ramparts: for such is the fury of the artillery, that he who relies on the protection of only one wall or rampart, deceives himself. And as the bastions ((if you want them not to exceed their regular measurements, for then they would be terraces and castles)) are not made so that others can retire into them, they are lost quickly. And therefore it is a wise practice to leave these bastions outside, and fortify the entrances of the terraces, and cover their gates with revets, so that one does not go in or out of the gate in a straight line, and there is a ditch with a bridge over it from the revet to the gate. The gates are also fortified with shutters, so as to allow your men to reenter, when, after going out to fight, it happens that the enemy drives them back, and in the ensuing mixing of men, the enemy does not enter with them. And therefore, these things have also been found which the ancients called "cataracts", which, being let down, keep out the enemy but saves one's friends; for in such cases, one can not avail himself of anything else, neither bridges, or the gate, since both are occupied by the crowd.

BATTISTA: I have seen these shutters that you mention, made of small beams, in Germany, in the form of iron grids, while those of ours are made entirely of massive planks. I would want to know whence this difference arises, and which is stronger.

FABRIZIO: I will tell you again, that the methods and organizations of war in all the world, with respect to those of the ancients, are extinct; but in Italy, they are entirely lost, and if there is something more powerful, it results from the examples of the Ultramontanes. You may have heard, and these others can remember, how weakly things were built before King Charles of France crossed into Italy in the year one thousand four hundred ninety four (1494). The battlements were made a half arm length thin (wide), the places for the cross-bowmen and bombardiers (gunners) were made with a small aperture outside and a large one inside, and with many other defects, which I will omit, not to be tedious; for the defenses are easily taken away from slender battlements; the (places for) bombardiers built that way are easily opened (demolished). Now from the French, we have learned to make the battlements wide and large, and also to make the (places of the) bombardiers wide on the inside, and narrow it at the center of the wall, and then again widen it up to the outside edge: and this results in the artillery being able to demolish its defenses only with difficulty, The French, moreover, have many other arrangements such as these, which, because they have not been seen thus, have not been given consideration. Among which, is this method of the shutters made in the form of a grid, which is by far a better method than yours; for if you have to repair the shutters of a gate such as yours, lowering it if you are locked inside, and hence are unable to injure the enemy, so that they can attack it safely either in the dark or with a fire. But if it is made in the shape of a grid, you can, once it is lowered, by those weaves and intervals, to be able to defend it with lances, cross-bows, and every other kind of arms.

BATTISTA: I have also seen another Ultramontane custom in Italy, and it is this, making the carriages of the artillery with the spokes of the wheels bent toward the axles. I would like to know why they make them this way, as it seems to me they would be stronger straight, as those of our wheels.

FABRIZIO: Never believe that things which differ from the ordinary are made at home, but if you would believe that I should make them such as to be more beautiful, you would err; for where strength is necessary, no account is taken of beauty; but they all arise from being safer and stronger than ours. The reason is this. When the carriage is loaded, it either goes on a level, or inclines to the right or left side. When it goes level, the wheels equally sustain the weight, which, being divided equally between them, does not burden them much; when it inclines, it comes to have all the weight of the load upon that wheel on which it inclines. If its spokes are straight, they can easily collapse, since the wheel being inclined, the spokes also come to incline, and do not sustain the weight in a straight line. And, thus, when the carriage rides level and when they carry less weight, they come to be stronger; when the carriage rides inclined and when they carry more weight, they are weaker. The contrary happens to the bent spokes of the French carriages; for when the carriage inclines to one side, it points (leans straight) on them, since being ordinarily bent, they then come to be (more) straight (vertical), and can sustain all the weight strongly; and when the carriage goes level and they (the spikes) are bent, they sustain half the weight.

But let us return to our Cities and Fortresses. The French, for the greater security of their towns, and to enable them during sieges to put into and withdraw forces from them more easily, also employ, in addition to the things mentioned, another arrangement, of which I have not yet seen any example in Italy: and it is this, that they erect two pilasters at the outside point of a draw-bridge, and upon each of them they balance a beam so that half of it comes over the bridge, and the other half outside. Then they join small beams to the part outside, which are woven together from one beam to another in the shape of a grid, and on the inside they attach a chain to the end of each beam. When they want to close the bridge from the outside, therefore, they release the chains and allow all that gridded part to drop, which closes the bridge when it is lowered, and when they want to open it, they pull on the chains, and they (gridded beams) come to be raised; and they can be raised so that a man can pass under, but not a horse, and also so much that a horse with the man can pass under, and also can be closed entirely, for it is lowered and raised like a lace curtain. This arrangement is more secure than the shutters: for it can be impeded by the enemy so that it cannot come down only with difficulty, (and) it does not come down in a straight line like the shutters which can easily be penetrated. Those who want to build a City, therefore, ought to have all the things mentioned installed; and in addition, they should want at least one mile around the wall where either farming or building would not be allowed, but should be open field where no bushes, embankments, trees, or houses, should exist which would impede the vision, and which should be in the rear of a besieging enemy. It is to be noted that a town which has its ditches outside with its embankments higher than the ground, is very weak; for they provide a refuge for the enemy who assaults you, but does not impede him in attacking you, because they can be easily forced (opened) and give his artillery an emplacement.

But let us pass into the town. I do not want to waste much time in showing you that, in addition to the things mentioned previously, provisions for living and fighting supplies must also be included, for they are the things which everyone needs, and without them, every other provision is in vain. And, generally, two things ought to be done, provision yourself, and deprive the enemy of the opportunity to avail himself of the resources of your country. Therefore, any straw, grain, and cattle, which you cannot receive in your house, ought to be destroyed. Whoever defends a town ought to see to it that nothing is done in a tumultuous and disorganized manner, and have means to let everyone know what he has to do in any incident. The manner is this, that the women, children, aged, and the public stay at home, and leave the town free to the young and the brave: who armed, are distributed for defense, part being on the walls, part at the gates, part in the principal places of the City, in order to remedy those evils which might arise within; another part is not assigned to any place, but is prepared to help anyone requesting their help. And when matters are so organized, only with difficulty can tumults arise which disturb you. I want you to note also that in attacking and defending Cities, nothing gives the enemy hope of being able to occupy a town, than to know the inhabitants are not in the habit of looking for the enemy; for often Cities are lost entirely from fear, without any other action. When one assaults such a City, he should make all his appearances (ostentatious) terrible. On the other hand, he who is assaulted ought to place brave men, who are not afraid of thoughts, but by arms, on the side where the enemy (comes to) fight; for if the attempt proves vain, courage grows in the besieged, and then the enemy is forced to overcome those inside with his virtu and his reputation.

The equipment with which the ancients defended the towns were many, such as, Ballistas, Onagers, Scorpions, Arc-Ballistas, Large Bows, Slingshots;

and those with which they assaulted were also many, such as, Battering Rams, Wagons, Hollow Metal Fuses (Muscoli), Trench Covers (Plutei), Siege Machines (Vinee), Scythes, Turtles (somewhat similar to present day tanks). In place of these things, today there is the artillery, which serves both attackers and defenders, and, hence, I will not speak further about it. But let us return to our discussion, and come to the details of the siege (attack). One ought to take care not to be able to be taken by hunger, and not to be forced (to capitulate) by assaults. As to hunger, it has been said that it is necessary, before the siege arrives, to be well provided with food. But when it is lacking during a long siege, some extraordinary means of being provided by friends who want to save you, have been observed to be employed, especially if a river runs in the middle of the besieged City, as were the Romans, when their castle of Casalino was besieged by Hannibal, who, not being able to send them anything else by way of the river, threw great quantities of nuts into it, which being carried by the river without being able to be impeded, fed the Casalinese for some time. Some, when they were besieged, in order to show the enemy they had grain left over, and to make them despair of being able to besiege (defeat) them by hunger, have either thrown bread outside the walls, or have given a calf grain to eat, and then allowed it to be taken, so that when it was killed, and being found full of grain, gave signs of an abundance which they do not have. On the other hand, excellent Captains have used various methods to enfamish the enemy. Fabius allowed the Campanians to sow so that they should lack that grain which they were sowing. Dionysius, when he was besieged at Reggio, feigned wanting to make an accord with them, and while it was being drawn, had himself provided with food, and then when, by this method, had depleted them of grain, pressed them and starved them. Alexander the Great, when he wanted to capture Leucadia, captured all the surrounding castles, and allowed the men from them to take refuge in it (the City), and thus by adding a great multitude, he starved them. As to assaults, it has been said that one ought to guard against the first onrush, with which the Romans often occupied many towns, assaulting them all at once from every side, and they called it attacking the city by its crown: as did Scipio when he occupied new Carthage in Spain. If this onrush is withstood, then only with difficulty will you be overcome. And even if it should occur that the enemy had entered inside the city by having forced the walls, even the small terraces give you some remedy if they are not abandoned; for many armies have, once they have entered into a town, been repulsed or slain. The remedy is, that the towns people keep themselves in high places, and fight them from their houses and towers. Which thing, those who have entered in the City, have endeavored to win in two ways: the one, to open the gates of the City and make a way for the townspeople by which they can escape in safety: the other, to send out a (message) by voice signifying that no one would be harmed unless armed, and whoever would throw his arms on the ground, they would pardon. Which thing has made the winning of many Cities easy. In addition to this, Cities are easy to capture if you fall on them unexpectedly, which you can do when you find yourself with your army far away, so that they do not believe that you either want to assault them, or that you can do it without your presenting yourself, because of the distance from the place. Whence, if you assault them secretly and quickly, it will almost always happen that you will succeed in reporting the victory. I unwillingly discuss those things which have happened in our times, as I would burden you with myself and my (ideas), and I would not know what to say in discussing other things. None the less, concerning this matter, I can not but cite the example of Cesare Borgia, called the Duke Valentine, who, when he was at Nocera with his forces, under the pretext of going to harm Camerino, turned toward the State of Urbino, and occupied a State in one day and without effort, which some other, with great time and expense, would barely have occupied. Those who are besieged must also guard themselves from the deceit and cunning of the enemy, and, therefore, the besieged should not trust anything which they see the enemy doing continuously, but always believe they are being done by deceit, and can change to injure them. When Domitius Calvinus was besieging a town, he undertook habitually to circle the walls of the City every day with a good part of his forces. Whence the townspeople, believing he was doing this for exercise, lightened the guard: when Domitius became aware of this, he assaulted them, and destroyed them. Some Captains, when they heard beforehand that aid was to come to the besieged, have clothed their soldiers with the insignia of those who were to come, and having introduced them inside, have occupied the town. Chimon, the Athenian, one night set fire to a Temple that was outside the town, whence, when the townspeople arrived to succor it, they left the town to the enemy to plunder. Some have put to death those who left the besieged castle to blacksmith (shoe horses), and redressing their soldiers with the clothes of the blacksmiths, who then surrendered the town to him. The ancient Captains also employed various methods to despoil the garrisons of the towns they want to take. Scipio, when he was in Africa, and desiring to occupy several castles in which garrisons had been placed by Carthaginians, feigned several times wanting to assault them, but then from fear not only abstained, but drew away from them. Which Hannibal believing to be true, in order to pursue him with a larger force and be able to attack him more easily, withdrew all the garrisons from them: (and) Scipio becoming aware of this, sent Maximus, his Captain, to capture them. Pyrrhus, when he was waging war in Sclavonia, in one of the Chief Cities of that country, where a large force had been brought in to garrison it, feigned to be desperate of being able to capture it, and turning to other places, caused her, in order to succor them, to empty herself of the garrison, so that it became easy to be forced (captured). Many have polluted the water and diverted rivers to take a town, even though they then did not succeed. Sieges and surrenders are also easily accomplished, by dismaying them by pointing out an accomplished victory, or new help which is come to their disfavor. The ancient Captains sought to occupy towns by treachery, corrupting some inside, but have used different methods. Some have sent one of their men under the disguise of a fugitive, who gained authority and confidence with the enemy, which he afterward used for his own benefit. Many by this means have learned the procedures of the guards, and through this knowledge have taken the town. Some have blocked the gate so that it could not be locked with a cart or a beam under some pretext, and by this means, made the entry easy to the enemy. Hannibal persuaded one to give him a castle of the Romans, and that he should feign going on a hunt at night, to show his inability to go by day for fear of the enemy, and when he returned with the game, placed his men inside with it, and killing the guard, captured the gate. You also deceive the besieged by drawing them outside the town and distant from it, by feigning flight when they assault you. And many ((among whom was Hannibal)) have, in addition, allowed their quarters to be taken in order to have the opportunity of placing them in their midst, and take the town from them. They deceive also by feigning departure, as did Forminus, the Athenian, who having plundered the country of the Calcidians, afterwards received their ambassadors, and filled their City with promises of safety and good will, who, as men of little caution, were shortly after captured by Forminus. The besieged ought to look out for men whom they have among them that are suspect, but sometimes they may want to assure themselves of these by reward, as well as by punishment. Marcellus, recognizing that Lucius Bancius Nolanus had turned to favor Hannibal, employed so much humanity and liberality toward him, that, from an enemy, he made him a very good friend. The besieged ought to use more diligence in their guards when the enemy is distant, than when he is near. And they ought to guard those places better which they think can be attacked less; for many towns have been lost when the enemy assaulted them on a side from which they did not believe they would be assaulted. And this deception occurs for two reasons: either because the place is strong and they believe it is inaccessible, or because the enemy cunningly assaults him on one side with feigned uproars, and on the other silently with the real assaults. And, therefore, the besieged ought to have a great awareness of this, and above all at all times, but especially at night, have good guards at the walls, and place there not only men, but dogs;

and keep them ferocious and ready, which by smell, detect the presence of the enemy, and with their baying discover him. And, in addition to dogs, it has been found that geese have also saved a City, as happened to the Romans when the Gauls besieged the Capitol. When Athens was besieged by the Spartans, Alcibiades, in order to see if the guards were awake, arranged that when a light was raised at night, all the guards should rise, and inflicted a penalty on those who did not observe it. Hissicratus, the Athenian, slew a guard who was sleeping, saying he was leaving him as he had found him. Those who are besieged have had various ways of sending news to their friends, and in order not to send embassies by voice, wrote letters in cipher, and concealed them in various ways. The ciphers are according to the desires of whoever arranges them, the method of concealment is varied. Some have written inside the scabbard of a sword. Others have put these letters inside raw bread, and then baked it, and gave it as food to him who brought it. Others have placed them in the most secret places of the body. Others have put them in the collar of a dog known to him who brings it. Others have written ordinary things in a letter, and then have written with water (invisible ink) between one line and another, which afterwards by wetting or scalding (caused) the letter to appear. This method has been very astutely observed in our time, where some wanting to point out a thing which was to be kept secret to their friends who lived inside a town, and not wanting to trust it in person, sent communications written in the customary manner, but interlined as I mentioned above, and had them hung at the gates of a Temple; which were then taken and read by those who recognized them from the countersigns they knew. Which is a very cautious method, because whoever brings it can be deceived by you, and you do not run any danger. There are infinite other ways by which anyone by himself likewise can find and read them. But one writes with more facility to the besieged than the besieged do to friends outside, for the latter can not send out such letters except by one who leaves the town under the guise of a fugitive, which is a doubtful and dangerous exploit when the enemy is cautious to a point. But as to those that are sent inside, he who is sent can, under many pretexts, go into the camp that is besieged, and from here await a convenient opportunity to jump into the town.

But let us come to talk of present captures, and I say that, if they occur when you are being fought in your City, which is not arranged with ditches inside, as we pointed out a little while ago, when you do not want the enemy to enter by the breaks in the wall made by artillery ((as there is no remedy for the break which it makes)), it is necessary for you, while the artillery is battering, to dig a ditch inside the wall that is being hit, at least thirty arm lengths wide, and throw all (the earth) that is excavated toward the town, which makes embankments and the ditch deeper: and you must do this quickly, so that if the wall falls, the ditch will be excavated at least five or six arm lengths deep. While this ditch is being excavated, it is necessary that it be closed on each side by a block house. And if the wall is so strong that it gives you time to dig the ditches and erect the block houses, that part which is battered comes to be stronger than the rest of the City, for such a repair comes to have the form that we gave to inside ditches. But if the wall is weak and does not give you time, then there is need to show virtu, and oppose them with armed forces, and with all your strength. This method of repair was observed by the Pisans when you went to besiege them, and they were able to do this because they had strong walls which gave them time, and the ground firm and most suitable for erecting ramparts and making repairs. Which, had they not had this benefit, would have been lost. It would always be prudent, therefore, first to prepare yourself, digging the ditches inside your City and throughout all its circuit, as we devised a little while ago; for in this case, as the defenses have been made, the enemy is awaited with leisure and safety. The ancients often occupied towns with tunnels in two ways: either they dug a secret tunnel which came out inside the town, and through which they entered it, in the way in which the Romans took the City of the Veienti: or, by tunnelling they undermined a wall, and caused it to be ruined. This last method is more effective today, and causes Cities located high up to be weaker, for they can be undermined more easily, and then when that powder which ignites in an instant is placed inside those tunnels, it not only ruins the wall, but the mountains are opened, and the fortresses are entirely disintegrated into several parts. The remedy for this is to build on a plain, and make the ditch which girds your City so deep, that the enemy can not excavate further below it without finding water, which is the only enemy of these excavations. And even if you find a knoll within the town that you defend, you cannot remedy it otherwise than to dig many deep wells within your walls, which are as outlets to those excavations which the enemy might be able to arrange against it. Another remedy is to make an excavation opposite to where you learn he is excavating: which method readily impedes him, but is very difficult to foresee, when you are besieged by a cautious enemy. Whoever is besieged, above all, ought to take care not to be attacked in times of repose, as after having engaged in battle, after having stood guard, that is, at dawn, the evening between night and day, and, above all, at dinner time, in which times many towns have been captured, and many armies ruined by those inside. One ought, therefore, to be always on guard with diligence on every side, and in good part well armed. I do not want to miss telling you that what makes defending a City or an encampment difficult, is to have to keep all the forces you have in them disunited; for the enemy being able all together to assault you at his discretion, you must keep every place guarded on all sides, and thus he assaults you with his entire force, and you defend it with part of yours. The besieged can also be completely overcome, while those outside cannot unless repulsed; whence many who have been besieged either in their encampment or in a town, although inferior in strength, have suddenly issued forth with all their forces, and have overcome the enemy. Marcellus did this at Nola, and Caesar did this in Gaul, where his encampment being assaulted by a great number of Gauls, and seeing he could not defend it without having to divide this forces into several parts, and unable to stay within the stockade with the driving attack of the enemy, opened the encampment on one side, and turning to that side with all his forces, attacked them with such fury, and with such virtu, that he overcame and defeated them. The constancy of the besieged has also often displeased and dismayed the besieger. And when Pompey was affronting Caesar, and Caesar's army was suffering greatly from hunger, some of his bread was brought to Pompey, who, seeing it made of grass, commanded it not be shown to his army in order not to frighten it, seeing what kind of enemies he had to encounter. Nothing gave the Romans more honor in the war against Hannibal, as their constancy; for, in whatever more inimical and adverse fortune, they never asked for peace, (and) never gave any sign of fear: rather, when Hannibal was around Rome, those fields on which he had situated his quarters were sold at a higher price than they would ordinarily have been sold in other times; and they were so obstinate in their enterprises, that to defend Rome, they did not leave off attacking Capua, which was being besieged by the Romans at the same time Rome was being besieged.

I know that I have spoken to you of many things, which you have been able to understand and consider by yourselves; none the less, I have done this ((as I also told you today)) to be able to show you, through them, the better kind of training, and also to satisfy those, if there should be any, who had not had that opportunity to learn, as you have. Nor does it appear to me there is anything left for me to tell you other than some general rules, with which you should be very familiar: which are these. What benefits the enemy, harms you; and what benefits you, harm the enemy. Whoever is more vigilant in observing the designs of the enemy in war, and endures much hardship in training his army, will incur fewer dangers, and can have greater hope for victory. Never lead your soldiers into an engagement unless you are assured of their courage, know they are without fear, and are organized, and never make an attempt unless you see they hope for victory. It is better to defeat the enemy by hunger than with steel; in such victory fortune counts more than virtu. No proceeding is better than that which you have concealed from the enemy until the time you have executed it. To know how to recognize an opportunity in war, and take it, benefits you more than anything else. Nature creates few men brave, industry and training makes many. Discipline in war counts more than fury. If some on the side of the enemy desert to come to your service, if they be loyal, they will always make you a great acquisition; for the forces of the adversary diminish more with the loss of those who flee, than with those who are killed, even though the name of the fugitives is suspect to the new friends, and odious to the old. It is better in organizing an engagement to reserve great aid behind the front line, than to spread out your soldiers to make a greater front. He is overcome with difficulty, who knows how to recognize his forces and those of the enemy. The virtu of the soldiers is worth more than a multitude, and the site is often of more benefit than virtu. New and speedy things frighten armies, while the customary and slow things are esteemed little by them: you will therefore make your army experienced, and learn (the strength) of a new enemy by skirmishes, before you come to an engagement with him. Whoever pursues a routed enemy in a disorganized manner, does nothing but become vanquished from having been a victor. Whoever does not make provisions necessary to live (eat), is overcome without steel. Whoever trusts more in cavalry than in infantry, or more in infantry than in cavalry, must settle for the location. If you want to see whether any spy has come into the camp during the day, have no one go to his quarters. Change your proceeding when you become aware that the enemy has foreseen it. Counsel with many on the things you ought to do, and confer with few on what you do afterwards. When soldiers are confined to their quarters, they are kept there by fear or punishment; then when they are led by war, (they are led) by hope and reward. Good Captains never come to an engagement unless necessity compels them, or the opportunity calls them. Act so your enemies do not know how you want to organize your army for battle, and in whatever way you organize them, arrange it so that the first line can be received by the second and by the third. In a battle, never use a company for some other purpose than what you have assigned it to, unless you want to cause disorder. Accidents are remedied with difficulty, unless you quickly take the facility of thinking. Men, steel, money, and bread, are the sinews of war; but of these four, the first two are more necessary, for men and steel find find money and bread, but money and bread do not find men and steel. The unarmed rich man is the prize of the poor soldier. Accustom your soldiers to despise delicate living and luxurious clothing.

This is as much as occurs to me generally to remind you, and I know I could have told you of many other things in my discussion, as for example, how and in how many ways the ancients organized their ranks, how they dressed, and how they trained in many other things; and to give you many other particulars, which I have not judged necessary to narrate, as much because you are able to see them, as because my intention has not been to show you in detail how the ancient army was created, but how an army should be organized in these times, which should have more virtu than they now have. Whence it does not please me to discuss the ancient matters further than those I have judged necessary to such an introduction. I know I should have enlarged more on the cavalry, and also on naval warfare; for whoever defines the military, says, that it is an army on land and on the sea, on foot and on horseback. Of naval matters, I will not presume to talk, not because of not being informed, but because I should leave the talk to the Genoese and Venetians, who have made much study of it, and have done great things in the past. Of the cavalry, I also do not want to say anything other than what I have said above, this part being ((as I said)) less corrupted. In addition to this, if the infantry, who are the nerve of the army, are well organized, of necessity it happens that good cavalry be created. I would only remind you that whoever organizes the military in his country, so as to fill (the quota) of cavalry, should make two provisions: the one, that he should distribute horses of good breed throughout his countryside, and accustom his men to make a round-up of fillies, as you do in this country with calves and mules: the other, ((so that the round-up men find a buyer)) I would prohibit anyone to keep mules who did not keep a horse; so that whoever wanted to keep a mount only, would also be constrained to keep a horse; and, in addition, none should be able to dress in silk, except whoever keeps a horse. I understand this arrangement has been done by some Princes of our times, and to have resulted in an excellent cavalry being produced in their countries in a very brief time. About other things, how much should be expected from the cavalry, I will go back to what I said to you today, and to that which is the custom. Perhaps you will also desire to learn what parts a Captain ought to have. In this, I will satisfy you in a brief manner; for I would not knowingly select any other man than one who should know how to do all those things which we have discussed today. And these would still not be enough for him if he did not know how to find them out by himself, for no one without imagination was ever very great in his profession; and if imagination makes for honor in other things, it will, above all, honor you in this one. And it is to be observed, that every creation (imagination), even though minor, is celebrated by the writers, as is seen where they praised Alexander the Great, who, in order to break camp more secretly, did not give the signal with the trumpet, but with a hat on the end of a lance. He is also praised for having ordered his soldiers, when coming to battle with the enemy, to kneel with the left foot (knee) so that they could more strongly withstand the attack (of the enemy); which not only gave him victory, but also so much praise that all the statues erected in his honor show him in that pose.

But as it is time to finish this discussion, I want to return to the subject, and so, in part, escape that penalty which, in this town, custom decrees for those who do not return. If you remember well, Cosimo, you said to me that I was, on the one hand, an exalter of antiquity, and a censurer of those who did not imitate them in serious matters, and, on the other (hand), in matters of war in which I worked very hard, I did not imitate them, you were unable to discover the reason: to that I replied, that men who want to do something must first prepare themselves to know how to do it in order to be able afterwards to do it when the occasion permits it. whether or not I would know how to bring the army to the ancient ways, I would rather you be the judge, who have heard me discuss on this subject at length; whence you have been able to know how much time I have consumed on these thoughts, and I also believe you should be able to imagine how much desire there is in me to put them into effect. Which you can guess, if I was ever able to do it, or if ever the opportunity was given to me. Yet, to make you more certain, and for my greater justification, I would like also to cite you the reasons, and in part, will observe what I promised you, to show you the ease and the difficulty that are present in such imitation. I say to you, therefore, that no activity among men today is easier to restore to its ancient ways than the military; but for those only who are Princes of so large a State, that they are able to assemble fifteen or twenty thousand young men from among their own subjects. On the other hand, nothing is more difficult than this to those who do not have such a convenience. And, because I want you to understand this part better, you have to know that Captains who are praised are of two kinds. The one includes those, who, with an army (well) ordered through its own natural discipline, have done great things, such as were the greater part of the Roman Citizens, and others, who have led armies, who have not had any hardship in maintaining them good, and to see to it that they were safely led. The other includes those who not only had to overcome the enemy, but before they came to this, had been compelled to make their army good and well ordered, (and) who, without doubt, deserve greater praise that those others merited who with a army which was (naturally) good have acted with so much virtu. Such as these were Pelopidas, Epaminondas, Tullus Hostilius, Phillip of Macedonia father of Alexander, Cyrus King of the Persians, and Gracchus the Roman. All these had first to make the army good, and then fight with it. All of these were able to do so, as much by their prudence, as by having subjects capable of being directed in such practices. Nor would it have been possible for any of them to accomplish any praiseworthy deed, no matter how good and excellent they might have been, should they have been in an alien country, full of corrupt men, and not accustomed to sincere obedience. It is not enough, therefore, in Italy, to govern an army already trained, but it is necessary first to know how to do it, and then how to command it. And of these, there need to be those Princes, who because they have a large State and many subjects, have the opportunity to accomplish this. Of whom, I cannot be one, for I have never commanded, nor can I command except armies of foreigners, and men obligated to others and not to me. Whether or not it is possible to introduce into them (those Princes) some of the things we discussed today, I want to leave to your judgement. Would I make one of these soldiers who practice today carry more arms than is customary, and in addition, food for two or three days, and a shovel? Should I make him dig, or keep him many hours every day under arms in feigned exercises, so that in real (battles) afterward he could be of value to me? Would they abstain from gambling, lasciviousness, swearing, and insolence, which they do daily? Would they be brought to so much discipline, obedience, and respect, that a tree full of apples which should be found in the middle of an encampment, would be left intact, as is read happened many times in the ancient armies? What can I promise them, by which they well respect, love, or fear me, when, with a war ended, they no longer must come to me for anything? Of what can I make them ashamed, who are born and brought up without shame? By what Deity or Saints do I make them take an oath? By those they adore, or by those they curse? I do not know any whom they adore; but I well know that they curse them all. How can I believe they will observe the promises to those men, for whom they show their contempt hourly? How can those who deprecate God, have reverence for men? What good customs, therefore, is it possible to instill in such people? And if you should tell me the Swiss and the Spaniards are good, I should confess they are far better than the Italians: but if you will note my discussion, and the ways in which both proceeded, you will see that there are still many things missing among them (the Swiss and Spaniards) to bring them up to the perfection of the ancients. And the Swiss have been good from their natural customs, for the reasons I told you today, and the others (Spaniards) from necessity; for when they fight in a foreign country, it seems to them they are constrained to win or die, and as no place appeared to them where they might flee, they became good. But it is a goodness defective in many parts, for there is nothing good in them except that they are accustomed to await the enemy up to the point of the pike and of the sword. Nor would there be anyone suitable to teach them what they lack, and much less anyone who does not (speak) their language.

But let us turn to the Italians, who, because they have not wise Princes, have not produced any good army; and because they did not have the necessity that the Spaniards had, have not undertaken it by themselves, so that they remain the shame of the world. And the people are not to blame, but their Princes are, who have been castigated, and by their ignorance have received a just punishment, ignominously losing the State, (and) without any show of virtu. Do you want to see if what I tell you is true? Consider how many wars have been waged in Italy, from the passage of King Charles (of France) until today; and wars usually make men warlike and acquire reputations; these, as much as they have been great (big) and cruel, so much more have caused its members and its leaders to lose reputation. This necessarily points out, that the customary orders were not, and are not, good, and there is no one who know how to take up the new orders. Nor do you ever believe that reputation will be acquired by Italian arms, except in the manner I have shown, and by those who have large States in Italy, for this custom can be instilled in men who are simple, rough, and your own, but not to men who are malignant, have bad habits, and are foreigners. And a good sculptor will never be found who believes he can make a beautiful statue from a piece of marble poorly shaped, even though it may be a rough one. Our Italian Princes, before they tasted the blows of the ultramontane wars, believed it was enough for them to know what was written, think of a cautious reply, write a beautiful letter, show wit and promptness in his sayings and in his words, know how to weave a deception, ornament himself with gems and gold, sleep and eat with greater splendor than others, keep many lascivious persons around, conduct himself avariciously and haughtily toward his subjects, become rotten with idleness, hand out military ranks at his will, express contempt for anyone who may have demonstrated any praiseworthy manner, want their words should be the responses of oracles; nor were these little men aware that they were preparing themselves to be the prey of anyone who assaulted them. From this, then, in the year one thousand four hundred ninety four (1494), there arose the great frights, the sudden flights, and the miraculous (stupendous) losses: and those most powerful States of Italy were several times sacked and despoiled in this manner. But what is worse is, that those who remained persist in the same error, and exist in the same disorder: and they do not consider that those who held the State anciently, had done all those things we discussed, and that they concentrated on preparing the body for hardships and the mind not to be afraid of danger. Whence it happened that Caesar, Alexander, and all those excellent men and Princes, were the first among the combatants, went around on foot, and even if they did lose their State, wanted also to lose their lives; so that they lived and died with virtu. And if they, or part of them, could be accused of having too much ambition to rule, there never could be found in them any softness or anything to condemn, which makes men delicate and cowardly. If these things were to be read and believed by these Princes, it would be impossible that they would not change their way of living, and their countries not change in fortune. And as, in the beginning of our discussion, you complained of your organization, I tell you, if you had organized it as we discussed above, and it did not give a good account for itself, then you have reason to complain; but if it is not organized and trained as I have said, (the Army) it can have reason to complain of you, who have made an abortion, and not a perfect figure (organization). The Venetians also, and the Duke of Ferrara, begun it, but did not pursue it; which was due to their fault, and not of their men. And I affirm to now, that any of them who have States in Italy today, will begin in this way, he will be the Lord higher than any other in this Province; and it will happen to his State as happened to the Kingdom of the Macedonians, which, coming under Phillip, who had learned the manner of organizing the armies from Epaminondas, the Theban, became, with these arrangements and practices ((while the rest of Greece was in idleness, and attended to reciting comedies)) so powerful, that in a few years, he was able to occupy it completely, and leave such a foundation to his son, that he was able to make himself Prince of the entire world. Whoever disparages these thoughts, therefore, if he be a Prince, disparages his Principality, and if he be a Citizen, his City. And I complain of nature, which either ought to make me a recognizer of this, or ought to have given me the faculty to be able to pursue it. Nor, even today when I am old, do I think I can have the opportunity: and because of this, I have been liberal with you, who, being young and qualified, when the things I have said please you, could, at the proper time, in favor of your Princes, aid and counsel them. I do not want you to be afraid or mistrustful of this, because this country appears to be born (to be destined) to resuscitate the things which are dead, as has been observed with Poetry, Painting, and Sculpture. But as for waiting for me, because of my years, do not rely on it. And, truly, if in the past fortune had conceded to me what would have sufficed for such an enterprise, I believe I would, in a very brief time, have shown the world how much the ancient institutions were of value, and, without doubt, I would have enlarged it with glory, or would have lost it without shame.

The Art of War

by Baron De Jomini
Translated by Capt. G.H. Mendell, and Lieut. W.P. Craighill

Table of Contents

Translator's Preface

In the execution of any undertaking there are extremes on either hand which are alike to be avoided. The rule holds in a special manner in making a translation. There is, on the one side, the extreme of too rigid adherence, word for word and line for line, to the original, and on the other is the danger of using too free a pen. In either case the sense of the author may not be truly given. It is not always easy to preserve a proper mean between these extremes. The translators of Jomini's Summary of the Principles of the Art of War have endeavored to render their author into plain English, without mutilating or adding to his ideas, attempting no display and making no criticisms.

To persons accustomed to read for instruction in military matters, it is not necessary to say a word with reference to the merits of Jomini. To those not thus accustomed heretofore, but who are becoming more interested in such subjects, (and this class must include the great mass of the American public,) it is sufficient to say, and it may be said with entire truth, that General Jomini is admitted by all competent judges to be one of the ablest military critics and historians of this or any other day.

The translation now presented to the people has been made with the earnest hope and the sincere expectation of its proving useful. As the existence of a large, well-instructed standing army is deemed incompatible with our institutions, it becomes the more important that military information be as extensively diffused as possible among the people. If by the present work the translators shall find they have contributed, even in an inconsiderable degree, to this important object, they will be amply repaid for the care and labor expended upon it.

To those persons to whom the study of the art of war is a new one, it is recommended to begin at the article "Strategy," Chapter III., from that point to read to the end of the Second Appendix, and then to return to Chapters I. and II. It should be borne in mind that this subject, to be appreciated, must be studied, map in hand: this remark is especially true of strategy. An acquaintance with the campaigns of Napoleon I. is quite important, as they are constantly referred to by Jomini and by all other recent writers on the military art.

U.S. Military Academy, West Point, N.Y. January, 1862.

Definitions of the Branches of the Art of War

The art of war, as generally considered, consists of five purely military branches,—viz.: Strategy, Grand Tactics, Logistics, Engineering, and Tactics. A sixth and essential branch, hitherto unrecognized, might be termed *Diplomacy in its relation to War*. Although this branch is more naturally and intimately connected with the profession of a statesman than with that of a soldier, it cannot be denied that, if it be useless to a subordinate general, it is indispensable to every general commanding an army: it enters into all the combinations which may lead to a war, and has a connection with the various operations to be undertaken in this war; and, in this view, it should have a place in a work like this.

To recapitulate, the art of war consists of six distinct parts:—

1. Statesmanship in its relation to war.

2. Strategy, or the art of properly directing masses upon the theater of war, either for defense or for invasion.

3. Grand Tactics.

4. Logistics, or the art of moving armies.

5. Engineering,—the attack and defense of fortifications.

6. Minor Tactics.

It is proposed to analyze the principal combinations of the first four branches, omitting the consideration of tactics and of the art of engineering.

Familiarity with all these parts is not essential in order to be a good infantry, cavalry, or artillery officer; but for a general, or for a staff officer, this knowledge is indispensable.

Statesmanship in its Relation to War

Under this head are included those considerations from which a statesman concludes whether a war is proper, opportune, or indispensable, and determines the various operations necessary to attain the object of the war.

A government goes to war,—

To reclaim certain rights or to defend them;

To protect and maintain the great interests of the state, as commerce, manufactures, or agriculture;

To uphold neighboring states whose existence is necessary either for the safety of the government or the balance of power;

To fulfill the obligations of offensive and defensive alliances;

To propagate political or religious theories, to crush them out, or to defend them;

To increase the influence and power of the state by acquisitions of territory;

To defend the threatened independence of the state;

To avenge insulted honor; or,

From a mania for conquest.

It may be remarked that these different kinds of war influence in some degree the nature and extent of the efforts and operations necessary for the proposed end. The party who has provoked the war may be reduced to the defensive, and the party assailed may assume the offensive; and there may be other circumstances which will affect the nature and conduct of a war, as,—

1. A state may simply make war against another state.

2. A state may make war against several states in alliance with each other.

3. A state in alliance with another may make war upon a single enemy.

4. A state may be either the principal party or an auxiliary.

5. In the latter case a state may join in the struggle at its beginning or after it has commenced.

6. The theater of war may be upon the soil of the enemy, upon that of an ally, or upon its own.

7. If the war be one of invasion, it may be upon adjacent or distant territory: it may be prudent and cautious, or it may be bold and adventurous.

8. It may be a national war, either against ourselves or against the enemy.

9. The war may be a civil or a religious war.

War is always to be conducted according to the great principles of the art; but great discretion must be exercised in the nature of the operations to be undertaken, which should depend upon the circumstances of the case.

For example: two hundred thousand French wishing to subjugate the Spanish people, united to a man against them, would not maneuver as the same number of French in a march upon Vienna, or any other capital, to compel a peace; nor would a French army fight the guerrillas of Mina as they fought the Russians at Borodino; nor would a French army venture to march upon Vienna without considering what might be the tone and temper of the governments and communities between the Rhine and the Inn, or between the Danube and the Elbe. A regiment should always fight in nearly the same way; but commanding generals must be guided by circumstances and events.

To these different combinations, which belong more or less to statesmanship, may be added others which relate solely to the management of armies. The name Military Policy is given to them; for they belong exclusively neither to diplomacy nor to strategy, but are still of the highest importance in the plans both of a statesman and a general.

Article I: Offensive Wars to Reclaim Rights

When a state has claims upon another, it may not always be best to enforce them by arms. The public interest must be consulted before action.

The most just war is one which is founded upon undoubted rights, and which, in addition, promises to the state advantages commensurate with the sacrifices required and the hazards incurred. Unfortunately, in our times there are so many doubtful and contested rights that most wars, though apparently based upon bequests, or wills, or marriages, are in reality but wars of expediency. The question of the succession to the Spanish crown under Louis XIV. was very clear, since it was plainly settled by a solemn will, and was supported by family ties and by the general consent of the Spanish nation; yet it was stoutly contested by all Europe, and produced a general coalition against the legitimate legatee.

Frederick II., while Austria and France were at war, brought forward an old claim, entered Silesia in force and seized this province, thus doubling the power of Prussia. This was a stroke of genius; and, even if he had failed, he could not have been much censured; for the grandeur and importance of the enterprise justified him in his attempt, as far as such attempts can be justified.

In wars of this nature no rules can be laid down. To watch and to profit by every circumstance covers all that can be said. Offensive movements should be suitable to the end to be attained. The most natural step would be to occupy the disputed territory: then offensive operations may be carried on according to circumstances and to the respective strength of the parties, the object being to secure the cession of the territory by the enemy, and the means being to threaten him in the heart of his own country. Every thing depends upon the alliances the parties may be able to secure with other states, and upon their military resources. In an offensive movement, scrupulous care must be exercised not to arouse the jealousy of any other state which might come to the aid of the enemy. It is a part of the duty of a statesman to foresee this chance, and to obviate it by making proper explanations and giving proper guarantees to other states.

Article II: of Wars Defensive Politically, and Offensive in a Military Point of View

A state attacked by another which renews an old claim rarely yields it without a war: it prefers to defend its territory, as is always more honorable. But it may be advantageous to take the offensive, instead of awaiting the attack on the frontiers.

There are often advantages in a war of invasion: there are also advantages in awaiting the enemy upon one's own soil. A power with no internal dissensions, and under no apprehension of an attack by a third party, will always find it advantageous to carry the war upon hostile soil. This course will spare its territory from devastation, carry on the war at the expense of the enemy, excite the ardor of its soldiers, and depress the spirits of the adversary. Nevertheless, in a purely military sense, it is certain that an army operating in its own territory, upon a theater of which all the natural and artificial features are well known, where all movements are aided by a knowledge of the country, by the favor of the citizens, and the aid of the constituted authorities, possesses great advantages.

These plain truths have their application in all descriptions of war; but, if the principles of strategy are always the same, it is different with the political part of war, which is modified by the tone of communities, by localities, and by the characters of men at the head of states and armies. The fact of these modifications has been used to prove that war knows no rules. Military science rests upon principles which can never be safely violated in the presence of an active and skillful enemy, while the moral and political part of war presents these variations. Plans of operations are made as circumstances may demand: to execute these plans, the great principles of war must be observed.

For instance, the plan of a war against France, Austria, or Russia would differ widely from one against the brave but undisciplined bands of Turks, which cannot be kept in order, are not able to maneuver well, and possess no steadiness under misfortunes.

Article III: Wars of Expediency

The invasion of Silesia by Frederick II., and the war of the Spanish Succession, were wars of expediency.

There are two kinds of wars of expediency: first, where a powerful state undertakes to acquire natural boundaries for commercial and political reasons; secondly, to lessen the power of a dangerous rival or to prevent his aggrandizement. These last are wars of intervention; for a state will rarely singly attack a dangerous rival: it will endeavor to form a coalition for that purpose.

These views belong rather to statesmanship or diplomacy than to war.

Article IV: Of Wars with or without Allies

Of course, in a war an ally is to be desired, all other things being equal. Although a great state will more probably succeed than two weaker states in alliance against it, still the alliance is stronger than either separately. The ally not only furnishes a contingent of troops, but, in addition, annoys the enemy to a great degree by threatening portions of his frontier which otherwise would have been secure. All history teaches that no enemy is so insignificant as to be despised and neglected by any power, however formidable.

Article V: Wars of Intervention

To interfere in a contest already begun promises more advantages to a state than war under any other circumstances; and the reason is plain. The power which interferes throws upon one side of the scale its whole weight and influence; it interferes at the most opportune moment, when it can make decisive use of its resources.

There are two kinds of intervention: 1. Intervention in the internal affairs of neighboring states; 2. Intervention in external relations.

Whatever may be said as to the moral character of interventions of the first class, instances are frequent. The Romans acquired power by these interferences, and the empire of the English India Company was assured in a similar manner. These interventions are not always successful. While Russia has added to her power by interference with Poland, Austria, on the contrary, was almost ruined by her attempt to interfere in the internal affairs of France during the Revolution.

Intervention in the external relations of states is more legitimate, and perhaps more advantageous. It may be doubtful whether a nation has the right to interfere in the internal affairs of another people; but it certainly has a right to oppose it when it propagates disorder which may reach the adjoining states.

There are three reasons for intervention in exterior foreign wars,—viz.: 1, by virtue of a treaty which binds to aid; 2, to maintain the political equilibrium; 3, to avoid certain evil consequences of the war already commenced, or to secure certain advantages from the war not to be obtained otherwise.

History is filled with examples of powers which have fallen by neglect of these principles. "A state begins to decline when it permits the immoderate aggrandizement of a rival, and a secondary power may become the arbiter of nations if it throw its weight into the balance at the proper time."

In a military view, it seems plain that the sudden appearance of a new and large army as a third party in a well-contested war must be decisive. Much will depend upon its geographical position in reference to the armies already in the field. For example, in the winter of 1807 Napoleon crossed the Vistula and ventured to the walls of Königsberg, leaving Austria on his rear and having Russia in front. If Austria had launched an army of one hundred thousand men from Bohemia upon the Oder, it is probable that the power of Napoleon would have been ended; there is every reason to think that his army could not have regained the Rhine. Austria preferred to wait till she could raise four hundred thousand men. Two years afterward, with this force she took the field, and was beaten; while one hundred thousand men well employed at the proper time would have decided the fate of Europe.

There are several kinds of war resulting from these two different interventions:—

1. Where the intervention is merely auxiliary, and with a force specified by former treaties.

2. Where the intervention is to uphold a feeble neighbor by defending his territory, thus shifting the scene of war to other soil.

3. A state interferes as a principal party when near the theater of war,—which supposes the case of a coalition of several powers against one.

4. A state interferes either in a struggle already in progress, or interferes before the declaration of war.

When a state intervenes with only a small contingent, in obedience to treaty-stipulations, it is simply an accessory, and has but little voice in the main operations; but when it intervenes as a principal party, and with an imposing force, the case is quite different.

The military chances in these wars are varied. The Russian army in the Seven Years' War was in fact auxiliary to that of Austria and France: still, it was a principal party in the North until its occupation of Prussia. But when Generals Fermor and Soltikoff conducted the army as far as Brandenburg it acted solely in the interest of Austria: the fate of these troops, far from their base, depended upon the good or bad maneuvering of their allies.

Such distant excursions are dangerous, and generally delicate operations. The campaigns of 1799 and 1805 furnish sad illustrations of this, to which we shall again refer in Article XXIX., in discussing the military character of these expeditions.

It follows, then, that the safety of the army may be endangered by these distant interventions. The counterbalancing advantage is that its own territory cannot then be easily invaded, since the scene of hostilities is so distant; so that what may be a misfortune for the general may be, in a measure, an advantage to the state.

In wars of this character the essentials are to secure a general who is both a statesman and a soldier; to have clear stipulations with the allies as to the part to be taken by each in the principal operations; finally, to agree upon an objective point which shall be in harmony with the common interests. By the neglect of these precautions, the greater number of coalitions have failed, or have maintained a difficult struggle with a power more united but weaker than the allies.

The third kind of intervention, which consists in interfering with the whole force of the state and near to its frontiers, is more promising than the others. Austria had an opportunity of this character in 1807, but failed to profit by it: she again had the opportunity in 1813. Napoleon had just collected his forces in Saxony, when Austria, taking his front of operations in reverse, threw herself into the struggle with two hundred thousand men, with almost perfect certainty of success. She regained in two months the Italian empire and her influence in Germany, which had been lost by fifteen years of disaster. In this intervention Austria had not only the political but also the military chances in her favor,—a double result, combining the highest advantages.

Her success was rendered more certain by the fact that while the theater was sufficiently near her frontiers to permit the greatest possible display of force, she at the same time interfered in a contest already in progress, upon which she entered with the whole of her resources and at the time most opportune for her.

This double advantage is so decisive that it permits not only powerful monarchies, but even small states, to exercise a controlling influence when they know how to profit by it.

Two examples may establish this. In 1552, the Elector Maurice of Saxony boldly declared war against Charles V., who was master of Spain, Italy, and the German empire, and had been victorious over Francis I. and held France in his grasp. This movement carried the war into the Tyrol, and arrested the great conqueror in his career.

In 1706, the Duke of Savoy, Victor Amadeus, by declaring himself hostile to Louis XIV., changed the state of affairs in Italy, and caused the recall of the French army from the banks of the Adige to the walls of Turin, where it encountered the great catastrophe which immortalized Prince Eugene.

Enough has been said to illustrate the importance and effect of these opportune interventions: more illustrations might be given, but they could not add to the conviction of the reader.

Article VI: Aggressive Wars for Conquest and other Reasons

There are two very different kinds of invasion: one attacks an adjoining state; the other attacks a distant point, over intervening territory of great extent whose inhabitants may be neutral, doubtful, or hostile.

Wars of conquest, unhappily, are often prosperous,—as Alexander, Cæsar, and Napoleon during a portion of his career, have fully proved. However, there are natural limits in these wars, which cannot be passed without incurring great disaster. Cambyses in Nubia, Darius in Scythia, Crassus and the Emperor Julian among the Parthians, and Napoleon in Russia, furnish bloody proofs of these truths.—The love of conquest, however, was not the only motive with Napoleon: his personal position, and his contest with England, urged him to enterprises the aim of which was to make him supreme. It is true that he loved war and its chances; but he was also a victim to the necessity of succeeding in his efforts or of yielding to England. It might be said that he was sent into this world to teach generals and statesmen what they should avoid. His victories teach what may be accomplished by activity, boldness, and skill; his disasters, what might have been avoided by prudence.

A war of invasion without good reason—like that of Genghis Khan—is a crime against humanity; but it may be excused, if not approved, when induced by great interests or when conducted with good motives.

The invasions of Spain of 1808 and of 1823 differed equally in object and in results: the first was a cunning and wanton attack, which threatened the existence of the Spanish nation, and was fatal to its author; the second, while combating dangerous principles, fostered the general interests of the country, and was the more readily brought to a successful termination because its object met with the approval of the majority of the people whose territory was invaded.

These illustrations show that invasions are not necessarily all of the same character. The first contributed largely to the fall of Napoleon; the second restored the relation between France and Spain, which ought never to have been changed.

Let us hope that invasions may be rare. Still, it is better to attack than to be invaded; and let us remember that the surest way to check the spirit of conquest and usurpation is to oppose it by intervention at the proper time.

An invasion, to be successful, must, be proportioned in magnitude to the end to be attained and to the obstacles to be overcome.

An invasion against an exasperated people, ready for all sacrifices and likely to be aided by a powerful neighbor, is a dangerous enterprise, as was well proved by the war in Spain, (1808,) and by the wars of the Revolution in 1792, 1793, and 1794. In these latter wars, if France was better prepared than Spain, she had no powerful ally, and she was attacked by all Europe upon both land and sea.

Although the circumstances were different, the Russian invasion of Turkey developed, in some respects, the same symptoms of national resistance. The religious hatred of the Ottoman powerfully incited him to arms; but the same motive was powerless among the Greeks, who were twice as numerous as the Turks. Had the interests of the Greeks and Turks been harmonized, as were those of Alsace with France, the united people would have been stronger, but they would have lacked the element of religious fanaticism. The war of 1828 proved that Turkey was formidable only upon the frontiers, where her bravest troops were found, while in the interior all was weakness.

When an invasion of a neighboring territory has nothing to fear from the inhabitants, the principles of strategy shape its course. The popular feeling rendered the invasions of Italy, Austria, and Prussia so prompt. (These military points are treated of in Article XXIX.) But when the invasion is distant and extensive territories intervene, its success will depend more upon diplomacy than upon strategy. The first step to insure success will be to secure the sincere and devoted alliance of a state adjoining the enemy, which will afford reinforcements of troops, and, what is still more important, give a secure base of operations, depots of supplies, and a safe refuge in case of disaster. The ally must have the same interest in success as the invaders, to render all this possible.

Diplomacy, while almost decisive in distant expeditions, is not powerless in adjacent invasions; for here a hostile intervention may arrest the most brilliant successes. The invasions of Austria in 1805 and 1809 might have ended differently if Prussia had interfered. The invasion of the North of

Germany in 1807 was, so to speak, permitted by Austria. That of Rumelia in 1829 might have ended in disaster, had not a wise statesmanship by negotiation obviated all chance of intervention.

Article VII: Wars of Opinion

Although wars of opinion, national wars, and civil wars are sometimes confounded, they differ enough to require separate notice.

Wars of opinion may be intestine, both intestine and foreign, and, lastly, (which, however, is rare,) they may be foreign or exterior without being intestine or civil.

Wars of opinion between two states belong also to the class of wars of intervention; for they result either from doctrines which one party desires to propagate among its neighbors, or from dogmas which it desires to crush,—in both cases leading to intervention. Although originating in religious or political dogmas, these wars are most deplorable; for, like national wars, they enlist the worst passions, and become vindictive, cruel, and terrible.

The wars of Islamism, the Crusades, the Thirty Years' War, the wars of the League, present nearly the same characteristics. Often religion is the pretext to obtain political power, and the war is not really one of dogmas. The successors of Mohammed cared more to extend their empire than to preach the Koran, and Philip II., bigot as he was, did not sustain the League in France for the purpose of advancing the Roman Church. We agree with M. Ancelot that Louis IX., when he went on a crusade in Egypt, thought more of the commerce of the Indies than of gaining possession of the Holy Sepulcher.

The dogma sometimes is not only a pretext, but is a powerful ally; for it excites the ardor of the people, and also creates a party. For instance, the Swedes in the Thirty Years' War, and Philip II. in France, had allies in the country more powerful than their armies. It may, however, happen, as in the Crusades and the wars of Islamism, that the dogma for which the war is waged, instead of friends, finds only bitter enemies in the country invaded; and then the contest becomes fearful.

The chances of support and resistance in wars of political opinions are about equal. It may be recollected how in 1792 associations of fanatics thought it possible to propagate throughout Europe the famous declaration of the rights of man, and how governments became justly alarmed, and rushed to arms probably with the intention of only forcing the lava of this volcano back into its crater and there extinguishing it. The means were not fortunate; for war and aggression are inappropriate measures for arresting an evil which lies wholly in the human passions, excited in a temporary paroxysm, of less duration as it is the more violent. Time is the true remedy for all bad passions and for all anarchical doctrines. A civilized nation may bear the yoke of a factious and unrestrained multitude for a short interval; but these storms soon pass away, and reason resumes her sway. To attempt to restrain such a mob by a foreign force is to attempt to restrain the explosion of a mine when the powder has already been ignited: it is far better to await the explosion and afterward fill up the crater than to try to prevent it and to perish in the attempt.

After a profound study of the Revolution, I am convinced that, if the Girondists and National Assembly had not been threatened by foreign armaments, they would never have dared to lay their sacrilegious hands upon the feeble but venerable head of Louis XVI. The Girondists would never have been crushed by the Mountain but for the reverses of Dumouriez and the threats of invasion. And if they had been permitted to clash and quarrel with each other to their hearts' content, it is probable that, instead of giving place to the terrible Convention, the Assembly would slowly have returned to the restoration of good, temperate, monarchical doctrines, in accordance with the necessities and the immemorial traditions of the French.

In a military view these wars are fearful, since the invading force not only is met by the armies of the enemy, but is exposed to the attacks of an exasperated people. It may be said that the violence of one party will necessarily create support for the invaders by the formation of another and opposite one; but, if the exasperated party possesses all the public resources, the armies, the forts, the arsenals, and if it is supported by a large majority of the people, of what avail will be the support of the faction which possesses no such means? What service did one hundred thousand Vendeans and one hundred thousand Federalists do for the Coalition in 1793?

History contains but a single example of a struggle like that of the Revolution; and it appears to clearly demonstrate the danger of attacking an intensely-excited nation. However the bad management of the military operations was one cause of the unexpected result, and before deducing any certain maxims from this war, we should ascertain what would have been the result if after the flight of Dumouriez, instead of destroying and capturing fortresses, the allies had informed the commanders of those fortresses that they contemplated no wrong to France, to her forts or her brave armies, and had marched on Paris with two hundred thousand men. They might have restored the monarchy; and, again, they might never have returned, at least without the protection of an equal force on their retreat to the Rhine. It is difficult to decide this, since the experiment was never made, and as all would have depended upon the course of the French nation and the army. The problem thus presents two equally grave solutions. The campaign of 1793 gave one; whether the other might have been obtained, it is difficult to say. Experiment alone could have determined it.

The military precepts for such wars are nearly the same as for national wars, differing, however, in a vital point. In national wars the country should be occupied and subjugated, the fortified places besieged and reduced, and the armies destroyed; whereas in wars of opinion it is of less importance to subjugate the country; here great efforts should be made to gain the end speedily, without delaying for details, care being constantly taken to avoid any acts which might alarm the nation for its independence or the integrity of its territory.

The war in Spain in 1823 is an example which may be cited in favor of this course in opposition to that of the Revolution. It is true that the conditions were slightly different; for the French army of 1792 was made up of more solid elements than that of the Radicals of the Isla de Leon. The war of the Revolution was at once a war of opinion, a national war, and a civil war,—while, if the first war in Spain in 1808 was thoroughly a national war, that of 1823 was a partial struggle of opinions without the element of nationality; and hence the enormous difference in the results.

Moreover, the expedition of the Duke of Angoulême was well carried out. Instead of attacking fortresses, he acted in conformity to the above-mentioned precepts. Pushing on rapidly to the Ebro, he there divided his forces, to seize, at their sources, all the elements of strength of their enemies,—which they could safely do, since they were sustained by a majority of the inhabitants. If he had followed the instructions of the Ministry, to proceed methodically to the conquest of the country and the reduction of the fortresses between the Pyrenees and the Ebro, in order to provide a base of operations, he would perhaps have failed in his mission, or at least made the war a long and bloody one, by exciting the national spirit by an occupation of the country similar to that of 1807.

Emboldened by the hearty welcome of the people, he comprehended that it was a political operation rather than a military one, and that it behooved him to consummate it rapidly. His conduct, so different from that of the allies in 1793, deserves careful attention from all charged with similar missions. In three months the army was under the walls of Cadiz.

If the events now transpiring in the Peninsula prove that statesmanship was not able to profit by success in order to found a suitable and solid order of things, the fault was neither in the army nor in its commanders, but in the Spanish government, which, yielding to the counsel of violent reactionaries, was unable to rise to the height of its mission. The arbiter between two great hostile interests, Ferdinand blindly threw himself into the arms of the party which professed a deep veneration for the throne, but which intended to use the royal authority for the furtherance of its own ends, regardless of consequences. The nation remained divided in two hostile camps, which it would not have been impossible to calm and reconcile in time. These camps came anew into collision, as I predicted in Verona in 1823,—a striking lesson, by which no one is disposed to profit in that beautiful and unhappy land, although history is not wanting in examples to prove that violent reactions, any more than revolutions, are not elements with which to construct and consolidate. May God grant that from this frightful conflict may emerge a strong and respected monarchy, equally separated from all factions, and based upon a disciplined army as well as upon the general interests of the country,—a monarchy capable of rallying to its support this incomprehensible Spanish nation, which, with merits not less extraordinary than its faults, was always a problem for those who were in the best position to know it.

Article VIII: National Wars

National wars, to which we have referred in speaking of those of invasion, are the most formidable of all. This name can only be applied to such as are waged against a united people, or a great majority of them, filled with a noble ardor and determined to sustain their independence: then every step is disputed, the army holds only its camp-ground, its supplies can only be obtained at the point of the sword, and its convoys are everywhere threatened or captured.

The spectacle of a spontaneous uprising of a nation is rarely seen; and, though there be in it something grand and noble which commands our admiration, the consequences are so terrible that, for the sake of humanity, we ought to hope never to see it. This uprising must not be confounded with a national defense in accordance with the institutions of the state and directed by the government.

This uprising may be produced by the most opposite causes. The serfs may rise in a body at the call of the government, and their masters, affected by a noble love of their sovereign and country, may set them the example and take the command of them; and, similarly, a fanatical people may arm under the appeal of its priests; or a people enthusiastic in its political opinions, or animated by a sacred love of its institutions, may rush to meet the enemy in defense of all it holds most dear.

The control of the sea is of much importance in the results of a national invasion. If the people possess a long stretch of coast, and are masters of the sea or in alliance with a power which controls it, their power of resistance is quintupled, not only on account of the facility of feeding the insurrection and of alarming the enemy on all the points he may occupy, but still more by the difficulties which will be thrown in the way of his procuring supplies by the sea.

The nature of the country may be such as to contribute to the facility of a national defense. In mountainous countries the people are always most formidable; next to these are countries covered with extensive forests.

The resistance of the Swiss to Austria and to the Duke of Burgundy, that of the Catalans in 1712 and in 1809, the difficulties encountered by the Russians in the subjugation of the tribes of the Caucasus, and, finally, the reiterated efforts of the Tyrolese, clearly demonstrate that the inhabitants of mountainous regions have always resisted for a longer time than those of the plains,—which is due as much to the difference in character and customs as to the difference in the natural features of the countries.

Defiles and large forests, as well as rocky regions, favor this kind of defense; and the Bocage of La Vendée, so justly celebrated, proves that any country, even if it be only traversed by large hedges and ditches or canals, admits of a formidable defense.

The difficulties in the path of an army in wars of opinions, as well as in national wars, are very great, and render the mission of the general conducting them very difficult. The events just mentioned, the contest of the Netherlands with Philip II. and that of the Americans with the English, furnish evident proofs of this; but the much more extraordinary struggle of La Vendée with the victorious Republic, those of Spain, Portugal, and the Tyrol against Napoleon, and, finally, those of the Morea against the Turks, and of Navarre against the armies of Queen Christina, are still more striking illustrations.

The difficulties are particularly great when the people are supported by a considerable nucleus of disciplined troops. The invader has only an army: his adversaries have an army, and a people wholly or almost wholly in arms, and making means of resistance out of every thing, each individual of whom conspires against the common enemy; even the non-combatants have an interest in his ruin and accelerate it by every means in their power. He holds scarcely any ground but that upon which he encamps; outside the limits of his camp every thing is hostile and multiplies a thousandfold the difficulties he meets at every step.

These obstacles become almost insurmountable when the country is difficult. Each armed inhabitant knows the smallest paths and their connections; he finds everywhere a relative or friend who aids him; the commanders also know the country, and, learning immediately the slightest movement on the part of the invader, can adopt the best measures to defeat his projects; while the latter, without information of their movements, and not in a condition to send out detachments to gain it, having no resource but in his bayonets, and certain safety only in the concentration of his columns, is like a blind man: his combinations are failures; and when, after the most carefully-concerted movements and the most rapid and fatiguing marches, he thinks he is about to accomplish his aim and deal a terrible blow, he finds no signs of the enemy but his camp-fires: so that while, like Don Quixote, he is attacking windmills, his adversary is on his line of communications, destroys the detachments left to guard it, surprises his convoys, his depots, and carries on a war so disastrous for the invader that he must inevitably yield after a time.

In Spain I was a witness of two terrible examples of this kind. When Ney's corps replaced Soult's at Corunna, I had camped the companies of the artillery-train between Betanzos and Corunna, in the midst of four brigades distant from the camp from two to three leagues, and no Spanish forces had been seen within fifty miles; Soult still occupied Santiago de Compostela, the division Maurice-Mathieu was at Ferrol and Lugo, Marchand's at Corunna and Betanzos: nevertheless, one fine night the companies of the train—men and horses—disappeared, and we were never able to discover what became of them: a solitary wounded corporal escaped to report that the peasants, led by their monks and priests, had thus made away with them. Four months afterward, Ney with a single division marched to conquer the Asturias, descending the valley of the Navia, while Kellermann debouched from Leon by the Oviedo road. A part of the corps of La Romana which was guarding the Asturias marched behind the very heights which inclose the valley of the Navia, at most but a league from our columns, without the marshal knowing a word of it: when he was entering Gijon, the army of La Romana attacked the center of the regiments of the division Marchand, which, being scattered to guard Galicia, barely escaped, and that only by the prompt return of the marshal to Lugo. This war presented a thousand incidents as striking as this. All the gold of Mexico could not have procured reliable information for the French; what was given was but a lure to make them fall more readily into snares.

No army, however disciplined, can contend successfully against such a system applied to a great nation, unless it be strong enough to hold all the essential points of the country, cover its communications, and at the same time furnish an active force sufficient to beat the enemy wherever he may present himself. If this enemy has a regular army of respectable size to be a nucleus around which to rally the people, what force will be sufficient to be superior everywhere, and to assure the safety of the long lines of communication against numerous bodies?

The Peninsular War should be carefully studied, to learn all the obstacles which a general and his brave troops may encounter in the occupation or conquest of a country whose people are all in arms. What efforts of patience, courage, and resignation did it not cost the troops of Napoleon, Massena, Soult, Ney, and Suchet to sustain themselves for six years against three or four hundred thousand armed Spaniards and Portuguese supported by the regular armies of Wellington, Beresford, Blake, La Romana, Cuesta, Castaños, Reding, and Ballasteros!

If success be possible in such a war, the following general course will be most likely to insure it,—viz.: make a display of a mass of troops proportioned to the obstacles and resistance likely to be encountered, calm the popular passions in every possible way, exhaust them by time and patience, display courtesy, gentleness, and severity united, and, particularly, deal justly. The examples of Henry IV. in the wars of the League, of Marshal Berwick in Catalonia, of Suchet in Aragon and Valencia, of Hoche in La Vendée, are models of their kind, which may be employed according to circumstances with equal success. The admirable order and discipline of the armies of Diebitsch and Paskevitch in the late war were also models, and were not a little conducive to the success of their enterprises.

The immense obstacles encountered by an invading force in these wars have led some speculative persons to hope that there should never be any other kind, since then wars would become more rare, and, conquest being also more difficult, would be less a temptation to ambitious leaders. This reasoning is rather plausible than solid; for, to admit all its consequences, it would be necessary always to be able to induce the people to take up arms, and it would also be necessary for us to be convinced that there would be in the future no wars but those of conquest, and that all legitimate though secondary wars, which are only to maintain the political equilibrium or defend the public interests, should never occur again: otherwise, how could it be known when and how to excite the people to a national war? For example, if one hundred thousand Germans crossed the Rhine and entered France, originally with the intention of preventing the conquest of Belgium by France, and without any other ambitious project, would it be a case where the whole population—men, women, and children—of Alsace, Lorraine, Champagne, and Burgundy, should rush to arms? to make a Saragossa of every walled town, to bring about, by way of reprisals, murder, pillage, and incendiarism throughout the country? If all this be not done, and the Germans, in consequence of some success, should occupy these provinces, who can say that they might not afterward seek to appropriate a part of them, even though at first they had never contemplated it? The difficulty of answering these two questions would seem to argue in favor of national wars. But is there no means of repelling such an invasion without bringing about an uprising of the whole population and a war of extermination? Is there no mean between these contests between the people and the old regular method of war between permanent armies? Will it not be sufficient, for the efficient defense of the country, to organize a militia, or landwehr, which, uniformed and called by their governments into service, would regulate the part the people should take in the war, and place just limits to its barbarities?

I answer in the affirmative; and, applying this mixed system to the cases stated above, I will guarantee that fifty thousand regular French troops, supported by the National Guards of the East, would get the better of this German army which had crossed the Vosges; for, reduced to fifty thousand men by many detachments, upon nearing the Meuse or arriving in Argonne it would have one hundred thousand men on its hands. To attain this mean, we have laid it down as a necessity that good national reserves be prepared for the army; which will be less expensive in peace and will insure the defense of the country in war. This system was used by France in 1792, imitated by Austria in 1809, and by the whole of Germany in 1813.

I sum up this discussion by asserting that, without being a utopian philanthropist, or a condottieri, a person may desire that wars of extermination may be banished from the code of nations, and that the defenses of nations by disciplined militia, with the aid of good political alliances, may be sufficient to insure their independence.

As a soldier, preferring loyal and chivalrous warfare to organized assassination, if it be necessary to make a choice, I acknowledge that my prejudices are in favor of the good old times when the French and English Guards courteously invited each other to fire first,—as at Fontenoy,—preferring them to the frightful epoch when priests, women, and children throughout Spain plotted the murder of isolated soldiers.

Article IX: Civil Wars, and Wars of Religion

Intestine wars, when not connected with a foreign quarrel, are generally the result of a conflict of opinions, of political or religious sectarianism. In the Middle Ages they were more frequently the collisions of feudal parties. Religious wars are above all the most deplorable.

We can understand how a government may find it necessary to use force against its own subjects in order to crush out factions which would weaken the authority of the throne and the national strength; but that it should murder its citizens to compel them to say their prayers in French or Latin, or to recognize the supremacy of a foreign pontiff, is difficult of conception. Never was a king more to be pitied than Louis XIV., who persecuted a million of industrious Protestants, who had put upon the throne his own Protestant ancestor. Wars of fanaticism are horrible when mingled with exterior wars, and they are also frightful when they are family quarrels. The history of France in the times of the League should be an eternal lesson for nations and kings. It is difficult to believe that a people so noble and chivalrous in the time of Francis I. should in twenty years have fallen into so deplorable a state of brutality.

To give maxims in such wars would be absurd. There is one rule upon which all thoughtful men will be agreed: that is, to unite the two parties or sects to drive the foreigners from the soil, and afterward to reconcile by treaty the conflicting claims or rights. Indeed, the intervention of a third power in a religious dispute can only be with ambitious views.

Governments may in good faith intervene to prevent the spreading of a political disease whose principles threaten social order; and, although these fears are generally exaggerated and are often mere pretexts, it is possible that a state may believe its own institutions menaced. But in religious disputes this is never the case; and Philip II. could have had no other object in interfering in the affairs of the League than to subject France to his influence, or to dismember it.

Article X: Double Wars, and the Danger of Undertaking Two Wars at Once

The celebrated maxim of the Romans, not to undertake two great wars at the same time, is so well known and so well appreciated as to spare the necessity of demonstrating its wisdom.

A government maybe compelled to maintain a war against two neighboring states; but it will be extremely unfortunate if it does not find an ally to come to its aid, with a view to its own safety and the maintenance of the political equilibrium. It will seldom be the case that the nations allied against it will have the same interest in the war and will enter into it with all their resources; and, if one is only an auxiliary, it will be an ordinary war.

Louis XIV., Frederick the Great, the Emperor Alexander, and Napoleon, sustained gigantic struggles against united Europe. When such contests arise from voluntary aggressions, they are proof of a capital error on the part of the state which invites them; but if they arise from imperious and inevitable circumstances they must be met by seeking alliances, or by opposing such means of resistance as shall establish something like equality between the strength of the parties.

The great coalition against Louis XIV., nominally arising from his designs on Spain, had its real origin in previous aggressions which had alarmed his neighbors. To the combined forces of Europe he could only oppose the faithful alliance of the Elector of Bavaria, and the more equivocal one of the Duke of Savoy, who, indeed, was not slow in adding to the number of his enemies. Frederick, with only the aid of the subsidies of England, and fifty thousand auxiliaries from six different states, sustained a war against the three most powerful monarchies of Europe: the division and folly of his opponents were his best friends.

Both these wars, as well as that sustained by Alexander in 1812, it was almost impossible to avoid.

France had the whole of Europe on its hands in 1793, in consequence of the extravagant provocations of the Jacobins, and the Utopian ideas of the Girondists, who boasted that with the support of the English fleets they would defy all the kings in the world. The result of these absurd calculations was a frightful upheaval of Europe, from which France miraculously escaped.

Napoleon is, to a certain degree, the only modern sovereign who has voluntarily at the same time undertaken two, and even three, formidable wars,—with Spain, with England, and with Russia; but in the last case he expected the aid of Austria and Prussia, to say nothing of that of Turkey and Sweden, upon which he counted with too much certainty; so that the enterprise was not so adventurous on his part as has been generally supposed.

It will be observed that there is a great distinction between a war made against a single state which is aided by a third acting as an auxiliary, and two wars conducted at the same time against two powerful nations in opposite quarters, who employ all their forces and resources. For instance, the double contest of Napoleon in 1809 against Austria and Spain aided by England was a very different affair from a contest with Austria assisted by an auxiliary force of a given strength. These latter contests belong to ordinary wars.

It follows, then, in general, that double wars should be avoided if possible, and, if cause of war be given by two states, it is more prudent to dissimulate or neglect the wrongs suffered from one of them, until a proper opportunity for redressing them shall arrive. The rule, however, is not without exception: the respective forces, the localities, the possibility of finding allies to restore, in a measure, equality of strength between the parties, are circumstances which will influence a government so threatened. We now have fulfilled our task, in noting both the danger and the means of remedying it.

Military Policy

We have already explained what we understand by this title. It embraces the moral combinations relating to the operations of armies. If the political considerations which we have just discussed be also moral, there are others which influence, in a certain degree, the conduct of a war, which belong neither to diplomacy, strategy, nor tactics. We include these under the head of *Military Policy*.

Military policy may be said to embrace all the combinations of any projected war, except those relating to the diplomatic art and strategy; and, as their number is considerable, a separate article cannot be assigned to each without enlarging too much the limits of this work, and without deviating from my intention,—which is, not to give a treatise on theses subjects, but to point out their relations to military operations.

Indeed, in this class we may place the passions of the nation to be fought, their military system, their immediate means and their reserves, their financial resources, the attachment they bear to their government or their institutions, the character of the executive, the characters and military abilities of the commanders of their armies, the influence of cabinet councils or councils of war at the capital upon their operations, the system of war in favor with their staff, the established force of the state and its armament, the military geography and statistics of the state which is to be invaded, and, finally, the resources and obstacles of every kind likely to be met with, all of which are included neither in diplomacy nor in strategy.

There are no fixed rules on such subjects, except that the government should neglect nothing in obtaining a knowledge of these details, and that it is indispensable to take them into consideration in the arrangement of all plans. We propose to sketch the principal points which ought to guide in this sort of combinations.

Article XI: Military Statistics and Geography

By the first of these sciences we understand the most thorough knowledge possible of the elements of power and military resources of the enemy with whom we are called upon to contend; the second consists in the topographical and strategic description of the theater of war, with all the obstacles, natural or artificial, to be encountered, and the examination of the permanent decisive points which may be presented in the whole extent of the frontier or throughout the extent of the country. Besides the minister of war, the commanding general and his chief of staff should be afforded this information, under the penalty of cruel miscalculations in their plans, as happens frequently in our day, despite the great strides civilized nations have taken in statistical, diplomatic, geographical, and topographical sciences. I will cite two examples of which I was cognizant. In 1796, Moreau's army, entering the Black Forest, expected to find terrible mountains, frightful defiles and forests, and was greatly surprised to discover, after climbing the declivities of the plateau that slope to the Rhine, that these, with their spurs, were the only mountains, and that the country, from the sources of the Danube to Donauwerth, was a rich and level plain.

The second example was in 1813. Napoleon and his whole army supposed the interior of Bohemia to be very mountainous,—whereas there is no district in Europe more level, after the girdle of mountains surrounding it has been crossed, which may be done in a single march.

All European officers held the same erroneous opinions in reference to the Balkan and the Turkish force in the interior. It seemed that it was given out at Constantinople that this province was an almost impregnable barrier and the palladium of the empire,—an error which I, having lived in the Alps, did not entertain. Other prejudices, not less deeply rooted, have led to the belief that a people all the individuals of which are constantly armed would constitute a formidable militia and would defend themselves to the last extremity. Experience has proved that the old regulations which placed the elite of the Janissaries in the frontier-cities of the Danube made the population of those cities more warlike than the inhabitants of the interior. In fact, the projects of reform of the Sultan Mahmoud required the overthrow of the old system, and there was no time to replace it by the new: so that the empire was defenseless. Experience has constantly proved that a mere multitude of brave men armed to the teeth make neither a good army nor a national defense.

Let us return to the necessity of knowing well the military geography and statistics of an empire. These sciences are not set forth in treatises, and are yet to be developed. Lloyd, who wrote an essay upon them, in describing the frontiers of the great states of Europe, was not fortunate in his maxims and predictions. He saw obstacles everywhere; he represents as impregnable the

Austrian frontier on the Inn, between the Tyrol and Passau, where Napoleon and Moreau maneuvered and triumphed with armies of one hundred and fifty thousand men in 1800, 1805, and 1809.

But, if these sciences are not publicly taught, the archives of the European staff must necessarily possess many documents valuable for instruction in them,—at least for the special staff school. Awaiting the time when some studious officer, profiting by those published and unpublished documents, shall present Europe with a good military and strategic geography, we may, thanks to the immense progress of topography of late years, partially supply the want of it by the excellent charts published in all European countries within the last twenty years. At the beginning of the French Revolution topography was in its infancy: excepting the semi-topographical map of Cassini, the works of Bakenberg alone merited the name. The Austrian and Prussian staff schools, however, were good, and have since borne fruit. The charts published recently at Vienna, at Berlin, Munich, Stuttgart, and Paris, as well as those of the institute of Herder at Fribourg, promise to future generals immense resources unknown to their predecessors.

Military statistics is not much better known than geography. We have but vague and superficial statements, from which the strength of armies and navies is conjectured, and also the revenue supposed to be possessed by a state,—which is far from being the knowledge necessary to plan operations. Our object here is not to discuss thoroughly these important subjects, but to indicate them, as facilitating success in military enterprises.

Article XII: Other Causes which exercise an Influence upon the Success of a War

As the excited passions of a people are of themselves always a powerful enemy, both the general and his government should use their best efforts to allay them. We have nothing to add to what has been said on this point under the head of national wars.

On the other hand, the general should do every thing to electrify his own soldiers, and to impart to them the same enthusiasm which he endeavors to repress in his adversaries. All armies are alike susceptible of this spirit: the springs of action and means, only, vary with the national character. Military eloquence is one means, and has been the subject of many a treatise. The proclamations of Napoleon and of Paskevitch, the addresses of the ancients to their soldiers, and those of Suwaroff to men of still greater simplicity, are models of their different kinds. The eloquence of the Spanish Juntas, and the miracles of the Madonna del Pilar, led to the same results by very different means. In general, a cherished cause, and a general who inspires confidence by previous success, are powerful means of electrifying an army and conducing to victory. Some dispute the advantages of this enthusiasm, and prefer imperturbable coolness in battle. Both have unmistakable advantages and disadvantages. Enthusiasm impels to the performance of great actions: the difficulty is in maintaining it constantly; and, when discouragement succeeds it, disorder easily results.

The greater or less activity and boldness of the commanders of the armies are elements of success or failure, which cannot be submitted to rules. A cabinet and a commander ought to consider the intrinsic value of their troops, and that resulting from their organization as compared with that of the enemy. A Russian general, commanding the most solidly organized troops in Europe, need not fear to undertake any thing against undisciplined and unorganized troops in an open country, however brave may be its individuals. [Irregular troops supported by disciplined troops may be of the greatest value, in destroying convoys, intercepting communication, &c., and may—as in the case of the French in 1812—make a retreat very disastrous.] Concert in action makes strength; order produces this concert, and discipline insures order; and without discipline and order no success is possible. The Russian general would not be so bold before European troops having the same instruction and nearly the same discipline as his own. Finally, a general may attempt with a Mack as his antagonist what it would be madness to do with a Napoleon.

The action of a cabinet in reference to the control of armies influences the boldness of their operations. A general whose genius and hands are tied by an Aulic council five hundred miles distant cannot be a match for one who has liberty of action, other things being equal.

As to superiority in skill, it is one of the most certain pledges of victory, all other things being equal. It is true that great generals have often been beaten by inferior ones; but an exception does not make a rule. An order misunderstood, a fortuitous event, may throw into the hands of the enemy all the chances of success which a skillful general had prepared for himself by his maneuvers. But these are risks which cannot be foreseen nor avoided. Would it be fair on that account to deny the influence of science and principles in ordinary affairs? This risk even proves the triumph of the principles, for it happens that they are applied accidentally by the army against which it was intended to apply them, and are the cause of its success. But, in admitting this truth, it may be said that it is an argument against science; this objection is not well founded, for a general's science consists in providing for his side all the chances possible to be foreseen, and of course cannot extend to the caprices of destiny. Even if the number of battles gained by skillful maneuvers did not exceed the number due to accident, it would not invalidate my assertion.

If the skill of a general is one of the surest elements of victory, it will readily be seen that the judicious selection of generals is one of the most delicate points in the science of government and one of the most essential parts of the military policy of a state. Unfortunately, this choice is influenced by so many petty passions, that chance, rank, age, favor, party spirit, jealousy, will have as much to do with it as the public interest and justice. This subject is so important that we will devote to it a separate article.

Article XIII: Military Institutions

One of the most important points of the military policy of a state is the nature of its military institutions. A good army commanded by a general of ordinary capacity may accomplish great feats; a bad army with a good general may do equally well; but an army will certainly do a great deal more if its own superiority and that of the general be combined.

Twelve essential conditions concur in making a perfect army:—

1. To have a good recruiting-system;

2. A good organization;

8. A well-organized system of national reserves;

4. Good instruction of officers and men in drill and internal duties as well as those of a campaign;

5. A strict but not humiliating discipline, and a spirit of subordination and punctuality, based on conviction rather than on the formalities of the service;

6. A well-digested system of rewards, suitable to excite emulation;

7. The special arms of engineering and artillery to be well instructed;

8. An armament superior, if possible, to that of the enemy, both as to defensive and offensive arms;

9. A general staff capable of applying these elements, and having an organization calculated to advance the theoretical and practical education of its officers;

10. A good system for the commissariat, hospitals, and of general administration;

11. A good system of assignment to command, and of directing the principal operations of war;

12. Exciting and keeping alive the military spirit of the people.

To these conditions might be added a good system of clothing and equipment; for, if this be of less direct importance on the field of battle, it nevertheless has a bearing upon the preservation of the troops; and it is always a great object to economize the lives and health of veterans.

None of the above twelve conditions can be neglected without grave inconvenience. A fine army, well drilled and disciplined, but without national reserves, and unskillfully led, suffered Prussia to fall in fifteen days under the attacks of Napoleon. On the other hand, it has often been seen of how much advantage it is for a state to have a good army. It was the care and skill of Philip and Alexander in forming and instructing their phalanxes and rendering them easy to move, and capable of the most rapid maneuvers, which enabled the Macedonians to subjugate India and Persia with a handful of choice troops. It was the excessive love of his father for soldiers which procured for Frederick the Great an army capable of executing his great enterprises.

A government which neglects its army under any pretext whatever is thus culpable in the eyes of posterity, since it prepares humiliation for its standards and its country, instead of by a different course preparing for it success. We are far from saying that a government should sacrifice every thing to the army, for this would be absurd; but it ought to make the army the object of its constant care; and if the prince has not a military education it will be very difficult for him to fulfill his duty in this respect. In this case—which is, unfortunately, of too frequent occurrence—the defect must be supplied by wise institutions, at the head of which are to be placed a good system of the general staff, a good system of recruiting, and a good system of national reserves.

There are, indeed, forms of government which do not always allow the executive the power of adopting the best systems. If the armies of the Roman and French republics, and those of Louis XIV. and Frederick of Prussia, prove that a good military system and a skillful direction of operations may be found in governments the most opposite in principle, it cannot be doubted that, in the present state of the world, the form of government exercises a great influence in the development of the military strength of a nation and the value of its troops.

When the control of the public funds is in the hands of those affected by local interest or party spirit, they may be so over-scrupulous and penurious as to take all power to carry on the war from the executive, whom very many people seem to regard as a public enemy rather than as a chief devoted to all the national interests.

The abuse of badly-understood public liberties may also contribute to this deplorable result. Then it will be impossible for the most far-sighted administration to prepare in advance for a great war, whether it be demanded by the most important interests of the country at some future time, or whether it be immediate and necessary to resist sudden aggressions.

In the futile hope of rendering themselves popular, may not the members of an elective legislature, the majority of whom cannot be Richelieus, Pitts, or Louvois, in a misconceived spirit of economy, allow the institutions necessary for a large, well-appointed, and disciplined army to fall into decay? Deceived by the seductive fallacies of an exaggerated philanthropy, may they not end in convincing themselves and their constituents that the pleasures of peace are always preferable to the more statesmanlike preparations for war?

I am far from advising that states should always have the hand upon the sword and always be established on a war-footing: such a condition of things would be a scourge for the human race, and would not be possible, except under conditions not existing in all countries. I simply mean that civilized governments ought always to be ready to carry on a war in a short time,—that they should never be found unprepared. And the wisdom of their institutions may do as much in this work of preparation as foresight in their administration and the perfection of their system of military policy.

If, in ordinary times, under the rule of constitutional forms, governments subjected to all the changes of an elective legislature are less suitable than others for the creation or preparation of a formidable military power, nevertheless, in great crises these deliberative bodies have sometimes attained very different results, and have concurred in developing to the full extent the national strength. Still, the small number of such instances in history makes rather a list of exceptional cases, in which a tumultuous and violent assembly, placed under the necessity of conquering or perishing, has profited by the extraordinary enthusiasm of the nation to save the country and themselves at the same time by resorting to the most terrible measures and by calling to its aid an unlimited dictatorial power, which overthrew both liberty and law under the pretext of defending them. Here it is the dictatorship, or the absolute and monstrous usurpation of power, rather than the form of the deliberative assembly, which is the true cause of the display of energy. What happened in the Convention after the fall of Robespierre and the terrible Committee of Public Safety proves this, as well as the Chambers of 1815. Now, if the dictatorial power, placed in the hands of a few, has always been a plank of safety in great crises, it seems natural to draw the conclusion that countries controlled by elective assemblies must be politically and militarily weaker than pure monarchies, although in other respects they present decided advantages.

It is particularly necessary to watch over the preservation of armies in the interval of a long peace, for then they are most likely to degenerate. It is important to foster the military spirit in the armies, and to exercise them in great maneuvers, which, though but faintly resembling those of actual war, still are of decided advantage in preparing them for war. It is not less important to prevent them from becoming effeminate, which may be done by employing them in labors useful for the defense of the country.

The isolation in garrisons of troops by regiments is one of the worst possible systems, and the Russian and Prussian system of divisions and permanent corps d'armée seems to be much preferable. In general terms, the Russian army now may be presented as a model in many respects; and if in many points its customs would be useless and impracticable elsewhere, it must be admitted that many good institutions might well be copied from it.

As to rewards and promotion, it is essential to respect long service, and at the same time to open a way for merit. Three-fourths of the promotions in each grade should be made according to the roster, and the remaining fourth reserved for those distinguished for merit and zeal. On the contrary, in time of war the regular order of promotion should be suspended, or at least reduced to a third of the promotions, leaving the other two-thirds for brilliant conduct and marked services.

The superiority of armament may increase the chances of success in war: it does not, of itself, gain battles, but it is a great element of success. Every one can recall how nearly fatal to the French at Bylau and Marengo was their great inferiority in artillery. We may also refer to the great gain of the heavy French cavalry in the resumption of the cuirass, which they had for so long thrown aside. Every one knows the great advantage of the lance. Doubtless, as skirmishers lancers would not be more effectual than hussars, but when charging in line it is a very different affair. How many brave cavalry soldiers have been the victims of the prejudice they bore against the lance because it was a little more trouble to carry than a saber!

The armament of armies is still susceptible of great improvements; the state which shall take the lead in making them will secure great advantages. There is little left to be desired in artillery; but the offensive and defensive arms of infantry and cavalry deserve the attention of a provident government.

The new inventions of the last twenty years seem to threaten a great revolution in army organization, armament, and tactics. Strategy alone will remain unaltered, with its principles the same as under the Scipios and Cæsars, Frederick and Napoleon, since they are independent of the nature of the arms and the organization of the troops.

The means of destruction are approaching perfection with frightful rapidity. [It will be recollected that the author wrote this many years ago, since which time the inventive genius of the age has been attentively directed to the improvement of fire-arms. Artillery, which he regarded as almost perfect, has certainly undergone important improvements, and the improved efficiency of small arms is no less marked, while we hear nothing now of Perkins's steam-guns; and as yet no civilized army has been organized upon the plan the author suggests for depriving these destructive machines of their Efficiency.—Translators.] The Congreve rockets, the effect and direction of which it is said the Austrians can now regulate,—the shrapnel howitzers, which throw a stream of canister as far as the range of a bullet,—the Perkins steam-guns, which vomit forth as many balls as a battalion,—will multiply the chances of destruction, as though the hecatombs of Eylau, Borodino, Leipsic, and Waterloo were not sufficient to decimate the European races.

If governments do not combine in a congress to proscribe these inventions of destruction, there will be no course left but to make the half of an army consist of cavalry with cuirasses, in order to capture with great rapidity these machines; and the infantry, even, will be obliged to resume its armor of the Middle Ages, without which a battalion will be destroyed before engaging the enemy.

We may then see again the famous men-at-arms all covered with armor, and horses also will require the same protection.

While there is doubt about the realization of these fears, it is, however, certain that artillery and pyrotechny have made advances which should lead us to think of modifying the deep formation so much abused by Napoleon. We will recur to this in the chapter on Tactics.

We will here recapitulate, in a few words, the essential bases of the military policy which ought to be adopted by a wise government.

1. The prince should receive an education both political and military. He will more probably find men of administrative ability in his councils than good statesmen or soldiers; and hence he should be both of the latter himself.

2. If the prince in person does not lead his armies, it will be his first duty and his nearest interest to have his place well supplied. He must confide the glory of his reign and the safety of his states to the general most capable of directing his armies.

3. The permanent army should not only always be upon a respectable footing, but it should be capable of being doubled, if necessary, by reserves, which should always be prepared. Its instruction and discipline should be of a high character, as well as its organization; its armament should at least be as good as that of its neighbors, and superior if possible.

4. The matériel of war should also be upon the best footing, and abundant. The reserves should be stored in the depots and arsenals. National jealousy should not be allowed to prevent the adoption of all improvements in this matériel made in other countries.

5. It is necessary that the study of the military sciences should be encouraged and rewarded, as well as courage and zeal. The scientific military corps should be esteemed and honored: this is the only way of securing for the army men of merit and genius.

6. The general staff in times of peace should be employed in labors preparatory for all possible contingencies of war. Its archives should be furnished with numerous historical details of the past, and with all statistical, geographical, topographical, and strategic treatises and papers for the present and future. Hence it is essential that the chief of this corps, with a number of its officers, should be permanently stationed at the capital in time of peace, and the war-office should be simply that of the general staff, except that there should be a secret department for those documents to be concealed from the subalterns of the corps.

7. Nothing should be neglected to acquire a knowledge of the geography and the military statistics of other states, so as to know their material and moral capacity for attack and defense, as well as the strategic advantages of the two parties. Distinguished officers should be employed in these scientific labors, and should be rewarded when they acquit themselves with marked ability.

8. When a war is decided upon, it becomes necessary to prepare, not an entire plan of operations,—which is always impossible,—but a system of operations in reference to a prescribed aim; to provide a base, as well as all the material means necessary to guarantee the success of the enterprise.

9. The system of operations ought to be determined by the object of the war, the kind of forces of the enemy, the nature and resources of the country, the characters of the nations and of their chiefs, whether of the army or of the state. In fine, it should be based upon the moral and material means of attack or defense which the enemy may be able to bring into action; and it ought to take into consideration the probable alliances that may obtain in favor of or against either of the parties during the war.

10. The financial condition of a nation is to be weighed among the chances of a war. Still, it would be dangerous to constantly attribute to this condition the importance attached to it by Frederick the Great in the history of his times. He was probably right at his epoch, when armies were chiefly recruited by voluntary enlistment, when the last crown brought the last soldier; but when national levies are well organised money will no longer exercise the same influence,—at least for one or two campaigns. If England has proved that money will procure soldiers and auxiliaries, France has proved that love of country and honor are equally productive, and that, when necessary, war may be made to support war. France, indeed, in the fertility of her soil and the enthusiasm of her leaders, possessed sources of temporary power which cannot be adopted as a general base of a system; but the results of its efforts were none the less striking. Every year the numerous reports of the cabinet of London, and particularly of M. d'Yvernois, announced that France was about to break down for want of money, while Napoleon had 200,000,000

francs [There was a deficit in the finances of France at the fall of Napoleon. It was the result of his disasters, and of the stupendous efforts he was obliged to make. There was no deficit in 1811.] in the vaults of the Tuileries, all the while meeting the expenses of the government, including the pay of his armies.

A power might be overrunning with gold and still defend itself very badly. History, indeed, proves that the richest nation is neither the strongest nor the happiest. Iron weighs at least as much as gold in the scales of military strength. Still, we must admit that a happy combination of wise military institutions, of patriotism, of well-regulated finances, of internal wealth and public credit, imparts to a nation the greatest strength and makes it best capable of sustaining a long war.

A volume would be necessary to discuss all the circumstances under which a nation may develop more or less strength, either by its gold or iron, and to determine the cases when war may be expected to support war. This result can only be obtained by carrying the army into the territory of the enemy; and all countries are not equally capable of furnishing resources to an assailant.

We need not extend further the investigation of these subjects which are not directly connected with the art of war. It is sufficient for our purpose to indicate their relations to a projected war; and it will be for the statesman to develop the modifications which circumstances and localities may make in these relations.

Article XIV: The Command of Armies, and the Chief Control over Operations

Is it an advantage to a state to have its armies commanded in person by the monarch? Whatever may be the decision on this point, it is certain that if the prince possess the genius of Frederick, Peter the Great, or Napoleon, he will be far from leaving to his generals the honor of performing great actions which he might do himself; for in this he would be untrue to his own glory and to the well-being of the country.

As it is not our mission to discuss the question whether it is more fortunate for a nation to have a warlike or a peace-loving prince, (which is a philanthropic question, foreign to our subject,) we will only state upon this point that, with equal merit and chances in other respects, a sovereign will always have an advantage over a general who is himself not the head of a state. Leaving out of the question that he is responsible only to himself for his bold enterprises, he may do much by the certainty he has of being able to dispose of all the public resources for the attainment of his end. He also possesses the powerful accessory of his favor, of recompenses and punishments; all will be devoted to the execution of his orders, and to insure for his enterprises the greatest success; no jealousy will interfere with the execution of his projects, or at least its exhibition will be rare and in secondary operations. Here are, certainly, sufficient motives to induce a prince to lead his armies, if he possess military capacity and the contest be of a magnitude worthy of him. But if he possess no military ability, if his character be feeble, and he be easily influenced, his presence with the army, instead of producing good results, will open the way for all manner of intrigues. Each one will present his projects to him; and, as he will not have the experience necessary to estimate them according to their merits, he will submit his judgment to that of his intimates. His general, interfered with and opposed in all his enterprises, will be unable to achieve success, even if he have the requisite ability. It may be said that a sovereign might accompany the army and not interfere with his general, but, on the contrary, aid him with all the weight of his influence. In this case his presence might be productive of good results, but it also might lead to great embarrassment. If the army were turned and cut off from its communications, and obliged to extricate itself, sword in hand, what sad results might not follow from the presence of the sovereign at head-quarters!

When a prince feels the necessity of taking the field at the head of his armies, but lacks the necessary self-confidence to assume the supreme direction of affairs, the best course will be that adopted by the Prussian government with Blücher,—viz.; he should be accompanied by two generals of the best capacity, one of them a man of executive ability, the other a well-instructed staff officer. If this trinity be harmonious, it may yield excellent results, as in the case of the army of Silesia in 1813.

The same system might apply in the case where the sovereign judges it proper to intrust the command to a prince of his house, as has frequently happened since the time of Louis XIV. It has often occurred that the prince possessed only the titular command, and that an adviser, who in reality commanded, was imposed upon him. This was the case with the Duke of Orleans and Marsin at the famous battle of Turin, afterward with the Duke of Burgundy and Vendôme at the battle of Audenarde, and, I think, also at Ulm with the Archduke Ferdinand and Mack. This system is deplorable, since no one is responsible for what is done. It is known that at the battle of Turin the Duke of Orleans exhibited more sagacity than Marsin, and it became necessary for the latter to show full secret authority from the king before the prince would yield his judgment and allow the battle to be lost. So at Ulm the archduke displayed more skill and courage than Mack, who was to be his mentor.

If the prince possess the genius and experience of the Archduke Charles, he should be invested with the untrammeled command, and be allowed full selection of his instruments. If he have not yet acquired the same titles to command, he may then be provided with an educated general of the staff, and another general distinguished for his talent in execution; but in no case will it be wise to invest either of these counselors with more authority than a voice in consultation.

We have already said that if the prince do not conduct his armies in person, his most important duty will be to have the position of commander well filled,—which, unfortunately, is not always done. Without going back to ancient times, it will be sufficient to recall the more modern examples under Louis XIV. and Louis XV. The merit of Prince Eugene was estimated by his deformed figure, and this drove him (the ablest commander of his time) into the ranks of the enemy. After Louvois' death, Tallard, Marsin, and Villeroi filled the places of Turenne, Condé, and Luxembourg, and subsequently Soubise and Clermont succeeded Marshal Saxe. Between the fashionable selections made in the Saloons of the Pompadours and Dubarrys, and Napoleon's preference for mere soldiers, there are many gradations, and the margin is wide enough to afford the least intelligent government means of making rational nominations; but, in all ages, human weaknesses will exercise an influence in one way or another, and artifice will often carry off the prize from modest or timid merit, which awaits a call for its services. But, leaving out of consideration all these influences, it will be profitable to inquire in what respects this choice of a commander will be difficult, even when the executive shall be most anxious to make it a judicious one. In the first place, to make choice of a skillful general requires either that the person who makes the selection shall be a military man, able to form an intelligent opinion, or that he should be guided by the opinions of others, which opens the way to the improper influence of cliques. The embarrassment is certainly less when there is at hand a general already illustrious by many victories; but, outside of the fact that every general is not a great leader because he has gained a battle, (for instance, Jourdan, Scherer, and many others,) it is not always the case that a victorious general is at the disposition of the government. It may well happen that after a long period of peace, there may not be a single general in Europe who has commanded in chief. In this case, it will be difficult to decide whether one general is better than another. Those who have served long in peace will be at the head of their arms or corps, and will have the rank appropriate for this position; but will they always be the most capable of filling it? Moreover, the intercourse of the heads of a government with their subordinates is generally so rare and transient, that it is not astonishing they should experience difficulty in assigning men to their appropriate positions. The judgment of the prince, misled by appearances, may err, and, with the purest intentions, he may well be deceived in his selections.

One of the surest means of escaping this misfortune would seem to be in realizing the beautiful fiction of Fénélon in Telemachus, by finding a faithful, sincere, and generous Philocles, who, standing between the prince and all aspirants for the command, would be able, by means of his more direct relations to the public, to enlighten the monarch in reference to selections of individuals best recommended by their character and abilities. But will this faithful friend never yield to personal affections? Will he be always free from prejudice? Suwaroff was rejected by Potemkin on account of his appearance, and it required all the art of Catherine to secure a regiment for the man who afterward shed so much luster upon the Russian arms.

It has been thought that public opinion is the best guide; but nothing could be more dangerous. It voted Dumouriez to be a Cæsar, when he was ignorant of the great operations of war. Would it have placed Bonaparte at the head of the army of Italy, when he was known only by two directors? Still, it must be admitted that, if not infallible, public sentiment is not to be despised, particularly if it survive great crises and the experience of events.

The most essential qualities for a general will always be as follow:—First, *A high moral courage, capable of great resolutions*; Secondly, *A physical courage which takes no account of danger*. His scientific or military acquirements are secondary to the above-mentioned characteristics, though if great they will be valuable auxiliaries. It is not necessary that he should be a man of vast erudition. His knowledge may be limited, but it should be thorough, and he should be perfectly grounded in the principles at the base of the art of war. Next in importance come the qualities of his personal character. A man who is gallant, just, firm, upright, capable of esteeming merit in others instead of being jealous of it, and skillful in making this merit conduce to his own glory, will always be a good general, and may even pass for a great man. Unfortunately, the disposition to do justice to merit in others is not the most common quality: mediocre minds are always jealous, and inclined to surround themselves with persons of little ability, fearing the reputation of being led, and not realizing that the nominal commander of an army always receives almost all the glory of its success, even when least entitled to it.

The question has often been discussed, whether it is preferable to assign to the command a general of long experience in service with troops, or an officer of the staff, having generally but little experience in the management of troops. It is beyond question that war is a distinct science of itself, and that it is quite possible to be able to combine operations skillfully without ever having led a regiment against an enemy. Peter the Great, Condé, Frederick, and Napoleon are instances of it. It cannot, then, be denied that an officer from the staff may as well as any other prove to be a great general, but it will not be because he has grown gray in the duties of a quartermaster that he will be capable of the supreme command, but because he has a natural genius for war and possesses the requisite characteristics. So, also, a general from the ranks of the infantry or cavalry may be as capable of conducting a campaign as the most profound tactician. So this question does not admit of a definite answer either in the affirmative or negative, since almost all will depend upon the personal qualities of the individuals; but the following remarks will be useful in leading to a rational conclusion:—

1. A general, selected from the general staff, engineers, or artillery, who has commanded a division or a corps d'armée, will, with equal chances, be superior to one who is familiar with the service of but one arm or special corps.

2. A general from the line, who has made a study of the science of war, will be equally fitted for the command.

3. That the character of the man is above all other requisites in a commander-in-chief.

Finally, He will be a good general in whom are found united the requisite personal characteristics and a thorough knowledge of the principles of the art of war.

The difficulty of always selecting a good general has led to the formation of a good general staff, which being near the general may advise him, and thus exercise a beneficial influence over the operations. A well-instructed general staff is one of the most useful of organizations; but care must be observed to prevent the introduction into it of false principles, as in this case it might prove fatal.

Frederick, when he established the military school of Potsdam, never thought it would lead to the "right shoulder forward" of General Ruchel, [General Ruchel thought at the battle of Jena that he could save the army by giving the command to advance the right shoulder in order to form an oblique line.] and to the teaching that the oblique order is the infallible rule for gaining all battles. How true it is that there is but a step from the sublime to the ridiculous!

Moreover, there ought to exist perfect harmony between the general and his chief of staff; and, if it be true that the latter should be a man of recognized ability, it is also proper to give the general the choice of the men who are to be his advisers. To impose a chief of staff upon a general would be to create anarchy and want of harmony; while to permit him to select a cipher for that position would be still more dangerous; for if he be himself a man of little ability, indebted to favor or fortune for his station, the selection will be of vital importance. The best means to avoid these dangers is to give the general the option of several designated officers, all of undoubted ability.

It has been thought, in succession, in almost all armies, that frequent councils of war, by aiding the commander with their advice, give more weight and effect to the direction of military operations. Doubtless, if the commander were a Soubise, a Clermont, or a Mack, he might well find in a council of war opinions more valuable than his own; the majority of the opinions given might be preferable to his; but what success could be expected from operations conducted by others than those who have originated and arranged them? What must be the result of an operation which is but partially understood by the commander, since it is not his own conception?

I have undergone a pitiable experience as prompter at head-quarters, and no one has a better appreciation of the value of such services than myself; and it is particularly in a council of war that such a part is absurd. The greater the number and the higher the rank of the military officers who compose the council, the more difficult will it be to accomplish the triumph of truth and reason, however small be the amount of dissent.

What would have been the action of a council of war to which Napoleon proposed the movement of Arcola, the crossing of the Saint-Bernard, the maneuver at Ulm, or that at Gera and Jena? The timid would have regarded them as rash, even to madness, others would have seen a thousand difficulties of execution, and all would have concurred in rejecting them; and if, on the contrary, they had been adopted, and had been executed by any one but Napoleon, would they not certainly have proved failures?

In my opinion, councils of war are a deplorable resource, and can be useful only when concurring in opinion with the commander, in which case they may give him more confidence in his own judgment, and, in addition, may assure him that his lieutenants, being of his opinion, will use every means to insure the success of the movement. This is the only advantage of a council of war, which, moreover, should be simply consultative and have no further authority; but if, instead of this harmony, there should be difference of opinion, it can only produce unfortunate results.

Accordingly, I think it safe to conclude that the best means of organizing the command of an army, in default of a general approved by experience, is—

1st. To give the command to a man of tried bravery, bold in the fight, and of unshaken firmness in danger.

2d. To assign, as his chief of staff, a man of high ability, of open and faithful character, between whom and the commander there may be perfect harmony. The victor will gain so much glory that he can spare some to the friend who has contributed to his success. In this way Blücher, aided by Gneisenau and Muffling, gained glory which probably he would not have been able to do of himself. It is true that this double command is more objectionable than an undivided one when a state has a Napoleon, a Frederick, or a Suwaroff to fill it; but when there is no great general to lead the armies it is certainly the preferable system.

Before leaving this important branch of the subject, another means of influencing military operations—viz.: that of a council of war at the seat of government—deserves notice. Louvois for a long time directed from Paris the armies of Louis XIV., and with success. Carnot, also, from Paris directed the armies of the Republic: in 1793 he did well, and saved France; in 1794 his action was at first very unfortunate, but he repaired his faults afterward by chance; in 1796 he was completely at fault. It is to be observed, however, that both Louvois and Carnot individually controlled the armies, and that there was no council of war. The Aulic council, sitting in Vienna, was often intrusted with the duty of directing the operations of the armies; and there has never been but one opinion in Europe as to its fatal influence. Whether this opinion is right or wrong, the Austrian generals alone are able to decide. My own opinion is that the functions of such a body in this connection should be limited to the adoption of a general plan of operations. By this I do not mean a plan which should trace out the campaign in detail, restricting the generals and compelling them to give battle without regard to circumstances, but a plan which should determine the object of the campaign, the nature of the operations, whether offensive or defensive, the material means to be applied to these first enterprises, afterward for the reserves, and finally for the levies which may be necessary if the country be invaded. These points, it is true, should be discussed in a council of both generals and ministers, and to these points should the control of the council be limited; for if it should not only order the general in command to march to Vienna or to Paris, but should also have the presumption to indicate the manner in which he should maneuver to attain this object, the unfortunate general would certainly be beaten, and the whole responsibility of his reverses should fall upon the shoulders of those who, hundreds of miles distant, took upon themselves the duty of directing the army,—a duty so difficult for any one, even upon the scene of operations.

Article XV: The Military Spirit of Nations, and the Morale of Armies

The adoption of the best regulations for the organization of an army would be in vain if the government did not at the same time cultivate a military spirit in its citizens. It may well be the case in London, situated on an island and protected from invasion by its immense fleets, that the title of a rich banker should be preferred to a military decoration; but a continental nation imbued with the sentiments and habits of the tradesmen of London or the bankers of Paris would sooner or later fall a prey to its neighbors. It was to the union of the civic virtues and military spirit fostered by their institutions that the Romans were indebted for their grandeur; and when they lost these virtues, and when, no longer regarding the military service as an honor as well as a duty, they relinquished it to mercenary Goths and Gauls, the fall of the empire became inevitable. It is doubtless true that whatever increases the prosperity of the country should be neither neglected nor despised; it is also necessary to honor the branches of industry which are the first instruments of this prosperity; but they should always be secondary to the great institutions which make up the strength of states in encouraging the cultivation of the manly and heroic virtues. Policy and justice both agree on this point; for, whatever Boileau may say, it is certainly more glorious to confront death in the footsteps of the Cæsars than to fatten upon the public miseries by gambling on the vicissitudes of the national credit. Misfortune will certainly fall upon the land where the wealth of the tax-gatherer or the greedy gambler in stocks stands, in public estimation, above the uniform of the brave man who sacrifices his life, health, or fortune to the defense of his country.

The first means of encouraging the military spirit is to invest the army with all possible social and public consideration. The second means is to give the preference to those who have rendered services to the state, in filling any vacancies in the administrative departments of the government, or even to require a certain length of military service as a qualification for certain offices. A comparison of the ancient military institutions of Rome with those of Russia and Prussia, is a subject worthy of serious attention; and it would also be interesting to contrast them with the doctrines of modern theorists, who declare against the employment of officers of the army in other public functions, and who wish for none but rhetoricians in the important offices of administration. [For instance, in France, instead of excluding all officers from the privilege of the elective franchise, it should be given to all colonels; and the generals should be eligible to the legislature. The most venal deputies will not be those from military life.] It is true that many public employments demand a special course of study; but cannot the soldier, in the abundant leisure of peace, prepare himself for the career he would prefer after having fulfilled his debt to his country in the profession of arms? If these administrative offices were conferred upon officers retired from the army in a grade not lower than that of captain, would it not be a stimulant for officers to attain that rank, and would it not lead them, when in garrisons, to find their recreations elsewhere than in the theaters and public clubs?

It may be possible that this facility of transfer from the military to the civil service would be rather injurious than favorable to a high military spirit, and that to encourage this spirit it would be expedient to place the profession of the soldier above all others. This was the early practice of the Mamelukes and Janissaries. Their soldiers were bought at the age of about seven years, and were educated in the idea that they were to die by their standards. Even the English—so jealous of their rights—contract, in enlisting as soldiers, the obligation for the whole length of their lives, and the Russian, in enlisting for twenty-five years, does what is almost equivalent. In such armies, and in those recruited by voluntary enlistments, perhaps it would not be advisable to tolerate this fusion of military and civil offices; but where the military service is a temporary duty imposed upon the people, the case is different, and the old Roman laws which required a previous military service of ten years in any aspirant for the public employments, seem to be best calculated to preserve the military spirit,—particularly in this age, when the attainment of material comfort and prosperity appears to be the dominant passion of the people.

However this may be, still, in my opinion, under all forms of government, it will be a wise part to honor the military profession, in order to encourage the love of glory and all the warlike virtues, under the penalty of receiving the reproaches of posterity and suffering insult and dependency.

It is not sufficient to foster the military spirit among the people, but, more than that, it is necessary to encourage it in the army. Of what avail would it be if the uniform be honored in the land and it be regarded as a duty to serve in the army, while the military virtues are wanting? The forces would be numerous but without valor.

The enthusiasm of an army and its military spirit are two quite different things, and should not be confounded, although they produce the same effects. The first is the effect of passions more or less of a temporary character,—of a political or religious nature, for instance, or of a great love of country; while the latter, depending upon the skill of the commander and resulting from military institutions, is more permanent and depends less upon circumstances, and should be the object of the attention of every far-seeing government. [It is particularly important that this spirit should pervade the officers and non-commissioned officers: if they be capable, and the nation brave, there need be no fear for the men.] Courage should be recompensed and honored, the different grades in rank respected, and discipline should exist in the sentiments and convictions rather than in external forms only.

The officers should feel the conviction that resignation, bravery, and faithful attention to duty are virtues without which no glory is possible, no army is respectable, and that firmness amid reverses is more honorable than enthusiasm in success,—since courage alone is necessary to storm a position, while it requires heroism to make a difficult retreat before a victorious and enterprising enemy, always opposing to him a firm and unbroken front. A fine retreat should meet with a reward equal to that given for a great victory.

By inuring armies to labor and fatigue, by keeping them from stagnation in garrison in times of peace, by inculcating their superiority over their enemies, without depreciating too much the latter, by inspiring a love for great exploits,—in a word, by exciting their enthusiasm by every means in harmony with their tone of mind, by honoring courage, punishing weakness, and disgracing cowardice,—we may expect to maintain a high military spirit.

Effeminacy was the chief cause of the ruin of the Roman legions: those formidable soldiers, who had borne the casque, buckler, and cuirass in the times of the Scipios under the burning sun of Africa, found them too heavy in the cool climates of Germany and Gaul; and then the empire was lost.

I have remarked that it is not well to create a too great contempt for the enemy, lest the *morale* of the soldier should be shaken if he encounter an obstinate resistance. Napoleon at Jena, addressing Lannes' troops, praised the Prussian cavalry, but promised that they would contend in vain against the bayonets of his Egyptians.

The officers and troops must be warned against those sudden panics which often seize the bravest armies when they are not well controlled by discipline, and hence when they do not recognize that in order is the surest hope of safety. It was not from want of courage that one hundred thousand Turks were beaten at Peterwardein by Prince Eugene, and at Kagoul by Romanzoff: it was because, once repulsed in their disorderly charges, every one yielded to his personal feelings, and because they fought individually, but not in masses and in order. An army seized with panic is similarly in a state of demoralization; because when disorder is once introduced all concerted action on the part of individuals becomes impossible, the voice of the officers can no longer be heard, no maneuver for resuming the battle can be executed, and there is no resource but in ignominious flight.

Nations with powerful imaginations are particularly liable to panics; and nothing short of strong institutions and skillful leaders can remedy it. Even the French, whose military virtues when well led have never been questioned, have often performed some quick movements of this kind which were highly ridiculous. We may refer to the unbecoming panic which pervaded the infantry of Marshal Villars after having gained the battle of Friedlingen, in 1704. The same occurred to Napoleon's infantry after the victory of Wagram and when the enemy was in full retreat. A still more extraordinary case was the flight of the 97th semi-brigade, fifteen hundred strong, at the siege of Genoa, before a platoon of cavalry. Two days afterward these same men took

Fort Diamond by one of the most vigorous assaults mentioned in modern history.

Still, it would seem to be easy to convince brave men that death comes more quickly and more surely to those who fly in disorder than to those who remain together and present a firm front to the enemy, or who rally promptly when their lines have been for the instant broken.

In this respect the Russian army may be taken as a model by all others. The firmness which it has displayed in all retreats is due in equal degrees to the national character, the natural instincts of the soldiers, and the excellent disciplinary institutions. Indeed, vivacity of imagination is not always the cause of the introduction of disorder: the want of the habit of order often causes it, and the lack of precautions on the part of the generals to maintain this order contributes to it. I have often been astonished at the indifference of most generals on this point. Not only did they not deign to take the slightest precaution to give the proper direction to small detachments or scattered men, and fail to adopt any signals to facilitate the rallying in each division of the fractions which may be scattered in a momentary panic or in an irresistible charge of the enemy, but they were offended that any one should think of proposing such precautions. Still, the most undoubted courage and the most severe discipline will often be powerless to remedy a great disorder, which might be in a great degree obviated by the use of rallying-signals for the different divisions. There are, it is true, cases where all human resources are insufficient for the maintenance of order, as when the physical sufferings of the soldiers have been so great as to render them deaf to all appeals, and when their officers find it impossible to do any thing to organize them,—which was the case in the retreat of 1812. Leaving out these exceptional cases, good habits of order, good logistical precautions for rallying, and good discipline will most frequently be successful, if not in preventing disorder, at least in promptly remedying it.

It is now time to leave this branch, of which I have only desired to trace an outline, and to proceed to the examination of subjects which are purely military.

Definition of Strategy and the Fundamental Principle of War

The art of war, independently of its political and moral relations, consists of five principal parts, viz.: Strategy, Grand Tactics, Logistics, Tactics of the different arms, and the Art of the Engineer. We will treat of the first three branches, and begin by defining them. In order to do this, we will follow the order of procedure of a general when war is first declared, who commences with the points of the highest importance, as a plan of campaign, and afterward descends to the necessary details. Tactics, on the contrary, begins with details, and ascends to combinations and generalization necessary for the formation and handling of a great army.

We will suppose an army taking the field: the first care of its commander should be to agree with the head of the state upon the character of the war: then he must carefully study the theater of war, and select the most suitable base of operations, taking into consideration the frontiers of the state and those of its allies.

The selection of this base and the proposed aim will determine the zone of operations. The general will take a first objective point: he will select the line of operations leading to this point, either as a temporary or permanent line, giving it the most advantageous direction; namely, that which promises the greatest number of favorable opportunities with the least danger. An army marching on this line of operations will have a front of operations and a strategic front. The temporary positions which the corps d'armée will occupy upon this front of operations, or upon the line of defense, will be strategic positions.

When near its first objective point, and when it begins to meet resistance, the army will either attack the enemy or maneuver to compel him to retreat; and for this end it will adopt one or two strategic lines of maneuvers, which, being temporary, may deviate to a certain degree from the general line of operations, with which they must not be confounded.

To connect the strategic front with the base as the advance is made, lines of supply, depots, &c. will be established.

If the line of operations be long, and there be hostile troops in annoying proximity to it, these bodies may either be attacked and dispersed or be merely observed, or the operations against the enemy may be carried on without reference to them. If the second of these courses be pursued, a double strategic front and large detachments will be the result.

The army being almost within reach of the first objective point, if the enemy oppose him there will be a battle; if indecisive, the fight will be resumed; if the army gains the victory, it will secure its objective point or will advance to attain a second. Should the first objective point be the possession of an important fort, the siege will be commenced. If the army be not strong enough to continue its march, after detaching a sufficient force to maintain the siege, it will take a strategic position to cover it, as did the army of Italy in 1796, which, less than fifty thousand strong, could not pass Mantua to enter Austria, leaving twenty-five thousand enemies within its walls, and having forty thousand more in front on the double line of the Tyrol and Frioul.

If the army be strong enough to make the best use of its victory, or if it have no siege to make, it will operate toward a second and more important objective point.

If this point be distant, it will be necessary to establish an intermediate point of support. One or more secure cities already occupied will form an eventual base: when this cannot be done, a small strategic reserve may be established, which will protect the rear and also the depots by temporary fortifications. When the army crosses large streams, it will construct *têtes de pont*; and, if the bridges are within walled cities, earth-works will be thrown up to increase the means of defense and to secure the safety of the eventual base or the strategic reserve which may occupy these posts.

Should the battle be lost, the army will retreat toward its base, in order to be reinforced therefrom by detachments of troops, or, what is equivalent, to strengthen itself by the occupation of fortified posts and camps, thus compelling the enemy to halt or to divide his forces.

When winter approaches, the armies will either go into quarters, or the field will be kept by the army which has obtained decisive success and is desirous of profiting to the utmost by its superiority. These winter campaigns are very trying to both armies, but in other respects do not differ from ordinary campaigns, unless it be in demanding increased activity and energy to attain prompt success.

Such is the ordinary course of a war, and as such we will consider it, while discussing combinations which result from these operations.

Strategy embraces the following points, viz.:—

1. The selection of the theater of war, and the discussion of the different combinations of which it admits.

2. The determination of the decisive points in these combinations, and the most favorable direction for operations.

3. The selection and establishment of the fixed base and of the zone of operations.

4. The selection of the objective point, whether offensive or defensive.

5. The strategic fronts, lines of defense, and fronts of operations.

6. The choice of lines of operations leading to the objective point or strategic front.

7. For a given operation, the best strategic line, and the different maneuvers necessary to embrace all possible cases.

8. The eventual bases of operations and the strategic reserves.

9. The marches of armies, considered as maneuvers.

10. The relation between the position of depots and the marches of the army.

11. Fortresses regarded as strategical means, as a refuge for an army, as an obstacle to its progress: the sieges to be made and to be covered.

12. Points for intrenched camps, *tétes de pont*, etc..

13. The diversions to be made, and the large detachments necessary.

These points are principally of importance in the determination of the first steps of a campaign; but there are other operations of a mixed nature, such as passages of streams, retreats, surprises, disembarkations, convoys, winter quarters, the execution of which belongs to tactics, the conception and arrangement to strategy.

The maneuvering of an army upon the battle-field, and the different formations of troops for attack, constitute Grand Tactics. Logistics is the art of moving armies. It comprises the order and details of marches and camps, and of quartering and supplying troops; in a word, it is the execution of strategical and tactical enterprises.

To repeat. Strategy is the art of making war upon the map, and comprehends the whole theater of operations. Grand Tactics is the art of posting troops upon the battle-field according to the accidents of the ground, of bringing them into action, and the art of fighting upon the ground, in contradistinction to planning upon a map. Its operations may extend over a field of ten or twelve miles in extent. Logistics comprises the means and arrangements which work out the plans of strategy and tactics. Strategy decides where to act; logistics brings the troops to this point; grand tactics decides the manner of execution and the employment of the troops.

It is true that many battles have been decided by strategic movements, and have been, indeed, but a succession of them; but this only occurs in the exceptional case of a dispersed army: for the general case of pitched battles the above definition holds good.

Grand Tactics, in addition to acts of local execution, relates to the following objects:—

1. The choice of positions and defensive lines of battle.

2. The offensive in a defensive battle.

3. The different orders of battle, or the grand maneuvers proper for the attack of the enemy's line.

4. The collision of two armies on the march, or unexpected battles.

5. Surprises of armies in the open field.

6. The arrangements for leading troops into battle.

7. The attack of positions and intrenched camps.

8. *Coups de main.*

All other operations, such as relate to convoys, foraging-parties, skirmishes of advanced or rear guards, the attack of small posts, and any thing accomplished by a detachment or single division, may be regarded as details of war, and not included in the great operations.

The Fundamental Principle of War

It is proposed to show that there is one great principle underlying all the operations of war,—a principle which must be followed in all good combinations. It is embraced in the following maxims:—

1. To throw by strategic movements the mass of an army, successively, upon the decisive points of a theater of war, and also upon the communications of the enemy as much as possible without compromising one's own.

2. To maneuver to engage fractions of the hostile army with the bulk of one's forces.

3. On the battle-field, to throw the mass of the forces upon the decisive point, or upon that portion of the hostile line which it is of the first importance to overthrow.

4. To so arrange that these masses shall not only be thrown upon the decisive point, but that they shall engage at the proper times and with energy.

This principle has too much simplicity to escape criticism: one objection is that it is easy to recommend throwing the mass of the forces upon the decisive points, but that the difficulty lies in recognizing those points.

This truth is evident; and it would be little short of the ridiculous to enunciate such a general principle without accompanying it with all necessary explanations for its application upon the field. In Article XIX. these decisive points will be described, and in Articles from XVIII. to XXII. will be discussed their relations to the different combinations. Those students who, having attentively considered what is there stated, still regard the determination of these points as a problem without a solution, may well despair of ever comprehending strategy.

The general theater of operations seldom contains more than three zones,—the right, the left, and the center; and each zone, front of operations, strategic position, and line of defense, as well as each line of battle, has the same subdivisions,—two extremities and the center. A direction upon one of these three will always be suitable for the attainment of the desired end. A direction upon one of the two remaining will be less advantageous; while the third direction will be wholly inapplicable. In considering the object proposed in connection with the positions of the enemy and the geography of the country, it will appear that in every strategic movement or tactical maneuver the question for decision will always be, whether to maneuver to the right, to the left, or directly in front. The selection of one of these three simple alternatives cannot, surely, be considered an enigma. The art of giving the proper direction to the masses is certainly the basis of strategy, although it is not the whole of the art of war. Executive talent, skill, energy, and a quick apprehension of events are necessary to carry out any combinations previously arranged.

We will apply this great principle to the different cases of strategy and tactics, and then show, by the history of twenty celebrated campaigns, that, with few exceptions, the most brilliant successes and the greatest reverses resulted from an adherence to this principle in the one case, and from a neglect of it in the other.

Of Strategic Combinations

Article XVI: Of the System of Operations

War once determined upon, the first point to be decided is, whether it shall be offensive or defensive; and we will first explain what is meant by these terms. There are several phases of the offensive: if against a great state, the whole or a large portion of whose territory is attacked, it is an *invasion*; if a province only, or a line of defense of moderate extent, be assailed, it is the ordinary offensive; finally, if the offensive is but an attack upon the enemy's position, and is confined to a single operation, it is called the taking the *initiative*. In a moral and political view, the offensive is nearly always advantageous: it carries the war upon foreign soil, saves the assailant's country from devastation, increases his resources and diminishes those of his enemy, elevates the *morale* of his army, and generally depresses the adversary.

It sometimes happens that invasion excites the ardor and energy of the adversary,—particularly when he feels that the independence of his country is threatened.

In a military point of view, the offensive has its good and its bad side. Strategically, an invasion leads to deep lines of operations, which are always dangerous in a hostile country. All the obstacles in the enemy's country, the mountains, rivers, defiles, and forts, are favorable for defense, while the inhabitants and authorities of the country, so far from being the instruments of the invading army, are generally hostile. However, if success be obtained, the enemy is struck in a vital point: he is deprived of his resources and compelled to seek a speedy termination of the contest.

For a single operation, which we have called the taking the *initiative*, the offensive is almost always advantageous, particularly in strategy. Indeed, if the art of war consists in throwing the masses upon the decisive points, to do this it will be necessary to take the initiative. The attacking party knows what he is doing and what he desires to do; he leads his masses to the point where he desires to strike. He who awaits the attack is everywhere anticipated: the enemy fall with large force upon fractions of his force: he neither knows where his adversary proposes to attack him nor in what manner to repel him.

Tactically, the offensive also possesses advantages, but they are less positive, since, the operations being upon a limited field, the party taking the initiative cannot conceal them from the enemy, who may detect his designs and by the aid of good reserves cause them to fail.

The attacking party labors under the disadvantages arising from the obstacles to be crossed before reaching the enemy's line; on which account the advantages and disadvantages of the tactical offensive are about equally balanced.

Whatever advantages may be expected either politically or strategically from the offensive, it may not be possible to maintain it exclusively throughout the war; for a campaign offensive in the beginning may become defensive before it ends.

A defensive war is not without its advantages, when wisely conducted. It may be passive or active, taking the offensive at times. The passive defense is always pernicious; the active may accomplish great successes. The object of a defensive war being to protect, as long as possible, the country threatened by the enemy, all operations should be designed to retard his progress, to annoy him in his enterprises by multiplying obstacles and difficulties, without, however, compromising one's own army. He who invades does so by reason of some superiority; he will then seek to make the issue as promptly as possible: the defense, on the contrary, desires delay till his adversary is weakened by sending off detachments, by marches, and by the privations and fatigues incident to his progress.

An army is reduced to the defensive only by reverses or by a positive inferiority. It then seeks in the support of forts, and in natural or artificial barriers, the means of restoring equality by multiplying obstacles in the way of the enemy. This plan, when not carried to an extreme, promises many chances of success, but only when the general has the good sense not to make the defense passive: he must not remain in his positions to receive whatever blows may be given by his adversary; he must, on the contrary, redouble his activity, and be constantly upon the alert to improve all opportunities of assailing the weak points of the enemy. This plan of war may be called the defensive-offensive, and may have strategical as well as tactical advantages.. It combines the advantages of both systems; for one who awaits his adversary upon a prepared field, with all his own resources in hand, surrounded by all the advantages of being on his own ground, can with hope of success take the initiative, and is fully able to judge when and where to strike.

During the first three campaigns of the Seven Years' War Frederick was the assailant; in the remaining four his conduct was a perfect model of the defensive-offensive. He was, however, wonderfully aided in this by his adversaries, who allowed him all the time he desired, and many opportunities of taking the offensive with success. Wellington's course was mainly the same in Portugal, Spain, and Belgium, and it was the most suitable in his circumstances. It seems plain that one of the greatest talents of a general is to know how to use (it may be alternately) these two systems, and particularly to be able to take the initiative during the progress of a defensive war.

Article XVII: Of the Theater of Operations

The theater of a war comprises all the territory upon which the parties may assail each other, whether it belong to themselves, their allies, or to weaker states who may be drawn into the war through fear or interest. When the war is also maritime, the theater may embrace both hemispheres,—as has happened in contests between France and England since the time of Louis XIV. The theater of a war may thus be undefined, and must, not be confounded with the theater of operations of one or the other army. The theater of a continental war between France and Austria may be confined to Italy, or may, in addition, comprise Germany if the German States take part therein.

Armies may act in concert or separately: in the first case the whole theater of operations may be considered as a single field upon which strategy directs the armies for the attainment of a definite end. In the second case each army will have its own independent theater of operations. The *theater of operations* of an army embraces all the territory it may desire to invade and all that it may be necessary to defend. If the army operates independently, it should not attempt any maneuver beyond its own theater, (though it should leave it if it be in danger of being surrounded,) since the supposition is that no concert of action has been arranged with the armies operating on the other fields. If, on the contrary, there be concert of action, the theater of operations of each army taken singly is but a zone of operations of the general field, occupied by the masses for the attainment of a common object.

Independently of its topographical features, each theater upon which one or more armies operate is composed, for both parties, as follows:—

1. Of a fixed base of operations.

2. Of a principal objective point.

3. Of fronts of operations, strategic fronts, and lines of defense.

4. Of zones and lines of operations.

5. Of temporary strategic lines and lines of communications.

6. Of natural or artificial obstacles to be overcome or to oppose to the enemy.

7. Of geographical strategic points, whose occupation is important, either for the offensive or defensive.

8. Of accidental intermediate bases of operations between the objective point and the primary base.

9. Of points of refuge in case of reverse.

For illustration, let us suppose the case of France invading Austria with two or three armies, to be concentrated under one commander, and starting from Mayence, from the Upper Rhine, from Savoy or the Maritime Alps, respectively. The section of country which each of these armies traverses may be considered as a zone of the general field of operations. But if the army of Italy goes but to the Adige without concerted action with the army of the Rhine, then what was before but a zone becomes for that army a theater of operations.

In every case, each theater must have its own base, its own objective point, its zones and lines of operations connecting the objective point with the base, either in the offensive or the defensive.

It has been taught and published that rivers are lines of operations *par excellence*. Now, as such a line must possess two or three roads to move the army within the range of its operations, and at least one line of retreat, rivers have been called lines of retreat, and even lines of maneuver. It would be much more accurate to say that rivers are excellent lines of supply, and powerful auxiliaries in the establishment of a good line of operations, but never the line itself.

It has also been maintained that, could one create a country expressly to be a good theater of war, converging roads would be avoided, because they facilitate invasion. Every country has its capital, its rich cities for manufactures or trade; and, in the very nature of things, these points must be the centers of converging routes. Could Germany be made a desert, to be molded into a theater of war at the pleasure of an individual, commercial cities and centers of trade would spring up, and the roads would again necessarily converge to these points. Moreover, was not the Archduke Charles enabled to beat Jourdan in 1796 by the use of converging routes? Besides, these routes are more favorable for defense than attack, since two divisions retreating upon these radial lines can effect a junction more quickly than two armies which are pursuing, and they may thus united defeat each of the pursuing masses separately.

Some authors have affirmed that mountainous countries abound in strategic positions; others have maintained that, on the contrary, these points are more rare among the Alps than in the plains, but also that if more rare they are more important and more decisive.

Some authors have represented that high ranges of mountains are, in war, inaccessible barriers. Napoleon, on the contrary, in speaking of the Rhetian Alps, said that "an army could pass wherever a man could put his foot."

Generals no less experienced than himself in mountain-warfare have united with him in this opinion, in admitting the great difficulty of carrying on a defensive war in such localities unless the advantages of partisan and regular warfare can be combined, the first to guard the heights and to harass the enemy, the second to give battle at the decisive points,—the junctions of the large valleys.

These differences of opinion are here noticed merely to show the reader that, so far from the art having reached perfection, there are many points that admit of discussion.

The most important topographical or artificial features which make up the theater of a war will, in succeeding portions of this chapter, be examined as to their strategic value; but here it may be proper to remark that this value will depend much upon the spirit and skill of the general. The great leader who crossed the Saint-Bernard and ordered the passage of the Splugen was far from believing in the impregnability of these chains; but he was also far from thinking that a muddy rivulet and a walled inclosure could change his destiny at Waterloo.

Article XVIII: Bases of Operations

A base of operations is the portion of country from which the army obtains its reinforcements and resources, from which it starts when it takes the offensive, to which it retreats when necessary, and by which it is supported when it takes position to cover the country defensively.

The base of operations is most generally that of supply,—though not necessarily so, at least as far as food is concerned; as, for instance, a French army upon the Elbe might be subsisted from Westphalia or Franconia, but its real base would certainly be upon the Rhine.

When a frontier possesses good natural or artificial barriers, it may be alternately either an excellent base for offensive operations, or a line of defense when the state is invaded. In the latter case it will always be prudent to have a second base in rear; for, although an army in its own country will everywhere find a point of support, there is still a vast difference between those parts of the country without military positions and means, as forts, arsenals, and fortified depots, and those other portions where these military resources are found; and these latter alone can be considered as safe bases of operations. An army may have in succession a number of bases: for instance, a French army in Germany will have the Rhine for its first base; it may have others beyond this, wherever it has allies or permanent lines of defense; but if it is driven back across the Rhine it will have for a base either the Meuse or the Moselle: it might have a third upon the Seine, and a fourth upon the Loire.

These successive bases may not be entirely or nearly parallel to the first. On the contrary, a total change of direction may become necessary. A French army repulsed beyond the Rhine might find a good base on Béfort or Besançon, on Mézières or Sedan, as the Russian army after the evacuation of Moscow left the base on the north and east and established itself upon the line of the Oka and the southern provinces. These lateral bases perpendicular to the front of defense are often decisive in preventing the enemy from penetrating to the heart of the country, or at least in rendering it impossible for him to maintain himself there. A base upon a broad and rapid river, both banks being held by strong works, would be as favorable as could be desired.

The more extended the base, the more difficulty will there be in covering it; but it will also be more difficult to cut the army off from it. A state whose capital is too near the frontier cannot have so favorable a base in a defensive war as one whose capital is more retired.

A base, to be perfect, should have two or three fortified points of sufficient capacity for the establishment of depots of supply. There should be a *tête de pont* upon each of its unfordable streams.

All are now agreed upon these principles; but upon other points opinions have varied. Some have asserted that a perfect base is one parallel to that of the enemy. My opinion is that bases perpendicular to those of the enemy are more advantageous, particularly such as have two sides almost perpendicular to each other and forming a re-entrant angle, thus affording a double base if required, and which, by giving the control of two sides of the strategic field, assure two lines of retreat widely apart, and facilitate any change of the line of operations which an unforeseen turn of affairs may necessitate.

The quotations which follow are from my treatise on Great Military Operations:—

"The general configuration of the theater of war may also have a great influence upon the direction of the lines of operations, and, consequently, upon the direction of the bases.

"If every theater of war forms a figure presenting four faces more or less regular, one of the armies, at the opening of the campaign, may hold one of these faces,—perhaps two,—while the enemy occupies the other, the fourth being closed by insurmountable obstacles. The different ways of occupying this theater will lead to widely different combinations. To illustrate, we will cite the theater of the French armies in Westphalia from 1757 to 1762, and that of Napoleon in 1806. In the first case, the side A B was the North Sea, B D the line of the Weser and the base of Duke Ferdinand, C D the line of the Main and the base of the French army, A C the line of the Rhine, also guarded by French troops. The French held two faces, the North Sea being the third; and hence it was only necessary for them, by maneuvers, to gain the side B D to be masters of the four faces, including the base and the communications of the enemy. The French army, starting from its base C D and gaining the front of operations F G H, could cut off the allied army I from its base B D; the latter would be thrown upon the angle A, formed by the lines of the Rhine, the Ems, and the sea, while the army E could communicate with its bases on the Main and Rhine.

"The movement of Napoleon in 1806 on the Saale was similar. He occupied at Jena and Naumburg the line F G H, then marched by Halle and Dessau to force the Prussian army I upon the sea, represented by the side A B. The result is well known.

"The art, then, of selecting lines of operations is to give them such directions as to seize the communications of the enemy without losing one's own. The line F G H, by its extended position, and the bend on the flank of the enemy, always protects the communications with the base C D; and this is exactly the maneuvers of Marengo, Ulm, and Jena.

"When the theater of war does not border upon the sea, it is always bounded by a powerful neutral state, which guards its frontiers and closes one side of the square. This may not be an obstacle insurmountable like the sea; but generally it may be considered as an obstacle upon which it would be dangerous to retreat after a defeat: hence it would be an advantage to force the enemy upon it. The soil of a power which can bring into the field one hundred and fifty or two hundred thousand troops cannot be violated with impunity; and if a defeated army made the attempt, it would be none the less cut off from its base. If the boundary of the theater of war should be the territory of a weak state, it would be absorbed in this theater, and the square would be enlarged till it reached the frontiers of a powerful state, or the sea. The outline of the frontiers may modify the shape of the quadrilateral so as to make it approach the figure of a parallelogram or trapezoid. In either case, the advantage of the army which has control of two faces of the figure, and possesses the power of establishing upon them a double base, will be still more decided, since it will be able more easily to cut the enemy off from the shortened side,—as was the case with the Prussian army in 1806, with the side B D J of the parallelogram formed by the lines of the Rhine, the Oder, the North Sea, and the mountainous frontier of Franconia."

The selection of Bohemia as a base in 1813 goes to prove the truth of my opinion; for it was the perpendicularity of this base to that of the French army which enabled the allies to neutralize the immense advantages which the line of the Elbe would otherwise have afforded Napoleon, and turned the advantages of the campaign in their favor. Likewise, in 1812, by establishing their base perpendicularly upon the Oka and Kalouga, the Russians were able to execute their flank march upon Wiazma and Krasnoi.

If any thing further be required to establish these truths, it will only be necessary to consider that, if the base be perpendicular to that of the enemy, the front of operations will be parallel to his line of operations, and that hence it will be easy to attack his communications and line of retreat.

It has been stated that perpendicular bases are particularly favorable in the case of a double frontier, as in the last figures. Critics may object to this that it does not agree with what is elsewhere said in favor of frontiers which are salient toward the enemy, and against double lines of operations with equality of force. (Art. XXI.) The objection is not well founded; for the greatest advantage of a perpendicular base consists in the fact that it forms such a salient, which takes in reverse a portion of the theater of operations. On the other hand, a base with two faces by no means requires that both should be occupied in force: on the contrary, upon one of them it will be sufficient to have some fortified points garrisoned by small bodies, while the great bulk of the force rests upon the other face,—as was done in the campaigns of 1800 and 1806. The angle of nearly ninety degrees formed by the portion of the Rhine from Constance to Basel, and thence to Kehl, gave General Moreau one base parallel and another perpendicular to that of his antagonist. He threw two divisions by his left toward Kehl on the first base, to attract the attention of the enemy to that point, while he moved with nine divisions upon the extremity of the perpendicular face toward Schaffhausen, which carried him in a few days to the gates of Augsburg, the two detached divisions having already rejoined him.

In 1806, Napoleon had also the double base of the Rhine and Main, forming almost a right re-entrant angle. He left Mortier upon the first and parallel one, while with the mass of his forces he gained the extremity of the perpendicular base, and thus intercepted the Prussians at Gera and Naumburg by reaching their line of retreat.

If so many imposing facts prove that bases with two faces, one of them being almost perpendicular to that of the enemy, are the best, it is well to recollect that, in default of such a base, its advantages may be partially supplied by a change of strategic front, as will be seen in Article XX.

Another very important point in reference to the proper direction of bases relates to those established on the sea-coast. These bases may be favorable in some circumstances, but are equally unfavorable in others, as may be readily seen from what precedes. The danger which must always exist of an army being driven to the sea seems so clear, in the ease of the establishment of the base upon it, (which bases can only be favorable to naval powers,) that it is astonishing to hear in our day praises of such a base. Wellington, coming with a fleet to the relief of Spain and Portugal, could not have secured a better base than that of Lisbon, or rather of the peninsula of Torres-Vedras, which covers all the avenues to that capital on the land side. The sea and the Tagus not only protected both flanks, but secured the safety of his only possible line of retreat, which was upon the fleet.

Blinded by the advantages which the intrenched camp of Torres-Vedras secured for the English, and not tracing effects to their real causes, many generals in other respects wise contend that no bases are good except such as rest on the sea and thus afford the army facilities of supply and refuge with both flanks secured. Fascinated by similar notions, Colonel Carion-Nizas asserted that in 1813 Napoleon ought to have posted half of his army in Bohemia and thrown one hundred and fifty thousand men on the mouths of the Elbe toward Hamburg; forgetting that the first precept for a continental army is to establish its base upon the front farthest *from* the sea, so as to secure the benefit of all its elements of strength, from which it might find itself cut off if the base were established upon the coast.

An insular and naval power acting on the continent would pursue a diametrically opposite course, but resulting from the same principle, viz.: *to establish the base upon those points where it can be sustained by all the resources of the country, and at the same time insure a safe retreat.*

A state powerful both on land and sea, whose squadrons control the sea adjacent to the theater of operations, might well base an army of forty or fifty thousand men upon the coast, as its retreat by sea and its supplies could be well assured; but to establish a continental army of one hundred and fifty thousand men upon such a base, when opposed by a disciplined and nearly equal force, would be an act of madness.

However, as every maxim has its exceptions, there is a case in which it may be admissible to base a continental army upon the sea: it is, when your adversary is not formidable upon land, and when you, being master of the sea, can supply the army with more facility than in the interior. We rarely see these conditions fulfilled: it was so, however, during the Turkish war of 1828 and 1829. The whole attention of the Russians was given to Varna and Bourghas, while Shumla was merely observed; a plan which they could not have pursued in the presence of a European army (even with the control of the sea) without great danger of ruin.

Despite all that has been said by triflers who pretend to decide upon the fate of empires, this war was, in the main, well conducted. The army covered itself by obtaining the fortresses of Brailoff, Varna, and Silistria, and afterward by preparing a depot at Sizeboli. As soon as its base was well established it moved upon Adrianople, which previously would have been madness. Had the season been a couple of months longer, or had the army not come so great a distance in 1828, the war would have terminated with the first campaign.

Besides permanent bases, which are usually established upon our own frontiers, or in the territory of a faithful ally, there are eventual or temporary bases, which result from the operations in the enemy's country; but, as these are rather temporary points of support, they will, to avoid confusion, be discussed in Article XXIII.

Article XIX: Strategic lines and Points, Decisive Points of the Theater of War, and Objective Points of Operations

Strategic lines and points are of different kinds. Some receive this title simply from their position, which gives them all their importance: these are permanent geographical strategic points. Others have a value from the relations they bear to the positions of the masses of the hostile troops and to the enterprises likely to be directed against them: such are strategic points of maneuver, and are eventual. Finally, there are points which have only a secondary importance, and others whose importance is constant and immense: the latter are called *decisive* strategic points.

Every point of the theater of war which is of military importance, whether from its position as a center of communication, or from the presence of military establishments or fortifications, is a geographical strategic point.

A distinguished general affirms that such a point would not necessarily be a strategic point, unless situated favorably for a contemplated operation. I think differently; for a strategic point is such essentially and by nature, and, no matter how far distant it may be from the scene of the first enterprises, it may be included in the field by some unforeseen turn of events, and thus acquire its full importance. It would, then, be more accurate to state that all strategic points are not necessarily decisive points.

Lines are strategic either from their geographical position or from their relation to temporary maneuvers. The first class may be subdivided as follows,—viz.: geographic lines which by their permanent importance belong to the decisive points [I may be reproached with inaccuracy of expression,—since a line cannot be a *point*, and yet I apply to lines the name of decisive or objective points. It seems almost useless to remark that *objective* points are not geometric points, but that the name is a form of expression used to designate the object which an army desires to attain.] of the theater of war, and those which have value merely because they connect two strategic points.

To prevent confusion, we will elsewhere treat of strategic lines in their relations to maneuvers,—confining ourselves here to what relates to the *decisive and objective points* of the zone of operations upon which enterprises occur.

Although these are most intimately connected, since every objective point ought necessarily to be one of the decisive points of the theater of war, there is nevertheless a distinction between them; for all decisive points cannot be at the same time the objective of operations. We will, then, define the first, in order to be more easily guided in our selection of the second.

I think the name of *decisive strategic point* should be given to all those which are capable of exercising a marked influence either upon the result of the campaign or upon a single enterprise. All points whose geographical position and whose natural or artificial advantages favor the attack or defense of a front of operations or of a line of defense are included in this number; and large, well-located fortresses occupy in importance the first rank among them.

The decisive points of a theater of war are of several kinds. The first are the geographic points and lines whose importance is permanent and a consequence of the configuration of the country. For example, take the case of the French in Belgium: whoever is master of the line of the Meuse will have the greatest advantages in taking possession of the country; for his adversary, being outflanked and inclosed between the Meuse and the North

Sea, will be exposed to the danger of total ruin if he give battle parallel to that sea. [This only applies to continental armies, and not to the English, who, having their base on Antwerp or Ostend, would have nothing to fear from an occupation of the line of the Meuse.] Similarly, the valley of the Danube presents a series of important points which have caused it to be looked upon as the key of Southern Germany.

Those points the possession of which would give the control of the junction of several valleys and of the center of the chief lines of communication in a country are also *decisive geographic points*. For instance, Lyons is an important strategic point, because it controls the valleys of the Rhone and Saône, and is at the center of communications between France and Italy and between the South and East; but it would not be a *decisive* point unless well fortified or possessing an extended camp with *têtes de pont*. Leipsic is most certainly a strategic point, inasmuch as it is at the junction of all the communications of Northern Germany. Were it fortified and did it occupy both banks of the river, it would be almost the key of the country,—if a country has a key, or if this expression means more than a decisive point.

All capitals are strategic points, for the double reason that they are not only centers of communications, but also the seats of power and government.

In mountainous countries there are defiles which are the only routes of exit practicable for an army; and these may be decisive in reference to any enterprise in this country. It is well known how great was the importance of the defile of Bard, protected by a single small fort, in 1800.

The second kind of decisive points are accidental points of maneuver, which result from the positions of the troops on both sides.

When Mack was at Ulm, in 1805, awaiting the approach of the Russian army through Moravia, the decisive point in an attack upon him was Donauwerth or the Lower Lech; for if his adversaries gained it before him he was cut off from his line of retreat, and also from the army intended to support him. On the contrary, Kray, who, in 1800, was in the same position, expected no aid from Bohemia, but rather from the Tyrol and from the army of Mélas in Italy: hence the decisive point of attack upon him was not Donauwerth, but on the opposite side, by Schaffhausen, since this would take in reverse his front of operations, expose his line of retreat, cut him off from his supporting army as well as from his base, and force him upon the Main. In the same campaign the first objective point of Napoleon was to fall upon the right of Mélas by the Saint-Bernard, and to seize his line of communications: hence Saint-Bernard, Ivrea, and Piacenza were decisive points only by reason of the march of Mélas upon Nice.

It may be laid down as a general principle that the decisive points of maneuver are on that flank of the enemy upon which, if his opponent operates, he can more easily cut him off from his base and supporting forces without being exposed to the same danger. The flank opposite to the sea is always to be preferred, because it gives an opportunity of forcing the enemy upon the sea. The only exception to this is in the case of an insular and inferior army, where the attempt, although dangerous, might be made to cut it off from the fleet.

If the enemy's forces are in detachments, or are too much extended, the decisive point is his center; for by piercing that, his forces will be more divided, their weakness increased, and the fractions may be crushed separately.

The decisive point of a battle-field will be determined by,—

1. The features of the ground.
2. The relation of the local features to the ultimate strategic aim.
3. The positions occupied by the respective forces.

These considerations will be discussed in the chapter on battles.

Objective Points

There are two classes of objective points,—objective *points of maneuver,* and *geographical objective points.* A geographical objective point may be an important fortress, the line of a river, a front of operations which affords good lines of defense or good points of support for ulterior enterprises. *Objective points of maneuver,* in contradistinction to *geographical objectives,* derive their importance from, and their positions depend upon, the situation of the hostile masses.

In strategy, the object of the campaign determines the objective point. If this aim be offensive, the point will be the possession of the hostile capital, or that of a province whose loss would compel the enemy to make peace. In a war of invasion the capital is, ordinarily, the objective point. However, the geographical position of the capital, the political relations of the belligerents with their neighbors, and their respective resources, are considerations foreign in themselves to the art of fighting battles, but intimately connected with plans of operations, and may decide whether an army should attempt or not to occupy the hostile capital. If it be concluded not to seize the capital, the objective point might be a part of the front of operations or line of defense where an important fort is situated, the possession of which would render safe the occupation of the neighboring territory. For instance, if France were to invade Italy in a war against Austria, the first objective point would be the line of the Ticino and Po; the second, Mantua and the line of the Adige. In the defensive, the objective point, instead of being that which it is desirable to gain possession of, is that which is to be defended. The capital, being considered the seat of power, becomes the principal objective point of the defense; but there may be other points, as the defense of a first line and of the first base of operations. Thus, for a French army reduced to the defensive behind the Rhine, the first objective would be to prevent the passage of the river; it would endeavor to relieve the forts in Alsace if the enemy succeeded in effecting a passage of the river and in besieging them: the second objective would be to cover the first base of operations upon the Meuse or Moselle,—which might be attained by a lateral defense as well as one in front.

As to the objective points of *maneuvers*,—that is, those which relate particularly to the destruction or decomposition of the hostile forces,—their importance may be seen by what has already been said. The greatest talent of a general, and the surest hope of success, lie in some degree in the good choice of these points. This was the most conspicuous merit of Napoleon. Rejecting old systems, which were satisfied by the capture of one or two points or with the occupation of an adjoining province, he was convinced that the best means of accomplishing great results was to dislodge and destroy the hostile army,—since states and provinces fall of themselves when there is no organized force to protect them. To detect at a glance the relative advantages presented by the different zones of operations, to concentrate the mass of the forces upon that one which gave the best promise of success, to be indefatigable in ascertaining the approximate position of the enemy, to fall with the rapidity of lightning upon his center if his front was too much extended, or upon that flank by which he could more readily seize his communications, to outflank him, to cut his line, to pursue him to the last, to disperse and destroy his forces,—such was the system followed by Napoleon in his first campaigns. These campaigns proved this system to be one of the very best.

When these maneuvers were applied, in later years, to the long distances and the inhospitable regions of Russia, they were not so successful as in Germany: however, it must be remembered that, if this kind of war is not suitable to all capacities, regions, or circumstances, its chances of success are still very great, and it is based upon principle. Napoleon abused the system; but this does not disprove its real advantages when a proper limit is assigned to its enterprises and they are made in harmony with the respective conditions of the armies and of the adjoining states.

The maxims to be given on these important strategic operations are almost entirely included in what has been said upon decisive points, and in what will be stated in Article XXI. in discussing the choice of lines of operations.

As to the choice of objective points, every thing will generally depend upon the aim of the war and the character which political or other circumstances may give it, and, finally, upon the military facilities of the two parties.

In cases where there are powerful reasons for avoiding all risk, it may be prudent to aim only at the acquisition of partial advantages,—such as the capture of a few towns or the possession of adjacent territory. In other cases, where a party has the means of achieving a great success by incurring great dangers, he may attempt the destruction of the hostile army, as did Napoleon.

The maneuvers of Ulm and Jena cannot be recommended to an army whose only object is the siege of Antwerp. For very different reasons, they could not be recommended to the French army beyond the Niemen, five hundred leagues from its frontiers, because there would be much more to be lost by failure than a general could reasonably hope to gain by success.

There is another class of decisive points to be mentioned, which are determined more from political than from strategic considerations: they play a great part in most coalitions, and influence the operations and plans of cabinets. They may be called *political objective points*.

Indeed, besides the intimate connection between statesmanship and war in its preliminaries, in most campaigns some military enterprises are undertaken to carry out a political end, sometimes quite important, but often very irrational. They frequently lead to the commission of great errors in strategy. We cite two examples. First, the expedition of the Duke of York to Dunkirk, suggested by old commercial views, gave to the operations of the allies a divergent direction, which caused their failure: hence this objective point was bad in a military view. The expedition of the same prince to Holland in 1799—likewise due to the views of the English cabinet, sustained by the intentions of Austria on Belgium—was not less fatal; for it led to the march of the Archduke Charles from Zurich upon Manheim,—a step quite contrary to the interests of the allied armies at the time it was undertaken. These illustrations prove that political objective points should be subordinate to strategy, at least until after a great success has been attained.

This subject is so extensive and so complicated that it would be absurd to attempt to reduce it to a few rules. The only one which can be given has just been alluded to, and is, that either the political objective points should be selected according to the principles of strategy, or their consideration should be postponed till after the decisive events of the campaign. Applying this rule to the examples just given, it will be seen that it was at Cambray or in the heart of France that Dunkirk should have been conquered in 1793 and Holland delivered in 1799; in other words, by uniting all the strength of the allies for great attempts on the decisive points of the frontiers. Expeditions of this kind are generally included in grand diversions,—to be treated of in a separate article.

Article XX: Fronts of Operations, Strategic Fronts, Lines of Defense, and Strategic Positions

There are some parts of the military science that so closely resemble each other, and are so intimately allied, that they are frequently confounded, although they are decidedly distinct. Such are *fronts of operations, strategic fronts, lines of defense,* and *strategic positions.* It is proposed in this article to show the distinction between them and to expose their relations to each other.

Fronts of Operations and Strategic Fronts

When the masses of an army are posted in a zone of operations, they generally occupy strategic positions. The extent of the front occupied toward the enemy is called the *strategic front.* The portion of the theater of war from which an enemy can probably reach this front in two or three marches is called the *front of operations.*

The resemblance between these two fronts has caused many military men to confound them, sometimes under one name and sometimes under the other.

Rigorously speaking, however, the strategic front designates that formed by the actual positions occupied by the masses of the army, while the other embraces the space separating the two armies, and extends one or two marches beyond each extremity of the strategic front, and includes the ground upon which the armies will probably come in collision.

When the operations of a campaign are on the eve of commencing, one of the armies will decide to await the attack of the other, and will undertake to prepare a line of defense, which may be either that of the strategic front or more to the rear. Hence the strategic front and line of defense may coincide, as was the case in 1795 and 1796 upon the Rhine, which was then a line of defense for both Austrians and French, and at the same time their strategic front and front of operations. This occasional coincidence of these lines doubtless leads persons to confound them, while they are really very different. An army has not necessarily a line of defense, as, for example, when it invades: when its masses are concentrated in a single position, it has no strategic front, but it is never without a front of operations.

The two following examples will illustrate the difference between the different terms.

At the resumption of hostilities in 1813, Napoleon's front of operations extended at first from Hamburg to Wittenberg; thence it ran along the line of the allies toward Glogau and Breslau, (his right being at Löwenberg,) and followed along the frontier of Bohemia to Dresden. His forces were stationed on this grand front in four masses, whose strategic positions were interior and central and presented three different faces. Subsequently, he retired behind the Elbe. His real line of defense then extended only from Wittenberg to Dresden, with a bend to the rear toward Marienberg, for Hamburg and Magdeburg were beyond the strategic field, and it would have been fatal for him to have extended his operations to these points.

The other example is his position about Mantua in 1796. His front of operations here really extended from the mountains of Bergamo to the Adriatic Sea, while his real line of defense was upon the Adige, between Lake Garda and Legnago: afterward it was upon the Mincio, between Peschiera and Mantua, while his strategic front varied according to his positions.

The front of operations being the space which separates the two armies, and upon which they may fight, is ordinarily parallel to the base of operations. The strategic front will have the same direction, and ought to be perpendicular to the principal line of operations, and to extend far enough on either flank to cover this line well. However, this direction may vary, either on account of projects that are formed, or on account of the attacks of the enemy; and it quite frequently happens that it is necessary to have a front perpendicular to the base and parallel to the original line of operations. Such a change of strategic front is one of the most important of all grand maneuvers, for by this means the control of two faces of the strategic field may be obtained, thus giving the army a position almost as favorable as if it possessed a base with two faces. (See Article XVIII.)

The strategic front of Napoleon in his march on Eylau illustrates these points. His pivots of operations were at Warsaw and Thorn, which made the Vistula a temporary base: the front became parallel to the Narew, from whence he set out, supported by Sierock, Pultusk, and Ostrolenka, to maneuver by his right and throw the Russians on Elbing and the Baltic. In such cases, if a point of support in the new direction can be obtained, the strategic front gives the advantages referred to above. It ought to be borne in mind in such maneuvers that the army should always be sure of regaining its temporary base if necessary; in other words, that this base should be prolonged behind the strategic front, and should be covered by it. Napoleon, marching from the Narew by Allenstein upon Eylau, had behind his left Thorn, and farther from the front of the army the *tête de pont* of Praga and Warsaw; so that his communications were safe, while Benningsen, forced to face him and to make his line parallel to the Baltic, might be cut off from his base, and be thrown back upon the mouths of the Vistula. Napoleon executed another very remarkable change of strategic front in his march from Gera upon Jena and Naumburg in 1806. Moreau made another in moving by his right upon Augsburg and Dillingen, fronting the Danube and France, and thereby forcing Kray to evacuate the intrenched camp at Ulm.

The change of the strategic front to a position perpendicular to the base may be a temporary movement for an operation of a few days' duration, or it may be for an indefinite time, in order to profit by important advantages afforded by certain localities, to strike decisive blows, or to procure for the army a good line of defense and good pivots of operations, which would be almost equivalent to a real base.

It often happens that an army is compelled to have a double strategic front, either by the features of the theater of war, or because every line of offensive operations requires protection on its flanks. As an example of the first, the frontiers of Turkey and Spain may be cited. In order to cross the Balkan or the Ebro, an army would be obliged to present a double front,—in the first case, to face the valley of the Danube; in the second, to confront forces coming from Saragossa or Leon.

All extensive countries necessitate, to a greater or less degree, the same precaution. A French army in the valley of the Danube will require a double front as soon as the Austrians have thrown sufficient troops into the Tyrol or Bohemia to give rise to any anxiety. Those countries which present a narrow frontier to the enemy are the only exception, since the troops left on the frontier to harass the flanks of the enemy could themselves be cut off and captured. This necessity of double strategic fronts is one of the most serious inconveniences of an offensive war, since it requires large detachments, which are always dangerous. (See Article XXXVI.)

Of course, all that precedes relates to regular warfare. In a national or intestine war the whole country is the scene of hostilities. Nevertheless, each large fraction of an army having a defined aim would have its own strategic front determined by the features of the country and the positions occupied by the large bodies of the enemy. Thus, Suchet in Catalonia and Massena in Portugal each had a strategic front, while the front of some other corps of the army was not clearly defined.

Lines of Defense

Lines of defense are classified as strategical and tactical. Strategical lines of defense are subdivided into two classes: 1. Permanent lines of defense, which are a part of the defensive system of a state, such as the line of a fortified frontier; 2. Eventual lines of defense, which relate only to the temporary position of an army.

The frontier is a permanent line of defense when it presents a well-connected system of obstacles, natural and artificial, such as ranges of mountains, broad rivers, and fortresses. Thus, the range of the Alps between France and Piedmont is a line of defense, since the practicable passes are guarded by forts which would prove great obstacles in the way of an army, and since the outlets of the gorges in the valleys of Piedmont are protected by large fortresses. The Rhine, the Oder, and the Elbe may also be considered as permanent lines of defense, on account of the important forts found upon them.

Every river of any considerable width, every range of mountains, and every defile, having their weak points covered by temporary fortifications, may be regarded as *eventual lines of defense*, both strategic and tactical, since they may arrest for some time the progress of the enemy, or may compel him to deviate to the right or left in search of a weaker point,—in which case the advantage is evidently strategic. If the enemy attack in front, the lines present an evident tactical advantage, since it is always more difficult to drive an army from its position behind a river, or from a point naturally and artificially strong, than to attack it on an open plain. On the other hand, this advantage must not be considered unqualified, lest we should fall into the system of positions which has been the ruin of so many armies; for, whatever may be the facilities of a position for defense, it is quite certain that the party which remains in it passive and receiving all the attacks of his adversary will finally yield. [This does not refer to intrenched camps, which make a great difference. They are treated of in Article XXVII.] In addition to this, since a position naturally very strong [It is a question here of positions of camps, and not of positions for battle. The latter will be treated of in the chapter devoted to Grand Tactics, (Article XXX.)] is difficult of access it will be as difficult of egress, the enemy may be able with an inferior force to confine the army by guarding all the outlets. This happened to the Saxons in the camp of Pirna, and to Wurmser in Mantua.

Strategic Positions

There is a disposition of armies to which the name of strategic position may be applied, to distinguish from tactical positions or positions for battle.

Strategic positions are those taken for some time and which are intended to cover a much greater portion of the front of operations than would be covered in an actual battle. All positions behind a river or upon a line of defense, the divisions of the army being separated by considerable distances, are of this class, such as those of Napoleon at Rivoli, Verona, and Legnago to overlook the Adige. His positions in 1813 in Saxony and Silesia in advance of his line of defense were strategic. The positions of the Anglo-Prussian armies on the frontier of Belgium before the battle of Ligny, (1814,) and that of Massena on the Limmat and Aar in 1799, were also strategic. Even winter quarters, when compact and in face of the enemy and not protected by an armistice, are strategic positions,—for instance, Napoleon on the Passarge in 1807. The daily positions taken up by an army beyond the reach of the enemy, which are sometimes spread out either to deceive him or to facilitate movements, are of this class.

This class also includes positions occupied by an army to cover several points and positions held by the masses of an army for the purposes of observation. The different positions taken up on a line of defense, the positions of detachments on a double front of operations, the position of a detachment covering a siege, the main army in the meanwhile operating on another point, are all strategic. Indeed, all large detachments or fractions of an army may be considered as occupying strategic positions.

The maxims to be given on the preceding points are few, since fronts, lines of defense, and strategic positions generally depend upon a multitude of circumstances giving rise to infinite variety.

In every case, the first general rule is that the communications with the different points of the line of operations be thoroughly assured.

In the defense it is desirable that the strategic fronts and lines of defense should present both upon the flanks and front formidable natural or artificial obstacles to serve as points of support. The points of support on the strategic front are called *pivots of operations*, and are practical temporary bases, but quite different from pivots of maneuver. For example, in 1796 Verona was an excellent pivot of operations for all Napoleon's enterprises about Mantua for eight months. In 1813 Dresden was his pivot.

Pivots of maneuver are detachments of troops left to guard points which it is essential to hold, while the bulk of the army proceeds to the fulfillment of some important end; and when this is accomplished the pivot of maneuver ceases to exist. Thus, Ney's corps was the pivot of Napoleon's maneuver by Donauwerth and Augsburg to cut Mack from his line of retreat. A pivot of operations, on the contrary, is a material point of both strategical and tactical importance, serves as a point of support and endures throughout a campaign.

The most desirable quality of a line of defense is that it should be as short as possible, in order to be covered with facility by the army if it is compelled to take the defensive. It is also important that the extent of the strategic front should not be so great as to prevent the prompt concentration of the fractions of the army upon an advantageous point.

The same does not altogether apply to the front of operations; for if it be too contracted it would be difficult for an army on the offensive to make strategic maneuvers calculated to produce great results, since a short front could be easily covered by the defensive army. Neither should the front of operations be too extended. Such a front is unsuitable for offensive operations, as it would give the enemy, if not a good line of defense, at least ample space to escape from the results of a strategic maneuver even if well planned. Thus, the beautiful operations of Marengo, Ulm, and Jena could not have produced the same results upon a theater of the magnitude of that of the Russian War in 1812, since the enemy, even if cut off from his line of retreat, could have found another by adopting a new zone of operations.

The essential conditions for every strategic position are that it should be more compact than the forces opposed, that all fractions of the army should have sure and easy means of concentrating, free from the intervention of the enemy. Thus, for forces nearly equal, all central or interior positions would be preferable to exterior ones, since the front in the latter case would necessarily be more extended and would lead to a dangerous division of force. Great mobility and activity on the part of the troops occupying these positions will be a strong element of security or of superiority over the enemy, since it renders possible rapid concentration at different and successive points of the front.

An army should never long occupy any strategic point without making selection of one or two tactical positions, for the purpose of there concentrating all the disposable force, and giving battle to the enemy when he shall have unveiled his designs. In this manner Napoleon prepared the fields of Rivoli and Austerlitz, Wellington that of Waterloo, and the Archduke Charles that of Wagram.

When an army either camps or goes into quarters, the general should be careful that the front be not too extended. A disposition which might be called the strategic square seems best, presenting three nearly-equal faces, so that the distance to be passed over would be about equal for all the divisions in concentrating upon the common center to receive an attack.

Every strategic line of defense should always possess a tactical point upon which to rally for defense should the enemy cross the strategic front. For instance, an army guarding a bank of a river, not being able to occupy in force the whole line, ought always to have a position in rear of the center selected, upon which to collect all his divisions, so as to oppose them united to the enemy when he has succeeded in effecting a passage.

For an army entering a country with the purpose either of subjugation or of temporary occupation, it would always be prudent, however brilliant may have been its earlier successes, to prepare a line of defense as a refuge in case of reverse. This remark is made to complete the subject: the lines themselves are intimately connected with temporary bases, and will be discussed in a future article, (XXIII.)

Article XXI: Zones and Lines of Operations

A zone of operations is a certain fraction of the whole theater of war, which may be traversed by an army in the attainment of its object, whether it act singly or in concert with other and secondary armies. For example, in the plan of campaign of 1796, Italy was the zone of the right, Bavaria that of the center, Franconia that of the left army.

A zone of operations may sometimes present but a single *line of operations*, either on account of the configuration of the country, or of the small number of practicable routes for an army found therein. Generally, however, a zone presents several *lines of operations*, depending partly upon the plans of the campaign, partly upon the number of great routes of communication existing in the theater of operations.

It is not to be understood from this that every road is of itself a *line of operations*,—though doubtless it may happen that any good road in a certain turn of affairs may become for the time-being such a line; but as long as it is only traversed by detachments, and lies beyond the sphere of the principal enterprises, it cannot truly be called the real line of operations. Moreover, the existence of several routes leading to the same front of operations, and separated by one or two marches, would not constitute so many lines of operations, but, being the communications of the different divisions of the same army, the whole space bounded by them would constitute but a single line.

The term *zone of operations* is applied to a large fraction of the general theater of war; the term *lines of operations* will designate the part of this fraction embraced by the enterprises of the army. Whether it follow a single or several routes, the term *strategic lines* will apply to those important lines which connect the decisive points of the theater of operations either with each other or with the front of operations; and, for the same reason, we give this name to those lines which the army would follow to reach one of these decisive points, or to accomplish an important maneuver which requires a temporary deviation from the principal line of operations. *Lines of communications* designate the practicable routes between the different portions of the army occupying different positions throughout the zone of operations.

For example, in 1813, after the accession of Austria to the Grand Coalition, three allied armies were to invade Saxony, one Bavaria, and another Italy: so that Saxony, or rather the country between Dresden, Magdeburg, and Breslau, formed the zone of operations of the mass of the forces. This zone had three *lines of operations* leading to Leipsic as an objective: the first was the line of the army of Bohemia, leading from the mountains of Erzgebirge by Dresden and Chemnitz upon Leipsic; the second was the line of the army of Silesia, going from Breslau by Dresden or by Wittenberg upon Leipsic; the third was that of Bernadotte from Berlin by Dessau to the same objective point. Each of these armies marched upon two or more adjacent parallel routes, but it could not be said that there were as many lines of operations as roads. The principal line of operations is that followed by the bulk of the army, and upon which depots of provisions, munitions, and other supplies are echeloned, and over which, if compelled, it would retreat.

If the choice of a zone of operations involves no extensive combinations, since there can never be more than two or three zones on each theater, and the advantages generally result from the localities, it is somewhat different with lines of operations, as they are divided into different classes, according to their relations to the different positions of the enemy, to the communications upon the strategic field, and to the enterprises projected by the commander.

Simple lines of operations are those of an army acting from a frontier when it is not subdivided into large independent bodies.

Double lines of operations are those of two independent armies proceeding from the same frontier, or those of two nearly equal armies which are commanded by the same general but are widely separated in distance and for long intervals of time. [This definition has been criticized; and, as it has given rise to misapprehension, it becomes necessary to explain it. In the first place, it must be borne in mind that it is a question of *maneuver-lines*, (that is, of strategic combinations,) and not of great routes. It must also be admitted that an army marching upon two or three routes, near enough to each other to admit of the concentration of the different masses within forty-eight hours, would not have two or three lines of operations. When Moreau and Jourdan entered Germany with two armies of 70,000 men each, being independent of each other, there was a double line of operations; but a French army of which only a detachment starts from the Lower Rhine to march on the Main, while the five or six other corps set out from the Upper Rhine to march on Ulm, would not have a double line of operations in the sense in which I use the term to designate a maneuver. Napoleon, when he concentrated seven corps and set them in motion by Bamberg to march on Gera, while Mortier with a single corps marched on Cassel to occupy Hesse and flank the principal enterprise, had but a single general line of operations, with an accessory detachment. The territorial line was composed of two arms or radii, but the operation was not double.]

Interior lines of operations are those adopted by one or two armies to oppose several hostile bodies, and having such a direction that the general can concentrate the masses and maneuver with his whole force in a shorter period of time than it would require for the enemy to oppose to them a greater force. [Some German writers have said that I confound central positions with the line of operations,—in which assertion they are mistaken. An army may occupy a central position in the presence of two masses of the enemy, and not have interior lines of operations: these are two very different things. Others have thought that I would have done better to use the term *radii of operations* to express the idea of double lines. The reasoning in this case is plausible if we conceive the theater of operations to be a circle; but, as every radius is, after all, a line, it is simply a dispute about words.] *Exterior lines* lead to the opposite result, and are those formed by an army which operates at the same time on both flanks of the enemy, or against several of his masses.

Concentric lines of operations are those which depart from widely-separated points and meet at the same point, either in advance of or behind the base.

Divergent lines are those by which an army would leave a given point to move upon several distinct points. These lines, of course, necessitate a subdivision of the army.

There are also *deep lines*, which are simply *long lines*.

The term *maneuver-lines* I apply to momentary strategic lines, often adopted for a single temporary maneuver, and which are by no means to be confounded with the real *lines of operations*.

Secondary lines are those of two armies acting so as to afford each other mutual support,—as, in 1796, the army of the Sambre and Meuse was secondary to the army of the Rhine, and, in 1812, the army of Bagration was secondary to that of Barclay.

Accidental lines are those brought about by events which change the original plan and give a new direction to operations. These are of the highest importance. The proper occasions for their use are fully recognized only by a great and active mind.

There may be, in addition, *provisional* and *definitive lines of operations*. The first designate the line adopted by an army in a preliminary, decisive enterprise, after which it is at liberty to select a more advantageous or direct line. They seem to belong as much to the class of temporary or eventual strategic lines as to the class of lines of operations.

These definitions show how I differ from those authors who have preceded me. Lloyd and Bulow attribute to these lines no other importance than that arising from their relations to the depots of the army: the latter has even asserted that when an army is encamped near its depots it has no lines of operations.

The following example will disprove this paradox. Let us suppose two armies, the first on the Upper Rhine, the second in advance of Dusseldorf or any other point of this frontier, and that their large depots are immediately behind the river,—certainly the safest, nearest, and most advantageous position for them which could possibly be adopted. These armies will have an offensive or defensive object: hence they will certainly have lines of operations, arising from the different proposed enterprises.

1. Their defensive territorial line, starting from their positions, will extend to the second line which they are to cover, and they would both be cut off from this second line should the enemy establish himself in the interval which separates them from it. Even if Mélas [This assertion has been disputed. I think it is correct; for Mélas, confined between the Bormida, the Tanaro, and the Po, was unable to recruit for his army, barely able to maintain a communication by couriers with his base, and he certainly would have been obliged to cut his way out or to surrender in case he had not been reinforced.] had possessed a year's supplies in Alessandria, he would none the less have been cut off from his base of the Mincio as soon as the victorious enemy occupied the line of the Po.

2. Their line would be double, and the enemy's single if he concentrated his forces to defeat these armies successively; it would be a double exterior line, and the enemy's a double interior, if the latter divided his forces into two masses, giving them such directions as to enable him to concentrate all his forces before the two armies first referred to could unite.

Bulow would have been more nearly right had he asserted that an army on its own soil is less dependent on its primitive line of operations than when on foreign ground; for it finds in every direction points of support and some of the advantages which are sought for in the establishment of lines of operations; it may even lose its line of operations without incurring great danger; but that is no reason why it has no line of operations.

Observations upon the Lines of Operations in the Wars of the French Revolution

At the beginning of this terrible and ever-varying struggle, Prussia and Austria were the only avowed enemies of France, and Italy was included in the theater of war only for purposes of reciprocal observation, it being too remote for decisive enterprises in view of the end proposed. The real theater extended from Huningue to Dunkirk, and comprised three zones of operations,—the first reaching along the Rhine from Huningue to Landau, and thence to the Moselle; the center consisting of the interval between the Meuse and Moselle; the third and left was the frontier from Givet to Dunkirk.

When France declared war, in April, 1792, her intention was to prevent a union of her enemies; and she had then one hundred thousand men in the zones just described, while Austria had but thirty-five thousand in Belgium. It is quite impossible to understand why the French did not conquer this country, when no effectual resistance could have been made. Four months intervened between the declaration of war and the concentration of the allied troops. Was it not probable that an invasion of Belgium would have prevented that of Champagne, and have given the King of Prussia a conception of the strength of France, and induced him not to sacrifice his armies for the secondary object of imposing upon France another form of government?

When the Prussians arrived at Coblentz, toward the end of July, the French were no longer able to invade. This *rôle* was reserved for the allies; and it is well known how they acquitted themselves.

The whole force of the French was now about one hundred and fifteen thousand men. It was scattered over a frontier of one hundred and forty leagues and divided into five corps d'armée, and could not make a good defense; for to paralyze them and prevent their concentration it was only necessary to attack the center. Political reasons were also in favor of this plan of attack: the end proposed was political, and could only be attained by rapid and vigorous measures. The line between the Moselle and Meuse, which was the center, was less fortified than the rest of the frontier, and, besides, gave the allies the advantage of the excellent fortress of Luxembourg as a base. They wisely adopted this plan of attack; but the execution was not equal to the conception.

The court of Vienna had the greatest interest in the war, for family reasons, as well as on account of the dangers to which a reverse might subject her provinces. For some reason, difficult to understand, Austria co-operated only to the extent of thirty battalions: forty-five thousand men remained as an army of observation in Brisgau, on the Rhine, and in Flanders. Where were the imposing armies she afterward displayed? and what more useful disposition could have been made of them than to protect the flanks of the invading army? This remarkable conduct on the part of Austria, which cost her so much, may account for the resolution of Prussia to retire at a later period, and quit the field, as she did, at the very moment when she should have entered it. During the campaign the Prussians did not exhibit the activity necessary for success. They spent eight days uselessly in camp at Kons. If they had anticipated Dumouriez at the Little Islands, or had even made a more serious effort to drive him from them, they would still have had all the advantage of a concentrated force against several scattered divisions, and could have prevented their junction and overthrown them separately. Frederick the Great would have justified the remark of Dumouriez at

Grandpré,—that, if his antagonist had been the great king, he (Dumouriez) would already have been driven behind Châlons.

The Austrians in this campaign proved that they were still imbued with the false system of Daun and Lascy, of covering every point in order to guard every point.

The fact of having twenty thousand men in Brisgau while the Moselle and Sarre were uncovered, shows the fear they had of losing a village, and how their system led to large detachments, which are frequently the ruin of armies.

Forgetting that the surest hope of victory lies in presenting the strongest force, they thought it necessary to occupy the whole length of a frontier to prevent invasion,—which was exactly the means of rendering invasion upon every point feasible.

I will further observe that, in thin campaign, Dumouriez foolishly abandoned the pursuit of the allies in order to transfer the theater from the center to the extreme left of the general field. Moreover, he was unable to perceive the great results rendered possible by this movement, but attacked the army of the Duke of Saxe-Teschen in front, while by descending the Meuse to Namur he might have thrown it back upon the North Sea toward Meuport or Ostend, and have destroyed it entirely in a more successful battle than that of Jemmapes.

The campaign of 1793 affords a new instance of the effect of a faulty direction of operations. The Austrians were victorious, and recovered Belgium, because Dumouriez unskillfully extended his front of operations to the gates of Rotterdam. Thus far the conduct of the allies deserves praise: the desire of reconquering these rich provinces justified this enterprise, which, moreover, was judiciously directed against the extreme right of the long front of Dumouriez. But after the French had been driven back under the guns of Valenciennes, and were disorganized and unable to resist, why did the allies remain six months in front of a few towns and permit the Committee of Public Safety to organize new armies? When the deplorable condition of France and the destitution of the wreck of the army of Dampierre are considered, can the parades of the allies in front of the fortresses in Flanders be understood?

Invasions of a country whose strength lies mainly in the capital are particularly advantageous. Under the government of a powerful prince, and in ordinary wars, the most important point is the head-quarters of the army; but under a weak prince, in a republic, and still more in wars of opinion, the capital is generally the center of national power. [The capture of Paris by the allies decided the fate of Napoleon; but he had no army, and was attacked by all Europe, and the French people had, in addition, separated their cause from his. If he had possessed fifty thousand more old soldiers, he would have shown that the capital was at his head-quarters.] If this is ever doubtful, it was not so on this occasion. Paris was France, and this to such an extent that two-thirds of the nation had risen against the government which oppressed them. If, after having beaten the French army at Famars, the allies had left the Dutch and Hanoverians to observe what remained of it, while the English and the Austrians directed their operations upon the Meuse, the Sarre, and the Moselle, in concert with the Prussians and a part of the useless army of the Upper Rhine, a force of one hundred and twenty thousand men, with its flanks protected by other troops, could have been pushed forward. It is even probable that, without changing the direction of the war or running great risks, the Dutch and Hanoverians could have performed the duty of observing Maubeuge and Valenciennes, while the bulk of the army pursued the remains of Dampierre's forces. After gaining several victories, however, two hundred thousand men were engaged in carrying on a few sieges and were not gaining a foot of ground. While they threatened France with invasion, they placed fifteen or sixteen bodies of troops, defensively, to cover their own frontier! When Valenciennes and Mayence capitulated, instead of falling with all their forces upon the camp at Cambray, they flew off, excentrically, to Dunkirk on one side and Landau on the other.

It is not less astonishing that, after making the greatest efforts in the beginning of the campaign upon the right of the general field, they should have shifted them afterward to the extreme left, so that while the allies were operating in Flanders they were in no manner seconded or aided by the imposing army upon the Rhine; and when, in its turn, this army took up the offensive, the allies remained inactive upon the Sambre. Do not these false combinations resemble those of Soubise and Broglie in 1761, and all the operations of the Seven Years' War?

In 1794 the phase of affairs is wholly changed. The French from a painful defensive pass to a brilliant offensive. The combinations of this campaign were doubtless well considered; but it is wrong to represent them as forming a new system of war. To be convinced of this, it is only necessary to observe that the respective positions of the armies in this campaign and in that of 1757 were almost identical, and the direction of the operations is quite the same. The French had four corps, which constituted two armies, as the King of Prussia had four divisions, which composed two armies.

These two large bodies took a concentric direction leading on Brussels, as Frederick and Schwerin had adopted in 1757 on Prague. The only difference between the two plans is that the Austrian troops in Flanders were not so much scattered as those of Brown in Bohemia; but this difference is certainly not favorable to the plan of 1794. The position of the North Sea was also unfavorable for the latter plan. To outflank the Austrian right, Pichegru was thrown between the sea and the mass of the enemy,—a direction as dangerous and faulty as could be given to great operations. This movement was the same as that of Benningsen on the Lower Vistula which almost lost the Russian army in 1807. The fate of the Prussian army, cut off from its communications and forced upon the Baltic, is another proof of this truth.

If the Prince of Coburg had acted with ability, he could easily have made Pichegru suffer for this audacious maneuver, which was performed a month before Jourdan was prepared to follow it up.

The center of the grand Austrian army intended to act upon the offensive was before Landrecies; the army was composed of one hundred and six battalions and one hundred and fifty squadrons; upon its right flank Flanders was covered by the corps d'armée of Clairfayt, and upon the left Charleroi was covered by that of the Prince de Kaunitz. The gain of a battle before Landrecies opened its gates; and upon General Chapuis was found a plan of the diversion in Flanders: only *twelve battalions* were sent to Clairfayt. A long time afterward, and after the French were known to have been successful, the corps of the Duke of York marched to Clairfayt's relief; but what was the use of the remainder of the army before Landrecies, after it was obliged by a loss of force to delay invasion? The Prince of Coburg threw away all the advantages of his central position, by allowing the French to concentrate in Belgium and to beat all his large detachments in detail.

Finally, the army moved, leaving a division at Cateau, and a part having been sent to the Prince de Kaunitz at Charleroi. If, instead of dividing this grand army, it had been directed upon Turcoing, there would have been concentrated there one hundred battalions and one hundred and forty squadrons; and what must then have been the result of this famous diversion of Pichegru, cut off from his own frontiers and shut up between the sea and two fortresses?

The plan of invasion adopted by the French had not only the radical error of exterior lines: it also failed in execution. The diversion on Courtray took place on April 26, and Jourdan did not arrive at Charleroi till the 3d of June,—more than a month afterward. Here was a splendid opportunity for the Austrians to profit by their central position. If the Prussian army had maneuvered by its right and the Austrian army by its left,—that is, both upon the Meuse,—the state of affairs would have been different. By establishing themselves in the center of a line of scattered forces they could have prevented the junction of the different fractions. It may be dangerous in a battle to attack the center of a close line of troops when it can be simultaneously sustained by the wings and the reserves; but it is quite different on a line of three hundred miles in extent.

In 1795 Prussia and Spain retired from the coalition, and the principal theater of war was shifted from the Rhine to Italy,—which opened a new field of glory for the French arms. Their lines of operations in this campaign were double; they desired to operate by Dusseldorf and Manheim. Clairfayt, wiser than his predecessors, concentrated his forces alternately upon these points, and gained victories at Manheim and in the lines of Mayence so decisive that they caused the army of the Sambre and Meuse to recross the Rhine to cover the Moselle, and brought Pichegru back to Landau.

In 1796 the lines of operations on the Rhine were copied from those of 1757 and those in Flanders in 1794, but with different results. The armies of the Rhine, and of the Sambre and Meuse, set out from the extremities of the base, on routes converging to the Danube. As in 1794, they were exterior lines. The Archduke Charles, more skillful than the Prince of Coburg, profited by his interior lines by concentrating his forces at a point nearer than that expected by the French. He then seized the instant when the Danube covered the corps of Latour, to steal several marches upon Moreau and attack and overwhelm Jourdan: the battle of Wurzburg decided the fate of Germany and compelled the army of Moreau to retreat.

Bonaparte now commences in Italy his extraordinary career. His plan is to separate the Piedmontese and Austrian armies. He succeeds by the battle of Millesimo in causing them to take two exterior strategic lines, and beats them successively at Mondovi and Lodi. A formidable army is collected in the Tyrol to raise the siege of Mantua: it commits the error of marching there in two bodies separated by a lake. The lightning is not quicker than Napoleon. He raises the siege, abandons every thing before Mantua, throws the greater part of his force upon the first column, which debouches by Brescia, beats it and forces it back upon the mountains: the second column arrives upon the same ground, and is there beaten in its turn, and compelled to retire into the Tyrol to keep up its communications with the right. Wurmser, upon whom these lessons are lost, desires to cover the two lines of Roveredo and Vicenza; Napoleon, after having overwhelmed and thrown the first back upon the Lavis, changes direction by the right, debouches by the gorges of the Brenta upon the left, and forces the remnant of this fine army to take refuge in Mantua, where it is finally compelled to surrender.

In 1799 hostilities recommence: the French, punished for having formed two exterior lines in 1796, nevertheless, have three upon the Rhine and the Danube. The army on the left observes the Lower Rhine, that of the center marches upon the Danube, Switzerland, flanking Italy and Swabia, being occupied by a third army as strong as both the others. *The three armies could be concentrated only in the valley of the Inn*, eighty leagues from their base of operations. The archduke has equal forces: he unites them against the center, which he defeats at Stockach, and the army of Switzerland is compelled to evacuate the Grisons and Eastern Switzerland. The allies in turn commit the same fault: instead of following up their success on this central line, which cost them so dearly afterward, they formed a double line in Switzerland and on the Lower Rhine. The army of Switzerland is beaten at Zurich, while the other trifles at Manheim.

In Italy the French undertake a double enterprise, which leaves thirty-two thousand men uselessly employed at Naples, while upon the Adige, where the vital blows were to be given or received, their force is too weak and meets with terrible reverses. When the army of Naples returns to the North, it commits the error of adopting a strategic direction opposed to Moreau's, and Suwaroff, by means of his central position, from which he derives full profit, marches against this army and beats it, while some leagues from the other.

In 1800, Napoleon has returned from Egypt, and every thing is again changed, and this campaign presents a new combination of lines of operations; one hundred and fifty thousand men march upon the two flanks of Switzerland, and debouch, one upon the Danube and the other upon the Po. This insures the conquest of vast regions. Modern history affords no similar combination. The French armies are upon interior lines, affording reciprocal support, while the Austrians are compelled to adopt an exterior line, which renders it impossible for them to communicate. By a skillful arrangement of its progress, the army of the reserve cuts off the enemy from his line of operations, at the same time preserving its own relations with its base and with the army of the Rhine, which forms its secondary line.

The analysis of the memorable events just sketched shows clearly the importance of a proper selection of lines of maneuver in military operations. Indeed, discretion on this point may repair the disasters of defeat, destroy the advantages of an adversary's victory, render his invasion futile, or assure the conquest of a province.

By a comparison of the combinations and results of the most noted campaigns, it will be seen that the lines of operations which have led to success have been established in conformity to the fundamental principle already alluded to,—viz.: that *simple and interior lines enable a general to bring into action, by strategic movements, upon the important point, a stronger force than the enemy.* The student may also satisfy himself that those which have failed contained faults opposed to this principle. An undue number of lines divides the forces, and permits fractions to be overwhelmed by the enemy.

Maxims on Lines of Operations

From the analysis of all the events herein referred to, as well as from that of many others, the following maxims result:—

1. If the art of war consists in bringing into action upon the decisive point of the theater of operations the greatest possible force, the choice of the line of operations, being the primary means of attaining this end, may be regarded as the fundamental idea in a good plan of a campaign. Napoleon proved this by the direction he gave his armies in 1805 on Donauwerth and in 1806 on Gera,—maneuvers that cannot be too much studied by military men.

Of course, it is impossible to sketch in advance the whole campaign. The objective point will be determined upon in advance, the general plan to be followed to attain it, and the first enterprise to be undertaken for this end: what is to follow will depend upon the result of this first operation and the new phases it may develop.

2. The direction to be given to this line depends upon the geographical situation of the theater of operations, but still more upon the position of the hostile masses upon this strategic field. *In every case, however, it must be directed upon the center or upon one of the extremities. Only when the assailing forces are vastly preponderating would it be otherwise than a fatal error to act upon the center and the two extremities at the same time.* [The inferiority of an army does not depend exclusively upon the number of soldiers: their military qualities, their *morale*, and the ability of their commander are also very important elements.]

It may be laid down as a general principle, that, if the enemy divide his forces on an extended front, the best direction of the maneuver-line will be upon his center, but in every other case, when it is possible, the best direction will be upon one of the flanks, and then upon the rear of his line of defense or front of operations.

The advantage of this maneuver arises more from the opportunity it affords of taking the line of defense in reverse than from the fact that by using it the assailant has to contend with but a part of the enemy's force. Thus, the army of the Rhine in 1800, gaining the extreme left of the line of defense of the Black Forest, caused it to yield almost without an effort. This army fought two battles on the right bank of the Danube, which, although not decisive, yet, from the judicious direction of the line of operations, brought about the invasion of Swabia and Bavaria. The results of the march of the army of the reserve by the Saint-Bernard and Milan upon the extreme right of Mélas were still more brilliant.

3. Even when the extremity of the enemy's front of operations is gained, it is not always safe to act upon his rear, since by so doing the assailant in many cases will lose his own communications. To avoid this danger, the line of operations should have a geographic and strategic direction, such that the

army will always find either to its rear or to the right or left a safe line of retreat. In this case, to take advantage of either of these flank lines of retreat would require a change of direction of the line of operations, (Maxim 12.)

The ability to decide upon such a direction is among the most important qualities of a general. The importance of a direction is illustrated by these examples.

If Napoleon in 1800, after passing the Saint-Bernard, had marched upon Asti or Alessandria, and had fought at Marengo without having previously protected himself on the side of Lombardy and of the left bank of the Po, he would have been more thoroughly cut off from his line of retreat than Mélas from his; but, having in his possession the secondary points of Casale and Pavia on the side of the Saint-Bernard, and Savona and Tenda toward the Apennines, in case of reverse he had every means of regaining the Var or the Valais.

In 1806, if he had marched from Gera directly upon Leipsic, and had there awaited the Prussian army returning from Weimar, he would have been cut off from the Rhine as much as the Duke of Brunswick from the Elbe, while by falling back to the west in the direction of Weimar he placed his front before the three roads of Saalfeld, Schleiz, and Hof, which thus became well-covered lines of communication. If the Prussians had endeavored to cut him off from these lines by moving between Gera and Baireuth, they would have opened to him his most natural line,—the excellent road from Leipsic to Frankfort,—as well as the two roads which lead from Saxony by Cassel to Coblentz, Cologne, and even Wesel.

4. Two independent armies should not be formed upon the same frontier: such an arrangement could be proper only in the case of large coalitions, or where the forces at disposal are too numerous to act upon the same zone of operations; and even in this case it would be better to have all the forces under the same commander, who accompanies the principal army.

5. As a consequence of the last-mentioned principle, with equal forces on the same frontier, a single line of operations will be more advantageous than a double one.

6. It may happen, however, that a double line will be necessary, either from the topography of the seat of war, or because a double line has been adopted by the enemy, and it will be necessary to oppose a part of the army to each of his masses.

7. In this case, interior or central lines will be preferable to exterior lines, since in the former case the fractions of the army can be concentrated before those of the enemy, and may thus decide the fate of the campaign. [When the fractions of an army are separated from the main body by only a few marches, and particularly when they are not intended to act separately

throughout the campaign, these are central strategic positions, and not lines of operations.] Such an army may, by a well-combined strategic plan, unite upon and overwhelm successively the fractions of the adversary's forces. To be assured of success in these maneuvers, a body of observation is left in front of the army to be held in check, with instructions to avoid a serious engagement, but to delay the enemy as much as possible by taking advantage of the ground, continually falling back upon the principal army.

8. A double line is applicable in the case of a decided superiority of force, when each army will be a match for any force the enemy can bring against it. In this case this course will be advantageous,—since a single line would crowd the forces so much as to prevent them all from acting to advantage. However, it will always be prudent to support well the army which, by reason of the nature of its theater and the respective positions of the parties, has the most important duty to perform.

9 The principal events of modern wars demonstrate the truth of two other maxims. The first is, that two armies operating on interior lines and sustaining each other reciprocally, and opposing two armies superior in numbers, should not allow themselves to be crowded into a too contracted space, where the whole might be overwhelmed at once. This happened to Napoleon at Leipsic. [In the movements immediately preceding the battle of Leipsic, Napoleon, strictly speaking, had but a single line of operations, and his armies were simply in central strategic positions; but the principle is the same, and hence the example is illustrative of lines of operations.] The second is, that interior lines should not be abused by extending them too far, thus giving the enemy the opportunity of overcoming the corps of observation. This risk, however, may be incurred if the end pursued by the main forces is so decisive as to conclude the war,—when the fate of these secondary bodies would be viewed with comparative indifference.

10. For the same reason, two converging lines are more advantageous than two divergent. The first conform better to the principles of strategy, and possess the advantage of covering the lines of communication and supply; but to be free from danger they should be so arranged that the armies which pass over them shall not be separately exposed to the combined masses of the enemy, before being able to effect their junction.

11. Divergent lines, however, may be advantageous when the center of the enemy has been broken and his forces separated either by a battle or by a strategic movement,—in which case divergent operations would add to the dispersion of the enemy. Such divergent lines would be interior, since the pursuers could concentrate with more facility than the pursued.

12. It sometimes happens that an army is obliged to change its line of operations in the middle of a campaign. This is a very delicate and important

step, which may lead to great successes, or to equally great disasters if not applied with sagacity, and is used only to extricate an army from an embarrassing position. Napoleon projected several of these changes; for in his bold invasions he was provided with new plans to meet unforeseen events.

At the battle of Austerlitz, if defeated, he had resolved to adopt a line of operations through Bohemia on Passau or Ratisbon, which would have opened a new and rich country to him, instead of returning by Vienna, which route lay through an exhausted country and from which the Archduke Charles was endeavoring to cut him off. Frederick executed one of these changes of the line of operations after the raising of the siege of Olmutz.

In 1814 Napoleon commenced the execution of a bolder maneuver, but one which was favored by the localities. It was to base himself upon the fortresses of Alsace and Lorraine, leaving the route to Paris open to the allies. If Mortier and Marmont could have joined him, and had he possessed fifty thousand more men, this plan would have produced the most decisive results and have put the seal on his military career.

13. As before stated, the outline of the frontiers, and the geographical character of the theater of operations, exercise a great influence on the direction to be given to these lines, as well as upon the advantages to be obtained. Central positions, salient toward the enemy, like Bohemia and Switzerland, are the most advantageous, because they naturally lead to the adoption of interior lines and facilitate the project of taking the enemy in reverse. The sides of this salient angle become so important that every means should be taken to render them impregnable. In default of such central positions, their advantages may be gained by the relative directions of maneuver-lines. C D maneuvering upon the right of the front of the army A B, and H I upon the left flank of G F, will form two interior lines I K and C K upon an extremity of the exterior lines A B, F G, which they may overwhelm separately by combining upon them. Such was the result of the operations of 1796, 1800, and 1809.

14. The general configuration of the bases ought also to influence the direction to be given to the lines of operations, these latter being naturally dependent upon the former. It has already been shown that the greatest advantage that can result from a choice of bases is when the frontiers allow it to be assumed parallel to the line of operations of the enemy, thus affording the opportunity of seizing this line and cutting him from his base.

But if, instead of directing the operations upon the decisive point, the line of operations be badly chosen, all the advantages of the perpendicular base may be lost. The army E, having the double base A C and C D, if it marched toward F, instead of to the right toward G H, would lose all the strategic advantages of its base C D.

The great art, then, of properly directing lines of operations, is so to establish them in reference to the bases and to the marches of the army as to seize the communications of the enemy without imperiling one's own, and is the most important and most difficult problem in strategy.

15. There is another point which exercises a manifest influence over the direction to be given to the line of operations; it is when the principal enterprise of the campaign is to cross a large river in the presence of a numerous and well-appointed enemy. In this case, the choice of this line depends neither upon the will of the general nor the advantages to be gained by an attack on one or another point; for the first consideration will be to ascertain where the passage can be most certainly effected, and where are to be found the means for this purpose. The passage of the Rhine in 1795, by Jourdan, was near Dusseldorf, for the same reason that the Vistula in 1831 was crossed by Marshal Paskevitch near Ossiek,—viz., that in neither case was there the bridge-train necessary for the purpose, and both were obliged to procure and take up the rivers large boats, bought by the French in Holland, and by the Russians at Thorn and Dantzic. The neutrality of Prussia permitted the ascent of the river in both cases, and the enemy was not able to prevent it. This apparently incalculable advantage led the French into the double invasions of 1795 and 1796, which failed because the double line of operations caused the defeat of the armies separately. Paskevitch was wiser, and passed the Upper Vistula with only a small detachment and after the principal army had already arrived at Lowicz.

When an army is sufficiently provided with bridge-trains, the chances of failure are much lessened; but then, as always, it is necessary to select the point which may, either on account of its topography or the position of the enemy, be most advantageous. The discussion between Napoleon and Moreau on the passage of the Rhine in 1800 is one of the most curious examples of the different combinations presented by this question, which is both strategic and tactical.

Since it is necessary to protect the bridges, at least until a victory is gained, the point of passage will exercise an influence upon the directions of a few marches immediately subsequent to the passage. The point selected in every case for the principal passage will be upon the center or one of the flanks of the enemy.

A united army which has forced a passage upon the center of an extended line might afterward adopt two divergent lines to complete the dispersion of the enemy, who, being unable to concentrate, would not think of disturbing the bridges.

If the line of the river is so short that the hostile army is more concentrated, and the general has the means of taking up after the passage a front perpendicular to the river, it would be better to pass it upon one of the extremities, in order to throw off the enemy from the bridges. This will be referred to in the article upon the passage of rivers.

16. There is yet another combination of lines of operations to be noticed. It is the marked difference of advantage between a line at home and one in a hostile country. The nature of the enemy's country will also influence these chances. Let us suppose an army crosses the Alps or the Rhine to carry on war in Italy or Germany. It encounters states of the second rank; and, even if they are in alliance, there are always rivalries or collisions of interest which will deprive them of that unity and strength possessed by a single powerful state. On the other hand, a German army invading France would operate upon a line much more dangerous than that of the French in Italy, because upon the first could be thrown the consolidated strength of Franco, united in feeling and interest. An army on the defensive, with its line of operations on its own soil, has resources everywhere and in every thing: the inhabitants, authorities, productions, towns, public depots and arsenals, and even private stores, are all in its favor. It is not ordinarily so abroad.

Lines of operations in rich, fertile, manufacturing regions offer to the assailants much greater advantages than when in barren or desert regions, particularly when the people are not united against the invader. In provinces like those first named the army would find a thousand necessary supplies, while in the other huts and straw are about the only resources. Horses probably may obtain pasturage; but every thing else must be carried by the army,—thus infinitely increasing the embarrassments and rendering bold operations much more rare and dangerous. The French armies, so long accustomed to the comforts of Swabia and Lombardy, almost perished in 1806 in the bogs of Pultusk, and actually did perish in 1812 in the marshy forests of Lithuania.

17. There is another point in reference to these lines which is much insisted upon by some, but which is more specious than important. It is that on each side of the line of operations the country should be cleared of all enemies for a distance equal to the depth of this line: otherwise the enemy might threaten the line of retreat. This rule is everywhere belied by the events of war. The nature of the country, the rivers and mountains, the morale of the armies, the spirit of the people, the ability and energy of the commanders, cannot be estimated by diagrams on paper. It is true that no considerable bodies of the enemy could be permitted on the flanks of the line of retreat; but a compliance with this demand would deprive an army of every means of taking a step in a hostile country; and there is not a campaign in recent wars, or in those of Marlborough and Eugene, which does not contradict this assertion. Was not General Moreau at the gates of Vienna

when Fussen, Scharnitz, and all the Tyrol were in possession of the Austrians? Was not Napoleon at Piacenza when Turin, Genoa, and the Col-di-Tenda were occupied by the army of Mélas? Did not Eugene march by way of Stradella and Asti to the aid of Turin, leaving the French upon the Mincio but a few leagues from his base?

Observations upon Interior Lines—What Has Been Said Against Them

Some of my critics have disputed as to the meaning of words and upon definitions; others have censured where they but imperfectly understood; and others have, by the light of certain important events, taken it upon themselves to deny my fundamental principles, without inquiring whether the conditions of the case which might modify the application of these principles were such as were supposed, or without reflecting that, even admitting what they claimed to be true, a single exception cannot disprove a rule based upon the experience of ages and upon natural principles.

In opposition to my maxims upon interior lines, some have quoted the famous and successful march of the allies upon Leipsic. This remarkable event, at first glance, seems to stagger the faith of those who believe in principles. At best, however, it is but one of those exceptional cases from which nothing can be inferred in the face of thousands of opposed instances. Moreover, it is easy to show that, far from overthrowing the maxims it has been brought to oppose, it will go to establish their soundness. Indeed, the critics had forgotten that in case of a considerable numerical superiority I recommended double lines of operations as most advantageous, particularly when concentric and arranged to combine an effort against the enemy at the decisive moment. Now, in the allied armies of Schwarzenberg, Blücher, Bernadotte, and Benningsen, this case of decided superiority is found. The inferior army, to conform to the principles of this chapter, should have directed its efforts against one of the extremities of his adversary, and not upon the center as it did: so that the events quoted against me are doubly in my favor.

Moreover, if the central position of Napoleon between Dresden and the Oder was disastrous, it must be attributed to the misfortunes of Culm, Katzbach, and Dennewitz,—in a word, to faults of execution, entirely foreign to the principles in question.

What I propose is, to act offensively upon the most important point with the greater part of the forces, but upon the secondary points to remain on the defensive, in strong positions or behind a river, until the decisive blow is struck, and the operation ended by the total defeat of an essential part of the army. Then the combined efforts of the whole army may be directed upon other points. Whenever the secondary armies are exposed to a decisive shock during the absence of the mass of the army, the system is not understood; and this was what happened in 1813.

If Napoleon, after his victory at Dresden, had vigorously pursued the allies into Bohemia, he would have escaped the disaster at Culm, have threatened Prague, and perhaps have dissolved the Coalition. To this error may be added a fault quite as great,—that of fighting decisive battles when he was not present with the mass of his forces. At Katzbach his instructions were not obeyed. He ordered Macdonald to wait for Blücher, and to fall upon him when he should expose himself by bold movements. Macdonald, on the contrary, crossed his detachments over torrents which were hourly becoming more swollen, and advanced to meet Blücher. If he had fulfilled his instructions and Napoleon had followed up his victory, there is no doubt that his plan of operations, based upon interior strategic lines and positions and upon a concentric line of operations, would have met with the most brilliant success. The study of his campaigns in Italy in 1796 and in France in 1814 shows that he knew how to apply this system.

There is another circumstance, of equal importance, which shows the injustice of judging central lines by the fate of Napoleon in Saxony,—viz.: *that his front of operations was outflanked on the right, and even taken in reverse, by the geographical position of the frontiers of Bohemia.* Such a case is of rare occurrence. A central position with such faults is not to be compared to one without them. When Napoleon made the application of these principles in Italy, Poland, Prussia, and France, he was not exposed to the attack of a hostile enemy on his flanks and rear. Austria could have threatened him in 1807; but she was then at peace with him and unarmed. To judge of a system of operations, it must be supposed that accidents and chances are to be as much in favor of as against it,—which was by no means the case in 1813, either in the geographic positions or in the state of the respective forces. Independently of this, it is absurd to quote the reverses at Katzbach and Dennewitz, suffered by his lieutenants, as proof capable of destroying a principle the simplest application of which required these officers not to allow themselves to be drawn into a serious engagement. Instead of avoiding they sought collisions. Indeed, what advantage can be expected from the system of central lines, if the parts of the army which have been weakened in order to strike decisive blows elsewhere, shall themselves seek a disastrous contest, instead of being contented with being bodies of observation? [Footnote 18: I am well aware that it is not always possible to avoid a combat without running greater risks than would result from a check; but Macdonald might have fought Blücher to advantage if he had better understood Napoleon's instructions.] In this case it is the enemy who applies the principle, and not he who has the interior lines. Moreover, in the succeeding campaign, the defense of Napoleon in Champagne, from the battle of Brienne to that of Paris, demonstrates fully the truth of these maxims.

The analysis of these two celebrated campaigns raises a strategic question which it would be difficult to answer by simple assertions founded upon theories. It is, whether the system of central lines loses its advantages when the masses are very large. Agreeing with Montesquieu, that the greatest enterprises fail from the magnitude of the arrangements necessary to consummate them, I am disposed to answer in the affirmative. It is very clear to me that an army of one hundred thousand men, occupying a central zone against three isolated armies of thirty or thirty-five thousand men, would be more sure of defeating them successively than if the central mass were four hundred thousand strong against three armies of one hundred and thirty-five thousand each; and for several good reasons:—

1. Considering the difficulty of finding ground and time necessary to bring a very large force into action on the day of battle, an army of one hundred and thirty or one hundred and forty thousand men may easily resist a much larger force.

2. If driven from the field, there will be at least one hundred thousand men to protect and insure an orderly retreat and effect a junction with one of the other armies.

3. The central army of four hundred thousand men requires such a quantity of provisions, munitions, horses, and *matériel* of every kind, that it will possess less mobility and facility in shifting its efforts from one part of the zone to another; to say nothing of the impossibility of obtaining provisions from a region too restricted to support such numbers.

4. The bodies of observation detached from the central mass to hold in check two armies of one hundred and thirty-five thousand each must be very strong, (from eighty to ninety thousand each;) and, being of such magnitude, if they are drawn into a serious engagement they will probably suffer reverses, the effects of which might outweigh the advantages gained by the principal army.

I have never advocated exclusively either a concentric or eccentric system. All my works go to show the eternal influence of principles, and to demonstrate that operations to be successful must be applications of principles.

Divergent or convergent operations may be either very good or very bad: all depends on the situation of the respective forces. The eccentric lines, for instance, are good when applied to a mass starting from a given point, and acting in divergent directions to divide and separately destroy two hostile forces acting upon exterior lines. Such was the maneuver of Frederick which brought about, at the end of the campaign of 1767, the fine battles of Rossbach and Leuthen. Such were nearly all the operations of Napoleon, whose favorite maneuver was to unite, by closely-calculated marches, imposing masses on the center, and, having pierced the enemy's center or turned his front, to give them eccentric directions to disperse the defeated army. [It will not be thought strange that I sometimes approve of concentric, and at other times divergent, maneuvers, when we reflect that among the finest operations of Napoleon there are some in which he employed these two systems alternately within twenty-four hours; for example, in the movements about Ratisbon in 1809.]

On the other hand, concentric operations are good in two cases: 1. When they tend to concentrate a scattered army upon a point where it will be sure to arrive before the enemy; 2. When they direct to the same end the efforts of two armies which are in no danger of being beaten separately by a stronger enemy.

Concentric operations, which just now seem to be so advantageous, may be most pernicious,—which should teach us the necessity of detecting the principles upon which systems are based, and not to confound principles and systems; as, for instance, if two armies set out from a distant base to march convergently upon an enemy whose forces are on interior lines and more concentrated, it follows that the latter could effect a union before the former, and would inevitably defeat them; as was the case with Moreau and Jourdan in 1796, opposed to the Archduke Charles.

In starting from the same points, or from two points much less separated than Dusseldorf and Strasbourg, an army may be exposed to this danger. What was the fate of the concentric columns of Wurmser and Quasdanovitch, wishing to reach the Mincio by the two banks of Lake Garda? Can the result of the march of Napoleon and Grouchy on Brussels be forgotten? Leaving Sombref, they were to march concentrically on this city,—one by Quatre-Bras, the other by Wavre. Blücher and Wellington, taking an interior strategic line, effected a junction before them, and the terrible disaster of Waterloo proved to the world that the immutable principles of war cannot be violated with impunity.

Such events prove better than any arguments that a system which is not in accordance with the principles of war cannot be good. I lay no claim to the creation of these principles, for they have always existed, and were applied by Cæsar, Scipio, and the Consul Nero, as well as by Marlborough and Eugene; but I claim to have been the first to point them out, and to lay down the principal chances in their various applications.

Article XXII: Strategic Lines

Mention has already been made of strategic lines of maneuvers, which differ essentially from lines of operations; and it will be well to define them, for many confound them. We will not consider those strategic lines which have a great and permanent importance by reason of their position and their relation to the features of the country, like the lines of the Danube and the Meuse, the chains of the Alps and the Balkan. Such lines can best be studied by a detailed and minute examination of the topography of Europe; and an excellent model for this kind of study is found in the Archduke Charles's description of Southern Germany.

The term *strategic* is also applied to all communications which lead by the most direct or advantageous route from one important point to another, as well as from the strategic front of the army to all of its objective points. It will be seen, then, that a theater of war is crossed by a multitude of such lines, but that at any given time those only which are concerned in the projected enterprise have any real importance. This renders plain the distinction between the general line of operations of a whole campaign, and these *strategic* lines, which are temporary and change with the operations of the army.

Besides territorial strategic lines, there are *strategic lines of maneuvers.*

An army having Germany as its general field might adopt as its zone of operations the space between the Alps and the Danube, or that between the Danube and the Main, or that between the mountains of Franconia and the sea. It would have upon its zone a single line of operations, or, at most, a double concentric line, upon interior, or perhaps exterior, directions,—while it would have successively perhaps twenty strategic lines as its enterprises were developed: it would have at first one for each wing which would join the general line of operations. If it operated in the zone between the Danube and the Alps, it might adopt, according to events, the strategic line leading from Ulm on Donauwerth and Ratisbon, or that from Ulm to the Tyrol, or that which connects Ulm with Nuremberg or Mayence.

It may, then, be assumed that the definitions applied to lines of operations, as well as the maxims referring to them, are necessarily applicable to strategic lines. These may be *concentric*, to inflict a decisive blow, or *eccentric*, after victory. They are rarely *simple*, since an army does not confine its march to a single road; but when they are double or triple, or even quadruple, they should be *interior* if the forces be equal, or *exterior* in the case of great numerical superiority. The rigorous application of this rule may perhaps sometimes be remitted in detaching a body on an exterior line, even when the forces are equal, to attain an important result without running much risk; but this is an affair of detachments, and does not refer to the important masses.

Strategic lines cannot be interior when our efforts are directed against one of the extremities of the enemy's front of operations.

The maxims above given in reference to lines of operations holding good for strategic lines, it is not necessary to repeat them, or to apply them to particular examples; but there is one, however, which deserves mention,—viz.: that it is important generally, in the selection of these temporary strategic lines, not to leave the line of operations exposed to the assaults of the enemy. Even this may, however, be done, to extricate the army from great danger, or to attain a great success; but the operation must be of short duration, and care must have been taken to prepare a plan of safe retreat, by a sudden change of the line of operations, if necessary, as has already been referred to.

We will illustrate this by the campaign of Waterloo. The Prussian army was based upon the Rhine, its line of operations extended from Cologne and Coblentz on Luxembourg and Namur; Wellington's base was Antwerp, and his line of operations the short road to Brussels. The sudden attack by Napoleon on Flanders decided Blücher to receive battle parallel to the English base, and not to his own, about which he seemed to have no uneasiness. This was pardonable, because he could always have a good chance of regaining Wesel or Nimeguen, and even might seek a refuge in Antwerp in the last extremity; but if the army had not had its powerful maritime allies it would have been destroyed. Beaten at Ligny, and seeking refuge at Gembloux and then at Wavre, Blücher had but three strategic lines to choose from: that which led directly to Maestricht, that farther north on Venloo, or the one leading to the English army near Mont St. Jean. He audaciously took the last, and triumphed by the application of interior strategic lines,—which Napoleon here, perhaps for the first time in his life, neglected. It will readily be seen that the line followed from Gembloux by Wavre to Mont St. Jean was neither a line of operations of the Prussian army nor a line of battle, but a *strategic line of maneuver*, and was interior. It was bold, because he exposed fully his own natural line of operations. The fact that he sought a junction with the English made his movement accord with the principles of war.

A less successful example was that of Ney at Dennewitz. Leaving Wittenberg, and going in the direction of Berlin, he moved to the right to gain the extreme left of the allies, but in so doing he left his primitive line of retreat exposed to the attacks of an enemy superior in force. His object was to gain communication with Napoleon, whose intention was to join him by Herzberg or Luckau; but Ney should from the beginning have taken all logistic and tactical means of accomplishing this change of strategic line and of informing his army of it. He did nothing of this kind,—either from forgetfulness, or on account of the feeling of aversion he had to any thing like a retreat,—and the severe losses at Dennewitz were the result.

Napoleon in 1796 gave one of the best illustrations of these different combinations of strategic lines. His general line of operations extended from the Apennines to Verona. When he had driven Wurmser upon Roveredo and determined to pursue him into the Tyrol, he pushed on in the valley of the Adige to Trent and the Lavis, where he learned that Wurmser had moved by the Brenta on the Frioul, doubtless to take him in reverse. There were but three courses open to him,—to remain in the narrow valley of the Adige at great risk, to retreat by Verona to meet Wurmser, or the last,—which was sublime, but rash,—to follow him into the valley of the Brenta, which was encircled by rugged mountains whose two passages might be held by the Austrians. Napoleon was not the man to hesitate between three such alternatives. He left Vaubois on the Lavis to cover Trent, and marched with the remainder of his forces on Bassano. The brilliant results of this bold step are well known. The route from Trent to Bassano was not the line of operations of the army, but a *strategic line of maneuver* still bolder than that of Blücher on Wavre. However, it was an operation of only three or four days' duration, at the end of which time Napoleon would either beat or be beaten at Bassano: in the first case, he would open direct communication with Verona and his line of operations; in the second, he could regain in great haste Trent, where, reinforced by Vaubois, he could fall back either upon Verona or Peschiera. The difficulties of the country, which made this march audacious in one respect, were favorable in another; for even if Wurmser had been victorious at Bassano he could not have interfered with the return to Trent, as there was no road to enable him to anticipate Napoleon. If Davidovitch on the Lavis had driven Vaubois from Trent, he might have embarrassed Napoleon; but this Austrian general, previously beaten at Roveredo, and ignorant of what the French army was doing for several days, and thinking it was all upon him, would scarcely have thought of resuming the offensive before Napoleon beaten at Bassano would have been on his retreat. Indeed, if Davidovitch had advanced as far as Roveredo, driving Vaubois before him, he would there have been surrounded by two French armies, who would have inflicted upon him the fate of Vandamme at Culm.

I have dwelt on this event to show that a proper calculation of time and distances, joined to great activity, may lead to the success of many adventures which may seem very imprudent. I conclude from this that it may be well sometimes to direct an army upon a route which exposes its line of operations, but that every measure must be taken to prevent the enemy from profiting by it, both by great rapidity of execution and by demonstrations which will deceive him and leave him in ignorance of what is taking place. Still, it is a very hazardous maneuver, and only to be adopted under an urgent necessity.

Article XXIII: Means of protecting a Line of Operations by Temporary Bases or Strategic Reserves

When a general enters a country offensively, he should form eventual or temporary bases,—which, of course, are neither so safe nor so strong as his own frontiers. A river with *têtes de ponts*, and one or two large towns secure from a *coup de main* to cover the depots of the army and to serve as points of assembling for the reserve troops, would be an excellent base of this kind. Of course, such a line could not be a temporary base if a hostile force were near the line of operations leading to the real base on the frontiers. Napoleon would have had a good real base on the Elbe in 1813 if Austria had remained neutral; but, she having joined his enemies, this line was taken in reverse, and became but a pivot of operations, favorable indeed for the execution of a single enterprise, but dangerous for a prolonged occupation, particularly in case of a serious reverse. As every army which is beaten in an enemy's country is exposed to the danger of being cut off from its own frontiers if it continues to occupy the country, these distant temporary bases are rather temporary points of support than real bases, and are in a measure eventual lines of defense. In general, we cannot expect to find in an enemy's country safe positions suitable even for a temporary base; and the deficiency must be supplied by a strategic reserve,—which is purely a modern invention. Its merits and demerits deserve notice.

Strategic Reserves

Reserves play an important part in modern warfare. From the executive, who prepares national reserves, down to the chief of a platoon of skirmishers, every commander now desires a reserve. A wise government always provides good reserves for its armies, and the general uses them when they come under his command. The state has its reserves, the army has its own, and every corps d'armée or division should not fail to provide one.

The reserves of an army are of two kinds,—those on the battle-field, and those which are intended to recruit and support the army: the latter, while organizing, may occupy important points of the theater of war, and serve even as strategic reserves; their positions will depend not only on their magnitude, but also on the nature of the frontiers and the distance from the base to the front of operations. Whenever an army takes the offensive, it should always contemplate the possibility of being compelled to act on the defensive, and by the posting of a reserve between the base and front of operations the advantage of an active reserve on the field of battle is gained: it can fly to the support of menaced points without weakening the active army. It is true that to form a reserve a number of regiments must be withdrawn from active service; but there are always reinforcements to arrive, recruits to be instructed, and convalescents to be used; and by organizing central depots for preparation of munitions and equipments, and by making them the rendezvous of all detachments going to and coming from the army, and adding to them a few good regiments to give tone, a reserve may be formed capable of important service.

Napoleon never failed to organize these reserves in his campaigns. Even in 1797, in his bold march on the Noric Alps, he had first Joubert on the Adige, afterward Victor (returning from the Roman States) in the neighborhood of Verona. In 1805 Ney and Augereau played the part alternately in the Tyrol and Bavaria, and Mortier and Marmont near Vienna.

In 1806 Napoleon formed like reserves on the Rhine, and Mortier used them to reduce Hesse. At the same time, other reserves were forming at Mayence under Kellermann, which took post, as fast as organized, between the Rhine and Elbe, while Mortier was sent into Pomerania. When Napoleon decided to push on to the Vistula in the same year, he directed, with much ostentation, the concentration of an army on the Elbe sixty thousand strong, its object being to protect Hamburg against the English and to influence Austria, whose disposition was as manifest as her interests.

The Prussians established a similar reserve in 1806 at Halle, but it was badly posted: if it had been established upon the Elbe at Wittenberg or Dessau, and had done its duty, it might have saved the army by giving Prince Hohenlohe and Blücher time to reach Berlin, or at least Stettin.

These reserves are particularly useful when the configuration of the country leads to double fronts of operations: they then fulfill the double object of observing the second front, and, in case of necessity, of aiding the operations of the main army when the enemy threatens its flanks or a reverse compels it to fall back toward this reserve.

Of course, care must be taken not to create dangerous detachments, and whenever these reserves can be dispensed with, it should be done, or the troops in the depots only be employed as reserves. It is only in distant invasions and sometimes on our own soil that they are useful: if the scene of hostilities be but five or six marches distant from the frontier, they are quite superfluous. At home they may generally be dispensed with: it is only in the case of a serious invasion, when new levies are organizing, that such a reserve, in an intrenched camp, under the protection of a fortress which serves as a great depot, will be indispensable.

The general's talents will be exercised in judging of the use of these reserves according to the state of the country, the length of the line of operations, the nature of the fortified points, and the proximity of a hostile state. He also decides upon their position, and endeavors to use for this purpose troops which will not weaken his main army so much as the withdrawal of his good troops.

These reserves ought to hold the most important points between the base and front of operations, occupy the fortified places if any have been reduced, observe or invest those which are held by the enemy; and if there be no fortress as a point of support, they should throw up intrenched camps or *têtes de ponts* to protect the depots and to increase the strength of their positions.

All that has been said upon pivots of operations is applicable to temporary bases and to strategic reserves, which will be doubly valuable if they possess such well-located pivots.

Article XXIV: The Old System of Wars of Position and the Modern System of Marches

By the system of positions is understood the old manner of conducting a methodical war, with armies in tents, with their supplies at hand, engaged in watching each other; one besieging a city, the other covering it; one, perhaps, endeavoring to acquire a small province, the other counteracting its efforts by occupying strong points. Such was war from the Middle Ages to the era of the French Revolution. During this revolution great changes transpired, and many systems of more or less value sprang up. War was commenced in 1792 as it had been in 1762: the French encamped near their strong places, and the allies besieged them. It was not till 1793, when assailed from without and within, that this system was changed. Thoroughly aroused, France threw one million men in fourteen armies upon her enemies. These armies had neither tents, provisions, nor money. On their marches they bivouacked or were quartered in towns; their mobility was increased and became a means of success. Their tactics changed also: the troops were put in columns, which were more easily handled than deployed lines, and, on account of the broken character of the country of Flanders and the Vosges, they threw out a part of their force as skirmishers to protect and cover the columns. This system, which was thus the result of circumstances, at first met with a success beyond all expectation: it disconcerted the methodical Austrian and Prussian troops as well as their generals. Mack, to whom was attributed the success of the Prince of Coburg, increased his reputation by directing the troops to extend their lines to oppose an open order to the fire of skirmishers. It had never occurred to the poor man that while the skirmishers made the noise the columns carried the positions.

The first generals of the Republic were fighting-men, and nothing more. The principal direction of affairs was in the hands of Carnot and of the Committee of Public Safety: it was sometimes judicious, but often bad. Carnot was the author of one of the finest strategic movements of the war. In 1793 he sent a reserve of fine troops successively to the aid of Dunkirk, Maubeuge, and Landau, so that this small force, moving rapidly from point to point, and aided by the troops already collected at these different points, compelled the enemy to evacuate France.

The campaign of 1794 opened badly. It was the force of circumstances, and not a premeditated plan, which brought about the strategic movement of the army of the Moselle on the Sambre; and it was this which led to the success of Fleurus and the conquest of Belgium.

In 1795 the mistakes of the French were so great that they were imputed to treachery. The Austrians, on the contrary, were better commanded by Clairfayt, Chateler, and Schmidt than they had been by Mack and the Prince of Coburg. The Archduke Charles, applying the principle of interior lines, triumphed over Moreau and Jourdan in 1796 by a single march.

Up to this time the fronts of the French armies had been large,—either to procure subsistence more easily, or because the generals thought it better to put all the divisions in line, leaving it to their commanders to arrange them for battle. The reserves were small detachments, incapable of redeeming the day even if the enemy succeeded in overwhelming but a single division. Such was the state of affairs when Napoleon made his *début* in Italy. His activity from the beginning worsted the Austrians and Piedmontese: free from useless incumbrances, his troops surpassed in mobility all modern armies. He conquered the Italian peninsula by a series of marches and strategic combats. His march on Vienna in 1797 was rash, but justified by the necessity of overcoming the Archduke Charles before he could receive reinforcements from the Rhine.

The campaign of 1800, still more characteristic of the man, marked a new era in the conception of plans of campaign and lines of operations. He adopted bold objective points, which looked to nothing less than the capture or destruction of whole armies. The orders of battle were less extended, and the more rational organization of armies in large bodies of two or three divisions was adopted. The system of modern strategy was here fully developed, and the campaigns of 1805 and 1806 were merely corollaries to the great problem solved in 1800. Tactically, the system of columns and skirmishers was too well adapted to the features of Italy not to meet with his approval.

It may now be a question whether the system of Napoleon is adapted to all capacities, epochs, and armies, or whether, on the contrary, there can be any return, in the light of the events of 1800 and 1809, to the old system of wars of position. After a comparison of the marches and camps of the Seven Years' War with those of the *seven weeks' war,*—as Napoleon called the campaign of 1806,—or with those of the three months which elapsed from the departure of the army from Boulogne in 1805 till its arrival in the plains of Moravia, the reader may easily decide as to the relative merits of the two systems.

The system of Napoleon was *to march twenty-five miles a day, to fight, and then to camp in quiet.* He told me that he knew no other method of conducting a war than this.

It may be said that the adventurous character of this great man, his personal situation, and the tone of the French mind, all concurred in urging him to undertakings which no other person, whether born upon a throne, or a general under the orders of his government, would ever dare to adopt. This is probably true; but between the extremes of very distant invasions, and wars of position, there is a proper mean, and, without imitating his impetuous audacity, we may pursue the line he has marked out. It is probable that the old system of wars of positions will for a long time be proscribed, or that, if adopted, it will be much modified and improved.

If the art of war is enlarged by the adoption of the system of marches, humanity, on the contrary, loses by it; for these rapid incursions and bivouacs of considerable masses, feeding upon the regions they overrun, are not materially different from the devastations of the barbarian hordes between the fourth and thirteenth centuries. Still, it is not likely that the system will be speedily renounced; for a great truth has been demonstrated by Napoleon's wars,—viz.: that remoteness is not a certain safeguard against invasion,—that a state to be secure must have a good system of fortresses and lines of defense, of reserves and military institutions, and, finally, a good system of government. Then the people may everywhere be organized as militia, and may serve as reserves to the active armies, which will render the latter more formidable; and the greater the strength of the armies the more necessary is the system of rapid operations and prompt results.

If, in time, social order assumes a calmer state,—if nations, instead of fighting for their existence, fight only for their interests, to acquire a natural frontier or to maintain the political equilibrium,—then a new right of nations may be agreed upon, and perhaps it will be possible to have armies on a less extensive scale. Then also we may see armies of from eighty to one hundred thousand men return to a mixed system of war,—a mean between the rapid incursions of Napoleon and the slow system of positions of the last century. Until then we must expect to retain this system of marches, which has produced so great results; for the first to renounce it in the presence of an active and capable enemy would probably be a victim to his indiscretion.

The science of marches now includes more than details, like the following, viz.: the order of the different arms in column, the time of departure and arrival, the precautions to be observed in the march, and the means of communication between the columns, all of which is a part of the duties of the staff of an army. Outside and beyond these very important details, there is a science of marches in the great operations of strategy. For instance, the march of Napoleon by the Saint-Bernard to fall upon the communications of Mélas, those made in 1805 by Donauwerth to cut off Mack, and in 1806 by Gera to turn the Prussians, the march of Suwaroff from Turin to the Trebbia to meet Macdonald, that of the Russian army on Taroutin, then upon Krasnoi, were decisive operations, not because of their relation to Logistics, but on account of their strategic relations.

Indeed, these skillful marches are but applications of the great principle of throwing the mass of the forces upon the decisive point; and this point is to be determined from the considerations given in Article XIX. What was the passage of the Saint-Bernard but a line of operations directed against an extremity of the strategic front of the enemy, and thence upon his line of retreat? The marches of Ulm and Jena were the same maneuvers; and what was Blücher's march at Waterloo but an application of interior strategic lines?

From this it may be concluded that all strategic movements which tend to throw the mass of the army successively upon the different points of the front of operations of the enemy, will be skillful, as they apply the principle of overwhelming a smaller force by a superior one. The operations of the French in 1793 from Dunkirk to Landau, and those of Napoleon in 1796, 1809, and 1814, are models of this kind.

One of the most essential points in the science of modern marches, is to so combine the movements of the columns as to cover the greatest strategic front, when beyond the reach of the enemy, for the triple object of deceiving him as to the objective in view, of moving with ease and rapidity, and of procuring supplies with more facility. However, it is necessary in this case to have previously arranged the means of concentration of the columns in order to inflict a decisive blow.

This alternate application of extended and concentric movements is the true test of a great general.

There is another kind of marches, designated as *flank marches*, which deserves notice. They have always been held up as very dangerous; but nothing satisfactory has ever been written about them. If by the term *flank marches* are understood tactical maneuvers made upon the field of battle in view of the enemy, it is certain that they are very delicate operations, though sometimes successful; but if reference is made to ordinary strategic marches, I see nothing particularly dangerous in them, unless the most common precautions of Logistics be neglected. In a strategic movement, the two hostile armies ought to be separated by about two marches, (counting the distance which separates the advanced guards from the enemy and from their own columns.) In such a case there could be no danger in a strategic march from one point to another.

There are, however, two cases where such a march would be altogether inadmissible: the first is where the system of the line of operations, of the strategic lines, and of the front of operations is so chosen as to present the flank to the enemy during a whole operation. This was the famous project of marching upon Leipsic, leaving Napoleon and Dresden on the flank, which would, if carried out, have proved fatal to the allies. It was modified by the Emperor Alexander upon the solicitations of the author.

The second case is where the line of operations is very long, (as was the case with Napoleon at Borodino,) and particularly if this line affords but a single suitable route for retreat: then every flank movement exposing this line would be a great fault.

In countries abounding in secondary communications, flank movements are still less dangerous, since, if repulsed, safety may be found in a change of the line of operations. The physical and moral condition of the troops and the more or less energetic characters of the commanders will, of course, be elements in the determination of such movements.

The often-quoted marches of Jena and Ulm were actual flank maneuvers; so was that upon Milan after the passage of the Chiusella, and that of Marshal Paskevitch to cross the Vistula at Ossiek; and their successful issue is well known.

A tactical maneuver by the flank in the presence of the enemy is quite a different affair. Ney suffered for a movement of this kind at Dennewitz, and so did Marmont at Salamanca and Frederick at Kolin.

Nevertheless, the celebrated maneuver of Frederick at Leuthen was a true flank movement, but it was covered by a mass of cavalry concealed by the heights, and applied against an army which lay motionless in its camp; and it was so successful because at the time of the decisive shock Daun was taken in flank, and not Frederick.

In the old system of marching in column at platoon distance, where line of battle could be formed to the right or left without deployment, (by a right or left into line,) movements parallel to the enemy's line were not *flank marches,* because the flank of the column was the real front of the line of battle.

The famous march of Eugene within view of the French army, to turn the lines of Turin, was still more extraordinary than that of Leuthen, and no less successful.

In these different battles, the maneuvers were tactical and not strategic. The march of Eugene from Mantua to Turin was one of the greatest strategic operations of the age; but the case above referred to was a movement made to turn the French camp the evening before the battle.

Article XXV: Depots of Supplies, and their Relation to Marches

The subject most nearly connected with the system of marches is the commissariat, for to march quickly and for a long distance food must be supplied; and the problem of supporting a numerous army in an enemy's country is a very difficult one. It is proposed to discuss the relation between the commissariat and strategy.

It will always be difficult to imagine how Darius and Xerxes subsisted their immense armies in Thrace, where now it would be a hard task to supply thirty thousand men. During the Middle Ages, the Greeks, barbarians, and more lately the Crusaders, maintained considerable bodies of men in that country. Cæsar said that war should support war, and he is generally believed to have lived at the expense of the countries he overran.

The Middle Ages were remarkable for the great migrations of all kinds, and it would be interesting to know the numbers of the Huns, Vandals, Goths, and Mongols who successively traversed Europe, and how they lived during their marches. The commissariat arrangements of the Crusaders would also be an interesting subject of research.

In the early periods of modern history, it is probable that the armies of Francis I., in crossing the Alps into Italy, did not carry with them large stores of provisions; for armies of their magnitude, of forty or fifty thousand men, could easily find provisions in the rich valleys of the Ticino and Po.

Under Louis XIV. and Frederick II. the armies were larger; they fought on their own frontiers, and lived from their storehouses, which were established as they moved. This interfered greatly with operations, restricting the troops within a distance from the depots dependent upon the means of transportation, the rations they could carry, and the number of days necessary for wagons to go to the depots and return to camp.

During the Revolution, depots of supply were abandoned from necessity. The large armies which invaded Belgium and Germany lived sometimes in the houses of the people, sometimes by requisitions laid upon the country, and often by plunder and pillage. To subsist an army on the granaries of Belgium, Italy, Swabia, and the rich banks of the Rhine and Danube, is easy,—particularly if it marches in a number of columns and does not exceed one hundred or one hundred and twenty thousand men; but this would be very difficult in some other countries, and quite impossible in Russia, Sweden, Poland, and Turkey. It may readily be conceived how great may be the rapidity and impetuosity of an army where every thing depends only on the strength of the soldiers' legs. This system gave Napoleon great advantages; but he abused it by applying it on too large a scale and to countries where it was impracticable.

A general should be capable of making all the resources of the invaded country contribute to the success of his enterprises: he should use the local authorities, if they remain, to regulate the assessments so as to make them uniform and legal, while he himself should see to their fulfillment. If the authorities do not remain, he should create provisional ones of the leading men, and endow them with extraordinary powers. The provisions thus acquired should be collected at the points most convenient for the operations of the army. In order to husband them, the troops may be quartered in the towns and villages, taking care to reimburse the inhabitants for the extra charge thus laid upon them. The inhabitants should also be required to furnish wagons to convey the supplies to the points occupied by the troops.

It is impossible to designate precisely what it will be prudent to undertake without having previously established these depots, as much depends upon the season, country, strength of the armies, and spirit of the people; but the following may be considered as general maxims:—

1. That in fertile and populous regions not hostile, an army of one hundred to one hundred and twenty thousand men, when so far distant from the enemy as to be able safely to recover a considerable extent of country, may draw its resources from it, during the time occupied by any single operation.

As the first operation never requires more than a month, during which time the great body of the troops will be in motion, it will be sufficient to provide, by depots of provisions, for the eventual wants of the army, and particularly for those of the troops obliged to remain at a particular point. Thus, the army of Napoleon, while half of it was besieging Ulm, would need bread until the surrender of the city; and if there had been a scarcity the operation might have failed.

2. During this time every effort should be made to collect the supplies obtained in the country, and to form depots, in order to subserve the wants of the army after the success of the operation, whether it take a position to recruit or whether it undertake a new enterprise.

3. The depots formed either by purchase or forced requisitions should be echeloned as much as possible upon three different lines of communication, in order to supply with more facility the wings of the army, and to extend as much as possible the area from which successive supplies are to be drawn, and, lastly, in order that the depots should be as well covered as possible. To this end, it would be well to have the depots on lines converging toward the principal line of operations, which will be generally found in the center. This arrangement has two real advantages: first, the depots are less exposed to the attempts of the enemy, as his distance from them is thereby increased; secondly, it facilitates the movements of the army in concentrating upon a single point of the line of operations to the rear, with a view of retaking the initiative from the enemy, who may have temporarily assumed the offensive and gained some advantage.

4. In thinly-settled and unproductive regions the army will lack its most necessary supplies: it will be prudent, in this case, not to advance too far from its depots, and to carry with it sufficient provisions to enable it, if compelled to do so, to fall back upon its lines of depots.

5. In national wars where the inhabitants fly and destroy every thing in their path, as was the case in Spain, Portugal, Russia, and Turkey, it is impossible to advance unless attended by trains of provisions and without having a sure base of supply near the front of operations. Under these circumstances a war of invasion becomes very difficult, if not impossible.

6. It is not only necessary to collect large quantities of supplies, but it is indispensable to have the means of conveying them with or after the army; and this is the greatest difficulty, particularly on rapid expeditions. To facilitate their transportation, the rations should consist of the most portable articles,—as biscuit, rice, etc..: the wagons should be both light and strong, so as to pass over all kinds of roads. It will be necessary to collect all the vehicles of the country, and to insure good treatment to their owners or drivers; and these vehicles should be arranged in parks at different points, so as not to take the drivers too far from their homes and in order to husband the successive resources. Lastly, the soldier must he habituated to carry with him several days' rations of bread, rice, or even of flour.

7. The vicinity of the sea is invaluable for the transportation of supplies; and the party which is master on this element can supply himself at will. This advantage, however, is not absolute in the case of a large continental army; for, in the desire to maintain communications with its depots, it may be drawn into operations on the coast, thus exposing itself to the greatest risks if the enemy maneuver with the mass of his forces upon the extremity opposite the sea. If the army advance too far from the coast, there will be

danger of its communications being intercepted; and this danger increases with the progress of the army.

8. A continental army using the sea for transportation should base itself on the land, and have a reserve of provisions independent of its ships, and a line of retreat prepared on the extremity of its strategic front opposed to the sea.

9. Navigable streams and canals, when parallel to the line of operations of the army, render the transportation of supplies much easier, and also free the roads from the incumbrances of the numerous vehicles otherwise necessary. For this reason, lines of operations thus situated are the most favorable. The water-communications themselves are not in this case the lines of operations, as has been asserted: on the contrary, it is essential that the troops should be able to move at some distance from the river, in order to prevent the enemy from throwing back the exterior flank upon the river,—which might be as dangerous as if it were the sea.

In the enemy's country the rivers can scarcely ever be used for transportation, since the boats will probably be destroyed, and since a small body of men may easily embarrass the navigation. To render it sure, it is necessary to occupy both banks,—which is hazardous, as Mortier experienced at Dirnstein. In a friendly country the advantages of rivers are more substantial.

10. In default of bread or biscuit, the pressing wants of an army may be fed by cattle on the hoof; and these can generally be found, in populous countries, in numbers to last for some little time. This source of supply will, however, be soon exhausted; and, in addition, this plan leads to plunder. The requisitions for cattle should be well regulated; and the best plan of all is to supply the army with cattle purchased elsewhere.

I will end this article by recording a remark of Napoleon which may appear whimsical, but which is still not without reason. He said that in his first campaigns the enemy was so well provided that when his troops were in want of supplies he had only to fall upon the rear of the enemy to procure every thing in abundance. This is a remark upon which it would be absurd to found a system, but which perhaps explains the success of many a rash enterprise, and proves how much actual war differs from narrow theory.

Article XXVI: The Defense of Frontiers by Forts and Intrenched Lines.—Wars of Sieges

Forts serve two principal purposes: first, to cover the frontiers; secondly, to aid the operations of the campaign.

The defense of frontiers is a problem generally somewhat indeterminate. It is not so for those countries whose borders are covered with great natural obstacles, and which present but few accessible points, and these admitting of defense by the art of the engineer. The problem here is simple; but in open countries it is more difficult. The Alps and the Pyrenees, and the lesser ranges of the Crapacks, of Riesengebirge, of Erzgebirge, of the Böhmerwald, of the Black Forest, of the Vosges, and of the Jura, are not so formidable that they cannot be made more so by a good system of fortresses.

Of all these frontiers, that separating France and Piedmont was best covered. The valleys of the Stura and Suza, the passes of Argentine, of Mont-Genèvre, and of Mont-Cenis,—the only ones considered practicable,—were covered by masonry forts; and, in addition, works of considerable magnitude guarded the issues of the valleys in the plains of Piedmont. It was certainly no easy matter to surmount these difficulties.

These excellent artificial defenses will not always prevent the passage of an army, because the small works which are found in the gorges may be carried, or the enemy, if he be bold, may find a passage over some other route hitherto deemed impracticable. The passage of the Alps by Francis I.,—which is so well described by Gaillard,—Napoleon's passage of the Saint-Bernard, and the Splugen expedition, prove that there is truth in the remark of Napoleon, *that an army can pass wherever a titan can set his foot,*—a maxim not strictly true, but characteristic of the man, and applied by him with great success.

Other countries are covered by large rivers, either as a first line or as a second. It is, however, remarkable that such lines, apparently so well calculated to separate nations without interfering with trade and communication, are generally not part of the real frontier. It cannot be said that the Danube divides Bessarabia from the Ottoman empire as long as the Turks have a foothold in Moldavia. The Rhine was never the real frontier of France and Germany; for the French for long periods held points upon the right bank, while the Germans were in possession of Mayence, Luxembourg, and the *têtes de ponts* of Manheim and Wesel on the left bank.

If, however, the Danube, the Rhine, Rhone, Elbe, Oder, Vistula, Po, and Adige be not exterior lines of the frontier, there is no reason why they should not be fortified as lines of permanent defense, wherever they permit the use of a system suitable for covering a front of operations.

An example of this kind is the Inn, which separates Bavaria from Austria: flanked on the south by the Tyrolese Alps, on the north by Bohemia and the Danube, its narrow front is covered by the three fortified places of Passau, Braunau, and Salzburg. Lloyd, with some poetic license, compares this frontier to two impregnable bastions whose curtain is formed of three fine forts and whose ditch is one of the most rapid of rivers. He has exaggerated these advantages; for his epithet of "impregnable" was decidedly disproved by the bloody events of 1800, 1805, and 1809.

The majority of the European states have frontiers by no means so formidable as that of the Alps and the Inn, being generally open, or consisting of mountains with practicable passes at a considerable number of points. We propose to give a set of general maxims equally applicable to all cases.

When the topography of a frontier is open, there should be no attempt to make a complete line of defense by building too many fortresses, requiring armies to garrison them, and which, after all, might not prevent an enemy from penetrating the country. It is much wiser to build fewer works, and to have them properly located, not with the expectation of absolutely preventing the ingress of the enemy, but to multiply the impediments to his progress, and, at the same time, to support the movements of the army which is to repel him.

If it be rare that a fortified place of itself absolutely prevents the progress of an army, it is, nevertheless, an embarrassment, and compels the army to detach a part of its force or to make *détours* in its march; while, on the other hand, it imparts corresponding advantages to the army which holds it, covers his depots, flanks, and movements, and, finally, is a place of refuge in case of need.

Fortresses thus exercise a manifest influence over military operations; and we now propose to examine their relations to strategy.

The first point to be considered is their location; the second lies in the distinction between the cases where an army can afford to pass the forts without a siege, and those where it will be necessary to besiege; the third point is in reference to the relations of an army to a siege which it proposes to cover.

As fortresses properly located favor military operations, in the same degree those which are unfortunately placed are disadvantageous. They are an incubus upon the army which is compelled to garrison them and the state whose men and money are wasted upon them. There are many in Europe in this category. It is bad policy to cover a frontier with fortresses very close together. This system has been wrongly imputed to Vauban, who, on the contrary, had a controversy with Louvois about the great number of points the latter desired to fortify. The maxims on this point are as follow:—

1. The fortified places should be in echelon, on three lines, and should extend from the frontiers toward the capital. [The memorable campaign of 1829 is evidence of the value of such a system. If the Porte had possessed masonry forts in the defiles of the Balkan and a good fortress toward Faki, the Russians would not have reached Adrianople, and the affair would not have been so simple.] There should be three in the first line, as many in the second, and a large place in the third, near the center of the state. If there be

four fronts, this would require, for a complete system, from twenty-four to thirty places.

It will be objected that this number is large, and that even Austria has not so many. It must be recollected that France has more than forty upon only a third of its frontiers, (from Besançon to Dunkirk,) and still has not enough on the third line in the center of the country. A Board convened for the purpose of considering the system of fortresses has decided quite recently that more were required. This does not prove that there were not already too many, but that certain points in addition should be fortified, while those on the first line, although too much crowded, may be maintained since they are already in existence. Admitting that France has two fronts from Dunkirk to Basel, one from Basel to Savoy, one from Savoy to Nice, in addition to the totally distinct line of the Pyrenees and the coast-line, there are six fronts, requiring forty to fifty places. Every military man will admit that this is enough, since the Swiss and coast fronts require fewer than the northeast. The system of arrangement of these fortresses is an important element of their usefulness. Austria has a less number, because she is bordered by the small German states, which, instead of being hostile, place their own forts at her disposal. Moreover, the number above given is what was considered necessary for a state having four fronts of nearly equal development. Prussia, being long and narrow, and extending from Königsberg almost to the gates of Metz, should not be fortified upon the same system as France, Spain, or Austria. Thus the geographical position and extent of states may either diminish or increase the number of fortresses, particularly when maritime forts are to be included.

2. Fortresses should always occupy the important strategic points already designated in Article XIX. As to their tactical qualities, their sites should not be commanded, and egress from them should be easy, in order to increase the difficulty of blockading them.

3. Those which possess the greatest advantages, either as to their own defense or for seconding the operations of an army, are certainly those situated on great rivers and commanding both banks. Mayence, Coblentz, and Strasbourg, including Kehl, are true illustrations and models of this kind. Places situated at the confluence of two great rivers command three different fronts, and hence are of increased importance. Take, for instance, Modlin. Mayence, when it had on the left bank of the Main the fort of Gustavusburg, and Cassel on the right, was the most formidable place in Europe, but it required a garrison of twenty-five thousand men: so that works of this extent must be few in number.

4. Large forts, when encompassing populous and commercial cities, are preferable to small ones,—particularly when the assistance of the citizens can

be relied on for their defense. Metz arrested the whole power of Charles V, and Lille for a whole year delayed Eugene and Marlborough. Strasbourg has many times proved the security of French armies. During the last wars these places were passed without being besieged by the invading forces, because all Europe was in arms against France; but one hundred and fifty thousand Germans having in their front one hundred thousand French could not penetrate to the Seine with impunity, leaving behind them these well-fortified points.

5. Formerly the operations of war were directed against towns, camps, and positions; recently they have been directed only against organized armies, leaving out of consideration all natural or artificial obstacles. The exclusive use of either of these systems is faulty: the true course is a mean between these extremes. Doubtless, it will always be of the first importance to destroy and disorganize all the armies of the enemy in the field, and to attain this end it may be allowable to pass the fortresses; but if the success be only partial it will be unwise to push the invasion too far. Here, also, very much depends upon the situation and respective strength of the armies and the spirit of the nations.

If Austria were the sole antagonist of France, she could not follow in the footsteps of the allies in 1814; neither is it probable that fifty thousand French will very soon risk themselves beyond the Noric Alps, in the very heart of Austria, as Napoleon did in 1797. [Still, Napoleon was right in taking the offensive in the Frioul, since the Austrians were expecting a reinforcement from the Rhine of twenty thousand men, and of course it was highly important to beat the Archduke Charles before this force joined him. In view of the circumstances of the case, Napoleon's conduct was in accordance with the principles of war.] Such events only occur under exceptional circumstances.

6. It may be concluded from what precedes,—1st, that, while fortified places are essential supports, abuse in their application may, by dividing an army, weaken it instead of adding to its efficiency; 2d, that an army may, with the view of destroying the enemy, pass the line of these forts,—always, however, leaving a force to observe them; 3d, that an army cannot pass a large river, like the Danube or the Rhine, without reducing at least one of the fortresses on the river, in order to secure a good line of retreat. Once master of this place, the army may advance on the offensive, leaving detachments to besiege other places; and the chances of the reduction of those places increase as the army advances, since the enemy's opportunities of hindering the siege are correspondingly diminished.

7. While large places are much the most advantageous among a friendly people, smaller works are not without importance, not to arrest an enemy,

who might mask them, but as they may materially aid the operations of an army in the field. The fort of Königstein in 1813 was as useful to the French as the fortress of Dresden, because it procured a *tête de pont* on the Elbe.

In a mountainous country, small, well-located forts are equal in value to fortified places, because their province is to close the passes, and not to afford refuge to armies: the little fort of Bard, in the valley of Aosta, almost arrested Napoleon's army in 1800.

8. It follows that each frontier should have one or two large fortresses as places of refuge, besides secondary forts and small posts to facilitate military operations. Walled cities with a shallow ditch may be very useful in the interior of a country, to contain depots, hospitals, etc., when they are strong enough to resist the attacks of any small bodies that may traverse the vicinity. They will be particularly serviceable if they can be defended by the militia, so as not to weaken the active army.

9. Large fortified places which are not in proper strategic positions are a positive misfortune for both the army and state.

10. Those on the sea-coast are of importance only in a maritime war, except for depots: they may even prove disastrous for a continental army, by holding out to it a delusive promise of support. Benningsen almost lost the Russian armies by basing them in 1807 on Königsberg,—which he did because it was convenient for supply. If the Russian army in 1812, instead of concentrating on Smolensk, had supported itself on Dunaburg and Riga, it would have been in danger of being forced into the sea and of being cut off from all its bases.

The relations between sieges and the operations of active armies are of two kinds. An invading army may pass by fortified places without attacking them, but it must leave a force to invest them, or at least to watch them; and when there are a number of them adjacent to each other it will be necessary to leave an entire corps d'armée, under a single commander, to invest or watch them as circumstances may require. When the invading army decides to attack a place, a sufficient force to carry on the siege will be assigned to this duty; the remainder may either continue its march or take a position to cover the siege.

Formerly the false system prevailed of encircling a city by a whole army, which buried itself in lines of circumvallation and contravallation. These lines cost as much in labor and expense as the siege itself. The famous case of the lines of Turin, which were fifteen miles in length, and, though guarded by seventy-eight thousand French, were forced by Prince Eugene with forty thousand men in 1706, is enough to condemn this ridiculous system.

Much as the recital of the immense labors of Cæsar in the investment of Alise may excite our admiration, it is not probable that any general in our times will imitate his example. Nevertheless, it is very necessary for the investing force to strengthen its position by detached works commanding the routes by which the garrison might issue or by which the siege might be disturbed from without. This was done by Napoleon at Mantua, and by the Russians at Varna.

Experience has proved that the best way to cover a siege is to beat and pursue as far as possible the enemy's forces which could interfere. If the besieging force is numerically inferior, it should take up a strategic position covering all the avenues by which succor might arrive; and when it approaches, as much of the besieging force as can be spared should unite with the covering force to fall upon the approaching army and decide whether the siege shall continue or not.

Bonaparte in 1796, at Mantua, was a model of wisdom and skill for the operations of an army of observation.

Intrenched Lines

Besides the lines of circumvallation and contravallation referred to above, there is another kind, which is more extended than they are, and is in a measure allied to permanent fortifications, because it is intended to protect a part of the frontiers.

As a fortress or an intrenched camp may, as a temporary refuge for an army, be highly advantageous, so to the same degree is the system of intrenched lines absurd. I do not now refer to lines of small extent closing a narrow gorge, like Fussen and Scharnitz, for they may be regarded as forts; but I speak of extended lines many leagues in length and intended to wholly close a part of the frontiers. For instance, those of Wissembourg, which, covered by the Lauter flowing in front, supported by the Rhine on the right and the Vosges on the left, seemed to fulfill all the conditions of safety; and yet they were forced on every occasion when they were assailed.

The lines of Stollhofen, which on the right of the Rhine played the same part as those of Wissembourg on the left, were equally unfortunate; and those of the Queich and the Kinzig had the same fate.

The lines of Turin, (1706,) and those of Mayence, (1795,) although intended as lines of circumvallation, were analogous to the lines in question in their extent and in the fate which befell them. However well they may be supported by natural obstacles, their great extent paralyzes their defenders, and they are almost always susceptible of being turned. To bury an army in intrenchments, where it may be outflanked and surrounded, or forced in front even if secure from a flank attack, is manifest folly; and it is to be hoped that we shall never see another instance of it. Nevertheless, in our chapter on Tactics we will treat of their attack and defense.

It may be well to remark that, while it is absurd to use these extended lines, it would be equally foolish to neglect the advantages to be derived from detached works in increasing the strength of a besieging force, the safety of a position, or the defense of a defile.

Article XXVII: The Connection of Intrenched Camps and Têtes de Ponts with Strategy

It would be out of place here to go into details as to the sites of ordinary camps and upon the means of covering them by advanced guards, or upon the advantages of field-fortifications in the defense of posts. Only fortified camps enter into the combinations of grand tactics, and even of strategy; and this they do by the temporary support they afford an army.

It may be seen by the example of the camp of Buntzelwitz, which saved Frederick in 1761, and by those of Kehl and Dusseldorf in 1796, that such a refuge may prove of the greatest importance. The camp of Ulm, in 1800, enabled Kray to arrest for a whole month the army of Moreau on the Danube; and Wellington derived great advantages from his camp of Torres-Vedras. The Turks were greatly assisted in defending the country between the Danube and the Balkan Mountains by the camp of Shumla.

The principal rule in this connection is that camps should be established on strategic points which should also possess tactical advantages. If the camp of Drissa was useless to the Russians in 1812, it was because it was not in a proper position in reference to their defensive system, which should have rested upon Smolensk and Moscow. Hence the Russians were compelled to abandon it after a few days.

The maxims which have been given for the determination of the great decisive strategic points will apply to all intrenched camps, because they ought only to be placed on such points. The influence of these camps is variable: they may answer equally well as points of departure for an offensive operation, as *têtes de ponts* to assure the crossing of a large river, as protection for winter quarters, or as a refuge for a defeated army.

However good may be the site of such a camp, it will always be difficult to locate it so that it may not be turned, unless, like the camp of Torres-Vedras, it be upon a peninsula backed by the sea. Whenever it can be passed either by the right or the left, the army will be compelled to abandon it or run the risk of being invested in it. The camp of Dresden was an important support to Napoleon for two months; but as soon as it was outflanked by the allies it had not the advantages even of an ordinary fortress; for its extent led to the sacrifice of two corps within a few days for want of provisions.

Despite all this, these camps, when only intended to afford temporary support to an army on the defensive, may still fulfill this end, even when the enemy passes by them, provided they cannot be taken in reverse,—that is, provided all their faces are equally safe from a *coup de main*. It is also important that they be established close to a fortress, where the depots may be safe, or which may cover the front of the camp nearest to the line of retreat.

In general terms, such a camp on a river, with a large *tête de pont* on the other side to command both banks, and near a large fortified city like Mayence or Strasbourg, is of undoubted advantage; but it will never be more than a temporary refuge, a means of gaining time and of collecting reinforcements. When the object is to drive away the enemy, it will be necessary to leave the camp and carry on operations in the open country.

The second maxim as to these camps is, that they are particularly advantageous to an army at home or near its base of operations. If a French army occupied an intrenched camp on the Elbe, it would be lost when the space between the Rhine and Elbe was held by the enemy; but if it were invested in an intrenched camp near Strasbourg, it might with a little assistance resume its superiority and take the field, while the enemy in the interior of France and between the relieving force and the intrenched army would have great difficulty in recrossing the Rhine.

We have heretofore considered these camps in a strategic light; but several German generals have maintained that they are suitable to cover places or to prevent sieges,—which appears to me to be a little sophistical. Doubtless, it will be more difficult to besiege a place when an army is encamped on its glacis; and it maybe said that the forts and camps are a mutual support; but, according to my view, the real and principal use of intrenched camps is always to afford, if necessary, a temporary refuge for an army, or the means of debouching offensively upon a decisive point or beyond a large river. To bury an army in such a camp, to expose it to the danger of being outflanked and cut off, simply to retard a siege, would be folly. The example of Wurmser, who prolonged the defense of Mantua, will be cited in opposition to this; but did not his army perish? And was this sacrifice really useful? I do not think so; for, the place having been once relieved and revictualed, and the siege-train having fallen into the hands of the Austrians, the siege was necessarily changed into a blockade, and the town could only be taken by reason of famine; and, this being the case, Wurmser's presence ought rather to have hastened than retarded its surrender.

The intrenched camp of the Austrians before Mayence in 1795 would, indeed, have prevented the siege of the place, if the French had possessed the means of carrying on a siege, as long as the Rhine had not been crossed; but as soon as Jourdan appeared on the Lahn, and Moreau in the Black Forest, it became necessary to abandon the camp and leave the place to its own means of defense. It would only be in the event of a fortress occupying a point such that it would be impossible for an army to pass it without taking it, that an intrenched camp, with the object of preventing an attack upon it, would be established; and what place in Europe is upon such a site?

So far from agreeing with these German authors, on the contrary, it seems to me that a very important question in the establishment of these camps near fortified places on a river, is whether they should be on the same bank as the place, or upon the other. When it is necessary to make a choice, by reason of the fact that the place cannot be located to cover both banks, I should decidedly prefer the latter.

To serve as a refuge or to favor a debouch, the camp should be on the bank of the river toward the enemy; and in this, case the principal danger to be feared is that the enemy might take the camp in reverse by passing the river at some other point; and if the fortress were upon the same bank us the camp, it would be of little service; while if upon the other bank, opposite to the camp, it would be almost impossible to take the latter in reverse. For instance, the Russians, who could not hold for twenty-four hours their camp of Drissa, would have defied the enemy for a long time if there had been a fortification on the right bank of the Dwina, covering the rear of the camp. So Moreau for three months, at Kehl, withstood all the efforts of the Archduke Charles; while if Strasbourg had not been there upon the opposite bank his camp would easily have been turned by a passage of the Rhine.

Indeed, it would be desirable to have the protection of the fortified place upon the other bank too; and a place holding both banks would fulfill this condition. The fortification of Coblentz, recently constructed, seems to introduce a new epoch. This system of the Prussians, combining the advantages of intrenched camps and permanent works, deserves attentive consideration; but, whatever may be its defects, it is nevertheless certain that it would afford immense advantages to an army intended to operate on the Rhine. Indeed, the inconvenience of intrenched camps on large rivers is that they are only very useful when beyond the river; and in this case they are exposed to the dangers arising from destruction of bridges (as happened to Napoleon at Essling,)—to say nothing of the danger of losing their provisions and munitions, or even of a front attack against which the works might not avail. The system of detached permanent works of Coblentz has the advantage of avoiding these dangers, by protecting the depots on the same bank as the army, and in guaranteeing to the army freedom from attack at least until the bridges be re-established. If the city were upon the right bank of the Rhine, and there were only an intrenched camp of field-works on the left bank, there would be no certainty of security either for the depots or the army. So, if Coblentz were a good ordinary fortress without detached forts, a large army could not so readily make it a place of refuge, nor would there be such facilities for debouching from it in the presence of an enemy. The fortress of Ehrenbreitstein, which is intended to protect Coblentz on the right bank, is so difficult of access that it would be quite easy to blockade it, and the egress of a force of any magnitude might be vigorously disputed.

Much has been recently said of a new system used by the Archduke Maximilian to fortify the intrenched camp of Linz,—by masonry towers. As I only know of it by hearsay and the description by Captain Allard in the *Spectateur Militaire*, I cannot discuss it thoroughly. I only know that the system of towers used at Genoa by the skillful Colonel Andreis appeared to me to be useful, but still susceptible of improvements,—which the archduke seems to have added. We are told that the towers of Linz, situated in ditches and covered by the glacis, have the advantage of giving a concentrated horizontal fire and of being sheltered from the direct shot of the enemy. Such towers, if well flanked and connected by a parapet, may make a very advantageous camp,—always, however, with some of the inconveniences of closed lines. If the towers are isolated, and the intervals carefully covered by field-works, (to be thrown up when required,) they will make a camp preferable to one covered by ordinary redoubts, but not so advantageous as afforded by the large detached forts of Coblentz. These towers number thirty-two, eight of which are on the left bank, with a square fort commanding the Perlingsberg. Of these twenty-four on the right bank, some seven or eight are only half-towers. The circumference of this line is about twelve miles. The towers are between five hundred and six hundred yards apart, and will be connected, in case of war, by a palisaded covered way. They are of masonry, of three tiers of guns, with a barbette battery which is the principal defense, mounting eleven twenty-four pounders. Two howitzers are placed in the upper tier. Those towers are placed in a wide and deep ditch, the *déblais* of which forms a high glacis which protects the tower from direct shot; but I should think it would be difficult to protect the artillery from direct fire.

Some say that this has cost about three-fourths of what a complete bastioned enceinte, necessary to make Linz a fortress of the first rank, would have cost; others maintain that it has not cost more than a quarter as much as a bastioned work, and that it subserves, besides, an entirely different object. If these works are to resist a regular siege, they are certainly very defective; but, regarded as an intrenched camp to give refuge and an outlet upon both banks of the Danube for a large army, they are appropriate, and would be of great importance in a war like that of 1809, and, if existing then, would probably have saved the capital.

To complete a grand system, it would perhaps have been better to encircle Linz with a regular bastioned line, and then to have built seven or eight towers between the eastern salient and the mouth of the Traun, within a direct distance of about two and a half miles, so as to have included for the camp only the curved space between Linz, the Traun, and the Danube. Then the double advantage of a fortress of the first rank and a camp under its guns would have been united, and, even if not quite so large, would have answered for a large army, particularly if the eight towers on the left bank and the fort of Perlingsberg had been preserved.

Têtes De Ponts

têtes de ponts are the most important of all field-works. The difficulties of crossing a river, particularly a large one, in the face of the enemy, demonstrate abundantly the immense utility of such works, which can be less easily dispensed with than intrenched camps, since if the bridges are safe an army is insured from the disastrous events which may attend a rapid retreat across a large river.

têtes de ponts are doubly advantageous when they are as it were *keeps* for a large intrenched camp, and will be triply so if they also cover the bank opposite to the location of the camp, since then they will mutually support each other. It is needless to state that these works are particularly important in an enemy's country and upon all fronts where there are no permanent works. It may be observed that the principal difference between the system of intrenched camps and that of *têtes de ponts* is that the best intrenched camps are composed of detached and closed works, while *têtes de ponts* usually consist of contiguous works not closed. An intrenched line to admit of defense must be occupied in force throughout its whole extent, which would generally require a large army; if, on the contrary, the intrenchments are detached closed works, a comparatively small force can defend them.

The attack and defense of these works will be discussed in a subsequent part of this volume.

Article XXVIII: Strategic Operations in Mountains

A mountainous country presents itself, in the combinations of war, under four different aspects. It may be the whole theater of the war, or it may be but a zone; it may be mountainous throughout its whole extent, or there may be a line of mountains, upon emerging from which the army may debouch into large and rich plains.

If Switzerland, the Tyrol, the Noric provinces, some parts of Turkey and Hungary, Catalonia and Portugal, be excepted, in the European countries the mountains are in single ranges. In these cases there is but a difficult defile to cross,—a temporary obstacle, which, once overcome, is an advantage rather than an objection. In fact, the range once crossed and the war carried into the plains, the chain of mountains may be regarded as an eventual base, upon which the army may fall back and find a temporary refuge. The only essential precaution to be observed is, not to allow the enemy to anticipate the army on this line of retreat. The part of the Alps between France and Italy, and the Pyrenees, (which are not so high, though equally broad,) are of this nature. The mountains of Bohemia and of the Black Forest, and the Vosges, belong to this class. In Catalonia the mountains cover the whole country as far as the Ebro: if the war were limited to this province, the combinations would not be the same as if there were but a line of mountains. Hungary in this respect differs little from Lombardy and Castile; for if the Crapacks in the eastern and northern part are as marked a feature as the Pyrenees, they are still but a temporary obstacle, and an army overcoming it, whether debouching in the basin of the Waag, of the Neytra, or of the Theiss, or in the fields of Mongatsch, would have the vast plains between the Danube and the Theiss for a field of operations. The only difference would be in the roads, which in the Alps, though few in number, are excellent, while in Hungary there are none of much value. In its northern part, this chain, though not so high, becomes broader, and would seem to belong to that class of fields of operations which are wholly mountainous; but, as its evacuation may be compelled by decisive operations in the valleys of the Waag or the Theiss, it must be regarded as a temporary barrier. The attack and defense of this country, however, would be a strategic study of the most interesting character.

When an extremely mountainous country, such as the Tyrol or Switzerland, is but a zone of operations, the importance of these mountains is secondary, and they must be observed like a fortress, the armies deciding the great contests in the valleys. It will, of course, be otherwise if this be the whole field.

It has long been a question whether possession of the mountains gave control of the valleys, or whether possession of the valleys gave control of the mountains. The Archduke Charles, a very intelligent and competent judge, has declared for the latter, and has demonstrated that the valley of the Danube is the key of Southern Germany. However, in this kind of questions much depends upon the relative forces and their arrangement in the country.

If sixty thousand French were advancing on Bavaria in presence of an equal force of Austrians, and the latter should throw thirty thousand men into the Tyrol, intending to replace them by reinforcements on its arrival on the Inn, it would be difficult for the French to push on as far as this line, leaving so large a force on its flanks masters of the outlets of Scharnitz, Fussen, Kufstein, and Lofers. But if the French force were one hundred and twenty thousand men, and had gained such successes as to establish its superiority over the army in its front, then it might leave a sufficient detachment to mask the passes of the Tyrol and extend its progress as far as Linz,—as Moreau did in 1800.

Thus far we have considered these mountainous districts as only accessory zones. If we regard them as the principal fields of operations, the strategic problem seems to be more complicated. The campaigns of 1799 and 1800 are equally rich in instruction on this branch of the art. In my account of them I have endeavored to bring out their teachings by a historical exposition of the events; and I cannot do better than refer my readers to it.

When we consider the results of the imprudent invasion of Switzerland by the French Directory, and its fatal influence in doubling the extent of the theater of operations and making it reach from the Texel to Naples, we cannot too much applaud the wisdom of France and Austria in the transactions which had for three centuries guaranteed the neutrality of Switzerland. Every one will be convinced of this by carefully studying the interesting campaigns of the Archduke Charles, Suwaroff, and Massena in 1799, and those of Napoleon and Moreau in 1800. The first is a model for operations upon an entirely mountainous field; the second is a model for wars in which the fate of mountainous countries is decided on the plains.

I will here state some of the deductions which seem to follow from this study.

When a country whose whole extent is mountainous is the principal theater of operations, the strategic combinations cannot be entirely based upon maxims applicable in an open country.

Transversal maneuvers to gain the extremity of the front of operations of the enemy here become always very difficult, and often impossible. In such a country a considerable army can be maneuvered only in a small number of valleys, where the enemy will take care to post advanced guards of sufficient strength to delay the army long enough to provide means for defeating the enterprise; and, as the ridges which separate these valleys will be generally crossed only by paths impracticable for the passage of an army, transversal marches can only be made by small bodies of light troops.

The important natural strategic points will be at the junction of the larger valleys or of the streams in those valleys, and will be few in number; and, if the defensive army occupy them with the mass of its forces, the invader will generally be compelled to resort to direct attacks to dislodge it.

However, if great strategic maneuvers in these cases be more rare and difficult, it by no means follows that they are less important. On the contrary, if the assailant succeed in gaining possession of one of these centers of communication between the large valleys upon the line of retreat of the enemy, it will be more serious for the latter than it would be in an open country; since the occupation of one or two difficult defiles will often be sufficient to cause the ruin of the whole army.

If the attacking party have difficulties to overcome, it must be admitted that the defense has quite as many, on account of the necessity of covering all the outlets by which an attack in force may be made upon the decisive points, and of the difficulties of the transversal marches which it would be compelled to make to cover the menaced points. In order to complete what I have said upon this kind of marches and the difficulties of directing them, I will refer to what Napoleon did in 1805 to cut off Mack from Ulm. If this operation was facilitated by the hundred roads which cross Swabia in all directions, and if it would have been impracticable in a mountainous country, for want of transversal routes, to make the long circuit from Donauwerth by Augsburg to Memmingen, it is also true that Mack could by these same hundred roads have effected his retreat with much greater facility than if he had been entrapped in one of the valleys of Switzerland or of the Tyrol, from which there was but a single outlet.

On the other hand, the general on the defensive may in a level country concentrate a large part of his forces; for, if the enemy scatter to occupy all the roads by which the defensive army may retire, it will be easy for the latter to crush these isolated bodies; but in a very mountainous country, where there are ordinarily but one or two principal routes into which other valleys open, even from the direction of the enemy, the concentration of forces becomes more difficult, since serious inconveniences may result if even one of these important valleys be not observed.

Nothing can better demonstrate the difficulty of strategic defense in mountainous regions than the perplexity in which we are involved when we attempt simply to give advice in such cases,—to say nothing of laying down maxims for them. If it were but a question of the defense of a single definite front of small extent, consisting of four or five converging valleys, the common junction of which is at a distance of two or three short marches from the summits of the ranges, it would be easier of solution. It would then be sufficient to recommend the construction of a good fort at the narrowest and least-easily turned point of each of these valleys. Protected by these forts, a few brigades of infantry should be stationed to dispute the passage, while half the army should be held in reserve at the junction, where it would be in position either to sustain the advanced guards most seriously threatened, or to fall upon the assailant with the whole force when he debouches. If to this be added good instructions to the commanders of the advanced guards, whether in assigning them the best point for rendezvous when their line of forts is pierced, or in directing them to continue to act in the mountains upon the flank of the enemy, the general on the defensive may regard himself as invincible, thanks to the many difficulties which the country offers to the assailant. But, if there be other fronts like this upon the right and left, all of which are to be defended, the problem is changed: the difficulties of the defense increase with the extent of the fronts, and this system of a cordon of forts becomes dangerous,—while it is not easy to adopt a better one.

We cannot be better convinced of these truths than by the consideration of the position of Massena in Switzerland in 1799. After Jourdan's defeat at Stockach, he occupied the line from Basel by Schaffhausen and Rheineck to Saint-Gothard, and thence by La Furca to Mont-Blanc. He had enemies in front of Basel, at Waldshut, at Schaffhausen, at Feldkirch, and at Chur; Bellegarde threatened the Saint-Gothard, and the Italian army menaced the Simplon and the Saint-Bernard. How was he to defend such a circumference? and how could he leave open one of these great valleys, thus risking every thing? From Rheinfelden to the Jura, toward Soleure, it was but two short marches, and there was the mouth of the trap in which the French army was placed. This was, then, the pivot of the defense. But how could he leave Schaffhausen unprotected? how abandon Rheineck and the Saint-Gothard? how open the Valais and the approach by Berne, without surrendering the whole of Switzerland to the Coalition? And if he covered each point even by a brigade, where would be his army when he would need it to give battle to an approaching force? It is a natural system on a level theater to concentrate the masses of an army; but in the mountains such a course would surrender the keys of the country, and, besides, it is not easy to say where an inferior army could be concentrated without compromising it.

After the forced evacuation of the line of the Rhine and Zurich, it seemed that the only strategic point for Massena to defend was the line of the Jura. He was rash enough to stand upon the Albis,—a line shorter than that of the Rhine, it is true, but exposed for an immense distance to the attacks of the Austrians. If Bellegarde, instead of going into Lombardy by the Valtellina, had marched to Berne or made a junction with the archduke, Massena would have been ruined. These events seem to prove that if a country covered with high mountains be favorable for defense in a tactical point of view, it is different in a strategic sense, because it necessitates a division of the troops.

This can only be remedied by giving them greater mobility and by passing often to the offensive.

General Clausewitz, whose logic is frequently defective, maintains, on the contrary, that, movements being the most difficult part in this kind of war, the defensive party should avoid them, since by such a course he might lose the advantages of the local defenses. He, however, ends by demonstrating that a passive defense must yield under an active attack,—which goes to show that the initiative is no less favorable in mountains than in plains. If there could be any doubt on this point, it ought to be dispelled by Massena's campaign in Switzerland, where he sustained himself only by attacking the enemy at every opportunity, even when he was obliged to seek him on the Grimsel and the Saint-Gothard. Napoleon's course was similar in 1796 in the Tyrol, when he was opposed to Wurmser and Alvinzi.

As for detailed strategic maneuvers, they may be comprehended by reading the events of Suwaroff's expedition by the Saint-Gothard upon the Muttenthal. While we must approve his maneuvers in endeavoring to capture Lecourbe in the valley of the Reuss, we must also admire the presence of mind, activity, and unyielding firmness which saved that general and his division. Afterward, in the Schachenthal and the Muttenthal, Suwaroff was placed in the same position as Lecourbe had been, and extricated himself with equal ability. Not less extraordinary was the ten days' campaign of General Molitor, who with four thousand men was surrounded in the canton of Glaris by more than thirty thousand allies, and yet succeeded in maintaining himself behind the Linth after four admirable fights. These events teach us the vanity of all theory in details, and also that in such a country a strong and heroic will is worth more than all the precepts in the world. After such lessons, need I say that one of the principal rules of this kind of war is, not to risk one's self in the valleys without securing the heights? Shall I say also that in this kind of war, more than in any other, operations should be directed upon the communications of the enemy? And, finally, that good temporary bases or lines of defense at the confluence of the great valleys, covered by strategic reserves, combined with great mobility and frequent offensive movements, will be the best means of defending the country?

I cannot terminate this article without remarking that mountainous countries are particularly favorable for defense when the war is a national one, in which the whole people rise up to defend their homes with the obstinacy which enthusiasm for a holy cause imparts: every advance is then dearly bought. But to be successful it is always necessary that the people be sustained by a disciplined force, more or less numerous: without this they must finally yield, like the heroes of Stanz and of the Tyrol.

The offensive against a mountainous country also presents a double case: it may either be directed upon a belt of mountains beyond which are extensive plains, or the whole theater may be mountainous.

In the first case there is little more to be done than this,—viz.: make demonstrations upon the whole line of the frontier, in order to lead the enemy to extend his defense, and then force a passage at the point which promises the greatest results. The problem in such a case is to break through a cordon which is strong less on account of the numbers of the defenders than from their position, and if broken at one point the whole line is forced. The history of Bard in 1800, and the capture of Leutasch and Scharnitz in 1805 by Ney, (who threw fourteen thousand men on Innspruck in the midst of thirty thousand Austrians, and by seizing this central point compelled them to retreat in all directions,) show that with brave infantry and bold commanders these famous mountain-ranges can generally be forced.

The history of the passage of the Alps, where Francis I. turned the army which was awaiting him at Suza by passing the steep mountains between Mont-Cenis and the valley of Queyras, is an example of those *insurmountable* obstacles which can always be surmounted. To oppose him it would have been necessary to adopt a system of cordon; and we have already seen what is to be expected of it. The position of the Swiss and Italians at Suza was even less wise than the cordon-system, because it inclosed them in a contracted valley without protecting the lateral issues. Their strategic plan ought to have been to throw troops into these valleys to defend the defiles, and to post the bulk of the army toward Turin or Carignano.

When we consider the *tactical* difficulties of this kind of war, and the immense advantages it affords the defense, we may be inclined to regard the concentration of a considerable force to penetrate by a single valley as an extremely rash maneuver, and to think that it ought to be divided into as many columns as there are practicable passes. In my opinion, this is one of the most dangerous of all illusions; and to confirm what I say it is only necessary to refer to the fate of the columns of Championnet at the battle of Fossano. If there be five or six roads on the menaced front, they should all, of course, be threatened; but the army should cross the chain in not more than two masses, and the routes which these follow should not be divergent; for if they were, the enemy might be able to defeat them separately. Napoleon's passage of the Saint-Bernard was wisely planned. He formed the bulk of his army on the center, with a division on each flank by Mont-Cenis and the Simplon, to divide the attention of the enemy and flank his march.

The invasion of a country entirely covered with mountains is a much greater and more difficult task than where a dénouement may be accomplished by a decisive battle in the open country; for fields of battle for the deployment of large masses are rare in a mountainous region, and the war becomes a succession of partial combats. Here it would be imprudent, perhaps, to penetrate on a single point by a narrow and deep valley, whose outlets might be closed by the enemy and thus the invading army be endangered: it might penetrate by the wings on two or three lateral lines, whose outlets should not be too widely separated, the marches being so arranged that the masses may debouch at the junction of the valleys at nearly the same instant. The enemy should be driven from all the ridges which separate these valleys.

Of all mountainous countries, the tactical defense of Switzerland would be the easiest, if all her inhabitants were united in spirit; and with their assistance a disciplined force might hold its own against a triple number.

To give specific precepts for complications which vary infinitely with localities, the resources and the condition of the people and armies, would be absurd. History, well studied and understood, is the best school for this kind of warfare. The account of the campaign of 1799 by the Archduke Charles, that of the campaigns which I have given in my History of the Wars of the Revolution, the narrative of the campaign of the Grisons by Ségur and Mathieu Dumas, that of Catalonia by Saint-Cyr and Suchet, the campaign of the Duke de Rohan in Valtellina, and the passage of the Alps by Gaillard, (Francis I.,) are good guides in this study.

Article XXIX: Grand Invasions and Distant Expeditions

There are several kinds of distant expeditions. The first are those which are merely auxiliary and belong to wars of intervention. The second are great continental invasions, through extensive tracts of country, which may be either friendly, neutral, doubtful, or hostile. The third are of the same nature, but made partly on land, partly by sea by means of numerous fleets. The fourth class comprises those beyond the seas, to found, defend, or attack distant colonies. The fifth includes the great descents, where the distance passed over is not very great, but where a powerful state is attacked.

As to the first, in a strategic point of view, a Russian army on the Rhine or in Italy, in alliance with the German States, would certainly be stronger and more favorably situated than if it had reached either of these points by passing over hostile or even neutral territory; for its base, lines of operations, and eventual points of support will be the same as those of its allies; it may find refuge behind their lines of defense, provisions in their depots, and munitions in their arsenals;—while in the other case its resources would be upon the Vistula or the Niemen, and it might afford another example of the sad fate of many of these great invasions.

In spite of the important difference between a war in which a state is merely an auxiliary, and a distant invasion undertaken for its own interest and with its own resources, there are, nevertheless, dangers in the way of these auxiliary armies, and perplexity for the commander of all the armies,—particularly if he belong to the state which is not a principal party; as may be learned from the campaign of 1805. General Koutousoff advanced on the Inn to the boundaries of Bavaria with thirty thousand Russians, to effect a junction with Mack, whose army in the mean time had been destroyed, with the exception of eighteen thousand men brought back from Donauwerth by Kienmayer. The Russian general thus found himself with fifty thousand men exposed to the impetuous activity of Napoleon with one hundred and fifty thousand, and, to complete his misfortune, he was separated from his own frontiers by a distance of about seven hundred and fifty miles. His position would have been hopeless if fifty thousand men had not arrived to reinforce him. The battle of Austerlitz—due to a fault of Weyrother—endangered the Russian army anew, since it was so far from its base. It almost became the victim of a distant alliance; and it was only peace that gave it the opportunity of regaining its own country.

The fate of Suwaroff after the victory of Novi, especially in the expedition to Switzerland, and that of Hermann's corps at Bergen in Holland, are examples which should be well studied by every commander under such circumstances. General Benningsen's position in 1807 was less disadvantageous, because, being between the Vistula and the Niemen, his communications with his base were preserved and his operations were in no respect dependent upon his allies. We may also refer to the fate of the French in Bohemia and Bavaria in 1742, when Frederick the Great abandoned them and made a separate peace. In this case the parties were allies rather than auxiliaries; but in the latter relation the political ties are never woven so closely as to remove all points of dissension which may compromise military operations. Examples of this kind have been cited in Article XIX., on political objective points.

History alone furnishes us instruction in reference to distant invasions across extensive territories. When half of Europe was covered with forests, pasturages, and flocks, and when only horses and iron were necessary to transplant whole nations from one end of the continent to the other, the Goths, Huns, Vandals, Normans, Arabs, and Tartars overran empires in succession. But since the invention of powder and artillery and the organization of formidable standing armies, and particularly since civilization and statesmanship have brought nations closer together and have taught them the necessity of reciprocally sustaining each other, no such events have taken place.

Besides these migrations of nations, there were other expeditions in the Middle Ages, which were of a more military character, as those of Charlemagne and others. Since the invention of powder there have been scarcely any, except the advance of Charles VIII. to Naples, and of Charles XII. into the Ukraine, which can be called distant invasions; for the campaigns of the Spaniards in Flanders and of the Swedes in Germany were of a particular kind. The first was a civil war, and the Swedes were only auxiliaries to the Protestants of Germany; and, besides, the forces concerned in both were not large. In modern times no one but Napoleon has dared to transport the armies of half of Europe from the Rhine to the Volga; and there is little danger that he will be imitated.

Apart from the modifications which result from great distances, all invasions, after the armies arrive upon the actual theater, present the same operations as all other wars. As the chief difficulty arises from these great distances, we should recall our maxims on deep lines of operations, strategic reserves, and eventual bases, as the only ones applicable; and here it is that their application is indispensable, although even that will not avert all danger. The campaign of 1812, although so ruinous to Napoleon, was a model for a distant invasion. His care in leaving Prince Schwarzenberg and Reynier on the Bug, while Macdonald, Oudinot, and Wrede guarded the Dwina, Victor covered Smolensk, and Augereau was between the Oder and Vistula, proves that he had neglected no humanly possible precaution in order to base himself safely; but it also proves that the greatest enterprises may fail simply on account of the magnitude of the preparations for their success.

If Napoleon erred in this contest, it was in neglecting diplomatic precautions; in not uniting under one commander the different bodies of troops on the Dwina and Dnieper; in remaining ten days too long at Wilna; in giving the command of his right to his brother, who was unequal to it; and in confiding to Prince Schwarzenberg a duty which that general could not perform with the devotedness of a Frenchman. I do not speak now of his error in remaining in Moscow after the conflagration, since then there was no remedy for the misfortune; although it would not have been so great if the retreat had taken place immediately. He has also been accused of having too much despised distances, difficulties, and men, in pushing on as far as the Kremlin. Before passing judgment upon him in this matter, however, we ought to know the real motives which induced him to pass Smolensk, instead of wintering there as he had intended, and whether it would have been possible for him to remain between that city and Vitebsk without having previously defeated the Russian army.

It is doubtless true that Napoleon neglected too much the resentment of Austria, Prussia, and Sweden, and counted too surely upon a *dénouement* between Wilna and the Dwina. Although he fully appreciated the bravery of the Russian armies, he did not realize the spirit and energy of the people. Finally, and chiefly, instead of procuring the hearty and sincere concurrence of a military state, whose territories would have given him a sure base for his attack upon the colossal power of Russia, he founded his enterprise upon the co-operation of a brave and enthusiastic but fickle people, and besides, he neglected to turn to the greatest advantage this ephemeral enthusiasm.

The fate of all such enterprises makes it evident that the capital point for their success, and, in fact, the only maxim to be given, is "never to attempt them without having secured the hearty and constant alliance of a respectable power near enough the field of operations to afford a proper base, where supplies of every kind may be accumulated, and which may also in case of reverse serve as a refuge and afford new means of resuming the offensive." As to the precautions to be observed in these operations, the reader is referred to Articles XXI. and XXII., on the safety of deep lines of operations and the establishment of eventual bases, as giving all the military means of lessening the danger; to these should be added a just appreciation of distances, obstacles, seasons, and countries,—in short, accuracy in calculation and moderation in success, in order that the enterprise may not be carried too far. We are far from thinking that any purely military maxims can insure the success of remote invasions: in four thousand years only five or six have been successful, and in a hundred instances they have nearly ruined nations and armies.

Expeditions of the third class, partly on land, partly by sea, have been rare since the invention of artillery, the Crusades being the last in date of occurrence; and probably the cause is that the control of the sea, after having been held in succession by several secondary powers, has passed into the hands of England, an insular power, rich in ships, but without the land-forces necessary for such expeditions.

It is evident that from both of these causes the condition of things now is very different from that existing when Xerxes marched to the conquest of Greece, followed by four thousand vessels of all dimensions, or when Alexander marched from Macedonia over Asia Minor to Tyre, while his fleet coasted the shore.

Nevertheless, if we no longer see such invasions, it is very true that the assistance of a fleet of men-of-war and transports will always be of immense value to any army on shore when the two can act in concert. Still, sailing-ships are an uncertain resource, for their progress depends upon the winds,—which may be unfavorable: in addition, any kind of fleet is exposed to great dangers in storms, which are not of rare occurrence.

The more or less hostile tone of the people, the length of the line of operations, and the great distance of the principal objective point, are the only points which require any deviation from the ordinary operations of war.

Invasions of neighboring states, if less dangerous than distant ones, are still not without great danger of failure. A French army attacking Cadiz might find a tomb on the Guadalquivir, although well based upon the Pyrenees and possessing intermediate bases upon the Ebro and the Tagus. Likewise, the army which in 1809 besieged Komorn in the heart of Hungary might have been destroyed on the plains of Wagram without going as far as the Beresina. The antecedents, the number of disposable troops, the successes already gained, the state of the country, will all be elements in determining the extent of the enterprises to be undertaken; and to be able to proportion them well to his resources, in view of the attendant circumstances, is a great talent in a general. Although diplomacy does not play so important a part in these invasions as in those more distant, it is still of importance; since, as stated in Article VI., there is no enemy, however insignificant, whom it would not be useful to convert into an ally. The influence which the change of policy of the Duke of Savoy in 1706 exercised over the events of that day, and the effects of the stand taken by Maurice of Saxony in 1551, and of Bavaria in 1813, prove clearly the importance of securing the strict neutrality of all states adjoining the theater of war, when their co-operation cannot be obtained.

Epitome of Strategy

The task which I undertook seems to me to have been passably fulfilled by what has been stated in reference to the strategic combinations which enter ordinarily into a plan of campaign. We have seen, from the definition at the beginning of this chapter, that, in the most important operations in war, *strategy* fixes the direction of movements, and that we depend upon *tactics* for their execution. Therefore, before treating of these mixed operations, it will be well to give here the combinations of grand tactics and of battles, as well as the maxims by the aid of which the application of the fundamental principle of war may be made.

By this method these operations, half strategic and half tactical, will be better comprehended as a whole; but, in the first place, I will give a synopsis of the contents of the preceding chapter.

From the different articles which compose it, we may conclude that the manner of applying the general principle of war to all possible theaters of operations is found in what follows:—

1. In knowing how to make the best use of the advantages which the reciprocal directions of the two bases of operations may afford, in accordance with Article XVIII.

2. In choosing, from the three zones ordinarily found in the strategic field, that one upon which the greatest injury can be done to the enemy with the least risk to one's self.

3. In establishing well, and giving a good direction to, the lines of operations; adopting for defense the concentric system of the Archduke Charles in 1796 and of Napoleon in 1814; or that of Soult in 1814, for retreats parallel to the frontiers.

On the offensive we should follow the system which led to the success of Napoleon in 1800, 1805, and 1806, when he directed his line upon the extremity of the strategic front; or we might adopt his plan which was successful in 1796, 1809, and 1814, of directing the line of operations upon the center of the strategic front: all of which is to be determined by the respective positions of the armies, and according to the maxims presented in Article XXI.

4. In selecting judicious eventual lines of maneuver, by giving them such directions as always to be able to act with the greater mass of the forces, and to prevent the parts of the enemy from concentrating or from affording each other mutual support.

5. In combining, in the same spirit of centralization, all strategic positions, and all large detachments made to cover the most important strategic points of the theater of war.

6. In imparting to the troops the greatest possible mobility and activity, so as, by their successive employment upon points where it may be important to act, to bring superior force to bear upon fractions of the hostile army.

The system of rapid and continuous marches multiplies the effect of an army, and at the same time neutralizes a great part of that of the enemy's, and is often sufficient to insure success; but its effect will be quintupled if the marches be skillfully directed upon the decisive strategic points of the zone of operations, where the severest blows to the enemy can be given.

However, as a general may not always be prepared to adopt this decisive course to the exclusion of every other, he must then be content with attaining a part of the object of every enterprise, by rapid and successive employment of his forces upon isolated bodies of the enemy, thus insuring their defeat. A general who moves his masses rapidly and continually, and gives them proper directions, may be confident both of gaining victories and of securing great results therefrom.

The oft-cited operations of 1809 and 1814 prove these truths most satisfactorily, as also does that ordered by Carnot in 1793, already mentioned in Article XXIV., and the details of which may be found in Volume IV. of my History of the Wars of the Revolution. Forty battalions, carried successively from Dunkirk to Menin, Maubeuge, and Landau, by reinforcing the armies already at those points, gained four victories and saved France. The whole science of marches would have been found in this wise operation had it been directed upon the decisive strategic point. The Austrian was then the principal army of the Coalition, and its line of retreat was upon Cologne: hence it was upon the Meuse that a general effort of the French would have inflicted the most severe blow. The Committee of Public Safety provided for the most pressing danger, and the maneuver contains half of the strategic principle; the other half consists in giving to such efforts the most decisive direction, as Napoleon did at Ulm, at Jena, and at Ratisbon. The whole of strategy is contained in these four examples.

It is superfluous to add that one of the great ends of strategy is to be able to assure real advantages to the army by preparing the theater of war most favorable for its operations, if they take place in its own country, by the location of fortified places, of intrenched camps, and of *têtes de ponts*, and by the opening of communications in the great decisive directions: these constitute not the least interesting part of the science. We have already seen how we are to recognize these lines and these decisive points, whether permanent or temporary. Napoleon has afforded instruction on this point by the roads of the Simplon and Mont-Cenis; and Austria since 1815 has profited by it in the roads from the Tyrol to Lombardy, the Saint-Gothard, and the Splugen, as well as by different fortified places projected or completed.

Grand Tactics and Battles

Battles are the actual conflicts of armies contending about great questions of national policy and of strategy. Strategy directs armies to the decisive points of a zone of operations, and influences, in advance, the results of battles; but tactics, aided by courage, by genius and fortune, gains victories.

Grand tactics is the art of making good combinations preliminary to battles, as well as during their progress. The guiding principle in tactical combinations, as in those of strategy, is to bring the mass of the force in hand against a part of the opposing army, and upon that point the possession of which promises the most important results.

Battles have been stated by some writers to be the chief and deciding features of war. This assertion is not strictly true, as armies have been destroyed by strategic operations without the occurrence of pitched battles, by a succession of inconsiderable affairs. It is also true that a complete and decided victory may give rise to results of the same character when there may have been no grand strategic combinations.

The results of a battle generally depend upon a union of causes which are not always within the scope of the military art: the nature of the order of battle adopted, the greater or less wisdom displayed in the plan of the battle, as well as the manner of carrying out its details, the more or less loyal and enlightened co-operation of the officers subordinate to the commander-in-chief, the cause of the contest, the proportions and quality of the troops, their greater or less enthusiasm, superiority on the one side or the other in artillery or cavalry, and the manner of handling these arms; but it is the *morale* of armies, as well as of nations, more than any thing else, which makes victories and their results decisive. Clausewitz commits a grave error in asserting that a battle not characterized by a maneuver to turn the enemy cannot result in a complete victory. At the battle of Zama, Hannibal, in a few brief hours, saw the fruits of twenty years of glory and success vanish before his eyes, although Scipio never had a thought of turning his position. At Rivoli the turning-party was completely beaten; nor was the maneuver more successful at Stockach in 1799, or at Austerlitz in 1805. As is evident from Article XXXII., I by no means intend to discourage the use of that maneuver, being, on the contrary, a constant advocate of it; but it is very important to know how to use it skillfully and opportunely, and I am, moreover, of opinion that if it be a general's design to make himself master of his enemy's communications while at the same time holding his own, he would do better to employ strategic than tactical combinations to accomplish it.

There are three kinds of battles: 1st, defensive battles, or those fought by armies in favorable positions taken up to await the enemy's attack; 2d, offensive battles, where one army attacks another in position; 3d, battles fought unexpectedly, and resulting from the collision of two armies meeting on the march. We will examine in succession the different combinations they present.

Article XXX: Positions and Defensive Battles

When an army awaits an attack, it takes up a position and forms its line of battle. From the general definitions given at the beginning of this work, it will appear that I make a distinction between *lines of battle* and *orders of battle*,—things which have been constantly confounded. I will designate as a *line of battle* the position occupied by battalions, either deployed or in columns of attack, which an army will take up to hold a camp and a certain portion of ground where it will await attack, having no particular project in view for the future: it is the right name to give to a body of troops formed with proper tactical intervals and distances upon one or more lines, as will be more fully explained in Article XLIII. On the contrary, I will designate as an *order of battle* an arrangement of troops indicating an intention to execute a certain maneuver; as, for example, the parallel order, the oblique order, the perpendicular order.

This nomenclature, although new, seems necessary to keeping up a proper distinction between two things which should by no means be confounded. [It is from no desire to make innovations that I have modified old terms or made new. In the development of a science, it is wrong for the same word to designate two very different things; and, if we continue to apply the term *order of battle* to the disposition of troops in line, it must be improper to designate certain important maneuvers by the terms *oblique order of battle, concave order of battle*, and it becomes necessary to use instead the terms *oblique system of battle*, etc. I prefer the method of designation I have adopted. The *order of battle* on paper may take the name *plan of organization*, and the ordinary formation of troops upon the ground will then be called *line of battle*.] From the nature of the two things, it is evident that the *line of battle* belongs especially to defensive arrangements; because an army awaiting an attack without knowing what or where it will be must necessarily form a rather indefinite and objectless line of battle. *order of battle*, on the contrary, indicating an arrangement of troops formed with an intention of fighting while executing some maneuver previously determined upon, belongs more particularly to offensive dispositions. However, it is by no means pretended that the line of battle is exclusively a defensive arrangement; for a body of troops may in this formation very well proceed to the attack of a position, while an army on the defensive may use the oblique order or any other. I refer above only to ordinary cases.

Without adhering strictly to what is called the system of a war of positions, an army may often find it proper to await the enemy at a favorable point, strong by nature and selected beforehand for the purpose of there fighting a defensive battle. Such a position may be taken up when the object is to cover an important objective point, such as a capital, large depots, or a decisive strategic point which controls the surrounding country, or, finally, to cover a siege.

There are two kinds of positions,—the *strategic*, which has been discussed in Article XX., and the *tactical*. The latter, again, are subdivided. In the first place, there are intrenched positions occupied to await the enemy under cover of works more or less connected,—in a word, intrenched camps. Their relations to strategic operations have been treated in Article XXVII., and their attack and defense are discussed in Article XXXV. Secondly, we have positions naturally strong, where armies encamp for the purpose of gaining a few days' time. Third and last are open positions, chosen in advance to fight on the defensive. The characteristics to be sought in these positions vary according to the object in view: it is, however, a matter of importance not to be carried away by the mistaken idea, which prevails too extensively, of giving the preference to positions that are very steep and difficult of access,—quite suitable places, probably, for temporary camps, but not always the best for battle-grounds. A position of this kind, to be really strong, must be not only steep and difficult of access, but should be adapted to the end had in view in occupying it, should offer as many advantages as possible for the kind of troops forming the principal strength of the army, and, finally, the obstacles presented by its features should be more disadvantageous for the enemy than for the assailed. For example, it is certain that Massena, in taking the strong position of the Albis, would have made a great error if his chief strength had been in cavalry and artillery; whilst it was exactly what was wanted for his excellent infantry. For the same reason, Wellington, whose whole dependence was in the fire of his troops, made a good choice of position at Waterloo, where all the avenues of approach were well swept by his guns. The position of the Albis was, moreover, rather a strategic position, that of Waterloo being simply a battle-ground.

The rules to be generally observed in selecting tactical positions are the following:—

1. To have the communications to the front such as to make it easier to fall upon the enemy at a favorable moment than for him to approach the line of battle.

2. To give the artillery all its effect in the defense.

3. To have the ground suitable for concealing the movements of troops between the wings, that they may be massed upon any point deemed the proper one.

4. To be able to have a good view of the enemy's movements.

5. To have an unobstructed line of retreat.

6. To have the flanks well protected, either by natural or artificial obstacles, so as to render impossible an attack upon their extremities, and to oblige the enemy to attack the center, or at least some point of the front.

This is a difficult condition to fulfill; for, if an army rests on a river, or a mountain, or an impenetrable forest, and the smallest reverse happens to it, a great disaster may be the result of the broken line being forced back upon the very obstacles which seemed to afford perfect protection. This danger—about which there can be no doubt—gives rise to the thought that points admitting an easy defense are better on a battle-field than insurmountable obstacles. [The park of Hougoumont, the hamlet of La Haye Sainte, and the rivulet of Papelotte were for Ney more serious obstacles than the famous position of Elchingen, where he forced a passage of the Danube, in 1805, upon the ruins of a burnt bridge. It may perhaps be said that the courage of the defenders in the two cases was not the same; but, throwing out of consideration this chance, it must be granted that the difficulties of a position, when properly taken advantage of, need not be insurmountable in order to render the attack abortive. At Elchingen the great height and steepness of the banks, rendering the fire almost ineffectual, were more disadvantageous than useful in the defense.]

7. Sometimes a want of proper support for the flanks is remedied by throwing a crotchet to the rear. This is dangerous; because a crotchet stuck on a line hinders its movements, and the enemy may cause great loss of life by placing his artillery in the angle of the two lines prolonged. A strong reserve in close column behind the wing to be guarded from assault seems better to fulfill the required condition than the crotchet; but the nature of the ground must always decide in the choice between the two methods. Full details on this point are given in the description of the battle of Prague, (Chapter II. of the Seven Years' War.)

8. We must endeavor in a defensive position not only to cover the flanks, but it often happens that there are obstacles on other points of the front, of

such a character as to compel an attack upon the center. Such a position will always be one of the most advantageous for defense,—as was shown at Malplaquet and Waterloo. Great obstacles are not essential for this purpose, as the smallest accident of the ground is sometimes sufficient: thus, the insignificant rivulet of Papelotte forced Ney to attack Wellington's center, instead of the left as he had been ordered.

When a defense is made of such a position, care must be taken to hold ready for movement portions of the wings thus covered, in order that they may take part in the action instead of remaining idle spectators of it.

The fact cannot be concealed, however, that all these means are but palliatives; and the best thing for an army standing on the defensive is to *know* how to take the offensive at a proper time, and *to take it*. Among the conditions to be satisfied by a defensive position has been mentioned that of enabling an easy and safe retreat; and this brings us to an examination of a question presented by the battle of Waterloo. Would an army with its rear resting upon a forest, and with a good road behind the center and each wing, have its retreat compromised, as Napoleon imagined, if it should lose the battle? My own opinion is that such a position would be more favorable for a retreat than an entirely open field; for a beaten army could not cross a plain without exposure to very great danger. Undoubtedly, if the retreat becomes a rout, a portion of the artillery left in battery in front of the forest would, in all probability, be lost; but the infantry and cavalry and a great part of the artillery could retire just as readily as across a plain. There is, indeed, no better cover for an orderly retreat than a forest,—this statement being made upon the supposition that there are at least two good roads behind the line, that proper measures for retreat have been taken before the enemy has had an opportunity to press too closely, and, finally, that the enemy is not permitted by a flank movement to be before the retreating army at the outlet of the forest, as was the case at Hohenlinden. The retreat would be the more secure if, as at Waterloo, the forest formed a concave line behind the center; for this re-entering would become a place of arms to receive the troops and give them time to pass off in succession on the main roads.

When discussing strategic operations, mention was made of the varying chances which the two systems, the *defensive* and the *offensive*, give rise to; and it was seen that especially in strategy the army taking the initiative has the great advantage of bringing up its troops and striking a blow where it may deem best, whilst the army which acts upon the defensive and awaits an attack is anticipated in every direction, is often taken unawares, and is always obliged to regulate its movements by those of the enemy. We have also seen that in tactics these advantages are not so marked, because in this case the operations occupy a smaller extent of ground, and the party taking the initiative cannot conceal his movements from the enemy, who, instantly observing, may at once counteract them by the aid of a good reserve. Moreover, the party advancing upon the enemy has against him all the disadvantages arising from accidents of ground that he must pass before reaching the hostile line; and, however flat a country it may be, there are always inequalities of the surface, such as small ravines, thickets, hedges, farm-houses, villages, etc.., which must either be taken possession of or be passed by. To these natural obstacles may also be added the enemy's batteries to be carried, and the disorder which always prevails to a greater or less extent in a body of men exposed to a continued fire either of musketry or artillery. Viewing the matter in the light of these facts, all must agree that in tactical operations the advantages resulting from taking the initiative are balanced by the disadvantages.

However undoubted these truths may be, there is another, still more manifest, which has been demonstrated by the greatest events of history. Every army which maintains a strictly defensive attitude must, if attacked, be at last driven from its position; whilst by profiting by all the advantages of the defensive system, and holding itself ready to take the offensive when occasion offers, it may hope for the greatest success. A general who stands motionless to receive his enemy, keeping strictly on the defensive, may fight ever so bravely, but he must give way when properly attacked. It is not so, however, with a general who indeed waits to receive his enemy, but with the determination to fall upon him offensively at the proper moment, to wrest from him and transfer to his own troops the moral effect always produced by an onward movement when coupled with the certainty of throwing the main strength into the action at the most important point,—a thing altogether impossible when keeping strictly on the defensive. In fact, a general who occupies a well-chosen position, where his movements are free, has the advantage of observing the enemy's approach; his forces, previously arranged in a suitable manner upon the position, aided by batteries placed so as to produce the greatest effect, may make the enemy pay very dearly for his advance over the space separating the two armies; and when the assailant, after suffering severely, finds himself strongly assailed at the moment when the victory seemed to be in his hands, the advantage will, in all probability, be his no longer, for the moral effect of such a counter-attack upon the part of an adversary supposed to be beaten is certainly enough to stagger the boldest troops.

A general may, therefore, employ in his battles with equal success either the offensive or defensive system; but it is indispensable,—1st, that, so far from limiting himself to a passive defense, he should know how to take the offensive at favorable moments; 2d, that his *coup-d'oeil* be certain and his coolness undoubted; 3d, that he be able to rely surely upon his troops; 4th, that, in retaking the offensive, he should by no means neglect to apply the general principle which would have regulated his order of battle had he done so in the beginning; 5th, that he strike his blows upon decisive points. These truths are demonstrated by Napoleon's course at Rivoli and Austerlitz, as well as by Wellington's at Talavera, at Salamanca, and at Waterloo.

Article XXXI: Offensive Battles, and Different Orders of Battle

We understand by offensive battles those which an army fights when assaulting another in position. [In every battle one party must be the assailant and the other assailed. Every battle is hence offensive for one party and defensive for the other.] An army reduced to the strategic defensive often takes the offensive by making an attack, and an army receiving an attack may, during the progress of the battle, take the offensive and obtain the advantages incident to it. History furnishes numerous examples of battles of each of these kinds. As defensive battles have been discussed in the preceding article, and the advantages of the defensive been pointed out, we will now proceed to the consideration of offensive movements.

It must be admitted that the assailant generally has a moral advantage over the assailed, and almost always acts more understandingly than the latter, who must be more or less in a state of uncertainty.

As soon as it is determined to attack the enemy, some order of attack must be adopted; and that is what I have thought ought to be called *order of battle*.

It happens also quite frequently that a battle must be commenced without a detailed plan, because the position of the enemy is not entirely known. In either case it should be well understood that there is in every battle-field a decisive point, the possession of which, more than of any other, helps to secure the victory, by enabling its holder to make a proper application of the principles of war: arrangements should therefore be made for striking the decisive blow upon this point.

The decisive point of a battle-field is determined, as has been already stated, by the character of the position, the bearing of different localities upon the strategic object in view, and, finally, by the arrangement of the contending forces. For example, suppose an enemy's flank to rest upon high ground from which his whole line might be attained, the occupation of this height seems most important, tactically considered; but it may happen that the height in question is very difficult of access, and situated exactly so as to be of the least importance, strategically considered. At the battle of Bautzen the left of the allies rested upon the steep mountains of Bohemia, which province was at that time rather neutral than hostile: it seemed that, tactically considered, the slope of these mountains was the decisive point to be held, when it was just the reverse, because the allies had but one line of retreat upon Reichenbach and Gorlitz, and the French, by forcing the right, which was in the plain, would occupy this line of retreat and throw the allies into the mountains, where they might have lost all their *matériel* and a great part of the personnel of their army. This course was also easier for them on account of the difference in the features of the ground, led to more important results, and would have diminished the obstacles in the future.

The following truths may, I think, be deduced from what has been stated: 1. The topographical key of a battle-field is not always the tactical key; 2. The decisive point of a battle-field is certainly that which combines strategic with topographical advantages; 3. When the difficulties of the ground are not too formidable upon the strategic point of the battle-field, this is generally the most important point; 4. It is nevertheless true that the determination of this point depends very much upon the arrangement of the contending forces. Thus, in lines of battle too much extended and divided the center will always be the proper point of attack; in lines well closed and connected the center is the strongest point, since, independently of the reserves posted there, it is easy to support it from the flanks: the decisive point in this case is therefore one of the extremities of the line. When the numerical superiority is considerable, an attack may be made simultaneously upon both extremities, but not when the attacking force is equal or inferior numerically to the enemy's. It appears, therefore, that all the combinations of a battle consist in so employing the force in hand as to obtain the most effective action upon that one of the three points mentioned which offers the greatest number of chances of success,—a point very easily determined by applying the analysis just mentioned.

The object of an offensive battle can only be to dislodge the enemy or to cut his line, unless it is intended by strategic maneuvers to ruin his army completely. An enemy is dislodged either by overthrowing him at some point of his line, or by outflanking him so as to take him in flank and rear, or by using both these methods at once; that is, attacking him in front while at the same time one wing is enveloped and his line turned.

To accomplish these different objects, it becomes necessary to make choice of the most suitable order of battle for the method to be used.

At least twelve orders of battle may be enumerated, viz.: 1. The simple parallel order; 2. The parallel order with a defensive or offensive crotchet; 3. The order reinforced upon one or both wings; 4. The order reinforced in the center; 5. The simple oblique order, or the oblique reinforced on the attacking wing; 6 and 7. The perpendicular order on one or both wings; 8.

The concave order; 9. The convex order; 10. The order by echelon on one or both wings; 11. The order by echelon on the center; 12. The order resulting from a strong combined attack upon the center and one extremity simultaneously.

Each of these orders may be used either by itself or, as has been stated, in connection with the maneuver of a strong column intended to turn the enemy's line. In order to a proper appreciation of the merits of each, it becomes necessary to test each by the application of the general principles which have been laid down. For example, it is manifest that the parallel order is worst of all, for it requires no skill to fight one line against another, battalion against battalion, with equal chances of success on either side: no tactical skill is needed in such a battle.

There is, however, one important case where this is a suitable order, which occurs when an army, having taken the initiative in great strategic operations, shall have succeeded in falling upon the enemy's communications and cutting off his line of retreat while covering its own; when the battle takes place between them, that army which has reached the rear of the other may use the parallel order, for, having effected the decisive maneuver previous to the battle, all its efforts should now be directed toward the frustration of the enemy's endeavor to open a way through for himself. Except for this single case, the parallel order is the worst of all. I do not mean to say that a battle cannot be gained while using this order, for one side or the other must gain the victory if the contest is continued; and the advantage will then be upon his side who has the best troops, who best knows when to engage them, who best manages his reserve and is most favored by fortune.

The parallel order with a crotchet upon the flank is most usually adopted in a defensive position. It may be also the result of an offensive combination; but then the crotchet is to the front, whilst in the case of defense it is to the rear. The battle of Prague is a very remarkable example of the danger to which such a crotchet is exposed if properly attacked.

The parallel order reinforced upon one wing, or upon the center, to pierce that of the enemy, is much more favorable than the two preceding ones, and is also much more in accordance with the general principles which have been laid down; although, when the contending forces are about equal, the part of the line which has been weakened to reinforce the other may have its own safety compromised if placed in line parallel to the enemy.

The oblique order is the best for an inferior force attacking a superior; for, in addition to the advantage of bringing the main strength of the forces against a single point of the enemy's line, it has two others equally important, since the weakened wing is not only kept back from the attack of the enemy, but performs also the double duty of holding in position the part of his line not attacked, and of being at hand as a reserve for the support, if necessary, of the engaged wing. This order was used by the celebrated Epaminondas at the battles of Leuctra and Mantinea. The most brilliant example of its use in modern times was given by Frederick the Great at the battle of Leuthen. (See Treatise on Grand Operations.)

The perpendicular order on one or both wings can only be considered an arrangement to indicate the direction along which the primary tactical movements might be made in a battle. Two armies will never long occupy the relative perpendicular positions; for if the army B were to take its first position on a line perpendicular to one or both extremities of the army A, the latter would at once change the front of a portion of its line; and even the army B, as soon as it extended itself to or beyond the extremity of A, must of necessity turn its columns either to the right or the left, in order to bring them near the enemy's line, and so take him in reverse, as at C, the result being two oblique lines. The inference is that one division of the assailing army would take a position perpendicular to the enemy's wing, whilst the remainder of the army would approach in front for the purpose of annoying him; and this would always bring us back to one of the oblique orders.

The attack on both wings, whatever be the form of attack adopted, may be very advantageous, but it is only admissible when the assailant is very decidedly superior in numbers; for, if the fundamental principle is to bring the main strength of the forces upon the decisive point, a weaker army would violate it in directing a divided attack against a superior force. This truth will be clearly demonstrated farther on.

The order concave in the center has found advocates since the day when Hannibal by its use gained the battle of Cannæ. This order may indeed be very good when the progress of the battle itself gives rise to it; that is, when the enemy attacks the center, this retires before him, and he suffers himself to be enveloped by the wings. But, if this order is adopted before the battle begins, the enemy, instead of falling on the center, has only to attack the wings, which present their extremities and are in precisely the same relative situation as if they had been assailed in flank. This order would, therefore, be scarcely ever used except against an enemy who had taken the convex order to fight a battle, as will be seen farther on.

An army will rarely form a semicircle, preferring rather a broken line with the center retired. If several writers may be believed, such an arrangement gave the victory to the English on the famous days of Crécy and Agincourt. This order is certainly better than a semicircle, since it does not so much present the flank to attack, whilst allowing forward movement by echelon and preserving all the advantages of concentration of fire. These advantages vanish if the enemy, instead of foolishly throwing himself upon the retired center, is content to watch it from a distance and makes his greatest effort upon one wing. Essling, in 1809, is an example of the advantageous use of a concave line; but it must not be inferred that Napoleon committed an error in attacking the center; for an army fighting with the Danube behind it and with no way of moving without uncovering its bridges of communication, must not be judged as if it had been free to maneuver at pleasure.

The convex order with the center salient answers for an engagement immediately upon the passage of a river when the wings must be retired and rested on the river to cover the bridges; also when a defensive battle is to be fought with a river in rear, which is to be passed and the defile covered, as at Leipsic; and, finally, it may become a natural formation to resist an enemy forming a concave line. If an enemy directs his efforts against the center or against a single wing, this order might cause the ruin of the whole army. [An attack upon the two extremities might succeed also in some cases, either when the force was strong enough to try it, or the enemy was unable to weaken his center to support the wings. As a rule, a false attack to engage the center, and a strong attack against one extremity, would be the best method to use against such a line.]

The French tried it at Fleurus in 1794, and were successful, because the Prince of Coburg, in place of making a strong attack upon the center or upon a single extremity, divided his attack upon five or six diverging lines, and particularly upon both wings at once. Nearly the same convex order was adopted at Essling, and during the second and third days of the famous battle of Leipsic. On the last occasion it had just the result that might have been expected.

The order by echelon upon the two wings is of the same nature as the perpendicular order being, however, better than that, because, the echelons being nearest each other in the direction where the reserve would be placed, the enemy would be less able, both as regards room and time, to throw himself into the interval of the center and make at that point a threatening counter-attack.

The order by echelon on the center may be used with special success against an army occupying a position too much cut up and too extended, because, its center being then somewhat isolated from the wings and liable to overthrow, the army thus cut in two would be probably destroyed. But, applying the test of the same fundamental principle, this order of attack would appear to be less certain of success against an army having a connected and closed line; for the reserve being generally near the center, and the wings being able to act either by concentrating their fire or by moving against the foremost echelons, might readily repulse them.

If this formation to some extent resembles the famous triangular wedge or *boar's head* of the ancients, and the column of Winkelried, it also differs from them essentially; for, instead of forming one solid mass,—an impracticable thing in our day, on account of the use of artillery,—it would have a large open space in the middle, which would render movements more easy. This formation is suitable, as has been said, for penetrating the center of a line too much extended, and might be equally successful against a line unavoidably immovable; but if the wings of the attacked line are brought at a proper time against the flanks of the foremost echelons, disagreeable consequences might result. A parallel order considerably reinforced on the center might perhaps be a much better arrangement, for the parallel line in this case would have at least the advantage of deceiving the enemy as to the point of attack, and would hinder the wings from taking the echelons of the center by the flank.

This order by echelons was adopted by Laudon for the attack of the intrenched camp of Buntzelwitz. (Treatise on Grand Operations, chapter xxviii.) In such a case it is quite suitable; for it is then certain that the defensive army being forced to remain within its intrenchments, there is no danger of its attacking the echelons in flank. But, this formation having the inconvenience of indicating to the enemy the point of his line which it is desired to attack, false attacks should be made upon the wings, to mislead him as to the true point of attack.

The order of attack in columns on the center and on one extremity at the same time is better than the preceding, especially in an attack upon an enemy's line strongly arranged and well connected. It may even be called the most reasonable of all the orders of battle. The attack upon the center, aided by a wing outflanking the enemy, prevents the assailed party falling upon the assailant and taking him in flank, as was done by Hannibal and Marshal Saxe. The enemy's wing which is hemmed in between the attacks on the center and at the extremity, having to contend with nearly the entire opposing force, will be defeated and probably destroyed. It was this maneuver which gave Napoleon his victories of Wagram and Ligny. This was what he wished to attempt at Borodino,—where he obtained only a partial success, on account of the heroic conduct of the Russian left and the division of Paskevitch in the famous central redoubt, and on account of the arrival of Baggavout's corps on the wing he hoped to outflank. He used it also at Bautzen,—where an unprecedented success would have been the result, but for an accident which interfered with the maneuver of the left wing intended to cut off the allies from the road to Wurschen, every arrangement having been made with that view.

It should be observed that these different orders are not to be understood precisely. A general who would expect to arrange his line of battle as regularly as upon paper or on a drill-ground would be greatly mistaken, and would be likely to suffer defeat. This is particularly true as battles are now fought. In the time of Louis XIV. or of Frederick, it was possible to form lines of battle almost as regular as geometrical figures, because armies camped under tents, almost always closely collected together, and were in presence of each other several days, thus giving ample time for opening roads and clearing spaces to enable the columns to be at regular distances from each other. But in our day,—when armies bivouac, when their division into several corps gives greater mobility, when they take position near each other in obedience to orders given them while out of reach of the general's eye, and often when there has been no time for thorough examination of the enemy's position,—finally, when the different arms of the service are intermingled in the line of battle,—under these circumstances, all orders of battle which must be laid out with great accuracy of detail are impracticable.

If every army were a solid mass, capable of motion as a unit under the influence of one man's will and as rapidly as thought, the art of winning battles would be reduced to choosing the most favorable order of battle, and a general could reckon with certainty upon the success of maneuvers arranged beforehand. But the facts are altogether different; for the great difficulty of the tactics of battles will always be to render certain the simultaneous entering into action of the numerous fractions whose efforts must combine to make such an attack as will give good ground to hope for victory: in other words, the chief difficulty is to cause these fractions to unite in the execution of the decisive maneuver which, in accordance with the original plan of the battle, is to result in victory.

Inaccurate transmission of orders, the manner in which they will be understood and executed by the subordinates of the general-in-chief, excess of activity in some, lack of it in others, a defective *coup-d'oeil militaire*,—every thing of this kind may interfere with the simultaneous entering into action of the different parts, without speaking of the accidental circumstances which may delay or prevent the arrival of a corps at the appointed place.

Hence result two undoubted truths: 1. The more simple a decisive maneuver is, the more sure of success will it be; 2. Sudden maneuvers seasonably executed during an engagement are more likely to succeed than those determined upon in advance, unless the latter, relating to previous strategic movements, will bring up the columns which are to decide the day upon those points where their presence will secure the expected result. Waterloo and Bautzen are proofs of the last. From the moment when Blücher and Bulow had reached the heights of Frichermont, nothing could have prevented the loss of the battle by the French, and they could then only fight to make the defeat less complete. In like manner, at Bautzen, as soon as Ney had reached Klix, the retreat of the allies during the night of the 20th of May could alone have saved them, for on the 21st it was too late; and, if Ney had executed better what he was advised to do, the victory would have been a very great one.

As to maneuvers for breaking through a line and calculations upon the co-operation of columns proceeding from the general front of the army, with the intention of effecting large detours around an enemy's flank, it may be stated that their result is always doubtful, since it depends upon such an accurate execution of carefully-arranged plans as is rarely seen. This subject will be considered in Art. XXXII.

Besides the difficulty of depending upon the exact application of an order of battle arranged in advance, it often happens that battles begin without even the assailant having a well-defined object, although the collision may have been expected. This uncertainty results either from circumstances prior to the battle, from ignorance of the enemy's position and plans, or from the fact that a portion of the army may be still expected to arrive on the field.

From these things many people have concluded that it is impossible to reduce to different systems the formations of orders of battle, or that the adoption of either of them can at all influence the result of an engagement,—an erroneous conclusion, in my opinion, even in the cases cited above. Indeed, in battles begun without any predetermined plan it is probable that at the opening of the engagement the armies will occupy lines nearly parallel and more or less strengthened upon some point; the party acting upon the defensive, not knowing in what quarter the storm will burst upon him, will hold a large part of his forces in reserve, to be used as occasion may require; the assailant must make similar efforts to have his forces well in hand; but as soon as the point of attack shall have been determined, the mass of his troops will be directed against the center or upon one wing of the enemy, or upon both at once.

There is nothing even in Napoleon's battles which disproves my assertion, although they are less susceptible than any others of being represented by lines accurately laid down. We see him, however, at Rivoli, at Austerlitz, and at Ratisbon, concentrating his forces toward the center to be ready at the favorable moment to fall upon the enemy. At the Pyramids he formed an oblique line of squares in echelon. At Leipsic, Essling, and Brienne he used a kind of convex order. At Wagram his order was to bring up two masses upon the center and right, while keeping back the left wing; and this he wished to repeat at Borodino and at Waterloo before the Prussians came up. At Eylau, although the collision was almost entirely unforeseen on account of the very unexpected return and offensive movement of the Russians, he outflanked their left almost perpendicularly, whilst in another direction he was endeavoring to break through the center; but these attacks were not simultaneous, that on the center being repulsed at eleven o'clock, whilst Davoust did not attack vigorously upon the left until toward one. At Dresden he attacked by the two wings, for the first time probably in his life, because his center was covered by a fortification and an intrenched camp, and, in addition, the attack of his left was combined with that of Vandamme upon the enemy's line of retreat. At Marengo, if we may credit Napoleon himself, the oblique order he assumed, resting his right at Castel Ceriole, saved him from almost inevitable defeat. Ulm and Jena were battles won by strategy before they were fought, tactics having but little to do with them. At Ulm there was not even a regular battle.

I think we may hence conclude that if it seems absurd to desire to mark out upon the ground orders of battle in such regular lines as would be used in tracing them on a sketch, a skillful general may nevertheless bear in mind the orders which have been indicated above, and may so combine his troops on the battle-field that the arrangement shall be similar to one of them. He should endeavor in all his combinations, whether deliberately arranged or adopted on the spur of the moment, to form a sound conclusion as to the important point of the battle-field; and this he can only do by observing well the direction of the enemy's line of battle, and not forgetting the direction in which strategy requires him to operate. He will then give his attention and efforts to this point, using a third of his force to keep the enemy in check or watch his movements, while throwing the other two-thirds upon the point the possession of which will insure him the victory. Acting thus, he will have satisfied all the conditions the science of grand tactics can impose upon him, and will have applied the principles of the art in the most perfect manner. The manner of determining the decisive point of a battle-field has been described in the preceding chapter, (Art. XIX.)

Having now explained the twelve orders of battle, it has occurred to me that this would be a proper place to reply to several statements made in the Memoirs of Napoleon published by General Montholon.

The great captain seems to consider the oblique order a modern invention, a theorist's fancy,—an opinion I can by no means share; for the oblique order is as old as Thebes and Sparta, and I have seen it used with my own eyes. This assertion of Napoleon's seems the more remarkable because Napoleon himself boasted of having used, at Marengo, the very order of which he thus denies the existence.

If we understand that the oblique order is to be applied in the rigid and precise manner inculcated by General Ruchel at the Berlin school. Napoleon was certainly right in regarding it as an absurdity; but I repeat that a line of battle never was a regular geometrical figure, and when such figures are used in discussing the combinations of tactics it can only be for the purpose of giving definite expression to an idea by the use of a known symbol. It is nevertheless true that every line of battle which is neither parallel nor perpendicular to the enemy's must be oblique of necessity. If one army attacks the extremity of another army, the attacking wing being reinforced by massing troops upon it while the weakened wing is kept retired from attack, the direction of the line must of necessity be a little oblique, since one end of it will be nearer the enemy than the other. The oblique order is so far from being a mere fancy that we see it used when the order is that by echelons on one wing.

As to the other orders of battle explained above, it cannot be denied that at Essling and Fleurus the general arrangement of the Austrians was a concave line, and that of the French a convex. In these orders parallel lines may be used as in the case of straight lines, and they would be classified as belonging to the parallel system when no part of the line was more strongly occupied or drawn up nearer to the enemy than another.

Laying aside for the present further consideration of these geometrical figures, it is to be observed that, for the purpose of fighting battles in a truly scientific manner, the following points must be attended to:—

1. An offensive order of battle should have for its object to force the enemy from his position by all reasonable means.

2. The maneuvers indicated by art are those intended to overwhelm one wing only, or the center and one wing at the same time. An enemy may also be dislodged by maneuvers for outflanking and turning his position.

3. These attempts have a much greater probability of success if concealed from the enemy until the very moment of the assault.

4. To attack the center and both wings at the same time, without having very superior forces, would be entirely in opposition to the rules of the art, unless one of these attacks can be made very strongly without weakening the line too much at the other points.

5. The oblique order has no other object than to unite at least half the force of the army in an overwhelming attack upon one wing, while the remainder is retired to the rear, out of danger of attack, being arranged either in echelon or in a single oblique line.

6 The different formations, convex, concave, perpendicular, or otherwise, may all be varied by having the lines of uniform strength throughout, or by massing troops at one point.

7. The object of the defense being to defeat the plans of the attacking party, the arrangements of a defensive order should be such as to multiply the difficulties of approaching the position, and to keep in hand a strong reserve,

well concealed, and ready to fall at the decisive moment upon a point where the enemy least expect to meet it.

8. It is difficult to state with precision what is the best method to use in forcing a hostile army to abandon its position. An order of battle would be perfect which united the double advantages of the fire of the arms and of the moral effect produced by an onset. A skillful mixture of deployed lines and columns, acting alternately as circumstances require, will always be a good combination. In the practical use of this system many variations must arise from differences in the *coup-d'oeil* of commanders, the *morale* of officers and soldiers, their familiarity with maneuvers and firings of all sorts, from varying localities, etc..

9. As it is essential in an offensive battle to drive the enemy from his position and to cut him up as much as possible, the best means of accomplishing this is to use as much material force as can be accumulated against him. It sometimes happens, however, that the direct application of main force is of doubtful utility, and better results may follow from maneuvers to outflank and turn that wing which is nearest the enemy's line of retreat. He may when thus threatened retire, when he would fight strongly and successfully if attacked by main force.

History is full of examples of the success of such maneuvers, especially when used against generals of weak character; and, although victories thus obtained are generally less decisive and the hostile army is but little demoralized, such incomplete successes are of sufficient importance not to be neglected, and a skillful general should know how to employ the means to gain them when opportunity offers, and especially should he combine these turning movements with attacks by main force.

10. The combination of these two methods—that is to say, the attack in front by main force and the turning maneuver—will render the victory more certain than the use of either separately; but, in all cases, too extended movements must be avoided, even in presence of a contemptible enemy.

11. The manner of driving an enemy from his position by main force is the following:—Throw his troops into confusion by a heavy and well-directed fire of artillery, increase this confusion by vigorous charges of cavalry, and follow up the advantages thus gained by pushing forward masses of infantry well covered in front by skirmishers and flanked by cavalry.

But, while we may expect success to follow such an attack upon the first line, the second is still to be overcome, and, after that, the reserve; and at this period of the engagement the attacking party would usually be seriously embarrassed, did not the moral effect of the defeat of the first line often occasion the retreat of the second and cause the general in command to lose his presence of mind. In fact, the attacking troops will usually be somewhat disordered, even in victory, and it will often be very difficult to replace them by those of the second line, because they generally follow the first line at such a distance as not to come within musket-range of the enemy; and it is always embarrassing to substitute one division for another in the heat of battle, at the moment when the enemy is putting forth all his strength in repelling the attack.

These considerations lead to the belief that if the general and the troops of the defensive army are equally active in the performance of their duty, and preserve their presence of mind, if their flanks and line of retreat are not threatened, the advantage will usually be on their side at the second collision of the battle; but to insure that result their second line and the cavalry must be launched against the victorious battalions of the adversary at the proper instant; for the loss of a few minutes may be irreparable, and the second line may be drawn into the confusion of the first.

12. From the preceding facts may be deduced the following truth: "that the most difficult as well as the most certain of all the means the assailant may use to gain the victory consists in strongly supporting the first line with the troops of the second line, and these with the reserve, and in a proper employment of masses of cavalry and of batteries, to assist in striking the decisive blow at the second line of the enemy; for here is presented the greatest of all the problems of the tactics of battles."

In this important crisis of battles, theory becomes an uncertain guide; for it is then unequal to the emergency, and can never compare in value with a natural talent for war, nor be a sufficient substitute for that intuitive *coup-d'oeil* imparted by experience in battles to a general of tried bravery and coolness.

The simultaneous employment of the largest number of troops of all arms combined, except a small reserve of each which should be always held in hand, [The great reserves must, of course, be also engaged when it is necessary; but it is always a good plan to keep back, as a final reserve, two or three battalions and five or six squadrons. Moreau decided the battle of Engen with four companies of infantry; and what Kellermann's cavalry accomplished at Marengo is known to every reader of history.] will, therefore, at the critical moment of the battle, be the problem which every skillful general will attempt to solve and to which he should give his whole attention. This critical moment is usually when the first line of the parties is broken, and all the efforts of both contestants are put forth,—on the one side to complete the victory, on the other to wrest it from the enemy. It is scarcely necessary to say that, to make this decisive blow more certain and effectual, a simultaneous attack upon the enemy's flank would be very advantageous.

13. In the defensive the fire of musketry can be much more effectively used than in the offensive, since when a position is to be carried it can be accomplished only by moving upon it, and marching and firing at the same time can be done only by troops as skirmishers, being an impossibility for the principal masses. The object of the defense being to break and throw into confusion the troops advancing to the attack, the fire of artillery and musketry will be the natural defensive means of the first line, and when the enemy presses too closely the columns of the second line and part of the cavalry must be launched against him. There will then be a strong probability of his repulse.

Article XXXII: Turning Maneuvers, and too extended Movement in Battles

We have spoken in the preceding article of maneuvers undertaken to turn an enemy's line upon the battle-field, and of the advantages which may be expected from them. A few words remain to be said as to the wide détours which these maneuvers sometimes occasion, causing the failure of so many plans seemingly well arranged.

It may be laid down as a principle that any movement is dangerous which is so extended as to give the enemy an opportunity, while it is taking place, of beating the remainder of the army in position. Nevertheless, as the danger depends very much upon the rapid and certain *coup-d'oeil* of the opposing general, as well as upon the style of warfare to which he is accustomed, it is not difficult to understand why so many maneuvers of this kind have failed against some commanders and succeeded against others, and why such a movement which would have been hazardous in presence of Frederick, Napoleon, or Wellington might have entire success against a general of limited capacity, who had not the tact to take the offensive himself at the proper moment, or who might himself have been in the habit of moving in this manner.

It seems, therefore, difficult to lay down a fixed rule on the subject. The following directions are all that can be given. Keep the mass of the force well in hand and ready to act at the proper moment, being careful, however, to avoid the danger of accumulating troops in too large bodies. A commander observing these precautions will be always prepared for any thing that may happen. If the opposing general shows little skill and seems inclined to indulge in extended movements, his adversary may be more daring.

A few examples drawn from history will serve to convince the reader of the truth of my statements, and to show him how the results of these extended movements depend upon the characters of the generals and the armies concerned in them.

In the Seven Years' War, Frederick gained the battle of Prague because the Austrians had left a feebly-defended interval of one thousand yards between their right and the remainder of their army,—the latter part remaining motionless while the right was overwhelmed. This inaction was the more extraordinary as the left of the Austrians had a much shorter distance to pass over in order to support their right than Frederick had to attack it; for the right was in the form of a crotchet, and Frederick was obliged to move on the arc of a large semicircle to reach it.

On the other hand, Frederick came near losing the battle of Torgau, because he made with his left a movement entirely too extended and disconnected (nearly six miles) with a view of turning the right of Marshal Daun. [For an account of these two battles, see Chapters II. and XXV. of the Treatise on Grand Military Operations.] Mollendorf brought up the right by a concentric movement to the heights of Siptitz, where he rejoined the king, whose line was thus reformed.

The battle of Rivoli is a noted instance in point. All who are familiar with that battle know that Alvinzi and his chief of staff Weyrother wished to surround Napoleon's little army, which was concentrated on the plateau of Rivoli. Their center was beaten,—while their left was piled up in the ravine of the Adige, and Lusignan with their right was making a wide *détour* to get upon the rear of the French army, where he was speedily surrounded and captured.

No one can forget the day of Stockach, where Jourdan conceived the unfortunate idea of causing an attack to be made upon a united army of sixty thousand men by three small divisions of seven thousand or eight thousand men, separated by distances of several leagues, whilst Saint-Cyr, with the third of the army, (thirteen thousand men,) was to pass twelve miles beyond the right flank and get in rear of this army of sixty thousand men, which could not help being victorious over these divided fractions, and should certainly have captured the part in their rear. Saint-Cyr's escape was indeed little less than a miracle.

We may call to mind how this same General Weyrother, who had desired to surround Napoleon at Rivoli, attempted the same maneuver at Austerlitz, in spite of the severe lesson he had formerly received. The left wing of the allied army, wishing to outflank Napoleon's right, to cut him off from Vienna, (where he did not desire to return,) by a circular movement of nearly six miles, opened an interval of a mile and a half in their line. Napoleon took advantage of this mistake, fell upon the center, and surrounded their left, which was completely shut up between Lakes Tellnitz and Melnitz.

Wellington gained the battle of Salamanca by a maneuver very similar to Napoleon's, because Marmont, who wished to cut off his retreat to Portugal, left an opening of a mile and a half in his line,—seeing which, the English general entirely defeated his left wing, that had no support.

If Weyrother had been opposed to Jourdan at Rivoli or at Austerlitz, he might have destroyed the French army, instead of suffering in each case a total defeat; for the general who at Stockach attacked a mass of sixty thousand men with four small bodies of troops so much separated as to be unable to give mutual aid would not have known how to take proper advantage of a wide detour effected in his presence. In the same way, Marmont was unfortunate in having at Salamanca an adversary whose chief merit was a rapid and practiced tactical *coup-d'oeil*. With the Duke of York or Moore for an antagonist, Marmont would probably have been successful.

Among the turning maneuvers which have succeeded in our day, Waterloo and Hohenlinden had the most brilliant results. Of these the first was almost altogether a strategic operation, and was attended with a rare concurrence of fortunate circumstances. As to Hohenlinden, we will search in vain in military history for another example of a single brigade venturing into a forest in the midst of fifty thousand enemies, and there performing such astonishing feats as Richepanse effected in the defile of Matenpoet, where he might have expected, in all probability, to lay down his arms.

At Wagram the turning wing under Davoust contributed greatly to the successful issue of the day; but, if the vigorous attack upon the center under Macdonald, Oudinot, and Bernadotte had not rendered opportune assistance, it is by no means certain that a like success would have been the result.

So many examples of conflicting results might induce the conclusion that no rule on this subject can be given; but this would be erroneous; for it seems, on the contrary, quite evident that, by adopting as a rule an order of battle well closed and well connected, a general will find himself prepared for any emergency, and little will be left to chance; but it is specially important for him to have a correct estimate of his enemy's character and his usual style of warfare, to enable him to regulate his own actions accordingly. In case of superiority in numbers or discipline, maneuvers may be attempted which would be imprudent were the forces equal or the commanders of the same capacity. A maneuver to outflank and turn a wing should be connected with other attacks, and opportunely supported by an attempt of the remainder of the army on the enemy's front, either against the wing turned or against the center. Finally, strategic operations to cut an enemy's line of communications before giving battle, and attack him in rear, the assailing army preserving its own line of retreat, are much more likely to be successful and effectual, and, moreover, they require no disconnected maneuver during the battle.

Article XXXIII: Unexpected Meeting of Two Armies on the March

The accidental and unexpected meeting of two armies on the march gives rise to one of the most imposing scenes in war.

In the greater number of battles, one party awaits his enemy in a position chosen in advance, which is attacked after a reconnoissance as close and accurate as possible. It often happens, however,—especially as war is now carried on,—that two armies approach each other, each intending to make an unexpected attack upon the other. A collision ensues unexpected by both armies, since each finds the other where it does not anticipate a meeting. One army may also be attacked by another which has prepared a surprise for it,—as happened to the French at Rossbach.

A great occasion of this kind calls into play all the genius of a skillful general and of the warrior able to control events. It is always possible to gain a battle with brave troops, even where the commander may not have great capacity; but victories like those of Lutzen, Luzzara, Eylau, Abensberg, can only be gained by a brilliant genius endowed with great coolness and using the wisest combinations.

There is so much chance in these accidental battles that it is by no means easy to lay down precise rules concerning them; but these are the very cases in which it is necessary to keep clearly before the mind the fundamental principles of the art and the different methods of applying them, in order to a proper arrangement of maneuvers that must be decided upon at the instant and in the midst of the crash of resounding arms.

Two armies marching, as they formerly did, with all their camp-equipage, and meeting unexpectedly, could do nothing better at first than cause their advanced guard to deploy to the right or left of the roads they are traversing. In each army the forces should at the same time be concentrated so that they may be thrown in a proper direction considering the object of the march. A grave error would be committed in deploying the whole army behind the advanced guard; because, even if the deployment were accomplished, the result would be nothing more than a badly-arranged parallel order, and if the enemy pressed the advanced guard with considerable vigor the consequence might be the rout of the troops which were forming. (See the account of the battle of Rossbach, Treatise on Grand Operations.)

In the modern system, when armies are more easily moved, marching upon several roads, and divided into masses which may act independently, these routs are not so much to be feared; but the principles are unchanged. The advanced guard must always be halted and formed, and then the mass of the troops concentrated in that direction which is best suited for carrying out the object of the march. Whatever maneuvers the enemy may then attempt, every thing will be in readiness to meet him.

Article XXXIV: of Surprises of Armies

I shall not speak here of surprises of small detachments,—the chief features in the wars of partisan or light troops, for which the light Russian and Turkish cavalry are so well adapted. I shall confine myself to an examination of the surprise of whole armies.

Before the invention of fire-arms, surprises were more easily effected than at present; for the reports of artillery and musketry firing are heard to so great a distance that the surprise of an army is now next to an impossibility, unless the first duties of field-service are forgotten and the enemy is in the midst of the army before his presence is known because there are no outposts to give the alarm. The Seven Years' War presents a memorable example in the surprise of Hochkirch. It shows that a surprise does not consist simply in falling upon troops that are sleeping or keeping a poor look-out, but that it may result from the combination of a sudden attack upon, and a surrounding of, one extremity of the army. In fact, to surprise an army it is not necessary to take it so entirely unawares that the troops will not even have emerged from their tents, but it is sufficient to attack it in force at the point intended, before preparations can be made to meet the attack.

As armies at the present day seldom camp in tents when on a march, prearranged surprises are rare and difficult, because in order to plan one it becomes necessary to have an accurate knowledge of the enemy's camp. At Marengo, at Lutzen, and at Eylau there was something like a surprise; but this term should only be applied to an entirely unexpected attack. The only great surprise to be cited is the case of Taroutin, in 1812, where Murat was attacked and beaten by Benningsen. To excuse his imprudence, Murat pretended that a secret armistice was in force; but there was really nothing of the kind, and he was surprised through his own negligence.

It is evident that the most favorable manner of attacking an army is to fall upon its camp just before daybreak, at the moment when nothing of the sort is expected. Confusion in the camp will certainly take place; and, if the assailant has an accurate knowledge of the locality and can give a suitable tactical and strategic direction to the mass of his forces, he may expect a complete success, unless unforeseen events occur. This is an operation by no means to be despised in war, although it is rare, and less brilliant than a great strategic combination which renders the victory certain even before the battle is fought.

For the same reason that advantage should be taken of all opportunities for surprising an adversary, the necessary precautions should be used to prevent such attacks. The regulations for the government of any well-organized army should point out the means for doing the last.

Article XXXV: Of the Attack by Main Force of Fortified Places, Intrenched Camps or Lines. —Of Coups de Main in General

There are many fortified places which, although not regular fortresses, are regarded as secure against *coups de main*, but may nevertheless be carried by escalade or assault, or through breaches not altogether practicable, but so steep as to require the use of ladders or some other means of getting to the parapet.

The attack of a place of this kind presents nearly the same combinations as that of an intrenched camp; for both belong to the class of *coups de main*.

This kind of attack will vary with circumstances: 1st, with the strength of the works; 2d, with the character of the ground on which they are built; 3d, with the fact of their being isolated or connected; 4th, with the morale of the respective parties. History gives us examples of all of these varieties.

For examples, take the intrenched camps of Kehl, Dresden, and Warsaw, the lines of Turin and Mayence, the intrenchments of Feldkirch, Scharnitz, and Assiette. Here I have mentioned several cases, each with varying circumstances and results. At Kehl (1796) the intrenchments were better connected and better constructed than at Warsaw. There was, in fact, a *tête de pont* nearly equal to a permanent fortification; for the archduke thought himself obliged to besiege it in form, and it would have been extremely hazardous for him to make an open attack upon it. At Warsaw the works were isolated, but of considerable relief, and they had as a keep a large city surrounded by loopholed walls, armed and defended by a number of desperate men.

Dresden, in 1813, had for a keep a bastioned enceinte, one front of which, however, was dismantled and had no other parapet than such as was suited to a field-work. The camp proper was protected by simple redoubts, at considerable distances apart, very poorly built, the keep giving it its sole strength. [The number of defenders at Dresden the first day (August 25) was twenty-four thousand, the next day, sixty-five thousand, and the third day, more than one hundred thousand.]

At Mayence and at Turin there were continuous lines of circumvallation; but if in the first case they were strong, they were certainly not so at Turin, where upon one of the important points there was an insignificant parapet with a command of three feet, and a ditch proportionally deep. In the latter case, also, the lines were between two fires, as they were attacked in rear by a strong garrison at the moment when Prince Eugene assailed them from without. At Mayence the lines were attacked in front, only a small detachment having succeeded in passing around the right flank.

The tactical measures to be taken in the attack of field-works are few in number. If it seems probable that a work may be surprised if attacked a little before day, it is altogether proper to make the attempt; but if this operation may be recommended in case of an isolated work, it is by no means to be expected that a large army occupying an intrenched camp will permit itself to be surprised,—especially as the regulations of all services require armies to stand to their arms at dawn. As an attack by main force seems likely to be the method followed in this case, the following simple and reasonable directions are laid down:—

1. Silence the guns of the work by a powerful artillery-fire, which at the same time has the effect of discouraging the defenders.

2. Provide for the troops all the materials necessary (such as fascines and short ladders) to enable them to pass the ditch and mount the parapet.

3. Direct three small columns upon the work to be taken, skirmishers preceding them, and reserves being at hand for their support.

4. Take advantage of every irregularity of the ground to get cover for the troops, and keep them sheltered as long as possible.

5. Give detailed instructions to the principal columns as to their duties when a work shall have been carried, and as to the manner of attacking the troops occupying the camp. Designate the bodies of cavalry which are to assist in attacking those troops if the ground permits. When all these arrangements are made, there is nothing more to be done but to bring up the troops to the attack as actively as possible, while a detachment makes an attempt at the gorge. Hesitancy and delay in such a case are worse than the most daring rashness.

Those gymnastic exercises are very useful which prepare soldiers for escalades and passing obstacles; and the engineers may with great advantage give their attention to providing means for facilitating the passage of the ditches of field-works and climbing their parapets.

Among all the arrangements in cases of this kind of which I have read, none are better than those for the assault of Warsaw and the intrenched camp of Mayence. Thielke gives a description of Laudon's dispositions for attacking the camp of Buntzelwitz, which, although not executed, is an excellent example for instruction. The attack of Warsaw may be cited as one of the finest operations of this sort, and does honor to Marshal Paskevitch and the troops who executed it. As an example not to be followed, no better can be given than the arrangements made for attacking Dresden in 1813.

Among attacks of this class may be mentioned the memorable assaults or escalades of Port Mahon in 1756, and of Berg-op-zoom in 1747,—both preceded by sieges, but still brilliant *coups de main*, since in neither case was the breach sufficiently large for a regular assault.

Continuous intrenched lines, although seeming to have a better interconnection than lines of detached works, are more easily carried, because they may be several leagues in extent, and it is almost impossible to prevent an enemy from breaking through them at some point. The capture of the lines of Mayence and Wissembourg, which are described in the History of the Wars of the Revolution, (Chapters XXI. and XXII.,) and that of the lines of Turin by Eugene of Savoy in 1706, are excellent lessons for study.

This famous event at Turin, which has been so often referred to, is so familiar to all readers that it is unnecessary to recall the details of it; but I cannot pass it by without remarking how easily the victory was bought and how little it should have been expected. The strategic plan was certainly admirable; and the march from the Adige through Piacenza to Asti by the right bank of the Po, leaving the French on the Mincio, was beautifully arranged, but its execution was exceedingly slow. When we examine the operations near Turin, we must confess that the victors owed more to their good fortune than to their wisdom. It required no great effort of genius upon the part of Prince Eugene to prepare the order he issued to his army; and he must have felt a profound contempt for his opponents to execute a march with thirty-five thousand allied troops of ten different nations between eighty thousand Frenchmen on the one side and the Alps on the other, and to pass around their camp for forty-eight hours by the most remarkable flank march that was ever attempted. The order for the attack was so brief and so devoid of instruction that any staff officer of the present day ought to write a better. Directing the formation of eight columns of infantry by brigade in two lines, giving them orders to carry the intrenchments and to make openings through them for the passage of the cavalry into the camp, make up the sum total of all the science exhibited by Eugene in order to carry out his rash undertaking It is true he selected the weak point of the intrenchment; for it was there so low that it covered only half the bodies of its defenders.

But I am wandering from my subject, and must return to the explanation of the measures most suitable for adoption in an attack on lines. If they have a sufficient relief to make it difficult to carry them by assault, and if on the other hand they may be outflanked or turned by strategic maneuvers, it is far better to pursue the course last indicated than to attempt a hazardous assault. If, however, there is any reason for preferring the attack by assault, it should be made upon one of the wings, because the center is the point most easily succored. There have been cases where an attack on the wing was expected by the defenders, and they have been deceived by a false attack made at that point, while the real attack took place at the center, and succeeded simply because unexpected. In these operations the locality and the character of the generals engaged must decide as to the proper course to be pursued.

The attack may be executed in the manner described for intrenched camps. It has sometimes happened, however, that these lines have had the relief and proportions of permanent works; and in this case escalade would be quite difficult, except of old earthen works whose slopes were worn away from the lapse of time and had become accessible for infantry of moderate activity. The ramparts of Ismail and Praga were of this character; so also was the citadel of Smolensk, which Paskevitch so gloriously defended against Ney, because he preferred making his stand at the ravines in front, rather than take shelter behind a parapet with an inclination of scarcely thirty degrees.

If one extremity of a line rests upon a river, it seems absurd to think of penetrating upon that wing, because the enemy collecting his forces, the mass of which would be near the center, might defeat the columns advancing between the center and the river and completely destroy them. This absurdity, however, has sometimes been successful; because the enemy driven behind his lines rarely thinks of making an offensive return upon the assailant, no matter how advantageous it might seem. A general and soldiers who seek refuge behind lines are already half conquered, and the idea of taking the offensive does not occur to them when their intrenchments are attacked. Notwithstanding these facts, I cannot advise such a course; and the general who would run such a risk and meet the fate of Tallard at Blenheim could have no just cause of complaint.

Very few directions can be given for the defense of intrenched camps and lines. The first is to be sure of having strong reserves placed between the center and each wing, or, to speak more accurately, on the right of the left wing and on the left of the right wing. With this arrangement succor can be easily and rapidly carried to a threatened point, which could not be done were there but one central reserve. It has been suggested that three reserves would not be too many if the intrenchment is very extensive; but I decidedly incline to the opinion that two are quite enough. Another recommendation may be given, and it is of great importance,—that the troops be made to understand they must by no means despair of finally defending a line which may be forced at one point; because, if a good reserve is at hand, it may take the offensive, attack the assailant, and succeed in driving him out of the work he may have supposed in his power.

Coups De Main

These are bold enterprises undertaken by a detachment of an army for the capture of posts of different strength or importance. [The distinction between the importance and the strength of a post must be observed; for it may be very strong and of very little importance, and vice aversá.] They partake of the nature both of surprises and attacks by main force, for both these methods may be employed in carrying an attempt of this sort to a successful issue. Although *coups de main* seem to be entirely tactical operations, their importance certainly depends on the relations of the captured posts to the strategic combinations in hand. It will become necessary, therefore, to say a few words with reference to coups de main in Article XXXVI., when speaking of detachments. However tiresome these repetitions may seem, I am obliged to state here the manner of executing such operations, as it is evidently a part of the subject of the attack of intrenchments.

I do not pretend to say that the rules of tactics apply to these operations; for their name, *coups de main*, implies that ordinary rules are not applicable to them. I desire only to call attention to them, and refer my readers to the different works, either historical or didactic, where they are mentioned.

I have previously stated that important results may often follow from these enterprises. The capture of Sizeboli in 1828, the unsuccessful attack of General Petrasch upon Kehl in 1796, the remarkable surprises of Cremona in 1702, of Gibraltar in 1704, and of Berg-op-zoom in 1814, as well as the escalades of Port Mahon and Badajos, give an idea of the different kinds of *coup de main*. Some are effected by surprise, others by open force. Skill, stratagems, boldness, on the part of the assailant, and fear excited among the assailed, are some of the things which have an influence upon the successful issue of *coups de main*.

As war is now waged, the capture of a post, however strong, is no longer of the same importance as formerly unless it has a direct influence upon the results of a great strategic operation.

The capture or destruction of a bridge defended by intrenchments, that of a large convoy, of a small fort closing important passes, like the two attacks which were made in 1799 upon the fort of Lucisteig in the Grisons; the capture of Leutasch and Scharnitz by Ney in 1805; finally, the capture of a post not even fortified, but used as a great depot of provisions and munitions much needed by the enemy;—such are the enterprises which will justify the risks to which a detachment engaging in them may be exposed.

Posts have been captured by filling up the ditches sometimes with fascines, sometimes with bags of wool; and manure has been used for the same purpose. Ladders are generally necessary, and should always be prepared. Hooks have been used in the hands and attached to the shoes of soldiers, to help them in climbing rocky heights which commanded the intrenchment. An entrance was effected through the sewers at Cremona by Prince Eugene.

In reading such facts, we must draw from them not rules, but hints; for what has been done once may be done again.

Of Several Mixed Operations, Which Are in Character Partly Strategical and Partly Tactical

Article XXXVI: Of Diversions and Great Detachments

The operations of the detachments an army may send out have so important a bearing on the success of a campaign, that the duty of determining their strength and the proper occasions for them is one of the greatest and most delicate responsibilities imposed upon a commander. If nothing is more useful in war than a strong detachment opportunely sent out and having a good *ensemble* of operations with the main body, it is equally certain that no expedient is more dangerous when inconsiderately adopted. Frederick the Great regarded it as one of the essential qualities of a general to know how to make his adversary send out many detachments, either with the view of destroying them in detail or of attacking the main body during their absence.

The division of armies into numerous detachments has sometimes been carried to so great an extent, and with such poor results, that many persons now believe it better to have none of them. It is undoubtedly much safer and more agreeable for an army to be kept in a single mass; but it is a thing at times impossible or incompatible with gaining a complete or even considerable success. The essential point in this matter is to send out as few detachments as possible.

There are several kinds of detachments.

1. There are large corps dispatched to a distance from the zone of operations of the main army, in order to make diversions of greater or less importance.

2. There are large detachments made in the zone of operations to cover important points of this zone, to carry on a siege, to guard a secondary base, or to protect the line of operations if threatened.

3. There are large detachments made upon the front of operations, in face of the enemy, to act in concert with the main body in some combined operation.

4. There are small detachments sent to a distance to try the effect of surprise upon isolated points, whose capture may have an important bearing upon the general operations of the campaign.

I understand by diversions those secondary operations carried out at a distance from the principal zone of operations, at the extremities of a theater of war, upon the success of which it is sometimes foolishly supposed the whole campaign depends. Such diversions are useful in but two cases, the first of which arises when the troops thus employed cannot conveniently act elsewhere on account of their distance from the real theater of operations, and the second is that where such a detachment would receive strong support from the population among which it was sent,—the latter case belonging rather to political than military combinations. A few illustrative examples may not be out of place here.

The unfortunate results for the allied powers of the Anglo-Russian expedition to Holland, and of that of the Archduke Charles toward the end of the last century, (which have been referred to in Article XIX.,) are well known.

In 1805, Napoleon was occupying Naples and Hanover. The allies intended an Anglo-Russian army to drive him out of Italy, while the combined forces of England, Russia, and Sweden should drive him from Hanover, nearly sixty thousand men being designed for these two widely-separated points. But, while their troops were collecting at the two extremities of Europe, Napoleon ordered the evacuation of Naples and Hanover, Saint-Cyr hastened to effect a junction with Massena in the Frioul, and Bernadotte, leaving Hanover, moved up to take part in the operations of Ulm and Austerlitz. After these astonishing successes, Napoleon had no difficulty in retaking Naples and Hanover. This is an example of the failure of diversions. I will give an instance where such an operation would have been proper.

In the civil wars of 1793, if the allies had sent twenty thousand men to La Vendée, they would have accomplished much more than by increasing the numbers of those who were fighting fruitlessly at Toulon, upon the Rhine, and in Belgium. Here is a case where a diversion would have been not only very useful, but decisive.

It has already been stated that, besides diversions to a distance and of small bodies, large corps are often detached in the zone of operations of the main army.

If the employment of these large corps thus detached for secondary objects is more dangerous than the diversions above referred to, it is no less true that they are often highly proper and, it may be, indispensable.

These great detachments are chiefly of two kinds. The first are permanent corps which must be sometimes thrown out in a direction opposite to the main line of operations, and are to remain throughout a campaign. The second are corps temporarily detached for the purpose of assisting in carrying out some special enterprise.

Among the first should be especially enumerated those fractions of an army that are detached either to form the strategic reserve, of which mention has been made, or to cover lines of operation and retreat when the configuration of the theater of the war exposes them to attack. For example, a Russian army that wishes to cross the Balkan is obliged to leave a portion of its forces to observe Shumla, Routchouk, and the valley of the Danube, whose direction is perpendicular to its line of operations. However successful it may be, a respectable force must always be left toward Giurgevo or Krajova, and even on the right bank of the river toward Routchouk.

This single example shows that it is sometimes necessary to have a double strategic front, and then the detachment of a considerable corps must be made to offer front to a part of the enemy's army in rear of the main army. Other localities and other circumstances might be mentioned where this measure would be equally essential to safety. One case is the double strategic front of the Tyrol and the Frioul for a French army passing the Adige. On whichever side it may wish to direct its main column, a detachment must be left on the other front sufficiently strong to hold in check the enemy threatening to cut the line of communications. The third example is the frontier of Spain, which enables the Spaniards to establish a double front,—one covering the road to Madrid, the other having Saragossa or Galicia as a base. To whichever side the invading army turns, a detachment must be left on the other proportioned in magnitude to the enemy's force in that direction.

All that can be said on this point is that it is advantageous to enlarge as much as possible the field of operations of such detachments, and to give them as much power of mobility as possible, in order to enable them by opportune movements to strike important blows. A most remarkable illustration of this truth was given by Napoleon in the campaign of 1797. Obliged as he was to leave a corps of fifteen thousand men in the valley of the Adige to observe the Tyrol while he was operating toward the Noric Alps, he preferred to draw this corps to his aid, at the risk of losing temporarily his line of retreat, rather than leave the parts of his army disconnected and exposed to defeat in detail. Persuaded that he could be victorious with his army united, he apprehended no particular danger from the presence of a few hostile detachments upon his communications.

Great movable and temporary detachments are made for the following reasons:—

1. To compel your enemy to retreat to cover his line of operations, or else to cover your own.

2. To intercept a corps and prevent its junction with the main body of the enemy, or to facilitate the approach of your own reinforcements.

3. To observe and hold in position a large portion of the opposing army, while a blow is struck at the remainder.

4. To carry off a considerable convoy of provisions or munitions, on receiving which depended the continuance of a siege or the success of any strategic enterprise, or to protect the march of a convoy of your own.

5. To make a demonstration to draw the enemy in a direction where you wish him to go, in order to facilitate the execution of an enterprise in another direction.

6. To mask, or even to invest, one or more fortified places for a certain time, with a view either to attack or to keep the garrison shut up within the ramparts.

7. To take possession of an important point upon the communications of an enemy already retreating.

However great may be the temptation to undertake such operations as those enumerated, it must be constantly borne in mind that they are always secondary in importance, and that the essential thing is to be successful at the decisive points. A multiplication of detachments must, therefore, be avoided. Armies have been destroyed for no other reason than that they were not kept together.

We will here refer to several of these enterprises, to show that their success depends sometimes upon good fortune and sometimes upon the skill of their designer, and that they often fail from faulty execution.

Peter the Great took the first step toward the destruction of Charles XII. by causing the seizure, by a strong detachment, of the famous convoy Lowenhaupt was bringing up. Villars entirely defeated at Denain the large detachment Prince Eugene sent out in 1709 under D'Albermale.

The destruction of the great convoy Laudon took from Frederick during the siege of Olmutz compelled the king to evacuate Moravia. The fate of the two detachments of Fouquet at Landshut in 1760, and of Fink at Maxen in 1759, demonstrates how difficult it is at times to avoid making detachments, and how dangerous they may be. To come nearer our own times, the disaster of Vandamme at Culm was a bloody lesson, teaching that a corps must not be thrust forward too boldly: however, we must admit that in this case the operation was well planned, and the fault was not so much in sending out the detachment as in not supporting it properly, as might easily have been done. That of Fink was destroyed at Maxen nearly on the same spot and for the same reason.

Diversions or demonstrations in the zone of operations of the army are decidedly advantageous when arranged for the purpose of engaging the enemy's attention in one direction, while the mass of the forces is collected upon another point where the important blow is to be struck. In such a case, care must be taken not only to avoid engaging the corps making the demonstration, but to recall it promptly toward the main body. We will mention two examples as illustrations of these facts.

In 1800, Moreau, wishing to deceive Kray as to the true direction of his march, carried his left wing toward Rastadt from Kehl, whilst he was really filing off his army toward Stockach; his left, having simply shown itself, returned toward the center by Fribourg in Brisgau.

In 1805, Napoleon, while master of Vienna, detached the corps of Bernadotte to Iglau to overawe Bohemia and paralyze the Archduke Ferdinand, who was assembling an army in that territory; in another direction he sent Davoust to Presburg to show himself in Hungary; but he withdrew them to Brunn, to take part in the event which was to decide the issue of the campaign, and a great and decisive victory was the result of his wise maneuvers. Operations of this kind, so far from being in opposition to the principles of the art of war, are necessary to facilitate their application.

It readily appears from what goes before that precise rules cannot be laid down for these operations, so varied in character, the success of which depends on so many minute details. Generals should run the risk of making detachments only after careful consideration and observation of all the surrounding circumstances. The only reasonable rules on the subject are these: send out as few detachments as possible, and recall thorn immediately when their duty is performed. The inconveniences necessarily attending them may be made as few as practicable, by giving judicious and carefully-prepared instructions to their commanders: herein lies the great talent of a good chief of staff.

One of the means of avoiding the disastrous results to which detachments sometimes lead is to neglect none of the precautions prescribed by tactics for increasing the strength of any force by posting it in good positions; but it is generally imprudent to engage in a serious conflict with too large a body of troops. In such cases ease and rapidity of motion will be most likely to insure safety. It seldom happens that it is right for a detachment to resolve to conquer or die in the position it has taken, whether voluntarily or by order.

It is certain that in all possible cases the rules of tactics and of field-fortification must be applied by detachments as well as by the army itself.

Since we have included in the number of useful cases of detachments those intended for *coups de main*, it is proper to mention a few examples of this kind to enable the reader to judge for himself. We may call to mind that one which was executed by the Russians toward the end of 1828 with the view of taking possession of Sizeboli in the Gulf of Bourghas. The capture of this feebly-fortified gulf, which the Russians rapidly strengthened, procured for them in case of success an essential *point d'appui* beyond the Balkan, where depots could be established in advance for the army intending to cross those mountains: in case of failure, no one was compromised,—not even the small corps which had been debarked, since it had a safe and certain retreat to the shipping.

In like manner, in the campaign of 1796, the *coup de main* attempted by the Austrians for the purpose of taking possession of Kehl and destroying the bridge whilst Moreau was returning from Bavaria, would have had very important consequences if it had not failed.

In attempts of this kind a little is risked to gain a great deal; and, as they can in no wise compromise the safety of the main army, they may be freely recommended.

Small bodies of troops thrown forward into the zone of the enemy's operations belong to the class of detachments that are judicious. A few hundred horsemen thus risked will be no great loss if captured; and they may be the means of causing the enemy great injury. The small detachments sent out by the Russians in 1807, 1812, and 1813 were a great hinderance to Napoleon's operations, and several times caused his plans to fail by intercepting his couriers.

For such expeditions officers should be selected who are bold and full of stratagems. They ought to inflict upon the enemy all the injury they can without compromising themselves. When an opportunity of striking a telling blow presents itself, they should not think for a moment of any dangers or difficulties in their path. Generally, however, address and presence of mind, which will lead them to avoid useless danger, are qualities more necessary for a partisan than cool, calculating boldness. For further information on this subject I refer my readers to Chapter XXXV. of the Treatise on Grand Operations, and to Article XLV. of this work, on light cavalry.

Article XXXVII: Passage of Rivers and Other Streams

The passage of a small stream, over which a bridge is already in place or might be easily constructed, presents none of the combinations belonging to grand tactics or strategy; but the passage of a large river, such as the Danube, the Rhine, the Po, the Elbe, the Oder, the Vistula, the Inn, the Ticino, etc., is an operation worthy the closest study.

The art of building military bridges is a special branch of military science, which is committed to pontoniers or sappers. It is not from this point of view that I propose to consider the passage of a stream, but as the attack of a military position and as a maneuver.

The passage itself is a tactical operation; but the determination of the point of passage may have an important connection with all the operations taking place within the entire theater of the war. The passage of the Rhine by General Moreau in 1800 is an excellent illustration of the truth of this remark. Napoleon, a more skillful strategist than Moreau, desired him to cross at Schaffhausen in order to take Kray's whole army in reverse, to reach Ulm before him, to cut him off from Austria and hurl him back upon the Main. Moreau, who had already a bridge at Basel, preferred passing, with greater convenience to his army, in front of the enemy, to turning his extreme left. The tactical advantages seemed to his mind much more sure than the strategical: he preferred the certainty of a partial success to the risk attending a victory which would have been a decisive one. In the same campaign Napoleon's passage of the Po is another example of the high strategic importance of the choice of the point of crossing. The army of the reserve, after the engagement of the Chiusella, could either march by the left bank of the Po to Turin, or cross the river at Crescentino and march directly to Genoa. Napoleon preferred to cross the Ticino, enter Milan, effect a junction with Moncey who was approaching with twenty thousand men by the Saint-Gothard pass, then to cross the Po at Piacenza, expecting to get before Mélas more certainly in that direction than if he came down too soon upon his line of retreat. The passage of the Danube at Donauwerth and Ingolstadt in 1805 was a very similar operation. The direction chosen for the passage was the prime cause of the destruction of Mack's army.

The proper strategic point of passage is easily determined by recollecting the principles laid down in Article XIX.; and it is here only necessary to remind the reader that in crossing a river, as in every other operation, there are permanent or geographical decisive points, and others which are relative or eventual, depending on the distribution of the hostile forces.

If the point selected combines strategic advantages with the tactical, no other point can be better; but if the locality presents obstacles exceedingly difficult to pass, another must be chosen, and in making the new selection care should be taken to have the direction of the movement as nearly as possible coincident with the true strategic direction. Independently of the general combinations, which exercise a great influence in fixing the point of passage, there is still another consideration, connected with the locality itself. The best position is that where the army after crossing can take its front of operations and line of battle perpendicular to the river, at least for the first marches, without being forced to separate into several corps moving upon different lines. This advantage will also save it the danger of fighting a battle with a river in rear, as happened to Napoleon at Essling.

Enough has been said with reference to the strategical considerations influencing the selection of the point of crossing a river. We will now proceed to speak of the passage itself. History is the best school in which to study the measures likely to insure the success of such operations. The ancients deemed the passage of the Granicus—which is a small stream—a wonderful exploit. So far as this point is concerned, the people of modern days can cite much greater.

The passage of the Rhine at Tholhuys by Louis XIV. has been greatly lauded; and it was really remarkable. In our own time, General Dedon has made famous the two passages of the Rhine at Kehl and of the Danube at Hochstadt in 1800. His work is a model as far as concerns the details; and in these operations minute attention to details is every thing. More recently, three other passages of the Danube, and the ever-famous passage of the Beresina, have exceeded every thing of the kind previously seen. The two first were executed by Napoleon at Essling and at Wagram, in presence of an army of one hundred and twenty thousand men provided with four hundred pieces of cannon, and at a point where the bed of the stream is broadest. General Pelet's interesting account of them should be carefully read. The third was executed by the Russian army at Satounovo in 1828, which, although not to be compared with the two just mentioned, was very remarkable on account of the great local difficulties and the vigorous exertions made to surmount them. The passage of the Beresina was truly wonderful. My object not being to give historical details on this subject, I direct my readers to the special narratives of these events. I will give several general rules to be observed.

1. It is essential to deceive the enemy as to the point of passage, that he may not accumulate an opposing force there. In addition to the strategic demonstrations, false attacks must be made near the real one, to divide the attention and means of the enemy. For this purpose half of the artillery should be employed to make a great deal of noise at the points where the passage is not to be made, whilst perfect silence should be preserved where the real attempt is to be made.

2. The construction of the bridge should be covered as much as possible by troops sent over in boats for the purpose of dislodging the enemy who

might interfere with the progress of the work; and these troops should take possession at once of any villages, woods, or other obstacles in the vicinity.

3. It is of importance also to arrange large batteries of heavy caliber, not only to sweep the opposite bank, but to silence any artillery the enemy might bring up to batter the bridge while building. For this purpose it is convenient to have the bank from which the passage is made somewhat higher than the other.

4. The proximity of a large island near the enemy's bank gives great facilities for passing over troops in boats and for constructing the bridge. In like manner, a smaller stream emptying into the larger near the point of passage is a favorable place for collecting and concealing boats and materials for the bridge.

5. It is well to choose a position where the river makes a re-entering bend, as the batteries on the assailant's side can cross their fire in front of the point where the troops are to land from the boats and where the end of the bridge is to rest, thus taking the enemy in front and flank when he attempts to oppose the passage.

6. The locality selected should be near good roads on both banks, that the army may have good communications to the front and rear on both banks of the river. For this reason, those points where the banks are high and steep should be usually avoided.

The rules for preventing a passage follow as a matter of course from those for effecting it, as the duty of the defenders is to counteract the efforts of the assailants. The important thing is to have the course of the river watched by bodies of light troops, without attempting to make a defense at every point. Concentrate rapidly at the threatened point, in order to overwhelm the enemy while a part only of his army shall have passed. Imitate the Duke of Vendôme at Cassano, and the Archduke Charles at Essling in 1809,—the last example being particularly worthy of praise, although the operation was not so decidedly successful as might have been expected.

In Article XXI. attention was called to the influence that the passage of a river, in the opening of a campaign, may have in giving direction to the lines of operations. We will now see what connection it may have with subsequent strategic movements.

One of the greatest difficulties to be encountered after a passage is to cover the bridge against the enemy's efforts to destroy it, without interfering too much with the free movement of the army. When the army is numerically very superior to the enemy, or when the river is passed just after a great victory gained, the difficulty mentioned is trifling; but when the campaign is just opening, and the two opposing armies are about equal, the case is very different.

If one hundred thousand Frenchmen pass the Rhine at Strasbourg or at Manheim in presence of one hundred thousand Austrians, the first thing to be done will be to drive the enemy in three directions,—first, before them as far as the Black Forest, secondly, by the right in order to cover the bridges on the Upper Rhine, and thirdly, by the left to cover the bridges of Mayence and the Lower Rhine. This necessity is the cause of an unfortunate division of the forces; but, to make the inconveniences of this subdivision as few as possible, the idea must be insisted on that it is by no means essential for the army to be separated into three equal parts, nor need these detachments remain absent longer than the few days required for taking possession of the natural point of concentration of the enemy's forces.

The fact cannot be concealed, however, that the case supposed is one in which the general finds his position a most trying one; for if he divides his army to protect his bridges he may be obliged to contend with one of his subdivisions against the whole of the enemy's force, and have it overwhelmed; and if he moves his army upon a single line, the enemy may divide his army and reassemble it at some unexpected point, the bridges may be captured or destroyed, and the general may find himself compromised before he has had time or opportunity to gain a victory.

The best course to be pursued is to place the bridges near a city which will afford a strong defensive point for their protection, to infuse all possible vigor and activity into the first operations after the passage, to fall upon the subdivisions of the enemy's army in succession, and to beat them in such a way that they will have no further desire of touching the bridges. In some cases eccentric lines of operations may be used. If the enemy has divided his one hundred thousand men into several corps, occupying posts of observation, a passage may be effected with one hundred thousand men at a single point near the center of the line of posts, the isolated defensive corps at this position may be overwhelmed, and two masses of fifty thousand men each may then be formed, which, by taking diverging lines of operations, can certainly drive off the successive portions of the opposing army, prevent them from reuniting, and remove them farther and farther from the bridges. But if, on the contrary, the passage be effected at one extremity of the enemy's strategic front, by moving rapidly along this front the enemy may be beaten throughout its whole extent,—in the same manner that Frederick tactically beat the Austrian line at Leuthen throughout its length,—the bridges will be secure in rear of the army, and remain protected during all the forward movements. It was in this manner that Jourdan, having passed the Rhine at Dusseldorf in 1795, on the extreme right of the Austrians, could have advanced in perfect safety toward the Main. He was driven away because the French, having a double and exterior line of operations, left one hundred and twenty thousand men inactive between Mayence and Basel, while Clairfayt repulsed Jourdan upon the Lahn. But this cannot diminish the importance of the advantages gained by passing a river upon one extremity of the enemy's strategic front. A commander-in-chief should either adopt this method, or that previously explained, of a central mass at the moment of passage, and the use of eccentric lines afterward, according to the circumstances of the case, the situation of the frontiers and bases of operations, as well as the positions of the enemy. The mention of these combinations, of which something has already been said in the article on lines of operations, does not appear out of place here, since their connection with the location of bridges has been the chief point under discussion.

It sometimes happens that, for cogent reasons, a double passage is attempted upon a single front of operations, as was the case with Jourdan and Moreau in 1796. If the advantage is gained of having in case of need a double line of retreat, there is the inconvenience, in thus operating on the two extremities of the enemy's front, of forcing him, in a measure, to concentrate on his center, and he may be placed in a condition to overwhelm separately the two armies which have crossed at different points. Such an operation will always lead to disastrous results when the opposing general has sufficient ability to know how to take advantage of this violation of principles.

In such a case, the inconveniences of the double passage may be diminished by passing over the mass of the forces at one of the points, which then becomes the decisive one, and by concentrating the two portions by interior lines as rapidly as possible, to prevent the enemy from destroying them separately. If Jourdan and Moreau had observed this rule, and made a junction of their forces in the direction of Donauwerth, instead of moving eccentrically, they would probably have achieved great successes in Bavaria, instead of being driven back upon the Rhine.

Article XXXVIII: Retreats and Pursuits

Retreats are certainly the most difficult operations in war. This remark is so true that the celebrated Prince de Ligne said, in his usual piquant style, that he could not conceive how an army ever succeeded in retreating. When we think of the physical and moral condition of an army in full retreat after a lost battle, of the difficulty of preserving order, and of the disasters to which disorder may lead, it is not hard to understand why the most experienced generals have hesitated to attempt such an operation.

What method of retreat shall be recommended? Shall the fight be continued at all hazards until nightfall and the retreat executed under cover of the darkness? or is it better not to wait for this last chance, but to abandon the field of battle while it can be done and a strong opposition still made to the pursuing army? Should a forced march be made in the night, in order to get as much start of the enemy as possible? or is it better to halt after a half-march and make a show of fighting again? Each of these methods, although entirely proper in certain cases, might in others prove ruinous to the whole army. If the theory of war leaves any points unprovided for, that of retreats is certainly one of them.

If you determine to fight vigorously until night, you may expose yourself to a complete defeat before that time arrives; and if a forced retreat must begin when the shades of night are shrouding every thing in darkness and obscurity, how can you prevent the disintegration of your army, which does not know what to do, and cannot see to do any thing properly? If, on the other hand, the field of battle is abandoned in broad daylight and before all possible efforts have been made to hold it, you may give up the contest at the very moment when the enemy is about to do the same thing; and this fact coming to the knowledge of the troops, you may lose their confidence,—as they are always inclined to blame a prudent general who retreats before the necessity for so doing may be evident to themselves. Moreover, who can say that a retreat commenced in the daylight in presence of an enterprising enemy may not become a rout?

When the retreat is actually begun, it is no less difficult to decide whether a forced march shall be made to get as much the start of the enemy as possible,—since this hurried movement might sometimes cause the destruction of the army, and might, in other circumstances, be its salvation. All that can be positively asserted on this subject is that, in general, with an army of considerable magnitude, it is best to retreat slowly, by short marches, with a well-arranged rear-guard of sufficient strength to hold the heads of the enemy's columns in check for several hours.

Retreats are of different kinds, depending upon the cause from which they result. A general may retire of his own accord before fighting, in order to draw his adversary to a position which he prefers to his present one. This is rather a prudent maneuver than a retreat. It was thus that Napoleon retired in 1805 from Wischau toward Brunn to draw the allies to a point which suited him as a battle-field. It was thus that Wellington retired from Quatre-Bras to Waterloo. This is what I proposed to do before the attack at Dresden, when the arrival of Napoleon was known. I represented the necessity of moving toward Dippoldiswalde to choose a favorable battle-field. It was supposed to be a retreat that I was proposing; and a mistaken idea of honor prevented a retrograde movement without fighting, which would have been the means of avoiding the catastrophe of the next day, (August 26, 1813.)

A general may retire in order to hasten to the defense of a point threatened by the enemy, either upon the flanks or upon the line of retreat. When an army is marching at a distance from its depots, in an exhausted country, it may be obliged to retire in order to get nearer its supplies. Finally, an army retires involuntarily after a lost battle, or after an unsuccessful enterprise.

These are not the only causes having an influence in retreats. Their character will vary with that of the country, with the distances to be passed over and the obstacles to be surmounted. They are specially dangerous in an enemy's country; and when the points at which the retreats begin are distant from the friendly country and the base of operations, they become painful and difficult.

From the time of the famous retreat of the Ten Thousand, so justly celebrated, until the terrible catastrophe which befell the French army in 1812, history does not make mention of many remarkable retreats. That of Antony, driven out of Media, was more painful than glorious. That of the Emperor Julian, harassed by the same Parthians, was a disaster. In more recent days, the retreat of Charles VIII. to Naples, when he passed by a corps of the Italian army at Fornovo, was an admirable one. The retreat of M. de Bellisle from Prague does not deserve the praises it has received. Those executed by the King of Prussia after raising the siege of Olmutz and after the surprise at Hochkirch were very well arranged; but they were for short distances. That of Moreau in 1796, which was magnified in importance by party spirit, was creditable, but not at all extraordinary. The retreat of Lecourbe from Engadin to Altorf, and that of Macdonald by Pontremoli after the defeat of the Trebbia, as also that of Suwaroff from the Muttenthal to Chur, were glorious feats of arms, but partial in character and of short duration. The retreat of the Russian army from the Niemen to Moscow—a space of two hundred and forty leagues,—in presence of such an enemy as Napoleon and such cavalry as the active and daring Murat commanded, was certainly admirable. It was undoubtedly attended by many favorable circumstances, but was highly deserving of praise, not only for the talent displayed by the generals who directed its first stages, but also for the admirable fortitude and soldierly bearing of the troops who performed it. Although the retreat from Moscow was a bloody catastrophe for Napoleon, it was also glorious for him and the troops who were at Krasnoi and the Beresina,—because the skeleton of the army was saved, when not a single man should have returned. In this ever-memorable event both parties covered themselves with glory.

The magnitude of the distances and the nature of the country to be traversed, the resources it offers, the obstacles to be encountered, the attacks to be apprehended, either in rear or in flank, superiority or inferiority in cavalry, the spirit of the troops, are circumstances which have a great effect in deciding the fate of retreats, leaving out of consideration the skillful arrangements which the generals may make for their execution.

A general falling back toward his native land along his line of magazines and supplies may keep his troops together and in good order, and may effect a retreat with more safety than one compelled to subsist his army in cantonments, finding it necessary to occupy an extended position. It would be absurd to pretend that a French army retiring from Moscow to the Niemen without supplies of provisions, in want of cavalry and draft horses, could effect the movement in the same good order and with the same steadiness as a Russian army, well provided with every thing necessary, marching in its own country, and covered by an immense number of light cavalry.

There are five methods of arranging a retreat:—

The first is to march in a single mass and upon one road.

The second consists in dividing the army into two or three corps, marching at the distance of a day's march from each other, in order to avoid confusion, especially in the *matériel*.

The third consists in marching upon a single front by several roads nearly parallel and having a common point of arrival.

The fourth consists in moving by constantly converging roads.

The fifth, on the contrary, consists in moving along diverging roads.

I have nothing to say as to the formation of rear-guards; but it is taken for granted that a good one should always be prepared and well sustained by a portion of the cavalry reserves. This arrangement is common to all kinds of retreats, but has nothing to do with the strategic relations of these operations.

An army falling back in good order, with the intention of fighting as soon as it shall have received expected reinforcements or as soon as it shall have reached a certain strategic position, should prefer the first method, as this particularly insures the compactness of the army and enables it to be in readiness for battle almost at any moment, since it is simply necessary to halt the heads of columns and form the remainder of the troops under their protection as they successively arrive. An army employing this method must not, however, confine itself to the single main road, if there are side-roads sufficiently near to be occupied which may render its movements more rapid and secure.

When Napoleon retired from Smolensk, he used the second method, having the portions of his army separated by an entire march. He made therein a great mistake, because the enemy was not following upon his rear, but moving along a lateral road which brought him in a nearly perpendicular direction into the midst of the separated French corps. The three fatal days of Krasnoi were the result. The employment of this method being chiefly to avoid incumbering the road, the interval between the departure of the several corps is sufficiently great when the artillery may readily file off. Instead of separating the corps by a whole march, the army would be better divided into two masses and a rear-guard, a half-march from each other. These masses, moving off in succession with an interval of two hours between the departure of their several army-corps, may file off without incumbering the road, at least in ordinary countries. In crossing the Saint-Bernard or the Balkan, other calculations would doubtless be necessary.

I apply this idea to an army of one hundred and twenty thousand or one hundred and fifty thousand men, having a rear-guard of twenty thousand or twenty-five thousand men distant about a half-march in rear. The army may be divided into two masses of about sixty thousand men each, encamped at a distance of three or four leagues from each other. Each of these masses will be subdivided into two or three corps, which may either move successively along the road or form in two lines across the road. In either case, if one corps of thirty thousand men moves at five A.M. and the other at seven, there will be no danger of interference with each other, unless something unusual should happen; for the second mass being at the same hours of the day about four leagues behind the first, they can never be occupying the same part of the road at the same time.

When there are practicable roads in the neighborhood, suitable at least for infantry and cavalry, the intervals may be diminished. It is scarcely necessary to add that such an order of march can only be used when provisions are plentiful; and the third method is usually the best, because the army is then marching in battle-order. In long days and in hot countries the best times for marching are the night and the early part of the day. It is one of the most difficult problems of logistics to make suitable arrangements of hours of departures and halts for armies; and this is particularly the case in retreats.

Many generals neglect to arrange the manner and times of halts, and great disorder on the march is the consequence, as each brigade or division takes the responsibility of halting whenever the soldiers are a little tired and find it agreeable to bivouac. The larger the army and the more compactly it marches, the more important does it become to arrange well the hours of departures and halts, especially if the army is to move at night. An ill-timed halt of part of a column may cause as much mischief as a rout.

If the rear-guard is closely pressed, the army should halt in order to relieve it by a fresh corps taken from the second mass, which will halt with this object in view. The enemy seeing eighty thousand men in battle-order will think it necessary to halt and collect his columns; and then the retreat should recommence at nightfall, to regain the space which has been lost.

The third method, of retreating along several parallel roads, is excellent when the roads are sufficiently near each other. But, if they are quite distant, one wing separated from the center and from the other wing may be compromised if the enemy attacks it in force and compels it to stand on the defensive. The Prussian army moving from Magdeburg toward the Oder, in 1806, gives an example of this kind.

The fourth method, which consists in following concentric roads, is undoubtedly the best if the troops are distant from each other when the retreat is ordered. Nothing can be better, in such a case, than to unite the forces; and the concentric retreat is the only method of effecting this.

The fifth method indicated is nothing else than the famous system of eccentric lines, which I have attributed to Bulow, and have opposed so warmly in the earlier editions of my works, because I thought I could not be mistaken either as to the sense of his remarks on the subject or as to the object of his system. I gathered from his definition that he recommended to a retreating army, moving from any given position, to separate into parts and pursue diverging roads, with the double object of withdrawing more readily from the enemy in pursuit and of arresting his march by threatening his flanks and his line of communications. I found great fault with the system, for the simple reason that a beaten army is already weak enough, without absurdly still further dividing its forces and strength in presence of a victorious enemy.

Bulow has found defenders who declare that I mistake his meaning, and that by the term *eccentric retreat* he did not understand a retreat made on several diverging roads, but one which, instead of being directed toward the center of the base of operations or the center of the country, should be eccentric to that focus of operations, and along the line of the frontier of the country.

I may possibly have taken an incorrect impression from his language, and in this case my criticism falls to the ground; for I have strongly recommended that kind of a retreat to which I have given the name of the parallel retreat. It is my opinion that an army, leaving the line which leads from the frontiers to the center of the state, with a view of moving to the right or the left, may very well pursue a course nearly parallel to the line of the frontiers, or to its front of operations and its base. It seems to me more rational to give the name of parallel retreat to such a movement as that described, designating as eccentric retreat that where diverging roads are followed, all leading from the strategic front.

However this dispute about words may result, the sole cause of which was the obscurity of Bulow's text, I find fault only with those retreats made along several diverging roads, under pretense of covering a greater extent of frontier and of threatening the enemy on both flanks.

By using these high-sounding words *flanks*, an air of importance may be given to systems entirely at variance with the principles of the art. An army in retreat is always in a bad state, either physically or morally; because a retreat can only be the result of reverses or of numerical inferiority. Shall such an army be still more weakened by dividing it? I find no fault with retreats executed in several columns, to increase the ease of moving, when these columns can support each other; but I am speaking of those made along diverging lines of operations. Suppose an army of forty thousand men retreating before another of sixty thousand. If the first forms four isolated divisions of about ten thousand men, the enemy may maneuver with two masses of thirty thousand men each. Can he not turn his adversary, surround, disperse, and ruin in succession all his divisions? How can they escape such a fate? By *concentration*. This being in direct opposition to a divergent system, the latter falls of itself.

I invoke to my support the great lessons of experience. When the leading divisions of the army of Italy were repulsed by Wurmser, Bonaparte collected them all together at Roverbella; and, although he had only forty thousand men, he fought and beat sixty thousand, because he had only to contend against isolated columns. If he had made a divergent retreat, what would have become of his army and his victories? Wurmser, after his first check, made an eccentric retreat, directing his two wings toward the extremities of the line of defense. What was the result? His right, although supported by the mountains of the Tyrol, was beaten at Trent. Bonaparte then fell upon the rear of his left, and destroyed that at Bassano and Mantua.

When the Archduke Charles gave way before the first efforts of the French armies in 1796, would he have saved Germany by an eccentric movement? Was not the salvation of Germany due to his concentric retreat? At last Moreau, who had moved with a very extended line of isolated divisions, perceived that this was an excellent system for his own destruction, if he stood his ground and fought or adopted the alternative of retreating. He concentrated his scattered troops, and all the efforts of the enemy were fruitless in presence of a mass which it was necessary to watch throughout the whole length of a line of two hundred miles. Such examples must put an end to further discussion. [Ten years after this first refutation of Bulow's idea, the concentric retreat of Barclay and Bagration saved the Russian army. Although it did not prevent Napoleon's first success, it was, in the end, the cause of his ruin.]

There are two cases in which divergent retreats are admissible, and then only as a last resource. First, when an army has experienced a great defeat in its own country, and the scattered fragments seek protection within the walls of fortified places. Secondly, in a war where the sympathies of the whole population are enlisted, each fraction of the army thus divided may serve as a nucleus of assembly in each province; but in a purely methodical war, with regular armies, carried on according to the principles of the art, divergent retreats are simply absurd.

There is still another strategical consideration as to the direction of a retreat,—to decide when it should be made perpendicularly to the frontier and toward the interior of the country, or when it should be parallel to the frontier. For example, when Marshal Soult gave up the line of the Pyrenees in 1814, he had to choose one of two directions for his retreat,—either by way of Bordeaux toward the interior of France, or by way of Toulouse parallel to the frontier formed by the Pyrenees. In the same way, when Frederick retired from Moravia, he marched toward Bohemia instead of returning to Silesia.

These parallel retreats are often to be preferred, for the reason that they divert the enemy from a march upon the capital of the state and the center of its power. The propriety of giving such a direction to a retreat must be determined by the configuration of the frontiers, the positions of the fortresses, the greater or less space the army may have for its marches, and the facilities for recovering its direct communications with the central portions of the state.

Spain is admirably suited to the use of this system. If a French army penetrates by way of Bayonne, the Spaniards may base themselves upon Pampeluna and Saragossa, or upon Leon and the Asturias; and in either case the French cannot move directly to Madrid, because their line of operations would be at the mercy of their adversary.

The frontier of the Turkish empire on the Danube presents the same advantages, if the Turks knew how to profit by them.

In France also the parallel retreat may be used, especially when the nation itself is not divided into two political parties each of which is striving for the possession of the capital. If the hostile army penetrates through the Alps, the French can act on the Rhone and the Saône, passing around the frontier as far as the Moselle on one side, or as far as Provence on the other. If the enemy enters the country by way of Strasbourg, Mayence, or Valenciennes, the same thing can be done. The occupation of Paris by the enemy would be impossible, or at least very hazardous, so long as a French army remained in good condition and based upon its circle of fortified towns. The same is the case for all countries having double fronts of operations. [In all these calculations I suppose the contending forces nearly equal. If the invading army is twice as strong as the defensive, it may be divided into two equal parts, one of which may move directly upon the capital, while the other may follow the army retiring along the frontier. If the armies are equal, this is impossible.]

Austria is perhaps not so fortunately situated, on account of the directions of the Rhetian and Tyrolean Alps and of the river Danube. Lloyd, however, considers Bohemia and the Tyrol as two bastions connected by the strong curtain of the river Inn, and regards this frontier as exceedingly well suited for parallel movements. This assertion was not well sustained by the events of the campaigns of 1800, 1805, and 1809; but, as the parallel method has not yet had a fair trial on that ground, the question is still an open one.

It seems to me that the propriety of applying the parallel method depends mainly upon the existing and the antecedent circumstances of each case. If a French army should approach from the Rhine by way of Bavaria, and should find allies in force upon the Lech and the Iser, it would be a very delicate operation to throw the whole Austrian army into the Tyrol and into Bohemia, with the expectation of arresting in this way the forward movement to Vienna. If half the Austrian army is left upon the Inn to cover the approaches to the capital, an unfortunate division of force is the consequence; and if it is decided to throw the whole army into the Tyrol, leaving the way to Vienna open, there would be great danger incurred if the enemy is at all enterprising. In Italy, beyond the Mincio, the parallel method would be of difficult application on the side of the Tyrol, as well as in Bohemia against an enemy approaching from Saxony, for the reason that the theater of operations would be too contracted.

In Prussia the parallel retreat may be used with great advantage against an army debouching from Bohemia upon the Elbe or the Oder, whilst its employment would be impossible against a French army moving from the Rhine, or a Russian army from the Vistula, unless Prussia and Austria were allies. This is a result of the geographical configuration of the country, which allows and even favors lateral movements: in the direction of its greatest dimension, (from Memel to Mayence;) but such a movement would be disastrous if made from Dresden to Stettin.

When an army retreats, whatever may be the motive of the operation, a pursuit always follows.

A retreat, even when executed in the most skillful manner and by an army in good condition, always gives an advantage to the pursuing army; and this is particularly the case after a defeat and when the source of supplies and reinforcements is at a great distance; for a retreat then becomes more difficult than any other operation in war, and its difficulties increase in proportion to the skill exhibited by the enemy in conducting the pursuit.

The boldness and activity of the pursuit will depend, of course, upon the character of the commanders and upon the *physique* and *morale* of the two armies. It is difficult to prescribe fixed rules for all cases of pursuits, but the following points must be recollected:—

1. It is generally better to direct the pursuit upon the flank of the retreating columns, especially when it is made in one's own country and where no danger is incurred in moving perpendicularly or diagonally upon the enemy's line of operations. Care must, however, be taken not to make too large a circuit; for there might then be danger of losing the retreating enemy entirely.

2. A pursuit should generally be as boldly and actively executed as possible, especially when it is subsequent to a battle gained; because the demoralized army may be wholly dispersed if vigorously followed up.

3. There are very few cases where it is wise to make a bridge of gold for the enemy, no matter what the old Roman proverb may say; for it can scarcely ever be desirable to pay an enemy to leave a country, unless in the case when an unexpected success shall have been gained over him by an army much inferior to his in numbers.

Nothing further of importance can be added to what has been said on the subject of retreats, as far as they are connected with grand combinations of strategy. We may profitably indicate several tactical measures which may render them more easy of execution.

One of the surest means of making a retreat successfully is to familiarize the officers and soldiers with the idea that an enemy may be resisted quite as well when coming on the rear as on the front, and that the preservation of order is the only means of saving a body of troops harassed by the enemy during a retrograde movement. Rigid discipline is at all times the best preservative of good order, but it is of special importance during a retreat. To enforce discipline, subsistence must be furnished, that the troops may not be obliged to straggle off for the purpose of getting supplies by marauding.

It is a good plan to give the command of the rear-guard to an officer of great coolness, and to attach to it staff officers who may, in advance of its movements, examine and select points suitable for occupation to hold the enemy temporarily in check. Cavalry can rally so rapidly on the main body that it is evidently desirable to have considerable bodies of such troops, as they greatly facilitate the execution of a slow and methodical retreat, and furnish the means of thoroughly examining the road itself and the neighborhood, so as to prevent an unexpected onset of the enemy upon the flanks of the retreating columns.

It is generally sufficient if the rear-guard keep the enemy at the distance of half a day's march from the main body. The rear-guard would run great risk of being itself cut off, if farther distant. When, however, there are defiles in its rear which are held by friends, it may increase the sphere of its operations and remain a full day's march to the rear; for a defile, when held, facilitates a retreat in the same degree that it renders it more difficult if in the power of the enemy. If the army is very numerous and the rear-guard proportionally large, it may remain a day's march in rear. This will depend, however, upon its strength, the nature of the country, and the character and strength of the pursuing force. If the enemy presses up closely, it is of importance not to permit him to do so with impunity, especially if the retreat is made in good order. In such a case it is a good plan to halt from time to time and fall unexpectedly upon the enemy's advanced guard, as the Archduke Charles did in 1796 at Neresheim, Moreau at Biberach, and Kleber at Ukerath. Such a maneuver almost always succeeds, on account of the surprise occasioned by an unexpected offensive return upon a body of troops which is thinking of little else than collecting trophies and spoils.

Passages of rivers in retreat are also operations by no means devoid of interest. If the stream is narrow and there are permanent bridges over it, the operation is nothing more than the passage of a defile; but when the river is wide and is to be crossed upon a temporary military bridge, it is a maneuver of extreme delicacy. Among the precautions to be taken, a very important one is to get the parks well advanced, so that they may be out of the way of the army; for this purpose it is well for the army to halt a half-day's march from the river. The rear-guard should also keep at more than the usual distance from the main body,—as far, in fact, as the locality and the respective forces opposed will permit. The army may thus file across the bridge without being too much hurried. The march of the rear-guard should be so arranged that it shall have reached a position in front of the bridge just as the last of the main body has passed. This will be a suitable moment for relieving the rear-guard by fresh troops strongly posted. The rear-guard will pass through the intervals of the fresh troops in position and will cross the river; the enemy, coming up and finding fresh troops drawn up to give him battle, will make no attempt to press them too closely. The new rear-guard will hold its position until night, and will then cross the river, breaking the bridges after it.

It is, of course, understood that as fast as the troops pass they form on the opposite bank and plant batteries, so as to protect the corps left to hold the enemy in check.

The dangers of such a passage in retreat, and the nature of the precautions which facilitate it, indicate that measures should always be taken to throw up intrenchments at the point where the bridge is to be constructed and the passage made. Where time is not allowed for the construction of a regular *tête de pont*, a few well-armed redoubts will be found of great value in covering the retreat of the last troops.

If the passage of a large river is so difficult when the enemy is only pressing on the rear of the column, it is far more so when the army is threatened both in front and rear and the river is guarded by the enemy in force.

The celebrated passage of the Beresina by the French is one of the most remarkable examples of such an operation. Never was an army in a more desperate condition, and never was one extricated more gloriously and skillfully. Pressed by famine, benumbed with cold, distant twelve hundred miles from its base of operations, assailed by the enemy in front and in rear, having a river with marshy banks in front, surrounded by vast forests, how could it hope to escape? It paid dearly for the honor it gained. The mistake of Admiral Tschitchagoff doubtless helped its escape; but the army performed heroic deeds, for which due praise should be given. We do not know whether to admire most the plan of operations which brought up the Russian armies from the extremities of Moldavia, from Moscow, and from Polotzk to the Beresina as to a rendezvous arranged in peace,—a plan which came near effecting the capture of their formidable adversary,—or the wonderful firmness of the lion thus pursued, who succeeded in opening a way through his enemies.

The only rules to be laid down are, not to permit your army to be closely pressed upon, to deceive the enemy as to the point of passage, and to fall headlong upon the corps which bars the way before the one which is following the rear of your column can come up. Never place yourself in a position to be exposed to such danger; for escape in such a case is rare.

If a retreating army should strive to protect its bridges either by regular *têtes de font,* or at least by lines of redoubts to cover the rear-guard, it is natural, also, that the enemy pursuing should use every effort to destroy the bridges. When the retreat is made down the bank of a river, wooden houses may be thrown into the stream, also fire-ships and mills,—a means the Austrians used in 1796 against Jourdan's army, near Neuwied on the Rhine, where they nearly compromised the army of the Sambre and the Meuse. The Archduke Charles did the same thing at Essling in 1809. He broke the bridge over the Danube, and brought Napoleon to the brink of ruin.

It is difficult to secure a bridge against attacks of this character unless there is time for placing a stockade above it. Boats may be anchored, provided with ropes and grappling-hooks to catch floating bodies and with means for extinguishing fire-boats.

Article XXXIX: Of Cantonments, either when on the March, or when established in Winter Quarters

So much has been written on this point, and its connection with my subject is so indirect, that I shall treat it very briefly.

To maintain an army in cantonments, in a war actively carried on, is generally difficult, however connected the arrangement may be, and there is almost always some point exposed to the enemy's attacks. A country where large towns abound, as Lombardy, Saxony, the Netherlands, Swabia, or old Prussia, presents more facilities for the establishment of quarters than one where towns are few; for in the former case the troops have not only convenient supplies of food, but shelters which permit the divisions of the army to be kept closely together. In Poland, Russia, portions of Austria and France, in Spain and in Southern Italy, it is more difficult to put an army into winter quarters.

Formerly, it was usual for each party to go into winter quarters at the end of October, and all the fighting after that time was of a partisan character and carried on by the advanced troops forming the outposts.

The surprise of the Austrian winter quarters in Upper Alsace in 1674, by Turenne, is a good example, from which may be learned the best method of conducting such an enterprise, and the precautions to be taken on the other side to prevent its success.

The best rules to be laid down on this subject seem to me to be the following. Establish the cantonments very compactly and connectedly and occupying a space as broad as long, in order to avoid having a too extended line of troops, which is always easily broken through and cannot be concentrated in time; cover them by a river, or by an outer line of troops in huts and with their position strengthened by field-works; fix upon points of assembly which may be reached by all the troops before the enemy can penetrate so far; keep all the avenues by which an enemy may approach constantly patrolled by bodies of cavalry; finally, establish signals to give warning if an attack is made at any point.

In the winter of 1807, Napoleon established his army in cantonments behind the Passarge in face of the enemy, the advanced guard alone being hutted near the cities of Gutstadt, Osterode, etc.. The army numbered more than one hundred and twenty thousand men, and much skill was requisite in feeding it and keeping it otherwise comfortable in this position until June. The country was of a favorable character; but this cannot be expected to be the case everywhere.

An army of one hundred thousand men may find it not very difficult to have a compact and well-connected system of winter quarters in countries where large towns are numerous. The difficulty increases with the size of the army. It must be observed, however, that if the extent of country occupied increases in proportion to the numbers in the army, the means of opposing an irruption of the enemy increase in the same proportion. The important point is to be able to assemble fifty thousand or sixty thousand men in twenty-four hours. With such an army in hand, and with the certainty of having it rapidly increased, the enemy may be held in check, no matter how strong he may be, until the whole army is assembled.

It must be admitted, however, that there will always be a risk in going into winter quarters if the enemy keeps his army in a body and seems inclined to make offensive movements; and the conclusion to be drawn from this fact is, that the only method of giving secure repose to an army in winter or in the midst of a campaign is to establish it in quarters protected by a river, or to arrange an armistice.

In the strategic positions taken up by an army in the course of a campaign, whether marching, or acting as an army of observation, or waiting for a favorable opportunity of taking the offensive, it will probably occupy quite compact cantonments. The selection of such positions requires great experience upon the part of a general, in order that he may form correct conclusions as to what he may expect the enemy to do. An army should occupy space enough to enable it to subsist readily, and it should also keep as much concentrated as possible, to be ready for the enemy should he show himself; and these two conditions are by no means easily reconciled. There is no better arrangement than to place the divisions of the army in a space nearly a square, so that in case of need the whole may be assembled at any point where the enemy may present himself. Nine divisions placed in this way, a half-day's march from each other, may in twelve hours assemble on the center. The same rules are to be observed in these cases as were laid down for winter quarters.

Article XL: Descents

These are operations of rare occurrence, and may be classed as among the most difficult in war when effected in presence of a well-prepared enemy.

Since the invention of gunpowder and the changes effected by it in navies, transports are so helpless in presence of the monstrous three-deckers of the present day, armed as they are with a hundred cannon, that an army can make a descent only with the assistance of a numerous fleet of ships of war which can command the sea, at least until the debarkation of the army takes place.

Before the invention of gunpowder, the transports were also the ships of war; they were moved along at pleasure by using oars, were light, and could skirt along the coasts; their number was in proportion to the number of troops to be embarked; and, aside from the danger of tempests, the operations of a fleet could be arranged with almost as much certainty as those of an army on land. Ancient history, for these reasons, gives us examples of more extensive debarkations than modern times.

Who does not recall to mind the immense forces transported by the Persians upon the Black Sea, the Bosporus, and the Archipelago,—the innumerable hosts landed in Greece by Xerxes and Darius,—the great expeditions of the Carthaginians and Romans to Spain and Sicily, that of Alexander into Asia Minor, those of Cæsar to England and Africa, that of Germanicus to the mouths of the Elbe,—the Crusades,—the expeditions of the Northmen to England, to France, and even to Italy?

Since the invention of cannon, the too celebrated Armada of Philip II. was the only enterprise of this kind of any magnitude until that set on foot by Napoleon against England in 1803. All other marine expeditions were of no great extent: as, for example, those of Charles V. and of Sebastian of Portugal to the coast of Africa; also the several descents of the French into the United States of America, into Egypt and St. Domingo, of the English to Egypt, Holland, Copenhagen, Antwerp, Philadelphia. I say nothing of Hoche's projected landing in Ireland; for that was a failure, and is, at the same time, an example of the difficulties to be apprehended in such attempts.

The large armies kept on foot in our day by the great states of the world prevent descents with thirty or forty thousand men, except against second-rate powers; for it is extremely difficult to find transportation for one hundred or one hundred and fifty thousand men with their immense trains of artillery, munitions, cavalry, etc..

We were, however, on the point of seeing the solution of the vast problem of the practicability of descents in great force, if it is true that Napoleon seriously contemplated the transportation of one hundred and sixty thousand veterans from Boulogne to the British Isles: unfortunately, his failure to execute this gigantic undertaking has left us entirely in the dark as to this grave question.

It is not impossible to collect fifty French ships-of-the-line in the Channel by misleading the English; this was, in fact, upon the point of being done; it is then no longer impossible, with a favorable wind, to pass over the flotilla in two days and effect a landing. But what would become of the army if a storm should disperse the fleet of ships of war and the English should return in force to the Channel and defeat the fleet or oblige it to regain its ports?

Posterity will regret, as the loss of an example to all future generations, that this immense undertaking was not carried through, or at least attempted. Doubtless, many brave men would have met their deaths; but were not those men mowed down more uselessly on the plains of Swabia, of Moravia, and of Castile, in the mountains of Portugal and the forests of Lithuania? What man would not glory in assisting to bring to a conclusion the greatest trial of skill and strength ever seen between two great nations? At any rate, posterity will find in the preparations made for this descent one of the most valuable lessons the present century has furnished for the study of soldiers and of statesmen. The labors of every kind performed on the coasts of France from 1803 to 1805 will be among the most remarkable monuments of the activity, foresight, and skill of Napoleon. It is recommended to the careful attention of young officers. But, while admitting the possibility of success for a great descent upon a coast so near as the English to Boulogne, what results should be expected if this armada had had a long sea-voyage to make? How could so many small vessels be kept moving, even for two days and nights? To what chances of ruin would not so many frail boats be exposed in navigating the open seas! Moreover, the artillery, munitions of war, equipments, provisions, and fresh water that must be carried with this multitude of men require immense labor in preparation and vast means of transportation.

Experience has shown clearly the difficulties attending such an expedition, even for thirty thousand men. From known facts, it is evident that a descent can be made with this number of men in four cases:—1st, against colonies or isolated possessions; 2d, against second-rate powers which cannot be immediately supported from abroad; 3d, for the purpose of effecting a temporary diversion, or to capture a position which it is important to hold for a time; 4th, to make a diversion, at once political and military, against a state already engaged in a great war, whose troops are occupied at a distance from the point of the descent.

It is difficult to lay down rules for operations of this character. About the only recommendations I can make are the following. Deceive the enemy as to the point of landing; choose a spot where the vessels may anchor in safety and the troops be landed together; infuse as much activity as possible into the operation, and take possession of some strong point to cover the development of the troops as they land; put on shore at once a part of the artillery, to give confidence and protection to the troops that have landed.

A great difficulty in such an operation is found in the fact that the transports can never get near the beach, and the troops must be landed in boats and rafts,—which takes time and gives the enemy great advantages. If the sea is rough, the men to be landed are exposed to great risks; for what can a body of infantry do, crowded in boats, tossed about by the waves, and ordinarily rendered unfit by sea-sickness for the proper use of their arms?

I can only advise the party on the defensive not to divide his forces too much by attempting to cover every point. It is an impossibility to line the entire coast with batteries and battalions for its defense; but the approaches to those places where large establishments are to be protected must be closed.

Signals should be arranged for giving prompt notice of the point where the enemy is landing, and all the disposable force should be rapidly concentrated there, to prevent his gaining a firm foothold.

The configuration of coasts has a great influence upon descents and their prosecution. There are countries where the coasts are steep and present few points of easy access for the ships and the troops to be landed: these few places may be more readily watched, and the descent becomes more difficult.

Finally, there is a strategical consideration connected with descents which may be usefully pointed out. The same principle which forbids a continental army from interposing the mass of its forces between the enemy and the sea requires, on the contrary, that an army landing upon a coast should always keep its principal mass in communication with the shore, which is at once its line of retreat and its base of supplies. For the same reason, its first care should be to make sure of the possession of one fortified harbor/ or at least of a tongue of land which is convenient to a good anchorage and may be easily strengthened by fortifications, in order that in case of reverse the troops may be re-embarked without hurry and loss.

Logistics; Or, the Practical Art of Moving Armies

Article XLI: A few Remarks on Logistics in General

Is logistics simply a science of detail? Or, on the contrary, is it a general science, forming one of the most essential parts of the art of war? or is it but a term, consecrated by long use, intended to designate collectively the different branches of staff duty,—that is to say, the different means of carrying out in practice the theoretical combinations of the art?

These questions will seem singular to those persons who are firmly convinced that nothing more remains to be said about the art of war, and believe it wrong to search out new definitions where every thing seems already accurately classified. For my own part, I am persuaded that good definitions lead to clear ideas; and I acknowledge some embarrassment in answering these questions which seem so simple.

In the earlier editions of this work I followed the example of other military writers, and called by the name of *logistics* the details of staff duties, which are the subject of regulations for field-service and of special instructions relating to the corps of quartermasters. This was the result of prejudices consecrated by time. The word *logistics* is derived, as we know, from the title of the *major général des logis*, (translated in German by *Quartiermeister*,) an officer whose duty it formerly was to lodge and camp the troops, to give direction to the marches of columns, and to locate them upon the ground. Logistics was then quite limited. But when war began to be waged without camps, movements became more complicated, and the staff officers had more extended functions. The chief of staff began to perform the duty of transmitting the conceptions of the general to the most distant points of the theater of war, and of procuring for him the necessary documents for arranging plans of operations. The chief of staff was called to the assistance of the general in arranging his plans, to give information of them to subordinates in orders and instructions, to explain them and to supervise their execution both in their *ensemble* and in their minute details: his duties were, therefore, evidently connected with all the operations of a campaign.

To be a good chief of staff, it became in this way necessary that a man should be acquainted with all the various branches of the art of war. If the term *logistics* includes all this, the two works of the Archduke Charles, the voluminous treatises of Guibert, Laroche-Aymon, Bousmard, and Ternay, all taken together, would hardly give even an incomplete sketch of what logistics is; for it would be nothing more nor less than the science of applying all possible military knowledge.

It appears from what has been said that the old term *logistics* is insufficient to designate the duties of staff officers, and that the real duties of a corps of such officers, if an attempt be made to instruct them in a proper manner for their performance, should be accurately prescribed by special regulations in accordance with the general principles of the art. Governments should take the precaution to publish well-considered regulations, which should define all the duties of staff officers and should give clear and accurate instructions as to the best methods of performing these duties.

The Austrian staff formerly had such a code of regulations for their government; but it was somewhat behind the times, and was better adapted to the old methods of carrying on war than the present. This is the only work of the kind I have seen. There are, no doubt, others, both public and secret; but I have no knowledge of their existence. Several generals—as, for instance, Grimoard and Thiebaut—have prepared manuals for staff officers, and the new royal corps of France has issued several partial sets of instructions; but there is nowhere to be found a complete manual on the subject.

If it is agreed that the old *logistics* had reference only to details of marches and camps, and, moreover, that the functions of staff officers at the present day are intimately connected with the most important strategical combinations, it must be admitted that logistics includes but a small part of the duties of staff officers; and if we retain the term we must understand it to be greatly extended and developed in signification, so as to embrace not only the duties of ordinary staff officers, but of generals-in-chief.

To convince my readers of this fact, I will mention the principal points that must be included if we wish to embrace in one view every duty and detail relating to the movements of armies and the undertakings resulting from such movements:—

1. The preparation of all the material necessary for setting the army in motion, or, in other words, for opening the campaign. Drawing up orders, instructions, and itineraries for the assemblage of the army and its subsequent launching upon its theater of operations.

2. Drawing up in a proper manner the orders of the general-in-chief for different enterprises, as well as plans of attack in expected battles.

3. Arranging with the chiefs of engineers and artillery the measures to be taken for the security of the posts which are to be used as depots, as well as those to be fortified in order to facilitate the operations of the army.

4. Ordering and directing reconnoissances of every kind, and procuring in this way, and by using spies, as exact information as possible of the positions and movements of the enemy.

5. Taking every precaution for the proper execution of movements ordered by the general. Arranging the march of the different columns, so that all may move in an orderly and connected manner. Ascertaining certainly that the means requisite for the ease and safety of marches are prepared. Regulating the manner and time of halts.

6. Giving proper composition to advanced guards, rear-guards, flankers, and all detached bodies, and preparing good instructions for their guidance. Providing all the means necessary for the performance of their duties.

7. Prescribing forms and instructions for subordinate commanders or their staff officers, relative to the different methods of drawing up the troops in columns when the enemy is at hand, as well as their formation in the most appropriate manner when the army is to engage in battle, according to the nature of the ground and the character of the enemy. [I refer here to general instructions and forms, which are not to be repeated every day: such repetition would be impracticable.]

8. Indicating to advanced guards and other detachments well-chosen points of assembly in case of their attack by superior numbers, and informing them what support they may hope to receive in case of need.

9. Arranging and superintending the march of trains of baggage, munitions, provisions, and ambulances, both with the columns and in their rear, in such manner that they will not interfere with the movements of the troops and will still be near at hand. Taking precautions for order and security, both on the march and when trains are halted and parked.

10. Providing for the successive arrival of convoys of supplies. Collecting all the means of transportation of the country and of the army, and regulating their use.

11. Directing the establishment of camps, and adopting regulations for their safety, good order, and police.

12. Establishing and organizing lines of operations and supplies, as well as lines of communications with these lines for detached bodies. Designating officers capable of organizing and commanding in rear of the army; looking out for the safety of detachments and convoys, furnishing them good instructions, and looking out also for preserving suitable means of communication of the army with its base.

13. Organizing depots of convalescent, wounded, and sickly men, movable hospitals, and workshops for repairs; providing for their safety.

14. Keeping accurate record of all detachments, either on the flanks or in rear; keeping an eye upon their movements, and looking out for their return to the main column as soon as their service on detachment is no longer necessary; giving them, when required, some center of action, and forming strategic reserves.

15. Organizing marching battalions or companies to gather up isolated men or small detachments moving in either direction between the army and its base of operations.

16. In case of sieges, ordering and supervising the employment of the troops in the trenches, making arrangements with the chiefs of artillery and engineers as to the labors to be performed by those troops and as to their management in sorties and assaults.

17. In retreats, taking precautionary measures for preserving order; posting fresh troops to support and relieve the rear-guard; causing intelligent officers to examine and select positions where the rear-guard may advantageously halt, engage the enemy, check his pursuit, and thus gain time; making provision in advance for the movement of trains, that nothing shall be left behind, and that they shall proceed in the most perfect order, taking all proper precautions to insure safety.

18. In cantonments, assigning positions to the different corps; indicating to each principal division of the army a place of assembly in case of alarm; taking measures to see that all orders, instructions, and regulations are implicitly observed.

An examination of this long list—which might easily be made much longer by entering into greater detail—will lead every reader to remark that these are the duties rather of the general-in-chief than of staff officers. This truth I announced some time ago; and it is for the very purpose of permitting the general-in-chief to give his whole attention to the supreme direction of the operations that he ought to be provided with staff officers competent to relieve him of details of execution. Their functions are therefore necessarily very intimately connected; and woe to an army where these authorities cease to act in concert! This want of harmony is often seen,—first, because generals are men and have faults, and secondly, because in every army there are found individual interests and pretensions, producing rivalry of the chiefs of staff and hindering them in performing their duties. [The chiefs of artillery, of engineers, and of the administrative departments all claim to have direct connection with the general-in-chief, and not with the chief of staff. There should, of course, be no hinderance to the freest intercourse between these high officers and the commander; but he should work with them in presence of the chief of staff, and send him all their correspondence: otherwise, confusion is inevitable.]

It is not to be expected that this treatise shall contain rules for the guidance of staff officers in all the details of their multifarious duties; for, in the first place, every different nation has staff officers with different names and rounds of duties,—so that I should be obliged to write new rules for each army; in the second place, these details are fully entered into in special books pertaining to these subjects.

I will, therefore, content myself with enlarging a little upon some of the first articles enumerated above:—

1. The measures to be taken by the staff officers for preparing the army to enter upon active operations in the field include all those which are likely to facilitate the success of the first plan of operations. They should, as a matter of course, make sure, by frequent inspections, that the *matériel* of all the arms of the service is in good order: horses, carriages, caissons, teams, harness, shoes, etc.. should be carefully examined and any deficiencies supplied. Bridge-trains, engineer-tool trains, *matériel* of artillery, siege-trains if they are to move, ambulances,—in a word, every thing which conies under the head of *matériel*,—should be carefully examined and placed in good order.

If the campaign is to be opened in the neighborhood of great rivers, gun-boats and flying bridges should be prepared, and all the small craft should be collected at the points and at the bank where they will probably be used. Intelligent officers should examine the most favorable points both for embarkations and for landings,—preferring those localities which present the greatest chances of success for a primary establishment on the opposite bank.

The staff officers will prepare all the itineraries that will be necessary for the movement of the several corps of the army to the proper points of assemblage, making every effort to give such direction to the marches that the enemy shall be unable to learn from them any thing relative to the projected enterprise.

If the war is to be offensive, the staff officers arrange with the chief engineer officers what fortifications shall be erected near the base of operations, when *têtes de ponts* or intrenched camps are to be constructed there. If the war is defensive, these works will be built between the first line of defense and the second base.

2. An essential branch of logistics is certainly that which relates to making arrangements of marches and attacks, which are fixed by the general and notice of them given to the proper persons by the chiefs of staff. The next most important qualification of a general, after that of knowing how to form good plans, is, unquestionably, that of facilitating the execution of his orders by their clearness of style. Whatever may be the real business of a chief of

staff, the greatness of a commander-in-chief will be always manifested in his plans; but if the general lacks ability the chief of staff should supply it as far as he can, having a proper understanding with the responsible chief.

I have seen two very different methods employed in this branch of the service. The first, which may be styled the old school, consists in issuing daily, for the regulation of the movements of the army, general instructions filled with minute and somewhat pedantic details, so much the more out of place as they are usually addressed to chiefs of corps, who are supposed to be of sufficient experience not to require the same sort of instruction as would be given to junior subalterns just out of school.

The other method is that of the detached orders given by Napoleon to his marshals, prescribing for each one simply what concerned himself, and only informing him what corps were to operate with him, either on the right or the left, but never pointing out the connection of the operations of the whole army. [I believe that at the passage of the Danube before Wagram, and at the opening of the second campaign of 1813, Napoleon deviated from his usual custom by issuing a general order.] I have good reasons for knowing that he did this designedly, either to surround his operations with an air of mystery, or for fear that more specific orders might fall into the hands of the enemy and assist him in thwarting his plans.

It is certainly of great importance for a general to keep his plans secret; and Frederick the Great was right when he said that if his night-cap knew what was in his head he would throw it into the fire. That kind of secrecy was practicable in Frederick's time, when his whole army was kept closely about him; but when maneuvers of the vastness of Napoleon's are executed, and war is waged as in our day, what concert of action can be expected from generals who are utterly ignorant of what is going on around them?

Of the two systems, the last seems to me preferable. A judicious mean may be adopted between the eccentric conciseness of Napoleon and the minute verbosity which laid down for experienced generals like Barclay, Kleist, and Wittgenstein precise directions for breaking into companies and reforming again in line of battle,—a piece of nonsense all the more ridiculous because the execution of such an order in presence of the enemy is impracticable. It would be sufficient, I think, in such cases, to give the generals special orders relative to their own corps, and to add a few lines in cipher informing them briefly as to the whole plan of the operations and the part they are to take individually in executing it. When a proper cipher is wanting, the order may be transmitted verbally by an officer capable of understanding it and repeating it accurately. Indiscreet revelations need then be no longer feared, and concert of action would be secured.

3. The army being assembled, and being in readiness to undertake some enterprise, the important thing will be to secure as much concert and precision of action as possible, whilst taking all the usual precaution's to gain accurate information of the route it is to pursue and to cover its movements thoroughly.

There are two kinds of marches,—those which are made out of sight of the enemy, and those which are made in his presence, either advancing or retiring. These marches particularly have undergone great changes in late years. Formerly, armies seldom came in collision until they had been several days in presence of each other, and the attacking party had roads opened by pioneers for the columns to move up parallel to each other. At present, the attack is made more promptly, and the existing roads usually answer all purposes. It is, however, of importance, when an army is moving, that pioneers and sappers accompany the advanced guard, to increase the number of practicable roads, to remove obstructions, throw small bridges over creeks, etc.., if necessary, and secure the means of easy communication between the different corps of the army.

In the present manner of marching, the calculation of times and distances becomes more complicated: the columns having each a different distance to pass over, in determining the hour of their departure and giving them instructions the following particulars must be considered:—1, the distances to be passed over; 2, the amount of *matériel* in each train; 3, the nature of the country; 4, the obstacles placed in the way by the enemy; 5, the fact whether or not it is important for the march to be concealed or open.

Under present circumstances, the surest and simplest method of arranging the movements of the great corps forming the wings of an army, or of all those corps not marching with the column attached to the general head-quarters, will be to trust the details to the experience of the generals commanding those corps,—being careful, however, to let them understand that the most exact punctuality is expected of them. It will then be enough to indicate to them the point to be reached and the object to be attained, the route to be pursued and the hour at which they will be expected to be in position. They should be informed what corps are marching either on the same roads with them or on side-roads to the right or left in order that they may govern themselves accordingly; they should receive whatever news there may be of the enemy, and have a line of retreat indicated to them. [Napoleon never did this, because he maintained that no general should ever think seriously of the possibility of being beaten. In many marches it is certainly a useless precaution; but it is often indispensable.]

All those details whose object it is to prescribe each day for the chiefs of corps the method of forming their columns and placing them in position are mere pedantry,—more hurtful than useful. To see that they march habitually according to regulation or custom is necessary; but they should be free to arrange their movements so as to arrive at the appointed place and time, at the risk of being removed from their command if they fail to do so without sufficient reason. In retreats, however, which are made along a single road by an army separated into divisions, the hours of departure and halts must be carefully regulated.

Each column should have its own advanced guard and flankers, that its march may be conducted with the usual precautions: it is convenient also, even when they form part of a second line, for the head of each column to be preceded by a few pioneers and sappers, provided with tools for removing obstacles or making repairs in case of accidents; a few of these workmen should also accompany each train: in like manner, a light trestle-bridge train will be found very useful.

4. The army on the march is often preceded by a general advanced guard, or, as is more frequent in the modern system, the center and each wing may have its special advanced guard. It is customary for the reserves and the center to accompany the head-quarters; and the general advanced guard, when there is one, will usually follow the same road: so that half the army is thus assembled on the central route. Under these circumstances, the greatest care is requisite to prevent obstructing the road. It happens sometimes, however, when the important stroke is to be made in the direction of one of the wings, that the reserves, the general head-quarters, and even the general advanced guard, may be moved in that direction: in this case, all the rules usually regulating the march of the center must be applied to that wing.

Advanced guards should be accompanied by good staff officers, capable of forming correct ideas as to the enemy's movements and of giving an accurate account of them to the general, thus enabling him to make his plans understandingly. The commander of the advanced guard should assist the general in the same way. A general advanced guard should be composed of light troops of all arms, containing some of the *élite* troops of the army as a main body, a few dragoons prepared to fight on foot, some horse-artillery, pontoniers, sappers, etc.., with light trestles and pontoons for passing small streams. A few good marksmen will not be out of place. A topographical officer should accompany it, to make a sketch of the country a mile or two on each side of the road. A body of irregular cavalry should always be attached, to spare the regular cavalry and to serve as scouts, because they are best suited to such service.

5. As the army advances and removes farther from its base, it becomes the more necessary to have a good line of operations and of depots which may keep up the connection of the army with its base. The staff officers will divide

the depots into departments, the principal depot being established in the town which can lodge and supply the greatest number of men: if there is a fortress suitably situated, it should be selected as the site of the principal depot.

The secondary depots may be separated by distances of from fifteen to thirty miles, usually in the towns of the country. The mean distance apart will be about twenty to twenty-five miles. This will give fifteen depots upon a line of three hundred miles, which should be divided into three or four brigades of depots. Each of these will have a commander and a detachment of troops or of convalescent soldiers, who regulate the arrangements for accommodating troops and give protection to the authorities of the country, (if they remain;) they furnish facilities for transmitting the mails and the necessary escorts; the commander sees that the roads and bridges are kept in good order. If possible, there should be a park of several carriages at each depot, certainly at the principal one in each brigade. The command of all the depots embraced within certain geographical limits should be intrusted to prudent and able general officers; for the security of the communications of the army often depends on their operations. [It may be objected that in some wars, as where the population is hostile, it may be very difficult, or impracticable, to organize lines of depots. In such cases they will certainly be exposed to great dangers; but these are the very cases where they are most necessary and should be most numerous. The line from Bayonne to Madrid was such a line, which resisted for four years the attacks of the guerrillas,—although convoys were sometimes seized. At one time the line extended as far as Cadiz.] These commands may sometimes become strategic reserves, as was explained in Art. XXIII.; a few good battalions, with the assistance of movable detachments passing continually between the army and the base, will generally be able to keep open the communications.

6. The study of the measures, partly logistical and partly tactical, to be taken by the staff officers in bringing the troops from the order of march to the different orders of battle, is very important, but requires going into such minute detail that I must pass it over nearly in silence, contenting myself with referring my readers to the numerous works specially devoted to this branch of the art of war.

Before leaving this interesting subject, I think a few examples should be given as illustrations of the great importance of a good system of logistics. One of these examples is the wonderful concentration of the French army in the plains of Gera in 1806; another is the entrance of the army upon the campaign of 1815.

In each of these cases Napoleon possessed the ability to make such arrangements that his columns, starting from points widely separated, were concentrated with wonderful precision upon the decisive point of the zone of operations; and in this way he insured the successful issue of the campaign. The choice of the decisive point was the result of a skillful application of the principles of strategy; and the arrangements for moving the troops give us an example of logistics which originated in his own closet. It has been long claimed that Berthier framed those instructions which were conceived with so much precision and usually transmitted with so much clearness; but I have had frequent opportunities of knowing that such was not the truth. The emperor was his own chief staff officer. Provided with a pair of dividers opened to a distance by the scale of from seventeen to twenty miles in a straight line, (which made from twenty-two to twenty-five miles, taking into account the windings of the roads,) bending over and sometimes stretched at full length upon his map, where the positions of his corps and the supposed positions of the enemy were marked by pins of different colors, he was able to give orders for extensive movements with a certainty and precision which were astonishing. Turning his dividers about from point to point on the map, he decided in a moment the number of marches necessary for each of his columns to arrive at the desired point by a certain day; then, placing pins in the new positions, and bearing in mind the rate of marching that he must assign to each column, and the hour of its setting out, he dictated those instructions which are alone enough to make any man famous.

Ney coming from the shores of Lake Constance, Lannes from Upper Swabia, Soult and Davoust from Bavaria and the Palatinate, Bernadotte and Augereau from Franconia, and the Imperial Guard from Paris, were all thus arranged in line on three parallel roads, to debouch simultaneously between Saalfeld, Gera, and Plauen, few persons in the army or in Germany having any conception of the object of these movements which seemed so very complicated.

In the same manner, in 1815, when Blücher had his army quietly in cantonments between the Sambre and the Rhine, and Wellington was attending *fêtes* in Brussels, both waiting a signal for the invasion of France, Napoleon, who was supposed to be at Paris entirely engrossed with diplomatic ceremonies, at the head of his guard, which had been but recently reformed in the capital, fell like a thunderbolt upon Charleroi and Blücher's quarters, his columns arriving from all points of the compass, with rare punctuality, on the 14th of June, in the plains of Beaumont and upon the banks of the Sambre. (Napoleon did not leave Paris until the 12th.)

The combinations described above were the results of wise strategic calculations, but their execution was undoubtedly a masterpiece of logistics. In order to exhibit more clearly the merit of these measures, I will mention, by way of contrast, two cases where faults in logistics came very near leading to fatal consequences. Napoleon having been recalled from Spain in 1809 by the fact of Austria's taking up arms, and being certain that this power intended war, he sent Berthier into Bavaria upon the delicate duty of concentrating the army, which was extended from Braunau as far as Strasbourg and Erfurt. Davoust was returning from the latter city, Oudinot from Frankfort; Massena, who had been on his way to Spain, was retiring toward Ulm by the Strasbourg route; the Saxons, Bavarians, and Wurtembergers were moving from their respective countries. The corps were thus separated by great distances, and the Austrians, who had been long concentrated, might easily break through this spider's web or brush away its threads. Napoleon was justly uneasy, and ordered Berthier to assemble the army at Ratisbon if the war had not actually begun on his arrival, but, if it had, to concentrate it in a more retired position toward Ulm.

The reason for this alternative order was obvious. If the war had begun, Ratisbon was too near the Austrian frontier for a point of assembly, as the corps might thus be thrown separately into the midst of two hundred thousand enemies; but by fixing upon Ulm as the point of rendezvous the army would be concentrated sooner, or, at any rate, the enemy would have five or six marches more to make before reaching-it,—which was a highly-important consideration as the parties were then situated.

No great talent was needed to understand this. Hostilities having commenced, however, but a few days after Berthier's arrival at Munich, this too celebrated chief of staff was so foolish as to adhere to a literal obedience of the order he had received, without conceiving its obvious intention: he not only desired the army to assemble at Ratisbon, but even obliged Davoust to return toward that city, when that marshal had had the good sense to fall back from Amberg toward Ingolstadt.

Napoleon, having, by good fortune, been informed by telegraph of the passage of the Inn twenty-four hours after its occurrence, came with the speed of lightning to Abensberg, just as Davoust was on the point of being surrounded and his army cut in two or scattered by a mass of one hundred and eighty thousand enemies. We know how wonderfully Napoleon succeeded in rallying his army, and what victories he gained on the glorious days of Abensberg, Siegberg, Landshut, Eckmühl, and Ratisbon, that repaired the faults committed by his chief of staff with his contemptible logistics.

We shall finish these illustrations with a notice of the events which preceded and were simultaneous with the passage of the Danube before the battle of Wagram. The measures taken to bring to a specified point of the island of Lobau the corps of the Viceroy of Italy from Hungary, that of Marmont from Styria, that of Bernadotte from Linz, are less wonderful than the famous imperial decree of thirty-one articles which regulated the details of the passage and the formation of the troops in the plains of Enzersdorf, in presence of one hundred and forty thousand Austrians and five hundred cannon, as if the operation had been a military *fête*. These masses were all assembled upon the island on the evening of the 4th of July; three bridges were immediately thrown over an arm of the Danube one hundred and fifty yards wide, on a very dark night and amidst torrents of rain; one hundred and fifty thousand men passed over the bridges, in presence of a formidable enemy, and were drawn up before mid-day in the plain, three miles in advance of the bridges which they covered by a change of front; the whole being accomplished in less time than might have been supposed necessary had it been a simple maneuver for instruction and after being several times repeated. The enemy had, it is true, determined to offer no serious opposition to the passage; but Napoleon did not know that fact, and the merit of his dispositions is not at all diminished by it.

Singularly enough, however, the chief of staff, although he made ten copies of the famous decree, did not observe that by mistake the bridge of the center had been assigned to Davoust, who had the right wing, whilst the bridge on the right was assigned to Oudinot, who was in the center. These two corps passed each other in the night, and, had it not been for the good sense of the men and their officers, a dreadful scene of confusion might have been the result. Thanks to the supineness of the enemy, the army escaped all disorder, except that arising from a few detachments following corps to which they did not belong. The most remarkable feature of the whole transaction is found in the fact that after such a blunder Berthier should have received the title of Prince of Wagram.

The error doubtless originated with Napoleon while dictating his decree; but should it not have been detected by a chief of staff who made ten copies of the order and whose duty it was to supervise the formation of the troops?

Another no less extraordinary example of the importance of good logistics was afforded at the battle of Leipsic. In fighting this battle, with a defile in rear of the army as at Leipsic, and in the midst of low ground, wooded, and cut up by small streams and gardens, it was highly important to have a number of small bridges, to prepare the banks for approaching them with ease, and to stake out the roads. These precautions would not have prevented the loss of a decisive battle; but they would have saved the lives of a considerable number of men, as well as the guns and carriages that were abandoned on account of the disorder and of there being no roads of escape. The unaccountable blowing up of the bridge of Lindenau was also the result of unpardonable carelessness upon the part of the staff corps, which indeed existed only in name, owing to the manner of Berthier's management of it. We must also agree that Napoleon, who was perfectly conversant with the logistical measures of an offensive campaign, had then never seriously thought what would be proper precautions in the event of defeat, and when the emperor was present himself no one thought of making any arrangement for the future unless by his direction.

To complete what I proposed when I commenced this article, it becomes necessary for me to add some remarks with reference to reconnoissances. They are of two kinds: the first are entirely topographical and statistical, and their object is to gain a knowledge of a country, its accidents of ground, its roads, defiles, bridges, etc.., and to learn its resources and means of every kind. At the present day, when the sciences of geography, topography, and statistics are in such an advanced state, these reconnoissances are less necessary than formerly; but they are still very useful, and it is not probable that the statistics of any country will ever be so accurate that they may be entirely dispensed with. There are many excellent books of instruction as to the art of making these reconnoissances, and I must direct the attention of my readers to them.

Reconnoissances of the other kind are ordered when it is necessary to gain information of the movements of the enemy. They are made by detachments of greater or less strength. If the enemy is drawn up in battle-order, the generals-in-chief or the chiefs of staff make the reconnoissance; if he is on the march, whole divisions of cavalry may be thrown out to break through his screen of posts.

Article XLII: of Reconnoissances and Other Means of Gaining Correct Information of the Movements of the Enemy

One of the surest ways of forming good combinations in war would be to order movements only after obtaining perfect information of the enemy's proceedings. In fact, how can any man say what he should do himself, if he is ignorant what his adversary is about? As it is unquestionably of the highest importance to gain this information, so it is a thing of the utmost difficulty, not to say impossibility; and this is one of the chief causes of the great difference between the theory and the practice of war.

From this cause arise the mistakes of those generals who are simply learned men without a natural talent for war, and who have not acquired that practical *coup-d'oeil* which is imparted by long experience in the direction of military operations. It is a very easy matter for a school-man to make a plan for outflanking a wing or threatening a line of communications upon a map, where he can regulate the positions of both parties to suit himself; but when he has opposed to him a skillful, active, and enterprising adversary, whose movements are a perfect riddle, then his difficulties begin, and we see an exhibition of the incapacity of an ordinary general with none of the resources of genius.

I have seen so many proofs of this truth in my long life, that, if I had to put a general to the test, I should have a much higher regard for the man who could form sound conclusions as to the movements of the enemy than for him who could make a grand display of theories,—things so difficult to put in practice, but so easily understood when once exemplified.

There are four means of obtaining information of the enemy's operations. The first is a well-arranged system of espionage; the second consists in reconnoissances made by skillful officers and light troops; the third, in questioning prisoners of war; the fourth, in forming hypotheses of probabilities. This last idea I will enlarge upon farther on. There is also a fifth method,—that of signals. Although this is used rather for indicating the presence of the enemy than for forming conclusions as to his designs, it may be classed with the others.

Spies will enable a general to learn more surely than by any other agency what is going on in the midst of the enemy's camps; for reconnoissances, however well made, can give no information of any thing beyond the line of the advanced guard. I do not mean to say that they should not be resorted to, for we must use every means of gaining information; but I do say that their results are small and not to be depended upon. Reports of prisoners are often useful, but it is generally dangerous to credit them. A skillful chief of staff will always be able to select intelligent officers who can so frame their questions as to elicit important information from prisoners and deserters.

The partisans who are sent to hang around the enemy's lines of operations may doubtless learn something of his movements; but it is almost impossible to communicate with them and receive the information they possess. An extensive system of espionage will generally be successful: it is, however, difficult for a spy to penetrate to the general's closet and learn the secret plans he may form: it is best for him, therefore, to limit himself to information of what he sees with his own eyes or hears from reliable persons. Even when the general receives from his spies information of movements, he still knows nothing of those which may since have taken place, nor of what the enemy is going finally to attempt. Suppose, for example, he learns that such a corps has passed through Jena toward Weimar, and that another has passed through Gera toward Naumburg: he must still ask himself the questions, Where are they going, and what enterprise are they engaged in? These things the most skillful spy cannot learn.

When armies camped in tents and in a single mass, information of the enemy's operations was certain, because reconnoitering-parties could be thrown forward in sight of the camps, and the spies could report accurately their movements; but with the existing organization into corps d'armée which either canton or bivouac, it is very difficult to learn any thing about them. Spies may, however, be very useful when the hostile army is commanded by a great captain or a great sovereign who always moves with the mass of his troops or with the reserves. Such, for example, were the Emperors Alexander and Napoleon. If it was known when they moved and what route they followed, it was not difficult to conclude what project was in view, and the details of the movements of smaller bodies needed not to be attended to particularly.

A skillful general may supply the defects of the other methods by making reasonable and well-founded hypotheses. I can with great satisfaction say that this means hardly ever failed me. Though fortune never placed me at the head of an army, I have been chief of staff to nearly a hundred thousand men, and have been many times called into the councils of the greatest sovereigns of the day, when the question under consideration was the proper direction to give to the combined armies of Europe; and I was never more than two or three times mistaken in my hypotheses and in my manner of solving the difficulties they offered. As I have said before, I have constantly noticed that, as an army can operate only upon the center or one extremity of its front of operations, there are seldom more than three or four suppositions that can possibly be made. A mind fully convinced of these truths and conversant with the principles of war will always be able to form a plan which will provide in advance for the probable contingencies of the future. I will cite a few examples which have come under my own observation.

In 1806, when people in France were still uncertain as to the war with Prussia, I wrote a memoir upon the probabilities of the war and the operations which would take place.

I made the three following hypotheses:—1st. The Prussians will await Napoleon's attack behind the Elbe, and will fight on the defensive as far as the Oder, in expectation of aid from Russia and Austria; 2d. Or they will advance upon the Saale, resting their left upon the frontier of Bohemia and defending the passes of the mountains of Franconia; 3d. Or else, expecting the French by the great Mayence road, they will advance imprudently to Erfurt.

I do not believe any other suppositions could be made, unless the Prussians were thought to be so foolish as to divide their forces, already inferior to the French, upon the two directions of Wesel and Mayence,—a useless mistake, since there had not been a French soldier on the first of these roads since the Seven Years' War.

These hypotheses having been made as above stated, if any one should ask what course Napoleon ought to pursue, it was easy to reply "that the mass of the French army being already assembled in Bavaria, it should be thrown upon the left of the Prussians by way of Grera and Hof, for the gordian knot of the campaign was in that direction, no matter what plan they should adopt."

If they advanced to Erfurt, he could move to Gera, cut their line of retreat, and press them back along the Lower Elbe to the North Sea. If they rested upon the Saale, he could attack their left by way of Hof and Gera, defeat them partially, and reach Berlin before them by way of Leipsic. If they stood fast behind the Elbe, he must still attack them by way of Gera and Hof.

Since Napoleon's direction of operations was so clearly fixed, what mattered it to him to know the details of their movements? Being certain of the correctness of these principles, I did not hesitate to announce, *a month before the war*, that Napoleon would attempt just what he did, and that if the Prussians passed the Saale battles would take place at Jena and Naumburg!

I relate this circumstance not from a feeling of vanity, for if that were my motive I might mention many more of a similar character. I have only been anxious to show that in war a plan of operations may be often arranged, simply based upon the general principles of the art, without much attention being of necessity given to the details of the enemy's movements.

Returning to our subject, I must state that the use of spies has been neglected to a remarkable degree in many modern armies. In 1813 the staff of Prince Schwarzenberg had not a single sou for expenditure for such services, and the Emperor Alexander was obliged to furnish the staff officers with funds from his own private purse to enable them to send agents into Lusatia for the purpose of finding out Napoleon's whereabouts. General Mack at Ulm, and the Duke of Brunswick in 1806, were no better informed; and the French generals in Spain often suffered severely, because it was impossible to obtain spies and to get information as to what was going on around them.

The Russian army is better provided than any other for gathering information, by the use of roving bodies of Cossacks; and history confirms my assertion.

The expedition of Prince Koudacheff, who was sent after the battle of Dresden to the Prince of Sweden, and who crossed the Elbe by swimming and marched in the midst of the French columns as far, nearly, as Wittenberg, is a remarkable instance of this class. The information furnished by the partisan troops of Generals Czernicheff, Benkendorf, Davidoff, and Seslawin was exceedingly valuable. We may recollect it was through a dispatch from Napoleon to the Empress Maria Louisa, intercepted near Châlons by the Cossacks, that the allies were informed of the plan he had formed of falling upon their communications with his whole disposable force, basing his operations upon the fortified towns of Lorraine and Alsace. This highly-important piece of information decided Blücher and Schwarzenberg to effect a junction of their armies, which the plainest principles of strategy had never previously brought to act in concert except at Leipsic and Brienne.

We know, also, that the warning given by Seslawin to General Doctoroff saved him from being crushed at Borovsk by Napoleon, who had just left Moscow in retreat with his whole army. Doctoroff did not at first credit this news,—which so irritated Seslawin that he effected the capture of a French officer and several soldiers of the guard from the French bivouacs and sent them as proofs of its correctness. This warning, which decided the march of Koutousoff to Maloi-Yaroslavitz, prevented Napoleon from taking the way by Kalouga, where he would have found greater facilities for refitting his army and would have escaped the disastrous days of Krasnoi and the Beresina. The catastrophe which befell him would thus have been lessened, though not entirely prevented.

Such examples, rare as they are, give us an excellent idea of what good partisan troops can accomplish when led by good officers.

I will conclude this article with the following summary:—

1. A general should neglect no means of gaining information of the enemy's movements, and, for this purpose, should make use of reconnoissances, spies, bodies of light troops commanded by capable officers, signals, and questioning deserters and prisoners.

2. By multiplying the means of obtaining information; for, no matter how imperfect and contradictory they may be, the truth may often be sifted from them.

3. Perfect reliance should be placed on none of these means.

4. As it is impossible to obtain exact information by the methods mentioned, a general should never move without arranging several courses of action for himself, based upon probable hypotheses that the relative situation of the armies enables him to make, and never losing sight of the principles of the art.

I can assure a general that, with such precautions, nothing very unexpected can befall him and cause his ruin,—as has so often happened to others; for, unless he is totally unfit to command an army, he should at least be able to form reasonable suppositions as to what the enemy is going to do, and fix for himself a certain line of conduct to suit each of these hypotheses. [I shall be accused, I suppose, of saying that no event in war can ever occur which may not be foreseen and provided for. To prove the falsity of this accusation, it is sufficient for me to cite the surprises of Cremona, Berg-op-zoom, and Hochkirch. I am still of the opinion, however, that such events even as these might always have been anticipated, entirely or in part, as at least within the limits of probability or possibility.] It cannot be too much insisted upon that the real secret of military genius consists in the ability to make these reasonable suppositions in any case; and, although their number is always small, it is wonderful how much this highly-useful means of regulating one's conduct is neglected.

In order to make this article complete, I must state what is to be gained by using a system of signals. Of these there are several kinds. Telegraphic signals may be mentioned as the most important of all. Napoleon owes his astonishing success at Ratisbon, in 1809, to the fact of his having established a telegraphic communication between the head-quarters of the army and France. He was still at Paris when the Austrian army crossed the Inn at Braunau with the intention of invading Bavaria and breaking through his line of cantonments. Informed, in twenty-four hours, of what was passing at a distance of seven hundred miles, he threw himself into his traveling-carriage, and a week later he had gained two victories under the walls of Ratisbon. Without the telegraph, the campaign would have been lost. This single fact is sufficient to impress us with an idea of its value.

It has been proposed to use portable telegraphs. Such a telegraphic arrangement, operated by men on horseback posted on high ground, could communicate the orders of the center to the extremities of a line of battle, as well as the reports of the wings to the head-quarters. Repeated trials of it were made in Russia; but the project was given up,—for what reason, however, I have not been able to learn. These communications could only be very brief, and in misty weather the method could not be depended upon. A vocabulary for such purposes could be reduced to a few short phrases, which might easily be represented by signs. I think it a method by no means useless, even if it should be necessary to send duplicates of the orders by officers capable of transmitting them with accuracy. There would certainly be a gain of rapidity. [When the above was written, the magnetic telegraph was not known.—Translators.] attempt of another kind was made in 1794, at the battle of Fleurus, where General Jourdan made use of the services of a balloonist to observe and give notice of the movements of the Austrians. I am not aware that he found the method a very useful one, as it was not again used; but it was claimed at the time that it assisted in gaining him the victory: of this, however, I have great doubts.

It is probable that the difficulty of having a balloonist in readiness to make an ascension at the proper moment, and of his making careful observations upon what is going on below, whilst floating at the mercy of the winds above, has led to the abandonment of this method of gaining information. By giving the balloon no great elevation, sending up with it an officer capable of forming correct opinions as to the enemy's movements, and perfecting a system of signals to be used in connection with the balloon, considerable advantages might be expected from its use. Sometimes the smoke of the battle, and the difficulty of distinguishing the columns, that look like liliputians, so as to know to which party they belong, will make the reports of the balloonists very unreliable. For example, a balloonist would have been greatly embarrassed in deciding, at the battle of Waterloo, whether it was Grouchy or Blücher who was seen coming up by the Saint-Lambert road; but this uncertainty need not exist where the armies are not so much mixed. I had ocular proof of the advantage to be derived from such observations when I was stationed in the spire of Gautsch, at the battle of Leipsic; and Prince Schwarzenberg's aid-de-camp, whom I had conducted to the same point, could not deny that it was at my solicitation the prince was prevailed upon to emerge from the marsh between the Pleisse and the Elster. An observer is doubtless more at his ease in a clock-tower than in a frail basket floating in mid-air; but steeples are not always at hand in the vicinity of battle-fields, and they cannot be transported at pleasure.

There is still another method of signaling, by the use of large fires kindled upon elevated points of the country. Before the invention of the telegraph, they afforded the means of transmitting the news of an invasion from one end of the country to the other. The Swiss have made use of them to call the militia to arms. They have been also used to give the alarm to winter quarters and to assemble the troops more rapidly. The signal-fires may be made still more useful if arranged so as to indicate to the corps of the army the direction of the enemy's threatening movements and the point where they should concentrate to meet him. These signals may also serve on sea-coasts to give notice of descents.

Finally, there is a kind of signals given to troops during an action, by means of military instruments. This method of signals has been brought to greater perfection in the Russian army than in any other I know of. While I am aware of the great importance of discovering a sure method of setting in motion simultaneously a large mass of troops at the will of the commander, I am convinced that it must be a long time before the problem is solved. Signals with instruments are of little use except for skirmishers. A movement of a long line of troops may be made nearly simultaneous by means of a shout begun at one point and passed rapidly from man to man; but these shouts seem generally to be a sort of inspiration, and are seldom the result of an order. I have seen but two cases of it in thirteen campaigns.

Of the Formation of Troops for Battle, and the Separate or Combined Use of the Three Arms

Article XLIII: Posting Troops in Line of Battle

Having explained in Article XXX. what is to be understood by the term *line of battle*, it is proper to add in what manner it is to be formed, and how the different troops are to be distributed in it.

Before the French Revolution, all the infantry, formed in regiments and brigades, was collected in a single battle-corps, drawn up in two lines, each of which had a right and a left wing. The cavalry was usually placed upon the wings, and the artillery—which at this period was very unwieldy—was distributed along the front of each line. The army camped together, marching by lines or by wings; and, as there were two cavalry wings and two infantry wings, if the march was by wings four columns were thus formed. When they marched by lines, (which was specially applicable to flank movements,) two columns were formed, unless, on account of local circumstances, the cavalry or a part of the infantry had camped in a third line,—which was rare.

This method simplified logistics very much, since it was only necessary to give such orders as the following:—"The army will move in such direction, by lines or by wings, by the right or by the left." This monotonous but simple formation was seldom deviated from; and no better could have been devised as war was carried on in those days.

The French attempted something new at Minden, by forming as many columns as brigades, and opening roads to bring them to the front in line,—a simple impossibility.

If the labor of staff officers was diminished by this method of camping and marching by lines, it must be evident that if such a system were applied to an army of one hundred thousand or one hundred and fifty thousand men, there would be no end to the columns, and the result would be the frequent occurrence of routs like that of Rossbach.

The French Revolution introduced the system of divisions, which broke up the excessive compactness of the old formation, and brought upon the field fractions capable of independent movement on any kind of ground. This change was a real improvement,—although they went from one extreme to the other, by returning nearly to the legionary formation of the Romans. These divisions, composed usually of infantry, artillery, and cavalry, maneuvered and fought separately. They were very much extended, either to enable them to subsist without the use of depots, or with an absurd expectation of prolonging the line in order to outflank that of the enemy. The seven or eight divisions of an army were sometimes seen marching on the same number of roads, ten or twelve miles distant from each other; the head-quarters was at the center, with no other support than five or six small regiments of cavalry of three hundred or four hundred men each, so that if the enemy concentrated the mass of his forces against one of these divisions and beat it, the line was pierced, and the general-in-chief, having no disposable infantry reserve, could do nothing but order a retreat to rally his scattered columns.

Bonaparte in his first Italian campaign remedied this difficulty, partly by the mobility of his army and the rapidity of his maneuvers, and partly by concentrating the mass of his divisions upon the point where the decisive blow was to fall. When he became the head of the government, and saw the sphere of his means and his plans constantly increasing in magnitude, he readily perceived that a stronger organization was necessary: he avoided the extremes of the old system and the new, while still retaining the advantages of the divisional system. Beginning with the campaign of 1800, he organized corps of two or three divisions, which he placed under the command of lieutenant-generals, and formed of them the wings, the center, and the reserve of his army. [Thus, the army of the Rhine was composed of a right wing of three divisions under Lecourbe, of a center of three divisions under Saint-Cyr, and of a left of two divisions under Saint-Suzanne, the general-in-chief having three divisions more as a reserve under his own immediate orders.]

This system was finally developed fully at the camp of Boulogne, where he organized permanent army corps under the command of marshals, who had under their orders three divisions of infantry, one of light cavalry, from thirty-six to forty pieces of cannon, and a number of sappers. Each corps was thus a small army, able at need to act independently as an army. The heavy cavalry was collected in a single strong reserve, composed of two divisions of cuirassiers, four of dragoons, and one of light cavalry. The grenadiers and the guard formed an admirable infantry reserve. At a later period—1812—the cavalry was also organized into corps of three divisions, to give greater unity of action to the constantly-increasing masses of this arm. This organization was as near perfection as possible; and the grand army, that brought about such great results, was the model which all the armies of Europe soon imitated.

Some military men, in their attempts to perfect the art, have recommended that the infantry division, which sometimes has to act independently, should contain three instead of two brigades, because this number will allow one for the center and each wing. This would certainly be an improvement; for if the division contains but two brigades there is an open space left in the center between the brigades on the wings: these brigades, having no common central support, cannot with safety act independently of each other. Besides this, with three brigades in a division, two may be engaged while the third is held in reserve,—a manifest advantage. But, if thirty brigades formed in ten divisions of three brigades are better than when formed in fifteen divisions of two brigades, it becomes necessary, in order to obtain this perfect divisional organization, to increase the numbers of the infantry by one-third, or to reduce the divisions of the army-corps from three to two,—which last would be a serious disadvantage, because the army-corps is much more frequently called upon to act independently than a division, and the subdivision into three parts is specially best for that. [Thirty brigades formed in fifteen divisions of two brigades each will have only fifteen brigades in the first line, while the same thirty brigades formed in ten divisions of three brigades each may have twenty brigades in the first line and ten in the second. But it then becomes necessary to diminish the number of divisions and to have but two in a corps,—which would be a faulty arrangement, because the corps is much more likely to be called upon for independent action than the division.]

What is the best organization to be given an army just setting out upon a campaign will for a long time to come be a problem in logistics; because it is extremely difficult to maintain the original organization in the midst of the operations of war, and detachments must be sent out continually.

The history of the grand army of Boulogne, whose organization seemed to leave nothing farther to be desired, proves the assertion just made. The center under Soult, the right under Davoust, the left under Ney, and the reserve under Lannes, formed together a regular and formidable battle-corps of thirteen divisions of infantry, without counting those of the guard and the grenadiers. Besides these, the corps of Bernadotte and Marmont detached to the right, and that of Augereau to the left, were ready for action on the flanks. But after the passage of the Danube at Donauwerth every thing was changed. Ney, at first reinforced to five divisions, was reduced to two; the battle-corps was divided partly to the right and partly to the left, so that this fine arrangement was destroyed.

It will always be difficult to fix upon a stable organization. Events are, however, seldom so complicated as those of 1805; and Moreau's campaign of 1800 proves that the original organization may sometimes be maintained, at least for the mass of the army. With this view, it would seem prudent to organize an army in four parts,—two wings, a center, and a reserve. The composition of these parts may vary with the strength of the army; but in order to retain this organization it becomes necessary to have a certain number of divisions out of the general line in order to furnish the necessary detachments. While these divisions are with the army, they may be attached to that part which is to receive or give the heaviest blows; or they may be employed on the flanks of the main body, or to increase the strength of the reserve. Bach of the four great parts of the army may be a single corps of three or four divisions, or two corps of two divisions each. In this last case there would be seven corps, allowing one for the reserve; but this last corps should contain three divisions, to give a reserve to each wing and to the center.

With seven corps, unless several more are kept out of the general line in order to furnish detachments, it may happen that the extreme corps may be detached, so that each wing might contain but two divisions, and from these a brigade might be occasionally detached to flank the march of the army, leaving but three brigades to a wing. This would be a weak order of battle.

These facts lead me to conclude that an organization of the line of battle in four corps of three divisions of infantry and one of light cavalry, with three or four divisions for detachments, would be more stable than one of seven corps, each of two divisions.

But, as every thing depends upon the strength of the army and of the units of which it is composed, as well as upon the character of the operations in which it may be engaged, the arrangement may be greatly varied. I cannot go into these details, and shall simply exhibit the principal combinations that may result from forming the divisions in two or three brigades and the corps in two or three divisions. I have indicated the formation of two infantry corps in two lines, either one behind the other, or side by side.

Note.—In all these formations the unit is the brigade in line; but these lines may be formed of deployed battalions, or of battalions in columns of attack by divisions of two companies. The cavalry attached to the corps will be placed on the flanks. The brigades might be so drawn up as to have one regiment in the first line and one in the second.

The question here presents itself, whether it is ever proper to place two corps one behind the other, as Napoleon often did, particularly at Wagram. I think that, except for the reserves, this arrangement may be used only in a position of expectation, and never as an order of battle; for it is much better for each corps to have its own second line and its reserve than to pile up several corps, one behind the other, under different commanders. However much one general may be disposed to support a colleague, he will always object to dividing up his troops for that purpose; and when in the general of the first line he sees not a colleague, but a hated rival, as too frequently happens, it is probable he will be very slow in furnishing the assistance which may be greatly needed. Moreover, a commander whose troops are spread out in a long line cannot execute his maneuvers with near so much facility as if his front was only half as great and was supported by the remainder of his own troops drawn up in rear.

In making our calculations, it is scarcely necessary to provide for the case of such immense masses being in the field as were seen from 1812 to 1815, when a single army contained fourteen corps varying in strength from two to five divisions. With such large numbers nothing better can be proposed than a subdivision into corps of three divisions each. Of these corps, eight would form the main body, and there would remain six for detachments and for strengthening any point of the main line that might require support. If this system be applied to an army of one hundred and fifty thousand men, it would be hardly practicable to employ divisions of two brigades each where Napoleon and the allies used corps.

If nine divisions form the main body,—that is, the wings and the center,—and six others form the reserve and detachments, fifteen divisions would be required, or thirty brigades,—which would make one hundred and eighty battalions, if each regiment contains three battalions. This supposition brings our army up to one hundred and forty-five thousand foot-soldiers and two hundred thousand in all. With regiments of two battalions there would be required one hundred and twenty battalions, or ninety-six thousand infantry; but if each regiment contains but two battalions, each battalion should be one thousand men strong, and this would increase the infantry to one hundred and twenty thousand men and the entire army to one hundred and sixty thousand men. These calculations show that the strength of the minor subdivisions must be carefully considered in arranging into corps and divisions. If an army does not contain more than one hundred thousand men, the formation by divisions is perhaps better than by corps. An example of this was Napoleon's army of 1800.

Having now endeavored to explain the best method of giving a somewhat permanent organization to the main body of an army, it will not be out of place for me to inquire whether this permanency is desirable, and if it is not advantageous to deceive the enemy by frequently changing the composition of corps and their positions.

I admit the advantage of thus deceiving the enemy; but it may be gained while still retaining a quite constant organization of the main body. If the divisions intended for detachments are joined to the wings and the center,—that is, if those parts contain each four divisions instead of three,—and if one or two divisions be occasionally added to the wing which is likely to bear the brunt of an engagement, each wing will be a corps properly of four divisions; but detachments will generally reduce it to three, and sometimes two, while it might, again, be reinforced by a portion of the reserve until it reached five divisions. The enemy would thus never know exactly the strength of the different parts of the line.

But I have dwelt sufficiently on these details. It is probable that, whatever be the strength and number of the subdivisions of an army, the organization into corps will long be retained by all the great powers of Europe, and calculations for the arrangement of the line of battle must be made upon that basis.

The distribution of the troops in the line of battle has changed in recent times, as well as the manner of arranging the line. Formerly it was usually composed of two lines, but now of two lines and one or more reserves. In recent [The term *recent* here refers to the later wars of Napoleon I.—Translators.] conflicts in Europe, when the masses brought into collision were very large, the corps were not only formed in two lines, but one corps was placed behind another, thus making four lines; and, the reserve being drawn up in the same manner, six lines of infantry were often the result, and several of cavalry. Such a formation may answer well enough as a preparatory one, but is by no means the best for battle, as it is entirely too deep.

The classical formation—if I may employ that term—is still two lines for the infantry. The greater or less extent of the battle-field and the strength of an army may necessarily produce greater depth at times; but these cases are the exceptions, because the formation of two lines and the reserves gives sufficient solidity, and enables a greater number of men to be simultaneously engaged.

When an army has a permanent advanced guard, it may be either formed in front of the line of battle or be carried to the rear to strengthen the reserve; [As the advanced guard is in presence of the enemy every day, and forms the rear-guard in retreat, it seems but fair at the hour of battle to assign it a position more retired than that in front of the line of battle.] but, as has been previously stated, this will not often happen with the present method of forming and moving armies. Each wing has usually its own advanced guard, and the advanced guard of the main or central portion of the army is naturally furnished by the leading corps: upon coming into view of the enemy, these advanced bodies return to their proper positions in line of battle. Often the cavalry reserve is almost entirely with the advanced guard; but this does not prevent its taking, when necessary, the place fixed for it in the line of battle by the character of the position or by the wishes of the commanding general.

From what has been stated above, my readers will gather that very great changes of army organization took place from the time of the revival of the art of war and the invention of gunpowder to the French Revolution, and that to have a proper appreciation of the wars of Louis XIV., of Peter the

Great, and of Frederick II., they should consider them from the stand-point of those days.

One portion of the old method may still be employed; and if, by way of example, it may not be regarded as a fundamental rule to post the cavalry on the wings, it may still be a very good arrangement for an army of fifty or sixty thousand men, especially when the ground in the center is not so suitable for the evolutions of cavalry as that near the extremities. It is usual to attach one or two brigades of light cavalry to each infantry corps, those of the center being placed in preference to the rear, whilst those of the wings are placed upon the flanks. If the reserves of cavalry are sufficiently numerous to permit the organization of three corps of this arm, giving one as reserve to the center and one to each wing, the arrangement is certainly a good one. If that is impossible, this reserve may be formed in two columns, one on the right of the left wing and the other on the left of the right wing. These columns may thus readily move to any point of the line that may be threatened. [This disposition of the cavalry, of course, is made upon the supposition that the ground is favorably situated for it. This is the essential condition of every well-arranged line of battle.]

The artillery of the present day has greater mobility, and may, as formerly, be distributed along the front, that of each division remaining near it. It may be observed, moreover, that, the organization of the artillery having been greatly improved, an advantageous distribution of it may be more readily made; but it is a great mistake to scatter it too much. Few precise rules can be laid down for the proper distribution of artillery. Who, for example, would dare to advise as a rule the filling up of a large gap in a line of battle with one hundred pieces of cannon in a single battery without adequate support, as Napoleon did successfully at Wagram? I do not desire to go here into much detail with reference to the use of this arm, but I will give the following rules:—

1. The horse-artillery should be placed on such ground that it can move freely in every direction.

2. Foot-artillery, on the contrary, and especially that of heavy caliber, will be best posted where protected by ditches or hedges from sudden charges of cavalry. It is hardly necessary for me to add—what every young officer should know already—that too elevated positions are not those to give artillery its greatest effect. Flat or gently-sloping ground is better.

3. The horse-artillery usually maneuvers with the cavalry; but it is well for each army-corps to have its own horse-artillery, to be readily thrown into any desired position. It is, moreover, proper to have horse-artillery in reserve, which may be carried as rapidly as possible to any threatened point. General Benningsen had great cause for self-congratulation at Eylau because he had

fifty light guns in reserve; for they had a powerful influence in enabling him to recover himself when his line had been broken through between the center and the left.

4. On the defensive, it is well to place some of the heavy batteries in front, instead of holding them in reserve, since it is desirable to attack the enemy at the greatest possible distance, with a view of checking his forward movement and causing disorder in his columns.

5. On the defensive, it seems also advisable to have the artillery not in reserve distributed at equal intervals in batteries along the whole line, since it is important to repel the enemy at all points. This must not, however, be regarded as an invariable rule; for the character of the position and the designs of the enemy may oblige the mass of the artillery to move to a wing or to the center.

6. In the offensive, it is equally advantageous to concentrate a very powerful artillery-fire upon a single point where it is desired to make a decisive stroke, with a view of shattering the enemy's line to such a degree that he will be unable to withstand an attack upon which the fate of the battle is to turn. I shall at another place have more to say as to the employment of artillery in battles.

Article XLIV: Formation and Employment of Infantry

Infantry is undoubtedly the most important arm of the service, since it forms four-fifths of an army and is used both in the attack and defense of positions. If we must admit that, next to the genius of the general, the infantry arm is the most valuable instrument in gaining a victory, it is no less true that most important aid is given by the cavalry and artillery, and that without their assistance the infantry might at times be very seriously compromised, and at others could achieve only partial success.

We shall not here introduce those old discussions about the shallow and the deep formations, although the question, which was supposed decided, is far from being settled absolutely. The war in Spain and the battle of Waterloo have again given rise to disputes as to the relative advantages of fire and the shallow order, and of columns of attack and the deep order. I will give my own opinion farther on.

There must, however, be no misconception on this subject. The question now is not whether Lloyd was right in wishing to add a fourth rank, armed with pikes, to the infantry formation, with the expectation of producing more effect by the shock when attacking, or opposing a greater resistance when attacked. Every officer of experience knows the difficulty of moving in an orderly manner several deployed battalions in three ranks at close order, and that a fourth rank would increase the disorder without adding any advantage. It is astonishing that Lloyd, who had seen service, should have insisted so much upon the material advantage to be gained by thus increasing the mass of a battalion; for it very rarely happens that such a collision between opposing troops takes place that mere weight decides the contest. If three ranks turn their backs to the enemy, the fourth will not check them. This increase in the number of ranks diminishes the front and the number of men firing upon the defensive, whilst in the offensive there is not near so much mobility as in the ordinary column of attack. It is much more difficult to move eight hundred men in line of battle in four ranks than in three: although in the former case the extent of front is less, the ranks cannot be kept properly closed.

Lloyd's proposal for remedying this diminution of front is so absurd that it is wonderful how a man of talents could have imagined it. He wishes to deploy twenty battalions, and leave between them one hundred and fifty yards, or an interval equal to their front. We may well ask what would befall those battalions thus separated. The cavalry may penetrate the intervals and scatter them like dust before the whirlwind.

But the real question now is, shall the line of battle consist of deployed battalions depending chiefly upon their fire, or of columns of attack, each battalion being formed in column on the central division and depending on its force and impetuosity?

I will now proceed to sum up the particulars bearing upon a decision of the question in hand.

There are, in fact, only five methods of forming troops to attack an enemy:—1, as skirmishers; 2, in deployed lines, either continuous or checkerwise; 3, in lines of battalions formed in column on the central divisions; 4, in deep masses; 5, in small squares.

The skirmishing-order is an accessory; for the duties of skirmishers are, not to form the line of battle, but to cover it by taking advantage of the ground, to protect the movements of columns, to fill up intervals, and to defend the skirts of a position.

These different manners of formation are, therefore, reducible to four: the shallow order, where the line is deployed in three ranks; the half-deep order, formed of a line of battalions in columns doubled on the center or in battalion squares; the mixed order, where regiments are partly in line and partly in column; finally, the deep order, composed of heavy columns of battalions deployed one behind the other.

In the three-rank formation, a battalion with four divisions [The word *division* being used to designate four or five regiments, as well as two companies of a battalion, there is danger of confusion in its use.] will have twelve ranks in such a column as shown above: there are in this way too many non-combatants, and the column presents too good a mark for the artillery. To remedy in part these inconveniences, it has been proposed, whenever infantry is employed in columns of attack, to form it in two ranks, to place only three divisions of a battalion one behind the other, and to spread out the fourth as skirmishers in the intervals of the battalions and upon the flanks: when the cavalry charges, these skirmishers may rally behind the other three divisions. Each battalion would thus have two hundred more men to fire, besides those thrown into the two front ranks from the third. There would be, also, an increase of the whole front. By this arrangement, while having really a depth of but six men, there would be a front of one hundred men, and four hundred men who could discharge their fire-arms, for each battalion. Force and mobility would both be obtained. [In the Russian army the skirmishers are taken from the third rank of each division,—which makes the column eight men in depth, instead of twelve, and gives more mobility. To facilitate rallying the skirmishers on the columns, it would be, perhaps, better to take the whole fourth division for that purpose, thus giving nine ranks, or three divisions of three ranks, against infantry, while against cavalry there would be twelve ranks.] A battalion of eight hundred men, formed in the ordinary manner in a column of four divisions, has about sixty files in each division, of which the first alone—and only two ranks of that—discharge their pieces. Bach battalion would deliver, therefore, one hundred and twenty shots at a volley, it would deliver four hundred.

While searching after methods of obtaining more fire when necessary, we must not forget that a column of attack is not intended to fire, and that its fire should be reserved until the last; for if it begins to fire while marching, the whole impulsive effect of its forward movement is lost. Moreover, this shallower order would only be advantageous against infantry, as the column of four divisions in three ranks—forming a kind of solid square—would be better against cavalry. The Archduke Charles found it advantageous at Essling, and particularly at Wagram, to adopt this last order, which was proposed by myself in my chapter on the General Principles of War, published in 1807. The brave cavalry of Bessières could make no impression upon these small masses.

To give more solidity to the column proposed, the skirmishers might, it is true, be recalled, and the fourth division reformed; but this would be a two-rank formation, and would offer much less resistance to a charge than the three-rank formation,—particularly on the flanks. If to remedy this inconvenience it is proposed to form squares, many military men believe that when in two ranks squares would not resist so well as columns. The English squares at Waterloo were, however, only in two ranks, and, notwithstanding the heroic efforts of the French cavalry, only one battalion was broken. I will observe, in conclusion, that, if the two-rank formation be used for the columns of attack, it will be difficult to preserve that in three ranks for deployed lines, as it is scarcely possible to have two methods of formation, or, at any rate, to employ them alternately in the same engagement. It is not probable that any European army, except the English, will undertake to use deployed lines in two ranks. If they do, they should never move except in columns of attack.

I conclude that the system employed by the Russians and Prussians, of forming columns of four divisions in three ranks, of which one may be employed as skirmishers when necessary, is more generally applicable than any other; whilst the other, of which mention has been made, would be suitable only in certain cases and would require a double formation.

There is a mixed order, which was used by Napoleon at the Tagliamento and by the Russians at Eylau, where, in regiments of three battalions, one was deployed to form the first line, and two others to the rear in columns. This arrangement—which belongs also to the half-deep order—is suitable for the offensive-defensive, because the first line pours a powerful fire upon the enemy, which must throw him into more or less confusion, and the troops formed in columns may debouch through the intervals and fall with advantage upon him while in disorder. This arrangement would probably be improved by placing the leading divisions of the two battalions of the wings upon the same line with the central deployed battalion. There would thus be a half-battalion more to each regiment in the first line,—a by no means unimportant thing for the delivery of fire. There may be reason to fear that, these divisions becoming actively engaged in firing, their battalions which are formed in column to be readily launched against the enemy may not be easily disengaged for that purpose. The order may be useful in many cases. I have therefore indicated it.

The order in very deep masses is certainly the most injudicious. In the later wars of Napoleon, twelve battalions were sometimes deployed and closed one upon the other, forming thirty-six ranks closely packed together. Such masses are greatly exposed to the destructive effects of artillery, their mobility and impulsion are diminished, while their strength is not increased. The use of such masses at Waterloo was one cause of the French being defeated. Macdonald's column was more fortunate at Wagram, but at a great sacrifice of life; and it is not probable that this column would have been victorious had it not been for the successes of Davoust and Oudinot on the left of the archduke's line.

When it is decided to risk such a mass, the precaution should certainly be taken of placing on each flank a battalion marching in file, so that if the enemy should charge the mass in flank it need not be arrested in its progress. Under the protection of these battalions, which may face toward the enemy, the column may continue its march to the point it is expected to reach: otherwise, this large mass, exposed to a powerful converging fire which it has no means of returning, will be thrown into confusion like the column at Fontenoy, or broken as was the Macedonian phalanx by Paulus Emilius.

Squares are good in plains and to oppose an enemy who has a superiority in cavalry. It is agreed that the regimental square is best for the defensive, and the battalion square for the offensive.

The figures may be perfect squares, or elongated to give a large front and pour a heavier column of fire in the direction of the enemy. A regiment of three battalions will thus form a long square, by wheeling the center battalion half to the right and half to the left.

In the Turkish wars squares were almost exclusively used, because hostilities were carried on in the vast plains of Bessarabia, Moldavia, or Wallachia, and the Turks had an immense force of cavalry. But if the seat of war be the Balkan Mountains or beyond them, and their irregular cavalry be replaced by an army organized according to the proportions usual in Europe, the importance of the square will disappear, and the Russian infantry will show its superiority in Rumelia.

However this may be, the order in squares by regiments or battalions seems suitable for every kind of attack, when the assailant has not the superiority in cavalry and maneuvers on level ground advantageous for the enemy's charges. The elongated square, especially when applied to a battalion of eight companies, three of which would march in front and one on each side, would be much better to make an attack than a deployed battalion. It would not be so good as the column proposed above; but there would be less unsteadiness and more impulsion than if the battalion marched in a deployed line. It would have the advantage, also, of being prepared to resist cavalry.

Squares may also be drawn up in echelons, so as entirely to unmask each other. All the orders of battle may be formed of squares as well as with deployed lines.

It cannot be stated with truth that any one of the formations described is always good or always bad; but there is one rule to the correctness of which every one will assent,—that a formation suitable for the offensive must possess the characteristics of *solidity, mobility,* and *momentum,* whilst for the defensive *solidity* is requisite, and also the power of delivering *as much fire as possible.*

This truth being admitted, it remains yet to be decided whether the bravest troops, formed in columns but unable to fire, can stand long in presence of a deployed line firing twenty thousand musket-balls in one round, and able to fire two hundred thousand or three hundred thousand in five minutes. In the later wars in Europe, positions have often been carried by Russian, French, and Prussian columns with their arms at a shoulder and without firing a shot. This was a triumph of *momentum* and the moral effect it produces; but under the cool and deadly fire of the English infantry the French columns did not succeed so well at Talavera, Busaco, Fuentes-de-Onore, Albuera, and Waterloo.

We must not, however, necessarily conclude from these facts that the advantage is entirely in favor of the shallow formation and firing; for when the French formed their infantry in those dense masses, it is not at all wonderful that the deployed and marching battalions of which they were composed, assailed on all sides by a deadly fire, should have been repulsed. Would the same result have been witnessed if they had used columns of attack formed each of a single battalion doubled on the center? I think not. Before deciding finally as to the superiority of the shallow order, with its facility for firing, over the half-deep order and its momentum, there should be several trials to see how a deployed line would stand an assault from a formation. These small columns have always succeeded wherever I have seen them tried.

Is it indeed an easy matter to adopt any other order when marching to attack a position? Can an immense deployed line be moved up into action while firing? I think no one will answer affirmatively. Suppose the attempt made to bring up twenty or thirty battalions in line, while firing either by file or by company, to the assault of a well-defended position: it is not very probable they would ever reach the desired point, or, if they did, it would be in about as good order as a flock of sheep.

What conclusions shall be drawn from all that has been said? 1. If the deep order is dangerous, the half-deep is excellent for the offensive. 2. The column of attack of single battalions is the best formation for carrying a position by assault; but its depth should be diminished as much as possible, that it may when necessary be able to deliver as heavy a column of fire as possible, and to diminish the effect of the enemy's fire: it ought also to be well covered by skirmishers and supported by cavalry. 3. The formation having the first line deployed and the second in columns is the best-suited to the defensive. 4. Either of them may be successful in the hands of a general of talent, who knows how to use his troops properly in the manner indicated in Articles XVI. and XXX.

Since this chapter was first written, numerous improvements have been made in the arms both of infantry and artillery, making them much more destructive. The effect of this is to incline men to prefer the shallower formations, even in the attack. We cannot, however, forget the lessons of experience; and, notwithstanding the use of rocket-batteries, shrapnel-shot, and the Perkins musket, I cannot imagine a better method of forming infantry for the attack than in columns of battalions. Some persons may perhaps desire to restore to infantry the helmets and breastplates of the fifteenth century, before leading them to the attack in deployed lines. But, if there is a general return to the deployed system, some better arrangement must be devised for marching to the attack than long, continuous lines, and either columns must be used with proper distances for deployment upon arriving near the enemy's position, or lines drawn up checkerwise, or the march must be by the flanks of companies,—all of which maneuvers are hazardous in presence of an enemy who is capable of profiting by the advantages on his side. A skillful commander will use either, or a combination of all, of these arrangements, according to circumstances.

Experience long ago taught me that one of the most difficult tactical problems is that of determining the best formation of troops for battle; but I have also learned that to solve this problem by the use of a single method is an impossibility.

In the first place, the topography of different countries is very various. In some, as Champagne, two hundred thousand men might be maneuvered in deployed lines. In others, as Italy, Switzerland, the valley of the Rhine, half of Hungary, it is barely possible to deploy a division of ten battalions. The degree of instruction of the troops, and their national characteristics, may also have an influence upon the system of formation.

Owing to the thorough discipline of the Russian army and its instruction in maneuvers of every kind, it may maintain in movements in long lines so much order and steadiness as to enable it to adopt a system which would be entirely out of the question for the French or Prussian armies of the present day. My long experience has taught me to believe that nothing is impossible; and I do not belong to the class of men who think that there can be but one type and one system for all armies and all countries.

To approximate as nearly as we can to the solution of the problem, it seems to me, we ought to find out:—1. The best method of moving when in sight of the enemy, but beyond his reach; 2. The best method of coming to close quarters with him; 3. The best defensive order.

In whatever manner we may settle these points, it seems desirable in all cases to exercise the troops—1. In marching in columns of battalions doubled on the center, with a view to deployment, if necessary, when coming into musket-range, or even to attack in column; 2. In marching in continuous deployed lines of eight or ten battalions; 3. In marching in deployed battalions arranged checkerwise,—as these broken lines are more easily moved than continuous lines; 4. In moving to the front by the flanks of companies; 5. In marching to the front in small squares, either in line or checkerwise; 6. In changing front while using these different methods of marching; 7. In changes of front executed by columns of companies at full distance, without deployment,—a more expeditious method than the others of changing front, and the one best suited to all kinds of ground.

Of all the methods of moving to the front, that by the flanks of companies would be the best if it was not somewhat dangerous. In a plain it succeeds admirably, and in broken ground is very convenient. It breaks up a line very much; but by accustoming the officers and privates to it, and by keeping the guides and color-bearers well aligned, all confusion can be avoided. The only objection to it is the danger to which the separated companies are exposed of being ridden down by cavalry. This danger may be avoided by having good cavalry scouts, and not using this formation too near the enemy, but only in getting over the first part of the large interval separating the two armies. At the least sign of the enemy's proximity the line could be reformed instantly, since the companies can come into line at a run. Whatever precautions may be taken, this maneuver should only be practiced with well-disciplined troops, never with militia or raw troops. I have never seen it tried in presence of an enemy,—but frequently at drills, where it has been found to succeed well, especially in changing front.

I have also seen attempts made to march deployed battalions in checkerwise order. They succeeded well; whilst marches of the same battalions in continuous lines did not. The French, particularly, have never been able to march steadily in deployed lines. This checkered order would be dangerous in case of an unexpected charge of cavalry. It may be employed in the first stages of the movement forward, to make it more easy, and the rear battalions would then come into line with the leading ones before reaching the enemy. Moreover, it is easy to form line at the moment of the charge, by leaving a small distance only between the leading and following battalions; for we must not forget that in the checkered order there are not two lines, but a single one, which is broken, to avoid the wavering and disorder observed in the marches of continuous lines.

It is very difficult to determine positively the best formation for making a serious and close attack upon an enemy. Of all the methods I have seen tried, the following seemed to succeed best. Form twenty-four battalions in two lines of battalions in columns doubled on the center ready for deployment: the first line will advance at charging-pace toward the enemy's line to within twice musket-range, and will then deploy at a run; the voltigeur-companies of each battalion will spread out in skirmishing-order, the remaining companies forming line and pouring in a continued fire by file; the second line of columns follows the first, and the battalions composing it pass at charging-step through the intervals of the first line. This maneuver was executed when no enemy was present; but it seems to me an irresistible combination of the advantages of firing and of the column.

Besides these lines of columns, there are three other methods of attacking in the half-deep order.

The first is that of lines composed of deployed battalions with others in column on the wings of those deployed. The deployed battalions and the leading divisions of those in column would open fire at half musket-range, and the assault would then be made. The second is that of advancing a deployed line and firing until reaching half musket-range, then throwing forward the columns of the second line through the intervals of the first.

Finally, a last method is that of advancing altogether in deployed lines, depending on the superiority of fire alone, until one or the other party takes to its heels,—a case not likely to happen.

I cannot affirm positively which of these methods is the best; for I have not seen them used in actual service. In fact, in real combats of infantry I have never seen any thing but battalions deployed commencing to fire by company, and finally by file, or else columns marching firmly against the enemy, who either retired without awaiting the columns, or repulsed them before an actual collision took place, or themselves moved out to meet the advance. I have seen *mêlées* of infantry in defiles and in villages, where the heads of columns came in actual bodily collision and thrust each other with the bayonet; but I never saw such a thing on a regular field of battle.

In whatever manner these discussions terminate, they are useful, and should be continued. It would be absurd to discard as useless the fire of infantry, as it would be to give up entirely the half-deep formation; and an army is ruined if forced to adhere to precisely the same style of tactical maneuvers in every country it may enter and against every different nation. It is not so much the mode of formation as the proper combined use of the different arms which will insure victory. I must, however, except very deep masses, as they should be entirely abandoned.

I will conclude this subject by stating that a most vital point to be attended to in leading infantry to the combat is to protect the troops as much as possible from the fire of the enemy's artillery, not by withdrawing them at inopportune moments, but by taking advantage of all inequalities and accidents of the ground to hide them from the view of the enemy. When the assaulting troops have arrived within musket-range, it is useless to calculate upon sheltering them longer: the assault is then to be made. In such cases covers are only suitable for skirmishers and troops on the defensive.

It is generally quite important to defend villages on the front of a position, or to endeavor to take them when held by an enemy who is assailed; but their importance should not be overestimated; for we must never forget the noted battle of Blenheim, where Marlborough and Eugene, seeing the mass of the French infantry shut up in the villages, broke through the center and captured twenty-four battalions which were sacrificed in defending these posts.

For like reasons, it is useful to occupy clumps of trees or brushwood, which may afford cover to the party holding them. They shelter the troops, conceal their movements, cover those of cavalry, and prevent the enemy from maneuvering in their neighborhood. The case of the park of Hougoumont at the battle of Waterloo is a fine example of the influence the possession of such a position, well chosen and strongly defended, may have in deciding the fate of a battle. At Hochkirch and Kolin the possession of the woods was very important.

Article XLV: Cavalry

The use a general should make of his cavalry depends, of course, somewhat upon its numerical strength as compared with that of the whole army, and upon its quality. Even cavalry of an inferior character may be so handled as to produce very great results, if set in action at proper moments.

The numerical proportion of cavalry to infantry in armies has varied greatly. It depends on the natural tastes of nations making their people more or less fit for good troopers. The number and quality of horses, also, have something to do with it. In the wars of the Revolution, the French cavalry, although badly organized and greatly inferior to the Austrian, performed wonders. In 1796 I saw what was pompously called the cavalry reserve of the army of the Rhine,—a weak brigade of barely fifteen hundred horses! Ten years later I saw the same reserve consisting of fifteen thousand or twenty thousand horses,—so much had ideas and means changed.

As a general rule, it may be stated that an army in an open country should contain cavalry to the amount of one-sixth its whole strength; in mountainous countries one-tenth will suffice.

The principal value of cavalry is derived from its rapidity and ease of motion. To these characteristics may be added its impetuosity; but we must be careful lest a false application be made of this last.

Whatever may be its importance in the *ensemble* of the operations of war, cavalry can never defend a position without the support of infantry. Its chief duty is to open the way for gaining a victory, or to render it complete by carrying off prisoners and trophies, pursuing the enemy, rapidly succoring a threatened point, overthrowing disordered infantry, covering retreats of infantry and artillery. An army deficient in cavalry rarely obtains a great victory, and finds its retreats extremely difficult.

The proper time and manner of bringing cavalry into action depend upon the ideas of the commander-in-chief, the plan of the battle, the enemy's movements, and a thousand other circumstances which cannot be mentioned here. I can only touch upon the principal things to be considered in its use.

All are agreed that a general attack of cavalry against a line in good order cannot be attempted with much hope of success, unless it be supported by infantry and artillery. At Waterloo the French paid dearly for having violated this rule; and the cavalry of Frederick the Great fared no better at Kunnersdorf. A commander may sometimes feel obliged to push his cavalry forward alone, but generally the best time for charging a line of infantry is when it is already engaged with opposing infantry. The battles of Marengo, Eylau, Borodino, and several others prove this.

There is one case in which cavalry has a very decided superiority over infantry,—when rain or snow dampens the arms of the latter and they cannot fire. Augereau's corps found this out, to their sorrow, at Eylau, and so did the Austrian left at Dresden.

Infantry that has been shaken by a fire of artillery or in any other way may be charged with success. A very remarkable charge of this kind was made by the Prussian cavalry at Hohenfriedberg in 1745. A charge against squares of good infantry in good order cannot succeed.

A general cavalry charge is made to carry batteries of artillery and enable the infantry to take the position more easily; but the infantry must then be at hand to sustain the cavalry, for a charge of this character has only a momentary effect, which must be taken advantage of before the enemy can return offensively upon the broken cavalry. The beautiful charge of the French upon Gosa at the battle of Leipsic, October 16, is a fine example of this kind. Those executed at Waterloo with the same object in view were admirable, but failed because unsupported. The daring charge of Ney's weak cavalry upon Prince Hohenlohe's artillery at Jena is an example of what may be done under such circumstances.

General charges are also made against the enemy's cavalry, to drive it from the field of battle and return more free to act against his infantry.

Cavalry may be successfully thrown against the flank or rear of an enemy's line at the moment of its being attacked in front by the infantry. If repulsed, it may rally upon the army at a gallop, and, if successful, it may cause the loss of the enemy's army. This operation is rarely attempted, but I see no reason why it should not be very good; for a body of cavalry well handled cannot be cut off even if it gets in rear of the enemy. This is a duty for which light cavalry is particularly fitted.

In the defensive, cavalry may also produce very valuable results by opportune dashes at a body of the enemy which has engaged the opposing line and either broken it through or been on the point of doing so. It may regain the advantages lost, change the face of affairs, and cause the destruction of an enemy flushed and disordered by his own success. This was proved at Eylau, where the Russians made a fine charge, and at Waterloo by the English cavalry. The special cavalry of a corps d'armée may charge at opportune moments, either to co-operate in a combined attack, or to take advantage of a false movement of the enemy, or to finish his defeat by pressing him while in retreat.

It is not an easy matter to determine the best mode of attacking, as it depends upon the object in view and other circumstances. There are but four methods of charging,—in columns, in lines at a trot, in lines at a gallop, and in open order,—all of which may be successfully used. In charges in line, the lance is very useful; in mêlées, the saber is much better: hence comes the idea of giving the lance to the front rank, which makes the first onslaught, and the saber to the second rank, which finishes the encounter usually in individual combats. Pistol-firing is of very little use except for outpost-duty, in a charge as foragers, or when light cavalry desires to annoy infantry and draw its fire previous to a charge. I do not know what the carbine is good for; since a body of cavalry armed with it must halt if they wish to fire with any accuracy, and they are then in a favorable condition for the enemy to attack. There are few marksmen who can with any accuracy fire a musket while on horseback and in rapid motion.

I have just said that all the methods of charging may be equally good. It must not be understood, however, that impetuosity always gives the advantage in a shock of cavalry against cavalry: the fast trot, on the contrary, seems to me the best gait for charges in line, because every thing depends, in such a case, upon the ensemble and good order of the movement,—things which cannot be obtained in charges at a fast gallop. Galloping is proper against artillery when it is important to get over the ground as rapidly as possible. In like manner, if the cavalry is armed with sabers, it may take the gallop at two hundred yards from the enemy's line if it stands firmly to receive the attack. But if the cavalry is armed with the lance, the fast trot is the proper gait, since the advantageous use of that weapon depends upon the preservation of good order: in a mêlée the lance is almost useless.

If the enemy advances at a fast trot, it does not seem prudent to gallop to meet him; for the galloping party will be much disordered, while the trotting party will not. The only advantage of the gallop is its apparent boldness and the moral effect it produces; but, if this is estimated at its true value by the enemy, it is reasonable to expect his firm and compact mass to be victorious over a body of horsemen galloping in confusion.

In their charges against infantry the Turks and Mamelukes showed the small advantage of mere impetuosity. No cavalry will penetrate where lancers or cuirassiers at a trot cannot. It is only when infantry is much disordered, or their fire poorly maintained, that there is any advantage in the impetuous gallop over the steady trot. To break good squares, cannon and lancers are required, or, better still, cuirassiers armed with lances. For charges in open order there are no better models for imitation than the Turks and the Cossacks.

Whatever method be adopted in charging, one of the best ways of using cavalry is to throw several squadrons opportunely upon the flanks of an enemy's line which is also attacked in front. That this maneuver may be completely successful, especially in charges of cavalry against cavalry, it should be performed at the very moment when the lines come in collision; for a minute too soon or too late its effect may be lost. It is highly important, therefore, that a cavalry commander should have a quick eye, sound judgment, and a cool head.

Much discussion has taken place about the proper manner of arming and organizing cavalry. The lance is the best arm for offensive purposes when a body of horsemen charge in line; for it enables them to strike an enemy who cannot reach them; but it is a very good plan to have a second rank or a reserve armed with sabers, which are more easily handled than the lance in hand-to-hand fighting when the ranks become broken. It would be, perhaps, better still to support a charge of lancers by a detachment of hussars, who can follow up the charge, penetrate the enemy's line, and complete the victory.

The cuirass is the best defensive armor. The lance and the cuirass of strong leather doubled seem to me the best armament for light cavalry, the saber and iron cuirass the best for heavy cavalry. Some military men of experience are inclined even to arm the cuirassiers with lances, believing that such cavalry, resembling very much the men-at-arms of former days, would bear down every thing before them. A lance would certainly suit them better than the musketoon; and I do not see why they should not have lances like those of the light cavalry.

Opinions will be always divided as to those amphibious animals called dragoons. It is certainly an advantage to have several battalions of mounted infantry, who can anticipate an enemy at a defile, defend it in retreat, or scour a wood; but to make cavalry out of foot-soldiers, or a soldier who is equally good on horse or on foot, is very difficult. This might have been supposed settled by the fate of the French dragoons when fighting on foot, had it not been seen that the Turkish cavalry fought quite as well dismounted as mounted. It has been said that the greatest inconvenience resulting from the use of dragoons consists in the fact of being obliged at one moment to make them believe infantry squares cannot resist their charges, and the next moment that a foot-soldier armed with his musket is superior to any horseman in the world. This argument has more plausibility than real force; for, instead of attempting to make men believe such contradictory statements, it would be much more reasonable to tell them that if brave cavalry may break a square, brave foot-soldiers may resist such a charge; that victory does not always depend upon the superiority of the arm, but upon a thousand other things; that the courage of the troops, the presence of mind of the commanders, the opportuneness of maneuvers, the effect of artillery and musketry fire, rain,—mud, even,—have been the causes of repulses or of victories; and, finally, that a brave man, whether on foot or mounted, will always be more than a match for a coward. By impressing these truths upon dragoons, they will believe themselves superior to their adversaries whether they fight on foot or on horseback. This is the case with the Turks and the Circassians, whose cavalry often dismount to fight on foot in a wood or behind a cover, musket in hand, like foot-soldiers.

It requires, however, fine material and fine commanders to bring soldiers to such perfection in knowledge of their duties.

The conviction of what brave men can accomplish, whether on foot or mounted, doubtless induced the Emperor Nicholas to collect the large number of fourteen or fifteen thousand dragoons in a single corps, while he did not consider Napoleon's unfortunate experiment with French dragoons, and was not restrained by the fear of often wanting a regiment of these troops at some particular point. It is probable that this concentration was ordered for the purpose of giving uniformity to the instruction of the men in their duties as foot and mounted soldiers, and that in war they were to be distributed to the different grand divisions of the army. It cannot be denied, however, that great advantages might result to the general who could rapidly move up ten thousand men on horseback to a decisive point and bring them into action as infantry. It thus appears that the methods of concentration and of distribution have their respective advantages and disadvantages. A judicious mean between the extremes would be to attach a strong regiment to each wing of the army and to the advanced guard, (or the rear-guard in a retreat,) and then to unite the remaining troops of this arm in divisions or corps.

Every thing that was said with reference to the formation of infantry is applicable to cavalry, with the following modifications:—

1. Lines deployed checkerwise or in echelons are much better for cavalry than full lines; whilst for infantry lines drawn up checkerwise are too much disconnected, and would be in danger if the cavalry should succeed in

penetrating and taking the battalions in flank. The checkerwise formation is only advantageous for infantry in preparatory movements before reaching the enemy, or else for lines of columns which can defend themselves in every direction against cavalry. Whether checkered or full lines be used, the distance between them ought to be such that if one is checked and thrown into confusion the others may not share it. It is well to observe that in the checkered lines the distance may be less than for full lines. In every case the second line should not be full. It should be formed in columns by divisions, or at least there should be left the spaces, if in line, of two squadrons, that may be in column upon the flank of each regiment, to facilitate the passage through of the troops which have been brought up.

2. When the order of columns of attack doubled on the center is used, cavalry should be formed in regiments and infantry only in battalions. The regiments should contain six squadrons, in order that, by doubling on the center into divisions, three may be formed. If there are only four squadrons, there can be but two lines.

3. The cavalry column of attack should never be formed *en masse* like that of infantry; but there should always be full or half squadron distance, that each may have room to disengage itself and charge separately. This distance will be so great only for those troops engaged. When they are at rest behind the line of battle, they may be closed up, in order to cover less ground and diminish the space to be passed over when brought into action. The masses should, of course, be kept beyond cannon-range.

4. A flank attack being much more to be apprehended by cavalry than in a combat of infantry with infantry, several squadrons should be formed in echelons by platoons on the flanks of a line of cavalry, which may form to the right or left, to meet an enemy coming in that direction.

5. For the same reason, it is important to throw several squadrons against the flanks of a line of cavalry which is attacked in front. Irregular cavalry is quite as good as the regular for this purpose, and it may be better.

6. It is also of importance, especially in cavalry, that the commander-in-chief increase the depth rather than the extent of the formation. For example, in a deployed division of two brigades it would not be a good plan for one brigade to form in a single line behind the other, but each brigade should have one regiment in the first line and one in the second. Each unit of the line will thus have its own proper reserve behind it,—an advantage not to be regarded as trifling; for in a charge events succeed each other so rapidly that it is impossible for a general to control the deployed regiments.

By adopting this arrangement, each general of brigade will be able to dispose of his own reserve; and it would be well, also, to have a general reserve for the whole division. This consideration leads me to think that five regiments would make a good division. The charge may then be made in line by brigades of two regiments, the fifth serving as a general reserve behind the center. Or three regiments may form the line, and two may be in column, one behind each wing. Or it may be preferable to use a mixed order, deploying two regiments and keeping the others in column. This is a good arrangement, because the three regiments, formed in columns by divisions behind the center and flanks of the line, cover those points, and can readily pass the line if it is beaten back. Cavalry deployed should be in checkered order rather than in full lines.

7. Two essential points are regarded as generally settled for all encounters of cavalry against cavalry. One is that the first line must sooner or later be checked; for, even upon the supposition of the first charge being entirely successful, it is always probable that the enemy will bring fresh squadrons to the contest, and the first line must at length be forced to rally behind the second. The other point is that, with troops and commanders on both sides equally good, the victory will remain with the party having the last squadrons in reserve in readiness to be thrown upon the flank of the enemy's line while his front is also engaged.

Attention to these truths will bring us to a just conclusion as to the proper method of forming a large mass of cavalry for battle.

Whatever order be adopted, care must be taken to avoid deploying large cavalry corps in full lines; for a mass thus drawn up is very unmanageable, and if the first line is checked suddenly in its career the second is also, and that without having an opportunity to strike a blow. This has been demonstrated many times. Take as an example the attack made by Nansouty in columns of regiments upon the Prussian cavalry deployed in front of Chateau-Thierry.

In opposing the formation of cavalry in more than two lines, I never intended to exclude the use of several lines checkerwise or in echelons, or of reserves formed in columns. I only meant to say that when cavalry, expecting to make a charge, is drawn up in lines one behind the other, the whole mass will be thrown into confusion as soon as the first line breaks and turns. [To disprove my statement, M. Wagner cites the case of the battle of Ramillies, where Marlborough, by a general charge of cavalry in fall lines, succeeded in beating the French drawn up checkerwise. Unless my memory deceives me, the allied cavalry was at first formed checkered in two lines; but the real cause of Marlborough's success was his seeing that Villeroi had paralyzed half his army behind Anderkirch and Gette, and his having the good sense to withdraw thirty-eight squadrons from this wing to reinforce his left, which in this way had twice as many cavalry as the French, and outflanked them. But I cheerfully admit that there may be many exceptions to a rule which I have not laid down more absolutely than all others relating to cavalry tactics,—a tactics, by the way, as changeable as the arm itself.]

With cavalry still more than with infantry the *morale* is very important. The quickness of eye and the coolness of the commander, and the intelligence and bravery of the soldier, whether in the *mêlée* or in the rally, will oftener be the means of assuring a victory than the adoption of this or that formation. When, however, a good formation is adopted and the advantages mentioned above are also present, the victory is more certain; and nothing can excuse the use of a vicious formation.

The history of the wars between 1812 and 1815 has renewed the old disputes upon the question whether regular cavalry will in the end get the better over an irregular cavalry which will avoid all serious encounters, will retreat with the speed of the Parthians and return to the combat with the same rapidity, wearing out the strength of its enemy by continual skirmishing. Lloyd has decided in the negative; and several exploits of the Cossacks when engaged with the excellent French cavalry seem to confirm his opinion. (When I speak of excellent French cavalry, I refer to its impetuous bravery, and not to its perfection; for it does not compare with the Russian or German cavalry either in horsemanship, organization, or in care of the animals.) We must by no means conclude it possible for a body of light cavalry deployed as skirmishers to accomplish as much as the Cossacks or other irregular cavalry. They acquire a habit of moving in an apparently disorderly manner, whilst they are all the time directing their individual efforts toward a common object. The most practiced hussars can never perform such service as the Cossacks, Tscherkesses, and Turks do instinctively.

Experience has shown that irregular charges may cause the defeat of the best cavalry in partial skirmishes; but it has also demonstrated that they are not to be depended upon in regular battles upon which the fate of a war may depend. Such charges are valuable accessories to an attack in line, but alone they can lead to no decisive results.

From the preceding facts we learn that it is always best to give cavalry a regular organization, and furnish them long weapons, not omitting, however, to provide, for skirmishing, etc.., an irregular cavalry armed with pistols, lances, and sabers.

Whatever system of organization be adopted, it is certain that a numerous cavalry, whether regular or irregular, must have a great influence in giving a turn to the events of a war. It may excite a feeling of apprehension at distant parts of the enemy's country, it can carry off his convoys, it can encircle his army, make his communications very perilous, and destroy the *ensemble* of his operations. In a word, it produces nearly the same results as a rising *en masse* of a population, causing trouble on the front, flanks, and rear of an army, and reducing a general to a state of entire uncertainty in his calculations.

Any system of organization, therefore, will be a good one which provides for great enlargement of the cavalry in time of war by the incorporation of militia; for they may, with the aid of a few good regular squadrons, be made excellent partisan soldiers. These militia would certainly not possess all the qualities of those warlike wandering tribes who live on horseback and seem born cavalry-soldiers; but they could in a measure supply the places of such. In this respect Russia is much better off than any of her neighbors, both on account of the number and quality of her horsemen of the Don, and the character of the irregular militia she can bring into the field at very short notice.

Twenty years ago I made the following statements in Chapter XXXV. of the Treatise on Grand Military Operations, when writing on this subject:—

"The immense advantages of the Cossacks to the Russian army are not to be estimated. These light troops, which are insignificant in the shock of a great battle, (except for falling upon the flanks,) are terrible in pursuits and in a war of posts. They are a most formidable obstacle to the execution of a general's designs,—because he can never be sure of the arrival and carrying out of his orders, his convoys are always in danger, and his operations uncertain. If an army has had only a few regiments of these half-regular cavalry-soldiers, their real value has not been known; but when their number increases to fifteen thousand or twenty thousand, their usefulness is fully recognized,—especially in a country where the population is not hostile to them.

"When they are in the vicinity, every convoy must be provided with a strong escort, and no movement can be expected to be undisturbed. Much unusual labor is thus made necessary upon the part of the opponent's regular cavalry, which is soon broken down by the unaccustomed fatigue.

"Volunteer hussars or lancers, raised at the time of war breaking out, may be nearly as valuable as the Cossacks, if they are well officered and move freely about from point to point."

In the Hungarians, Transylvanians, and Croats, Austria has resources possessed by few other states. The services rendered by mounted militia have proved, however, that this kind of cavalry may be very useful, if for no other purpose than relieving the regular cavalry of those occasional and extra duties to be performed in all armies, such as forming escorts, acting as orderlies, protecting convoys, serving on outposts, etc.. Mixed corps of regular and irregular cavalry may often be more really useful than if they were entirely composed of cavalry of the line,—because the fear of compromising a body of these last often restrains a general from pushing them forward in daring operations where he would not hesitate to risk his irregulars, and he may thus lose excellent opportunities of accomplishing great results.

Article XLVI: Employment of Artillery

Artillery is an arm equally formidable both in the offensive and defensive. As an offensive means, a great battery well managed may break an enemy's line, throw it into confusion, and prepare the way for the troops that are to make an assault. As a defensive means, it doubles the strength of a position, not only on account of the material injury it inflicts upon the enemy while at a distance, and the consequent moral effect upon his troops, but also by greatly increasing the peril of approaching near, and specially within the range of grape. It is no less important in the attack and defense of fortified places or intrenched camps; for it is one of the main reliances in modern systems of fortification.

I have already in a former portion of this book given some directions as to the distribution of artillery in a line of battle; but it is difficult to explain definitely the proper method of using it in the battle itself. It will not be right to say that artillery can act independently of the other arms, for it is rather an accessory. At Wagram, however, Napoleon threw a battery of one hundred pieces into the gap left by the withdrawal of Massena's corps, and thus held in check the Austrian center, notwithstanding their vigorous efforts to advance. This was a special case, and should not be often imitated.

I will content myself with laying down a few fundamental rules, observing that they refer to the present state of artillery service, (1838.) The recent discoveries not yet being fully tested, I shall say little with reference to them.

1. In the offensive, a certain portion of the artillery should concentrate its fire upon the point where a decisive blow is to be struck. Its first use is to shatter the enemy's line, and then it assists with its fire the attack of the infantry and cavalry.

2. Several batteries of horse-artillery should follow the offensive movements of the columns of attack, besides the foot-batteries intended for the same purpose. Too much foot-artillery should not move with an offensive column. It may be posted so as to co-operate with the column without accompanying it. When the cannoneers can mount the boxes, it may have greater mobility and be advanced farther to the front.

3. It has already been stated that half of the horse-artillery should be held in reserve, that it may be rapidly moved to any required point. [Greater mobility is now given to foot-artillery by mounting the men on the boxes.] For this purpose it should be placed upon the most open ground, whence it can

move readily in every direction. I have already indicated the best positions for the heavy calibers.

4. The batteries, whatever may be their general distribution along the defensive line, should give their attention particularly to those points where the enemy would be most likely to approach, either on account of the facility or the advantage of so doing. The general of artillery should therefore know the decisive strategic and tactical points of the battle-field, as well as the topography of the whole space occupied. The distribution of the reserves of artillery will be regulated by these.

5. Artillery placed on level ground or ground sloping gently to the front is most favorably situated either for point-blank or ricochet firing: a converging fire is the best.

6. It should be borne in mind that the chief office of all artillery in battles is to overwhelm the enemy's troops, and not to reply to their batteries. It is, nevertheless, often useful to fire at the batteries, in order to attract their fire. A third of the disposable artillery may be assigned this duty, but two-thirds at least should be directed against the infantry and cavalry of the enemy.

7. If the enemy advance in deployed lines, the batteries should endeavor to cross their fire in order to strike the lines obliquely. If guns can be so placed as to enfilade a line of troops, a most powerful effect is produced.

8. When the enemy advance in columns, they may be battered in front. It is advantageous also to attack them obliquely, and especially in flank and reverse. The moral effect of a reverse fire upon a body of troops is inconceivable; and the best soldiers are generally put to flight by it. The fine movement of Ney on Preititz at Bautzen was neutralized by a few pieces of Kleist's artillery, which took his columns in flank, checked them, and decided the marshal to deviate from the excellent direction he was pursuing. A few pieces of light artillery, thrown at all hazards upon the enemy's flank, may produce most important results, far overbalancing the risks run.

9. Batteries should always have supports of infantry or cavalry, and especially on their flanks. Cases may occur where the rule may be deviated from: Wagram is a very remarkable example of this.

10. It is very important that artillerists, when threatened by cavalry, preserve their coolness. They should fire first solid shot, next shells, and then grape, as long as possible. The infantry supports should, in such a case, form squares in the vicinity, to shelter the horses, and, when necessary, the cannoneers. When the infantry is drawn up behind the pieces, large squares of sufficient size to contain whatever they should cover are best; but when the infantry is on the flanks, smaller squares are better. Rocket-batteries may also be very efficient in frightening the horses.

11. When infantry threatens artillery, the latter should continue its fire to the last moment, being careful not to commence firing too soon. The cannoneers can always be sheltered from an infantry attack if the battery is properly supported. This is a case for the co-operation of the three arms; for, if the enemy's infantry is thrown into confusion by the artillery, a combined attack upon it by cavalry and infantry will cause its destruction.

12. The proportions of artillery have varied in different wars. Napoleon conquered Italy in 1800 with forty or fifty pieces,—whilst in 1812 he invaded Russia with one thousand pieces thoroughly equipped, and failed. These facts show that any fixed rule on the subject is inadmissible. Usually three pieces to a thousand combatants are allowed; but this allowance will depend on circumstances.

The relative proportions of heavy and light artillery vary also between wide limits. It is a great mistake to have too much heavy artillery, whose mobility must be much less than that of the lighter calibers. A remarkable proof of the great importance of having a strong artillery-armament was given by Napoleon after the battle of Eylau. The great havoc occasioned among his troops by the numerous guns of the Russians opened his eyes to the necessity of increasing his own. With wonderful vigor, he set all the Prussian arsenals to work, those along the Rhine, and even at Metz, to increase the number of his pieces, and to cast new ones in order to enable him to use the munitions previously captured. In three months he doubled the *matériel* and *personnel* of his artillery, at a distance of one thousand miles from his own frontiers,—a feat without a parallel in the annals of war.

13. One of the surest means of using the artillery to the best advantage is to place in command of it a general who is at once a good strategist and tactician. This chief should be authorized to dispose not only of the reserve artillery, but also of half the pieces attached to the different corps or divisions of the army. He should also consult with the commanding general as to the moment and place of concentration of the mass of his artillery in order to contribute most to a successful issue of the day, and he should never take the responsibility of thus massing his artillery without previous orders from the commanding general.

Article XLVII: Of the Combined Use of the Three Arms

To conclude this Summary in a proper manner, I ought to treat of the combined use of the three arms; but I am restrained from so doing by considering the great variety of points necessary to be touched upon if I should attempt to go into an examination of all the detailed operations that would arise in the application of the general rules laid down for each of the arms.

Several authors—chiefly German—have treated this subject very extensively, and their labors are valuable principally because they consist mainly of citations of numerous examples taken from the actual minor engagements of the later wars. These examples must indeed take the place of rules, since experience has shown that fixed rules on the subject cannot be laid down. It seems a waste of breath to say that the commander of a body of troops composed of the three arms should employ them so that they will give mutual support and assistance; but, after all, this is the only fundamental rule that can be established, for the attempt to prescribe for such a commander a special course of conduct in every case that may arise, when these cases may be infinitely varied, would involve him in an inextricable labyrinth of instructions. As the object and limits of this Summary do not allow me to enter upon the consideration of such details, I can only refer my readers to the best works which do treat of them.

I have said all I can properly say when I advise that the different arms be posted in conformity with the character of the ground, according to the object in view and the supposed designs of the enemy, and that they be used simultaneously in the manner best suited to them, care being taken to enable them to afford mutual support. A careful study of the events of previous wars, and especially experience in the operations of war, will give an officer correct ideas on these points, and the ability to use, at the right time and place, his knowledge of the properties of the three arms, either single or combined.

Conclusion

I am constrained to recapitulate the principal facts which may be regarded as fundamental in war. War in its *ensemble* is not a science, but an art. Strategy, particularly, may indeed be regulated by fixed laws resembling those of the positive sciences, but this is not true of war viewed as a whole. Among other things, combats may be mentioned as often being quite independent of scientific combinations, and they may become essentially dramatic, personal qualities and inspirations and a thousand other things frequently being the controlling elements. The passions which agitate the masses that are brought into collision, the warlike qualities of these masses, the energy and talent of their commanders, the spirit, more or less martial, of nations and epochs, [The well-known Spanish proverb, *He was brave on such a day*, may be applied to nations as to individuals. The French at Rossbach were not the same people as at Jena, nor the Prussians at Prentzlow as at Dennewitz.]—in a word, every thing that can be called the poetry and metaphysics of war,—will have a permanent influence on its results.

Shall I be understood as saying that there are no such things as tactical rules, and that no theory of tactics can be useful? What military man of intelligence would be guilty of such an absurdity? Are we to imagine that Eugene and Marlborough triumphed simply by inspiration or by the superior courage and discipline of their battalions? Or do we find in the events of Turin, Blenheim, and Ramillies maneuvers resembling those seen at Talavera, Waterloo, Jena, or Austerlitz, which were the causes of the victory in each case? When the application of a rule and the consequent maneuver have procured victory a hundred times for skillful generals, and always have in their favor the great probability of leading to success, shall their occasional failure be a sufficient reason for entirely denying their value and for distrusting the effect of the study of the art? Shall a theory be pronounced absurd because it has only three-fourths of the whole number of chances of success in its favor?

The *morale* of an army and its chief officers has an influence upon the fate of a war; and this seems to be due to a certain physical effect produced by the moral cause. For example, the impetuous attack upon a hostile line of twenty thousand brave men whose feelings are thoroughly enlisted in their cause will produce a much more powerful effect than the attack of forty thousand demoralized or apathetic men upon the same point.

Strategy, as has already been explained, is the art of bringing the greatest part of the forces of an army upon the important point of the theater of war or of the zone of operations.

Tactics is the art of using these masses at the points to which they shall have been conducted by well-arranged marches; that is to say, the art of making them act at the decisive moment and at the decisive point of the field of battle. When troops are thinking more of flight than of fight, they can no longer be termed active masses in the sense in which I use the term.

A general thoroughly instructed in the theory of war, but not possessed of military *coup-d'oeil*, coolness, and skill, may make an excellent strategic plan and be entirely unable to apply the rules of tactics in presence of an enemy: his projects will not be successfully carried out, and his defeat will be probable. If he be a man of character, he will be able to diminish the evil results of his failure, but if he lose his wits he will lose his army.

The same general may, on the other hand, be at once a good tactician and strategist, and have made all the arrangements for gaining a victory that his means will permit: in this case, if he be only moderately seconded by his troops and subordinate officers, he will probably gain a decided victory. If, however, his troops have neither discipline nor courage, and his subordinate officers envy and deceive him, [The unskillful conduct of a subordinate who is incapable of understanding the merit of a maneuver which has been ordered, and who will commit grave faults in its execution, may produce the same result of causing the failure of the plans of an excellent commander.] he will undoubtedly see his fine hopes fade away, and his admirable combinations can only have the effect of diminishing the disasters of an almost unavoidable defeat.

No system of tactics can lead to victory when the *morale* of an army is bad; and even when it may be excellent the victory may depend upon some occurrence like the rupture of the bridges over the Danube at Essling. Neither will victories be necessarily gained or lost by rigid adherence to or rejection of this or that manner of forming troops for battle.

These truths need not lead to the conclusion that there can be no sound rules in war, the observance of which, the chances being equal, will lead to success. It is true that theories cannot teach men with mathematical precision what they should do in every possible case; but it is also certain that they will always point out the errors which should be avoided; and this is a highly-important consideration, for these rules thus become, in the hands of skillful generals commanding brave troops, means of almost certain success.

The correctness of this statement cannot be denied; and it only remains to be able to discriminate between good rules and bad. In this ability consists the whole of a man's genius for war. There are, however, leading principles which assist in obtaining this ability. Every maxim relating to war will be good if it indicates the employment of the greatest portion of the means of action at the decisive moment and place. In Chapter III. I have specified all the strategic combinations which lead to such a result. As regards tactics, the principal thing to be attended to is the choice of the most suitable order of battle for the object in view. When we come to consider the action of masses on the field, the means to be used may be an opportune charge of cavalry, a strong battery put in position and unmasked at the proper moment, a column of infantry making a headlong charge, or a deployed division coolly and steadily pouring upon the enemy a fire, or they may consist of tactical maneuvers intended to threaten the enemy's flanks or rear, or any other maneuver calculated to diminish the confidence of the adversary. Each of these things may, in a particular case, be the cause of victory. To define the cases in which each should be preferred is simply impossible.

If a general desires to be a successful actor in the great drama of war, his first duty is to study carefully the theater of operations, that he may see clearly the relative advantages and disadvantages it presents for himself and his enemies. This being done, he can understandingly proceed to prepare his base of operations, then to choose the most suitable zone of operations for his main efforts, and, in doing so, keep constantly before his mind the principles of the art of war relative to lines and fronts of operations. The offensive army should particularly endeavor to cut up the opposing army by skillfully selecting objective points of maneuver; it will then assume, as the objects of its subsequent undertakings, geographical points of more or less importance, depending upon its first successes.

The defensive army, on the contrary, should endeavor, by all means, to neutralize the first forward movement of its adversary, protracting operations as long as possible while not compromising the fate of the war, and deferring a decisive battle until the time when a portion of the enemy's forces are either exhausted by labors, or scattered for the purpose of occupying invaded provinces, masking fortified places, covering sieges, protecting the line of operations, depots, etc..

Up to this point every thing relates to a first plan of operations; but no plan can provide with certainty for that which is uncertain always,—the character and the issue of the first conflict. If your lines of operations have been skillfully chosen and your movements well concealed, and if on the other hand your enemy makes false movements which permit you to fall on fractions of his army, you maybe successful in your campaign, without fighting general battles, by the simple use of your strategic advantages. But if the two parties seem about equally matched at the time of conflict, there will result one of those stupendous tragedies like Borodino, Wagram, Waterloo, Bautzen, and Dresden, where the precepts of grand tactics, as indicated in the chapter on that subject, must have a powerful influence.

If a few prejudiced military men, after reading this book and carefully studying the detailed and correct history of the campaigns of the great masters of the art of war, still contend that it has neither principles nor rules, I can only pity them, and reply, in the famous words of Frederick, that "a mule which had made twenty campaigns under Prince Eugene would not be a better tactician than at the beginning."

Correct theories, founded upon right principles, sustained by actual events of wars, and added to accurate military history, will form a true school of instruction for generals. If these means do not produce great men, they will at least produce generals of sufficient skill to take rank next after the natural masters of the art of war.

Supplement to the Summary of the Art of War

My Summary of the Art of War, published in 1836, to assist in the military instruction of the Hereditary Grand Duke of Russia, contained a concluding article that was never printed. I deem it expedient to give it now in the form of a supplement, and add a special article upon the means of acquiring a certain and ready strategic *coup-d'oeil*.

It is essential for the reader of my Summary to understand clearly that in the military science, as in every other, the study of details is easy for the man who has learned how to seize the fundamental features to which all others are secondary. I am about to attempt a development of these elements of the art; and my readers should endeavor to apprehend them clearly and to apply them properly.

I cannot too often repeat that the theory of the great combinations of war is in itself very simple, and requires nothing more than ordinary intelligence and careful consideration. Notwithstanding its simplicity, many learned military men have difficulty in grasping it thoroughly. Their minds wander off to accessory details, in place of fixing themselves on first causes, and they go a long way in search of what is just within their reach if they only would think so.

Two very different things must exist in a man to make him a general: *he must know how to arrange a good plan of operations, and how to carry it to a successful termination.* The first of these talents may be a natural gift, but it may also be acquired and developed by study. The second depends more on individual character, is rather a personal attribute, and cannot be created by study, although it may be improved.

It is particularly necessary for a monarch or the head of a government to possess the first of these talents, because in such case, although he may not have the ability to execute, he can arrange plans of operations and decide correctly as to the excellence or defects of those submitted to him by others. He is thus enabled to estimate properly the capacity of his generals, and when he finds a general producing a good plan, and having firmness and coolness, such a man may be safely trusted with the command of an army.

If, on the other hand, the head of a state is a man of executive ability, but not possessing the faculty of arranging wise military combinations, he will be likely to commit all the faults that have characterized the campaigns of many celebrated warriors who were only brave soldiers without being at all improved by study.

From the principles which I have laid down, and their application to several famous campaigns, my readers will perceive that the theory of the great combinations of war may be summed up in the following truths.

The science of strategy consists, in the first place, in knowing how to choose well a theater of war and to estimate correctly that of the enemy. To do this, a general must accustom himself to decide as to the importance of decisive points,—which is not a difficult matter when he is aided by the hints I have given on the subject, particularly in Articles from XVIII. to XXII.

The art consists, next, in a proper employment of the troops upon the theater of operations, whether offensive or defensive. (See Article XVII.) This employment of the forces should be regulated by two fundamental principles: the first being, *to obtain by free and rapid movements the advantage of bringing the mass of the troops against fractions of the enemy; the second, to strike in the most decisive direction,*—that is to say, in that direction where the consequences of his defeat may be most disastrous to the enemy, while at the same time his success would yield him no great advantages.

The whole science of great military combination is comprised in these two fundamental truths. Therefore, all movements that are disconnected or more extended than those of the enemy would be grave faults; so also would the occupation of a position that was too much cut up, or sending out a large detachment unnecessarily. On the contrary, every well-connected, compact system of operations would be wise; so also with central strategic lines, and every strategic position less extended than the enemy's.

The application of these fundamental principles is also very simple. If you have one hundred battalions against an equal number of the enemy's, you may, by their mobility and by taking the initiative, bring eighty of them to the decisive point while employing the remaining twenty to observe and deceive half of the opposing army. You will thus have eighty battalions against fifty at the point where the important contest is to take place. You will reach this point by rapid marches, by interior lines, or by a general movement toward one extremity of the hostile line. I have indicated the cases in which one or the other of these means is to be preferred. (See pages 114 and following.)

In arranging a plan of operations, it is important to remember "*that a strategic theater, as well as every position occupied by an army, has a center and two extremities.*" A theater has usually three zones,—a right, a left, and a central.

In choosing a zone of operations, select one,—1, that will furnish a safe and advantageous base; 2, in which the least risk will be run by yourself, while the enemy will be most exposed to injury; 3, bearing in mind the antecedent situations of the two parties, and, 4, the dispositions and inclinations of the powers whose territories are near the theater of war.

One of the zones will always be decidedly bad or dangerous, while the other two will be more or less suitable according to circumstances.

The zone and base being fixed upon, the object of the first attempts must be selected. This is choosing an objective of operations. There are two very different kinds: some, that are called *territorial or geographical objectives*, refer simply to an enemy's line of defense which it is desired to get possession of, or a fortress or intrenched camp to be captured; *the others, on the contrary, consist entirely in the destruction or disorganization of the enemy's forces, without giving attention to geographical points of any kind*. This was the favorite objective of Napoleon. [The objective may be in some degree *political*,—especially in cases of wars of intervention in the affairs of another country; but it then really becomes geographical.]

I can profitably add nothing to what I have already written on this point, (page 86;) *and, as the choice of the objective is by far the most important thing in a plan of operations*, I recommend the whole of Article XIX.

The objective being determined upon, the army will move toward it by one or two lines of operations, care being taken to conform to the fundamental principle laid down, and to avoid double lines, unless the character of the theater of war makes it necessary to use them, or the enemy is very inferior either in the number or the quality of his troops. Article XXI. treats this subject fully. If two geographical lines are used, it is essential to move the great mass of the forces along the most important of them, and to occupy the secondary line by detachments having a concentric direction, if possible, with the main body.

The army, being on its way toward the objective, before arriving in presence of the enemy and giving battle, occupies daily or temporary strategic positions: the front it embraces, or that upon which the enemy may attack, is its front of operations. There is an important consideration with reference to the direction of the front of operations and to changes it may receive, which I have dwelt upon in Article XX., (page 93.)

The fundamental principle requires, even when the forces are equal, that the front be less extensive than the enemy's,—especially if the front remains unchanged for some time. If your strategic positions are more closely connected than the enemy's, you can concentrate more rapidly and more easily than he can, and in this way the fundamental principle will be applied. If your positions are interior and central, the enemy cannot concentrate except by passing by the mass of your divisions or by moving in a circle around them: he is then exactly in a condition not to be able to apply the fundamental principle, while it is your most obvious measure.

But if you are very weak and the enemy very strong, a central position, that may be surrounded on all sides by forces superior at every point, is untenable, unless the enemy's corps are very far separated from each other, as was the case with the allied armies in the Seven Years' War; or unless the central zone has a natural barrier on one or two of its sides, like the Rhine, the Danube, or the Alps, which would prevent the enemy from using his forces simultaneously. In case of great numerical inferiority it is, nevertheless, wiser to maneuver upon one of the extremities than upon the center of the enemy's line, especially if his masses are sufficiently near to be dangerous to you.

It was stated above that strategy, besides indicating the decisive points of a theater of war, requires two things:—1st, that the principal mass of the force be moved against fractions of the enemy's, to attack them in succession; 2d, that the best direction of movement be adopted,—that is to say, one leading straight to the decisive points already known, and afterward upon secondary points.

It will be recollected that the allies had ten principal corps on the frontier of France from the Rhine to the North Sea.

The Duke of York was attacking Dunkirk. (No. 1.)

Marshal Freytag was covering the siege. (No. 2.)

The Prince of Orange was occupying an intermediate position at Menin. (No. 3.)

The Prince of Coburg, with the main army, was attacking Maubeuge, and was guarding the space between that place and the Scheldt by strong detachments. (No. 4.)

Clairfayt was covering the siege. (No. 5.)

Benjouski was covering Charleroi and the Meuse, toward Thuin and Charleroi, the fortifications of which were being rebuilt. (No. 6.)

Another corps was covering the Ardennes and Luxembourg. (No. 7.)

The Prussians were besieging Landau. (No. 8.)

The Duke of Brunswick was covering the siege in the Vosges. (No. 9.)

General Wurmser was observing Strasbourg and the army of the Rhine. (No. 10.)

The French, besides the detachments in front of each of the hostile corps, had five principal masses in the camps of Lille, Douai, Guise, Sarre Louis, and Strasbourg, (a, b, c, d, e.) A strong reserve, (g,) composed of the best troops drawn from the camps of the northern frontier, was intended to be thrown upon all the points of the enemy's line in succession, assisted by the troops already in the neighborhood, (i, k, l, m.)

This reserve; assisted by the divisions of the camp of Cassel near Dunkirk, commenced its operations by beating corps 1 and 2, under the Duke of York; then that of the Dutch, (No. 3,) at Menin; next that of Clairfayt, (5,) before Maubeuge; finally, joining the army of the Moselle toward Sarre Louis, it beat the Duke of Brunswick in the Vosges, and, with the assistance of the army of the Rhine, (f,) drove Wurmser from the lines of Wissembourg.

The general principle was certainly well applied, and every similar operation will be praiseworthy. But, as the Austrians composed half the allied forces, and they had their lines of retreat from the points 4, 5, and 6 upon the Rhine, it is evident that if the French had collected three of their large corps in order to move them against Benjouski at Thuin, (No. 6,) and then fallen upon the Prince of Coburg's left by the Charleroi road, they would have thrown the imperial army upon the North Sea, and would have obtained immense results.

The Committee of Public Safety deemed it a matter of great importance that Dunkirk should not be permitted to fell into the hands of the English. Besides this, York's corps, encamped on the downs, might be cut off and thrown upon the sea; and the disposable French masses for this object were at Douai, Lille, and Cassel: so that there were good reasons for commencing operations by attacking the English. The principal undertaking failed, because Houchard did not appreciate the strategic advantage he had, and did not know how to act on the line of retreat of the Anglo-Hanoverian army. He was guillotined, by way of punishment, although he saved Dunkirk; yet he failed to cut off the English as he might have done.

It will be observed that this movement of the French reserve along the whole front was the cause of five victories, neither of which had decisive results, *because the attacks were made in front,* and because, when the cities were relieved, the allied armies not being cut through, and the French reserve moving on to the different points in succession, none of the victories was pushed to its legitimate consequences. If the French had based themselves upon the five fortified towns on the Meuse, had collected one hundred thousand men by bold and rapid marches, had fallen upon the center of those separated corps, had crushed Benjouski, assailed the Prince of Coburg in his rear, beaten him, and pursued him vigorously as Napoleon pursued at Ratisbon, and as he wished to do at Ligny in 1815, the result would have been very different.

I have mentioned this example, as it illustrates very well the two important points to be attended to in the strategic management of masses of troops; that is, their employment at different points in succession and at decisive points. [The operations mentioned show the advantage of employing masses at the decisive point, not because it was done in 1793, but because it was not done.

If Napoleon had been in Carnot's place, he would have fallen with all his force upon Charleroi, whence be would have attacked the left of the Prince of Coburg and cut his line of retreat. Let any one compare the results of Carnot's half-skillful operations with the wise maneuvers of Saint-Bernard and Jena, and be convinced.]

Every educated military man will be impressed by the truths educed, and will be convinced that the excellence of maneuvers will depend upon their conforming to the principle already insisted upon; that is to say, the great part of the force must be moved against one wing or the center, according to the position of the enemy's masses. It is of importance in battles to calculate distances with still greater accuracy; for the results of movements on the battle-field following them more rapidly than in the case of strategic maneuvers, every precaution must be taken to avoid exposing any part of the line to a dangerous attack from the enemy, especially if he is compactly drawn up. Add to these things calmness during the action; the ability to choose positions for fighting battles in the manner styled the defensive with *offensive returns*, (Art. XXX.;) the simultaneous employment of the forces in striking the decisive blow, the faculty of arousing the soldiers and moving them forward at opportune moments; and we have mentioned every thing which can assist, as far as the general is concerned, in assuring victories, and every thing which will constitute him a skillful tactician.

It is almost always easy to determine the decisive point of a field of battle, but not so with the decisive moment; and it is precisely here that genius and experience are every thing, and mere theory of little value.

It is important, also, to consider attentively Article XLII., which explains how a general may make a small number of suppositions as to what the enemy may or can do, and as to what course of conduct he shall himself pursue upon those hypotheses. He may thus accustom himself to be prepared for any eventuality.

I must also call attention to Article XXVIII., upon great detachments. These are necessary evils, and, if not managed with great care, may prove ruinous to the best armies. The essential rules on this point are, to make as few detachments as possible, *to have them readily movable*, to draw them back to the main body as soon as practicable, and to give them good instructions for avoiding disasters.

I have nothing to say relative to the first two chapters on military policy; for they are themselves nothing more than a brief summary of this part of the art of war, which chiefly concerns statesmen, but should be thoroughly understood by military men. I will, however, invite special attention to Article XIV., relating to the command of armies or to the choice of generals-in-chief,—a subject worthy the most anxious care upon the part of a wise government; for upon it often depends the safety of the nation.

We may be confident that a good strategist will make a good chief of staff for an army; but for the command in chief is required a man of tried qualities, of high character and known energy. The united action of two such men as commander-in-chief and chief of staff, when a great captain of the first order cannot be had, may produce the most brilliant results.

Note upon the Means of Acquiring a Good Strategic Coup-d'oeil

The study of the principles of strategy can produce no valuable practical results if we do nothing more than keep them in remembrance, never trying to apply them, with map in hand, to hypothetical wars, or to the brilliant operations of great captains. By such exercises may be procured a rapid and certain strategic *coup-d'oeil*,—the most valuable characteristic of a good general, without which he can never put in practice the finest theories in the world.

When a military man who is a student of his art has become fully impressed by the advantages procured by moving a strong mass against successive fractions of the enemy's force, and particularly when he recognizes the importance of constantly directing the main efforts upon decisive points of the theater of operations, he will naturally desire to be able to perceive at a glance what are these decisive points. I have already, in Chapter III., page 70, of the preceding Summary, indicated the simple means by which this knowledge may be obtained. There is, in fact, one truth of remarkable simplicity which obtains in all the combinations of a methodical war. It is this:—*in every position a general may occupy, he has only to decide whether to operate by the right, by the left, or by the front.*

To be convinced of the correctness of this assertion, let us first take this general in his private office at the opening of the war. His first care will be to choose that zone of operations which will give him the greatest number of chances of success and be the least dangerous for him in case of reverse. As no theater of operations can have more than three zones, (that of the right, that of the center, and that of the left,) and as I have in Articles from XVII. to XXII. pointed out the manner of perceiving the advantages and dangers of these zones, the choice of a zone of operations will be a matter of no difficulty.

When the general has finally chosen a zone within which to operate with the principal portion of his forces, and when these forces shall be established in that zone, the army will have a front of operations toward the hostile army, which will also have one. Now, these fronts of operations will each have its right, left, and center. It only remains, then, for the general to decide upon which of these directions he can injure the enemy most,—for this will always be the best, especially if he can move upon it without endangering his own communications. I have dwelt upon this point also in the preceding Summary.

Finally, when the two armies are in presence of each other upon the field of battle where the decisive collision is to ensue, and are upon the point of coming to blows, they will each have a right, left, and center; and it remains for the general to decide still between these three directions of striking.

Let us take, as an illustration of the truths I have mentioned, the theater of operations, already referred to, between the Rhine and the North Sea.

Although this theater presents, in one point of view, four geographical sections,—viz.: the space between the Rhine and the Moselle, that between the Moselle and the Meuse, that between the Meuse and the Scheldt, and that between the last river and the sea,—it is nevertheless true that an army of which A A is the base and B B the front of operations will have only three general directions to choose from; for the two spaces in the center will form a single central zone, as it will always have one on the right and another on the left.

The army B B, wishing to take the offensive against the army CC, whose base was the Rhine, would have three directions in which to operate. If it maneuvered by the extreme right, descending the Moselle, (toward D,) it would evidently threaten the enemy's line of retreat toward the Rhine; but he, concentrating the mass of his forces toward Luxembourg, might fall upon the left of the army D and compel it to change front and fight a battle with its rear toward the Rhine, causing its ruin if seriously defeated.

If, on the contrary, the army B wished to make its greatest effort upon the left, (toward E,) in order to take advantage of the finely-fortified towns of Lille and Valenciennes, it would be exposed to inconveniences still more serious than before. For the army CC, concentrating in force toward Audenarde, might fall on the right of B, and, outflanking this wing in the battle, might throw it upon the impassable country toward Antwerp between the Scheldt and the sea,—where there would remain but two things for it to do: either to surrender at discretion, or cut its way through the enemy at the sacrifice of half its numbers.

It appears evident, therefore, that the left zone would be the most disadvantageous for army B, and the right zone would be inconvenient, although somewhat favorable in a certain point of view. The central zone remains to be examined. This is found to possess all desirable advantages, because the army B might move the mass of its force toward Charleroi with a view of cutting through the immense front of operations of the enemy, might overwhelm his center, and drive the right back upon Antwerp and the Lower Scheldt, without seriously exposing its own communications.

When the forces are chiefly concentrated upon the most favorable zone, they should, of course, have that direction of movement toward the enemy's front of operations which is in harmony with the chief object in view. For example, if you shall have operated by your right against the enemy's left, with the intention of cutting off the greater portion of his army from its base of the Rhine, you should certainly continue to operate in the same direction; for if you should make your greatest effort against the right of the enemy's front, while your plan was to gain an advantage over his left, your operations could not result as you anticipated, no matter how well they might be executed. If, on the contrary, you had decided to take the left zone, with the intention of crowding the enemy back upon the sea, you ought constantly to maneuver by your right in order to accomplish your object; for if you maneuvered by the left, yourself and not the enemy would be the party thrown back upon the sea in case of a reverse.

Applying these ideas to the theaters of the campaigns of Marengo, Ulm, and Jena, we find the same three zones, with this difference, that in those campaigns the central direction was not the best. In 1800, the direction of the left led straight to the left bank of the Po, on the line of retreat of Mélas; in 1805, the left zone was the one which led by the way of Donauwerth to the extreme right, and the line of retreat of Mack; in 1806, however, Napoleon could reach the Prussian line of retreat by the right zone, filing off from Bamberg toward Gera.

In 1800, Napoleon had to choose between a line of operations on the right, leading to the sea-shore toward Nice and Savona, that of the center, leading by Mont-Cenis toward Turin, and that of the left, leading to the line of communications of Mélas, by way of Saint-Bernard or the Simplon. The first two directions had nothing in their favor, and the right might have been very dangerous,—as, in fact, it proved to Massena, who was forced back to Genoa and there besieged. The decisive direction was evidently that by the left.

I have said enough to explain my ideas on this point.

The subject of battles is somewhat more complicated; for in the arrangements for these there are both strategical and tactical considerations to be taken into account and harmonized. A position for battle, being necessarily connected with the line of retreat and the base of operations, must have a well-defined strategic direction; but this direction must also depend somewhat upon the character of the ground and the stations of the troops of both parties to the engagement: these are tactical considerations. Although an army usually takes such a position for a battle as will keep its line of retreat behind it, sometimes it is obliged to assume a position parallel to this line. In such a case it is evident that if you fall with overwhelming force upon the wing nearest the line of retreat, the enemy may be cut off or destroyed, or, at least, have no other chance of escape than in forcing his way through your line.

At the battle of Leuthen Frederick overwhelmed the Austrian left, which was in the direction of their line of retreat; and for this reason the right wing was obliged to take refuge in Breslau, where it capitulated a few days later.

In such cases there is no cause for hesitation. The decisive point is that wing of the enemy which is nearest his line of retreat, and this line you must seize while protecting your own.

When an enemy has one or two lines of retreat perpendicular to and behind his position of battle, it will generally be best to attack the center, or that wing where the obstacles of the ground shall be the least favorable for the defense; for in such a case the first consideration is to gain the battle, without having in view the total destruction of the enemy. That depends upon the relative numerical strength, the *morale* of the two armies, and other circumstances, with reference to which no fixed rules can be laid down.

Finally, it happens sometimes that an army succeeds in seizing the enemy's line of retreat before fighting a battle, as Napoleon did at Marengo, Ulm, and Jena. The decisive point having in such case been secured by skillful marches before fighting, it only remains to prevent the enemy from forcing his way through your line. You can do nothing better than fight a parallel battle, as there is no reason for maneuvering against one wing more than the other. But for the enemy who is thus cut off the case is very different. He should certainly strike most heavily in the direction of that wing where he can hope most speedily to regain his proper line of retreat; and if he throws the mass of his forces there, he may save at least a large portion of them. All that he has to do is to determine whether this decisive effort shall be toward the right or the left.

It is proper for me to remark that the passage of a great river in the presence of a hostile army is sometimes an exceptional case to which the general rules will not apply. In these operations, which are of an exceedingly delicate character, the essential thing is to keep the bridges safe. If, after effecting the passage, a general should throw the mass of his forces toward the right or the left with a view of taking possession of some decisive point, or of driving his enemy back upon the river, whilst the latter was collecting all his forces in another direction to seize the bridges, the former army might be in a very critical condition in case of a reverse befalling it. The battle of Wagram is an excellent example in point,—as good, indeed, as could be desired. I have treated this subject in Article XXXVII.

A military man who clearly perceives the importance of the truths that have been stated will succeed in acquiring a rapid and accurate *coup-d'oeil*. It will be admitted, moreover, that a general who estimates them at their true value, and accustoms himself to their use, either in reading military history, or in hypothetical cases on maps, will seldom be in doubt, in real campaigns, what he ought to do; and even when his enemy attempts sudden and unexpected movements, he will always be ready with suitable measures for counteracting them, by constantly bearing in mind the few simple fundamental principles which should regulate all the operations of war.

Heaven forbid that I should pretend to lessen the dignity of the sublime art of war by reducing it to such simple elements! I appreciate thoroughly the difference between the directing principles of combinations arranged in the quiet of the closet, and that special talent which is indispensable to the individual who has, amidst the noise and confusion of battle, to keep a hundred thousand men co-operating toward the attainment of one single object. I know well what should be the character and talents of the general who has to make such masses move as one man, to engage them at the proper point simultaneously and at the proper moment, to keep them supplied with arms, provisions, clothing, and munitions. Still, although this special talent, to which I have referred, is indispensable, it must be granted that the ability to give wise direction to masses upon the best strategic points of a theater of operations is the most sublime characteristic of a great captain. How many brave armies, under the command of leaders who were also brave and possessed executive ability, have lost not only battles, but even empires, because they were moved imprudently in one direction when they should have gone in the other! Numerous examples might be mentioned; but I will refer only to Ligny, Waterloo, Bautzen, Dennewitz, Leuthen.

I will say no more; for I could only repeat what has already been said. To relieve myself in advance of the blame which will be ascribed to me for attaching too much importance to the application of the few maxims laid down in my writings, I will repeat what I was the first to announce:— "*that war is not an exact science, but a drama full of passion*; that the moral qualities, the talents, the executive foresight and ability, the greatness of character, of the leaders, and the impulses, sympathies, and passions of the masses, have a great influence upon it." I may be permitted also, after having written the detailed history of thirty campaigns and assisted in person in twelve of the most celebrated of them, to declare that I have not found a single case where these principles, correctly applied, did not lead to success.

As to the special executive ability and the well-balanced penetrating mind which distinguish the practical man from the one who knows only what others teach him, I confess that no book can introduce those things into a head where the germ does not previously exist by nature. I have seen many generals—marshals, even—attain a certain degree of reputation by talking largely of principles which they conceived incorrectly in theory and could not apply at all. I have seen these men intrusted with the supreme command of armies, and make the most extravagant plans, because they were totally deficient in good judgment and were filled with inordinate self-conceit. My works are not intended for such misguided persons as these, but my desire has been to facilitate the study of the art of war for careful, inquiring minds, by pointing out directing principles. Taking this view, I claim credit for having rendered valuable service to those officers who are really desirous of gaining distinction in the profession of arms.

Finally, I will conclude this short summary with one last truth:—

"The first of all the requisites for a man's success as a leader is, that he be perfectly brave. When a general is animated by a truly martial spirit and can communicate it to his soldiers, he may commit faults, but he will gain victories and secure deserved laurels."

Appendix: on the Formation of Troops for Battle

Happening to be in Paris, near the end of 1851, a distinguished person did me the honor to ask my opinion as to whether recent improvements in fire-arms would cause any great modifications in the manner of making war.

I replied that they would probably have an influence upon the details of tactics, but that, in great strategic operations and the grand combinations of battles, victory would, now as ever, result from the application of the principles which had led to the success of great generals in all ages,—of Alexander and Cæsar as well as of Frederick and Napoleon. My illustrious interlocutor seemed to be completely of my opinion.

The heroic events which have recently occurred near Sebastopol have not produced the slightest change in my opinion. This gigantic contest between two vast intrenched camps, occupied by entire armies and mounting two thousand guns of the largest caliber, is an event without precedent, which will have no equal in the future; for the circumstances which produced it cannot occur again.

Moreover, this contest of cannon with ramparts, bearing no resemblance to regular pitched battles fought in the center of a continent, cannot influence in any respect the great combinations of war, nor even the tactics of battles.

The bloody battles of the Alma and Inkermann, by giving evidence of the murderous effect of the new fire-arms, naturally led me to investigate the changes which it might be necessary to make on this account in the tactics for infantry.

I shall endeavor to fulfill this task in a few words, in order to complete what was published on this point twenty years ago in the Summary of the Art of War.

The important question of the influence of musketry-fire in battles is not new: it dates from the reign of Frederick the Great, and particularly from the battle of Mollwitz, which he gained (it was said) because his infantry-soldiers, by the use of cylindrical rammers in loading their muskets, were able to fire three shots per minute more than their enemies. [It is probable that Baron

Jomini here refers to iron, instead of cylindrical, ramrods. Before 1730, all European troops used wooden ramrods; and the credit of the invention of iron ones is attributed by some to the Prince of Anhalt, and by others to Prince Leopold of Dessau. The Prussians were the first to adopt the iron ramrod, and at the date of the battle of Mollwitz (1741) it had not been introduced into the Austrian service. Frederick did not adopt the cylindrical ramrod till 1777, thirty-six years after the battle of Mollwitz. The advantage of the cylindrical ramrod consisted in this,—that the soldier in loading saved the time necessary to turn the ramrod; but obviously this small economy of time could never have enabled him to load three times while the enemy loaded once,—all other things being equal.—Translators.] The discussion which arose at this epoch between the partisans of the shallow and deep orders of formation for troops is known to all military students.

The system of deployed lines in three ranks was adopted for the infantry; the cavalry, formed in two ranks, and in the order of battle, was deployed upon the wings, or a part was held in reserve.

The celebrated regulation for maneuvers of 1791 fixed the deployed as the only order for battle: it seemed to admit the use of battalion-columns doubled on the center only in partial combats,—such as an attack upon an isolated post, a village, a forest, or small intrenchments. [Columns by battalions closed in mass seemed only to be intended to use in long columns on the march, to keep them closed, in order to facilitate their deployment.]

The insufficient instruction in maneuvers of the troops of the Republic forced the generals, who were poor tacticians, to employ in battle the system of columns supported by numerous skirmishers. Besides this, the nature of the countries which formed the theaters of operations—the Vosges, Alps, Pyrenees, and the difficult country of La Vendée—rendered this the only appropriate system. How would it have been possible to attack the camps of Saorgio, Figueras, and Mont-Cenis with deployed regiments?

In Napoleon's time, the French generally used the system of columns, as they were nearly always the assailants.

In 1807, I published, at Glogau in Silesia, a small pamphlet with the title of "Summary of the General Principles of the Art of War," in which I proposed to admit for the attack the system of lines formed of columns of battalions by divisions of two companies; in other words, to march to the attack in lines of battalions closed in mass or at half-distance, preceded by numerous skirmishers, and the columns being separated by intervals that may vary between that necessary for the deployment of a battalion and the minimum of the front of one column.

What I had recently seen in the campaigns of Ulm, Austerlitz, Jena, and Eylau had convinced me of the difficulty, if not the impossibility, of marching an army in deployed lines in either two or three ranks, to attack an enemy in position. It was this conviction which led me to publish the pamphlet above referred to. This work attracted some attention, not only on account of the treatise on strategy, but also on account of what was said on tactics.

The successes gained by Wellington in Spain and at Waterloo with troops deployed in lines of two ranks were generally attributed to the murderous effect of the infantry-fire, and created doubt in some minds as to the propriety of the use of small columns; but it was not till after 1815 that the controversies on the best formation for battle wore renewed by the appearance of a pamphlet by the Marquis of Chambray.

In these discussions, I remarked the fatal tendency of the clearest minds to reduce every system of war to absolute forms, and to cast in the same mold all the tactical combinations a general may arrange, without taking into consideration localities, moral circumstances, national characteristics, or the abilities of the commanders. I had proposed to use lines of small columns, especially in the attack: I never intended to make it an exclusive system, particularly for the defense.

I had two opportunities of being convinced that this formation was approved of by the greatest generals of our times. The first was at the Congress of Vienna, in the latter part of 1814: the Archduke Charles observed "that he was under great obligations for the summary I had published in 1807, which General Walmoden had brought to him in 1808 from Silesia." At the beginning of the war of 1809, the prince had not thought it possible to apply the formation which I had proposed; but at the battle of Essling the contracted space of the field induced him to form a part of his army in columns by battalions, (the landwehr particularly,) and they resisted admirably the furious charges of the cuirassiers of General d'Espagne, which, in the opinion of the archduke, they could not have done if they had been deployed.

At the battle of Wagram, the greater part of the Austrian line was formed in the same way as at Essling, and after two days of terrible fighting the archduke abandoned the field of battle, not because his army was badly beaten, but because his left was outflanked and thrown back so as to endanger his line of retreat on Hungary. The prince was satisfied that the firm bearing of his troops was in part due to this mixture of small columns with deployed battalions.

The second witness is Wellington; although his evidence is, apparently, not so conclusive. Having been presented to him at the Congress of Verona in 1823, I had occasion to speak to him on the subject of the controversies to which his system of formation for battle (a system to which a great part of his success had been attributed) had given rise. He remarked that he was convinced the manner of the attack of the French upon him, in columns more or less deep, was very dangerous against a solid, well-armed infantry having confidence in its fire and well supported by artillery and cavalry. I observed to the duke that these deep columns were very different from the small columns which I proposed,—a formation which insures in the attack steadiness, force, and mobility, while deep masses afford no greater mobility and force than a deployed line, and are very much more exposed to the ravages of artillery.

I asked the illustrious general if at Waterloo he had not formed the Hanoverian, Brunswick, and Belgian troops in columns by battalions. He answered, "Yes; because I could not depend upon them so well as upon the English." I replied that this admission proved that he thought a line formed of columns by battalions was more firm than long deployed lines. He replied, "They are certainly good, also; but their use always depends upon the localities and the spirit of the troops. A general cannot act in the same manner under all circumstances."

To this illustrious evidence I might add that Napoleon himself, in the campaign of 1813, prescribed for the attack the formation of the infantry in columns by divisions of two companies in two ranks, as the most suitable,—which was identically what I had proposed in 1807.

The Duke of Wellington also admitted that the French columns at Waterloo, particularly those of their right wing, were not small columns of battalions, but enormous masses, much more unwieldy and much deeper.

If we can believe the Prussian accounts and plans of the battle, it would seem that Ney's four divisions were formed in but four columns, at least in their march to the attack of La Haye Sainte and the line extending from this farm to the Papelotte. I was not present; but several officers have assured me that at one time the troops were formed in columns by divisions of two brigades each, the battalions being deployed behind each other at six paces' interval.

This circumstance demonstrates how much is wanting in the military terms of the French. We give the same name of *division* to masses of four regiments and to fractions of a battalion of two companies each,—which is absurd. Let us suppose, for example, that Napoleon had directed on the 18th of June, 1815, the formation of the line in columns by divisions and by battalions, intending that the regulation of 1813 should be followed. His lieutenants might naturally have understood it very differently, and, according to their interpretation of the order, would have executed one of the following formations:—

1. Either the four divisions of the right wing would have been formed in four large masses, each one of eight or twelve battalions. (according to the

strength of the regiments,) [We suppose each regiment to consist of two battalions: if there should be three in each regiment, the deep column would then consist of twelve lines of either twenty-four or thirty-six ranks.]

2. Or each division would have been formed in eight or twelve columns of battalions by divisions of two platoons or companies, according to the system I have proposed.—

I do not mean to assert positively that this confusion of words led to the deep masses at Waterloo; but it might have done so; and it is important that in every language there should be two different terms to express two such different things as a *coup-d'oeil* of twelve battalions and a *coup-d'oeil* of a quarter of a battalion.

Struck with what precedes, I thought it proper to modify my Summary already referred to, which was too concise, and in my revision of it I devoted a chapter to the discussion of the advantages and disadvantages of the different formations for battle. I also added some considerations relative to a mixed system used at Eylau by General Benningsen, which consisted in forming a regiment of three battalions by deploying the central one, the other two being in column on the wings.

After these discussions, I drew the conclusions:—

1. That Wellington's system was certainly good for the defensive.

2. That the system of Benningsen might, according to circumstances, be as good for the offensive as for the defensive, since it was successfully used by Napoleon at the passage of the Tagliamento.

3. That the most skillful tactician would experience great difficulty in marching forty or fifty deployed battalions in two or three ranks over an interval of twelve or fifteen hundred yards, preserving sufficient order to attack an enemy in position with any chance of success, the front all the while being played upon by artillery and musketry.

I have never seen any thing of the kind in my experience. I regard it as impossible, and am convinced that such a line could not advance to the attack in sufficiently good order to have the force necessary for success.

Napoleon was in the habit of addressing his marshals in these terms:—"Take your troops up in good order, and make a vigorous assault upon the enemy." I ask, what means is there of carrying up to the assault of an enemy forty or fifty deployed battalions as a whole in good order? They will reach the enemy in detachments disconnected from each other, and the commander cannot exercise any control over the mass as a whole.

I saw nothing of this kind either at Ulm, Jena, Eylau, Bautzen, Dresden, Culm, or Leipsic; neither did it occur at Austerlitz, Friedland, Katzbach, or Dennewitz.

I am not aware that Wellington, in any of his battles, ever marched in deployed lines to the attack of an enemy in position. He generally awaited the attack. At Vittoria and Toulouse he gained the victory by maneuvers against the flanks; and at Toulouse Soult's right wing was beaten while descending the heights to attack. Even at Waterloo, what fate would have befallen the English army if, leaving the plateau of Mont Saint-Jean, it had marched in deployed order to attack Napoleon in position on the heights of La Belle Alliance?

I will be pardoned for these recapitulations, as they seem to be necessary to the solution of a question which has arisen since my Summary of the Art of War was written.

Some German generals, recognizing fully the advantages derived in 1813 from the system of columns of battalions, have endeavored to add to its value by dividing up the columns and increasing their number, so as to make them more shallow and to facilitate their deployment. With this view, they propose, instead of forming four divisions or companies one behind the other, to place them beside each other, not deployed, but in small columns. That is, if the battalion consists of four companies of two hundred and forty men each, each company is to be divided into four sections of sixty each: one of these sections will be dispersed as skirmishers, and the other three, in two ranks, will form a small column; so that the battalion, instead of forming one column, will form four, and the regiment of three battalions will form twelve small columns instead of three—

It is certain that it would be easier to march such a line against the enemy than if deployed; but these diminutive columns of sixty skirmishers and one hundred and eighty men in the ranks would never present the same order and solidity as a single column of a battalion. Still as the system has some advantages, it deserves a trial; and, indeed, it has already been practiced in Prussia and Austria.

The same formation applies equally to battalions of six or eight companies. In this case the battalion would not be formed by companies, but by divisions of two companies,—that is, in three or four columns, according to the number of companies.

Two serious inconveniences appear to me to attach to each of these formations. If vigorously charged by cavalry, these small subdivisions would be in great danger; and even in attacking the enemy's line, if driven back and pursued, disorder would be more likely to occur than in the columns of battalions. Still, either of them may be employed, according to circumstances, localities, and the *morale* of the troops. Experience alone can assign to each its proper value. I am not aware whether the Austrians applied these columns of companies at Custozza and Novara, or whether these maneuvers have only been practiced in their camps of instruction.

Be that as it may, there is another not less important question to be considered:—

"Will the adoption of the rifled small-arms and improved balls bring about any important changes in the formation for battle and the now recognized principles of tactics?"

If these arms aided the allies at the Alma and Inkermann, it was because the Russians were not provided with them; and it must not be forgotten that in a year or two all armies will alike be furnished with them, so that in future the advantage will not be confined to one side.

What change will it make in tactics?

Will whole armies be deployed as skirmishers, or will it not still be necessary to preserve either the formation of lines deployed in two or three ranks, or lines of battalions in columns?

Will battles become mere duels with the rifle, where the parties will fire upon each other, without maneuvering, until one or the other shall retreat or be destroyed?

What military man will reply in the affirmative?

It follows, therefore, that, to decide battles, maneuvers are necessary, and victory will fall to the general who maneuvers most skillfully; and he cannot maneuver except with deployed lines or lines of columns of battalions, either whole or subdivided into columns of one or two companies. To attempt to prescribe by regulation under what circumstances either of these systems is to be applied would be absurd.

If a general and an army can be found such that he can march upon the enemy in a deployed line of forty or fifty battalions, then let the shallow order be adopted, and the formation in columns be confined to the attack of isolated posts; but I freely confess that I would never accept the command of an army under this condition. The only point for a regulation for the formation for battle is to forbid the use of very deep columns, because they are heavy, and difficult to move and to keep in order. Besides, they are so much exposed to artillery that their destruction seems inevitable, and their great depth does not increase in any respect their chances of success.

If the organization of an army were left to me, I would adopt for infantry the formation in two ranks, and a regimental organization according with the formation for battle. I would then make each regiment of infantry to consist of three battalions and a depot. Each battalion should consist of six companies, so that when in column by division the depth would be three divisions or six ranks.

This formation seems most reasonable, whether it is desired to form the battalion in columns of attack by divisions on the center of each battalion, or on any other division.

The columns of attack, since the depth is only six ranks, would not be so much exposed to the fire of artillery, but would still have the mobility necessary to take the troops up in good order and launch them upon the enemy with great force. The deployment of these small columns could be executed with great ease and promptitude; and for the forming of a square a column of three divisions in depth would be preferable in several respects to one of four or six divisions.

In the Russian service each battalion consists of four companies of two hundred and fifty men each; each company being as strong as a division in the French organization. The maneuver of double column on the center is not practicable, since the center is here merely an interval separating the second and third companies. Hence the column must be simple, not on the center, but on one of the four companies. Something analogous to the double column on the center would be attained by forming the first and fourth companies behind the second and third respectively; but then the formation would be in two lines rather than in column; and this is the reason why I would prefer the organization of the battalion in six companies or three divisions.

By dividing each of the four companies into two platoons, making eight in all, the formation of *double column on the center* might be made on the fourth and fifth platoons as the leading division; but then each division would be composed of two platoons belonging to different companies, so that each captain would have half of the men of his company under the command of another officer, and half of his own division would be made up of another company.

Such an arrangement in the attack would be very inconvenient; for, as the captain is the real commander, father, and judge of the men of his own company, he can always obtain more from them in the way of duty than any stranger. In addition, if the double column should meet with a decided repulse, and it should be necessary to reform it in line, it would be difficult to prevent disorder, the platoons being obliged to run from one side to the other to find their companies. In the French system, where each battalion consists of eight companies, forming as many platoons at drill, this objection does not exist, since each company is conducted by its own captain. It is true that there will be two captains of companies in each division; but this will be rather an advantage than the reverse, since there will be a rivalry and emulation between the two captains and their men, which will lead to greater display of bravery: besides, if necessary, the senior captain is there, to command the division as a whole.

It is time to leave these secondary details and return to the important question at issue.

Since I have alluded to the system adopted by Wellington, it is proper to explain it so that it can be estimated at its true value in the light of historical events.

In Spain and Portugal, particularly, Wellington had under his command a mass of troops of the country, in which he placed but little confidence in regular formation in a pitched battle, on account of their want of instruction and discipline, but which were animated by a lively hatred of the French and formed bodies of skirmishers useful in harassing the enemy. Having learned by experience the effects of the fury and impetuosity of the French columns when led by such men as Massena and Ney, Wellington decided upon wise means of weakening this impetuosity and afterward securing a triumph over it. He chose positions difficult to approach, and covered all their avenues by swarms of Spanish and Portuguese riflemen, who were skilled in taking advantage of the inequalities of the ground; he placed a part of his artillery on the tactical crest of his position, and a part more to the rear, and riddled the advancing columns with a murderous artillery and musketry fire, while his excellent English infantry, sheltered from the fire, were posted a hundred paces in rear of the crest, to await the arrival of these columns; and when the latter appeared on the summit, wearied, out of breath, decimated in numbers, they were received with a general discharge of artillery and musketry and immediately charged by the infantry with the bayonet.

This system, which was perfectly rational and particularly applicable to Spain and Portugal, since he had there great numbers of this kind of troops and there was a great deal of rough ground upon which they could be useful as marksmen, needed some modifications to make it applicable to Belgium. At Waterloo the duke took his position on a plateau with a gentle slope like a glacis, where his artillery had a magnificent field of fire, and where it produced a terrible effect: both flanks of this plateau were well protected. Wellington, from the crest of the plateau, could discover the slightest movement in the French army, while his own were hidden; but, nevertheless, his system would not have prevented his losing the battle if a number of other circumstances had not come to his aid.

Every one knows more or less correctly the events of this terrible battle, which I have elsewhere impartially described. I demonstrated that its result was due neither to the musketry-fire nor to the use of deployed lines by the English, but to the following accidental causes, viz.:—

1. To the mud, which rendered the progress of the French in the attack painful and slow, and caused their first attacks to be less effective, and prevented their being properly sustained by the artillery.

2. To the original formation of very deep columns on the part of the French, principally on the right wing.

3. To the want of unity in the employment of the three arms: the infantry and cavalry made a number of charges alternating with each other, but they were in no case simultaneous.

4. Finally and chiefly, to the unexpected arrival of the whole Prussian army at the decisive moment on the right flank, if not the rear, of the French.

Every experienced military man will agree that, in spite of the mud and the firmness of the English infantry, if the mass of the French infantry had been thrown on the English in columns of battalions immediately after the great charge of cavalry, the combined army would have been broken and forced back on Antwerp. Independently of this, if the Prussians had not arrived, the English would have been compelled to retreat; and I maintain that this battle cannot justly be cited as proof of the superiority of musketry-fire over well-directed attacks in columns.

From all these discussions we may draw the following conclusions, viz.:—

1. That the improvements in fire-arms will not introduce any important change in the manner of taking troops into battle, but that it would be useful to introduce into the tactics of infantry the formation of columns by companies, and to have a numerous body of good riflemen or skirmishers, and to exercise the troops considerably in firing. Those armies which have whole regiments of light infantry may distribute them through the different brigades; but it would be preferable to detail sharp-shooters alternately in each company as they are needed, which would be practicable when the troops are accustomed to firing: by this plan the light-infantry regiments could be employed in the line with the others; and should the number of sharp-shooters taken from the companies be at any time insufficient, they could be reinforced by a battalion of light infantry to each division.

2. That if Wellington's system of deployed lines and musketry-fire be excellent for the defense, it would be difficult ever to employ it in an attack upon an enemy in position.

3. That, in spite of the improvements of fire-arms, two armies in a battle will not pass the day in firing at each other from a distance: it will always be necessary for one of them to advance to the attack of the other.

4. That, as this advance is necessary, success will depend, as formerly, upon the most skillful maneuvering according to the principles of grand tactics, which consist in this, viz.: in knowing how to direct the great mass of the

troops at the proper moment upon the decisive point of the battle-field, and in employing for this purpose the simultaneous action of the three arms.

5. That it would be difficult to add much to what has been said on this subject in Chapters IV. and V.; and that it would be unreasonable to define by regulation an absolute system of formation for battle.

6. That victory may with much certainty be expected by the party taking the offensive when the general in command possesses the talent of taking his troops into action in good order and of boldly attacking the enemy, adopting the system of formation best adapted to the ground, to the spirit and quality of his troops, and to his own character.

Finally, I will terminate this article with the following remark: That war, far from being an exact science, is a terrible and impassioned drama, regulated, it is true, by three or four general principles, but also dependent for its results upon a number of moral and physical complications.

Sketch of the Principal Maritime Expeditions

I have thought it proper to give here an account of the principal maritime expeditions, to be taken in connection with maxims on descents.

The naval forces of Egypt, Phoenicia, and Rhodes are the earliest mentioned in history, and of them the account is confused. The Persians conquered these nations, as well as Asia Minor, and became the most formidable power on both land and sea.

About the same time the Carthaginians, who were masters of the coast of Mauritania, being invited by the inhabitants of Cadiz, passed the straits, colonized Boetica and took possession of the Balearic Isles and Sardinia, and finally made a descent on Sicily.

The Greeks contended against the Persians with a success that could not have been expected,—although no country was ever more favorably situated for a naval power than Greece, with her fifty islands and her great extent of coast.

The merchant marine of Athens produced her prosperity, and gave her the naval power to which Greece was indebted for her independence. Her fleets, united with those of the islands, were, under Themistocles, the terror of the Persians and the rulers of the East. They never made grand descents, because their land-forces were not in proportion to their naval strength. Had Greece been a united government instead of a confederation of republics, and had the navies of Athens, Syracuse, Corinth, and Sparta been combined instead of fighting among each other, it is probable that the Greeks would have conquered the world before the Romans.

If we can believe the exaggerated traditions of the old Greek historians, the famous army of Xerxes had not less than four thousand vessels; and this number is astonishing, even when we read the account of them by Herodotus. It is more difficult to believe that at the same time, and by a concerted movement, five thousand other vessels landed three hundred thousand Carthaginians in Sicily, where they were totally defeated by Gelon on the same day that Themistocles destroyed the fleet of Xerxes at Salamis. Three other expeditions, under Hannibal, Imilcon, and Hamilcar, carried into Sicily from one hundred to one hundred and fifty thousand men: Agrigentum and Palermo were taken, Lilybæum was founded, and Syracuse besieged twice. The third time Androcles, with fifteen thousand men, landed in Africa, and made Carthage tremble. This contest lasted one year and a half.

Alexander the Great crossed the Hellespont with only fifty thousand men: his naval force was only one hundred and sixty sail, while the Persians had four hundred; and to save his fleet Alexander sent it back to Greece.

After Alexander's death, his generals, who quarreled about the division of the empire, made no important naval expedition.

Pyrrhus, invited by the inhabitants of Tarentum and aided by their fleet, landed in Italy with twenty-six thousand infantry, three thousand horses, and the first elephants which had been seen in Italy. This was two hundred and eighty years before the Christian era.

Conqueror of the Romans at Heraclea and Ascoli, it is difficult to understand why he should have gone to Sicily at the solicitation of the Syracusans to expel the Carthaginians. Recalled, after some success, by the Tarentines, he recrossed the straits, harassed by the Carthaginian fleet: then, reinforced by the Samnites or Calabrians, he, a little too late, concluded to march on Rome. He in turn was beaten and repulsed on Beneventum, when he returned to Epirus with nine thousand men, which was all that remained of his force.

Carthage, which had been prospering for a long time, profited by the ruin of Tyre and the Persian empire.

The Punic wars between Carthage and Rome, now the preponderating power in Italy, were the most celebrated in the maritime annals of antiquity. The Romans were particularly remarkable for the rapidity with which they improved and increased their marine. In the year 264 B.C. their boats or vessels were scarcely fit to cross to Sicily; and eight years after found Regulus conqueror at Ecnomos, with three hundred and forty large vessels, each with three hundred rowers and one hundred and twenty combatants, making in all one hundred and forty thousand men. The Carthaginians, it is said, were stronger by twelve to fifteen thousand men and fifty vessels.

The victory of Ecnomos—perhaps more extraordinary than that of Actium—was the first important step of the Romans toward universal empire. The subsequent descent in Africa consisted of forty thousand men; but the greater part of this force being recalled to Sicily, the remainder was overthrown, and Regulus, being made prisoner, became as celebrated by his death as by his famous victory.

The great fleet which was to avenge him was successful at Clypea, but was destroyed on its return by a storm; and its successor met the same fate at Cape Palinuro. In the year 249 B.C. the Romans were defeated at Drepanum, and lost twenty-eight thousand men and more than one hundred vessels. Another fleet, on its way to besiege Lilybæum, in the same year, was lost off Cape Pactyrus.

Discouraged by this succession of disasters, the Senate at first resolved to renounce the sea; but, observing that the power of Sicily and Spain resulted from their maritime superiority, it concluded to arm its fleets again, and in the year 242 Lutatius Catullus set out with three hundred galleys and seven hundred transports for Drepanum, and gained the battle in the Ægates Islands, in which the Carthaginians lost one hundred and twenty vessels. This victory brought to a close the first Punic war.

The second, distinguished by Hannibal's expedition to Italy, was less maritime in its character. Scipio, however, bore the Roman eagles to Cartagena, and by its capture destroyed forever the empire of the Carthaginians in Spain. Finally, he carried the war into Africa with a force inferior to that of Regulus; but still he succeeded in gaining the battle of Zama, imposing a shameful peace on Carthage and burning five hundred of her ships. Subsequently Scipio's brother crossed the Hellespont with twenty-five thousand men, and at Magnesia gained the celebrated victory which surrendered to the mercy of the Romans the kingdom of Antiochus and all Asia. This expedition was aided by a victory gained at Myonnesus in Ionia, by the combined fleets of Rome and Rhodes, over the navy of Antiochus.

From this time Rome had no rival, and she continued to add to her power by using every means to insure to her the empire of the sea. Paulus Emilius in the year 168 B.C. landed at Samothrace at the head of twenty-five thousand men, conquered Perseus, and brought Macedonia to submission.

Twenty years later, the third Punic war decided the fate of Carthage. The important port of Utica having been given up to the Romans, an immense fleet was employed in transporting to this point eighty thousand foot-soldiers and four thousand horses; Carthage was besieged, and the son of Paulus Emilius and adopted son of the great Scipio had the glory of completing the victory which Emilius and Scipio had begun, by destroying the bitter rival of his country.

After this triumph, the power of Rome in Africa, as well as in Europe, was supreme; but her empire in Asia was for a moment shaken by Mithridates. This powerful king, after seizing in succession the small adjacent states, was in command of not less than two hundred and fifty thousand men, and of a fleet of four hundred vessels, of which three hundred were decked. He defeated the three Roman generals who commanded in Cappadocia, invaded Asia Minor and massacred there at least eighty thousand Roman subjects, and even sent a large army into Greece.

Sylla landed in Greece with a reinforcement of twenty-five thousand Romans, and retook Athens; but Mithridates sent in succession two large armies by the Bosporus and the Dardanelles: the first, one hundred thousand strong, was destroyed at Chæronea, and the second, of eighty thousand men, met a similar fate at Orchomenus. At the same time, Lucullus, having collected all the maritime resources of the cities of Asia Minor, the islands, and particularly of Rhodes, was prepared to transport Sylla's army from Sestos to Asia; and Mithridates, from fear, made peace.

In the second and third wars, respectively conducted by Murena and Lucullus, there were no descents effected. Mithridates, driven step by step into Colchis, and no longer able to keep the sea, conceived the project of turning the Black Sea by the Caucasus, in order to pass through Thrace to assume the offensive,—a policy which it is difficult to understand, in view of the fact that he was unable to defend his kingdom against fifty thousand Romans.

Cæsar, in his second descent on England, had six hundred vessels, transporting forty thousand men. During the civil wars he transported thirty-five thousand men to Greece. Antony came from Brundusium to join him with twenty thousand men, and passed through the fleet of Pompey,—in which act he was as much favored by the lucky star of Cæsar as by the arrangements of his lieutenants.

Afterward Cæsar carried an army of sixty thousand men to Africa; they did not, however, go in a body, but in successive detachments.

The greatest armament of the latter days of the Roman republic was that of Augustus, who transported eighty thousand men and twelve thousand horses into Greece to oppose Antony; for, besides the numerous transports required for such an army, there were two hundred and sixty vessels of war to protect them. Antony was superior in force on land, but trusted the empire of the world to a naval battle: he had one hundred and seventy war-vessels, in addition to sixty of Cleopatra's galleys, the whole manned by twenty-two thousand choice troops, besides the necessary rowers.

Later, Germanicus conducted an expedition of one thousand vessels, carrying sixty thousand men, from the mouths of the Rhine to the mouths of the Ems. Half of this fleet was destroyed on its return by a storm; and it is difficult to understand why Germanicus, controlling both banks of the Rhine, should have exposed his army to the chances of the sea, when he could have reached the same point by land in a few days.

When the Roman authority extended from the Rhine to the Euphrates, maritime expeditions were rare; and the great contest with the races of the North of Europe, which began after the division of the empire, gave employment to the Roman armies on the sides of Germany and Thrace. The eastern fraction of the empire still maintained a powerful navy, which the possession of the islands of the Archipelago made a necessity, while at the same time it afforded the means.

The first five centuries of the Christian era afford but few events of interest in maritime warfare. The Vandals, having acquired Spain, landed in Africa, eighty thousand strong, under Genseric. They were defeated by Belisarius; but, holding the Balearic Isles and Sicily, they controlled the Mediterranean for a time.

At the very epoch when the nations of the East invaded Europe, the Scandinavians began to land on the coast of England. Their operations are little better known than those of the barbarians: they are hidden in the mysteries of Odin.

The Scandinavian bards attribute two thousand five hundred vessels to Sweden. Less poetical accounts assign nine hundred and seventy to the Danes and three hundred to Norway: these frequently acted in concert.

The Swedes naturally turned their attention to the head of the Baltic, and drove the Varangians into Russia. The Danes, more favorably situated with respect to the North Sea, directed their course toward the coasts of France and England.

If the account cited by Depping is correct, the greater part of these vessels were nothing more than fishermen's boats manned by a score of rowers. There were also *snekars*, with twenty banks or forty rowers. The largest had thirty-four banks of rowers. The incursions of the Danes, who had long before ascended the Seine and Loire, lead us to infer that the greater part of these vessels were very small.

However, Hengist, invited by the Briton Vortigern, transported five thousand Saxons to England in eighteen vessels,—which would go to show that there were then also large vessels, or that the marine of the Elbe was superior to that of the Scandinavians.

Between the years 527 and 584, three new expeditions, under Ida and Cridda, gained England for the Saxons, who divided it into seven kingdoms; and it was not until three centuries had elapsed (833) that they were again united under the authority of Egbert.

The African races, in their turn, visited the South of Europe. In 712, the Moors crossed the Straits of Gibraltar, under the lead of Tarik. They came, five thousand strong, at the invitation of Count Julian; and, far from meeting great resistance, they were welcomed by the numerous enemies of the Visigoths. This was the happy era of the Caliphs, and the Arabs might well pass for liberators in comparison with the tyrants of the North. Tarik's army, soon swelled to twenty thousand men, defeated Rodrigo at Jerez and reduced the kingdom to submission. In time, several millions of the inhabitants of

Mauritania crossed the sea and settled in Spain; and if their numerous migrations cannot be regarded as descents, still, they form one of the most curious and interesting scenes in history, occurring between the incursions of the Vandals in Africa and the Crusades in the East.

A revolution not less important, and one which has left more durable traces, marked in the North the establishment of the vast empire now known as Russia. The Varangian princes, invited by the Novgorodians, of whom Rurik was the chief, soon signalized themselves by great expeditions.

In 902, Oleg is said to have embarked eighty thousand men in two thousand boats on the Dnieper: they passed the falls of the river and debouched in the Black Sea, while their cavalry followed the banks. They proceeded to Constantinople, and forced Leo the Philosopher to pay tribute.

Forty years subsequently, Igor took the same route with a fleet said to have consisted of ten thousand boats. Near Constantinople his fleet, terrified by the effects of the Greek fire, was driven on the coast of Asia, where the force was disembarked. It was defeated, and the expedition returned home.

Not discouraged, Igor re-established his fleet and army and descended to the mouths of the Danube, where the Emperor Romanus I. sent to renew the tribute and ask for peace, (943.)

In 967, Svatoslav, favored by the quarrel of Nicephorus with the King of Bulgaria, embarked sixty thousand men, debouched into the Black Sea, ascended the Danube, and seized Bulgaria. Recalled by the Petchenegs, who were menacing Kiew, he entered into alliance with them and returned into Bulgaria, broke his alliance with the Greeks, and, being reinforced by the Hungarians, crossed the Balkan and marched to attack Adrianople. The throne of Constantine was held by Zimisces, who was worthy of his position. Instead of purchasing safety by paying tribute, as his predecessors had done, he raised one hundred thousand men, armed a respectable fleet, repulsed Svatoslav at Adrianople, obliged him to retreat to Silistria, and took by assault the capital of the Bulgarians. The Russian prince marched to meet him, and gave battle not far from Silistria, but was obliged to re-enter the place, where he sustained one of the most memorable sieges recorded in history.

In a second and still more bloody battle, the Russians performed prodigies of valor, but were again compelled to yield to numbers. Zimisces, honoring courage, finally concluded an advantageous treaty.

About this period the Danes were attracted to England by the hope of pillage; and we are told that Lothaire called their king, Ogier, to France to be avenged of his brothers. The first success of these pirates increased their fondness for this sort of adventure, and for five or six years their bands swarmed on the coasts of France and Britain and devastated the country.

Ogier, Hastings, Regner, and Sigefroi conducted them sometimes to the mouths of the Seine, sometimes to the mouths of the Loire, and finally to those of the Garonne. It is even asserted that Hastings entered the Mediterranean and ascended the Rhone to Avignon; but this is, to say the least, doubtful. The strength of their fleets is not known: the largest seems to have been of three hundred sail.

In the beginning of the tenth century, Rollo at first landed in England, but, finding little chance of success against Alfred, he entered into alliance with him, landed in Neustria in 911, and advanced from Rouen on Paris: other bodies marched from Nantes on Chartres. Repulsed here, Rollo overran and ravaged the neighboring provinces. Charles the Simple saw no better means of delivering his kingdom of this ever-increasing scourge than to offer Rollo the fine province of Neustria on condition that he would marry his daughter and turn Christian,—an offer which was eagerly accepted.

Thirty years later, Rollo's step-son, annoyed by the successors of Charles, called to his aid the King of Denmark. The latter landed in considerable force, defeated the French, took the king prisoner, and assured Rollo's son in the possession of Normandy.

During the same interval (838 to 950) the Danes exhibited even greater hostility toward England than to France, although they were much more assimilated to the Saxons than to the French in language and customs. Ivar, after pillaging the kingdom, established his family in Northumberland. Alfred the Great, at first beaten by Ivar's successors, succeeded in regaining his throne and in compelling the submission of the Danes.

The aspect of affairs changes anew: Sweyn, still more fortunate than Ivar, after conquering and devastating England, granted peace on condition that a sum of money should be paid, and returned to Denmark, leaving a part of his army behind him.

Ethelred, who had weakly disputed with Sweyn what remained of the Saxon power, thought he could not do better to free himself from his importunate guests than to order a simultaneous massacre of all the Danes in the kingdom, (1002.) But Sweyn reappeared in the following year at the head of an imposing force, and between 1003 and 1007 three successive fleets effected disembarkations on the coast, and unfortunate England was ravaged anew.

In 1012, Sweyn landed at the mouth of the Humber and again swept over the land like a torrent, and the English, tired of obedience to kings who could not defend them, recognized him as king of the North. His son, Canute the Great, had to contend with a rival more worthy of him, (Edmund Ironside.) Returning from Denmark at the head of a considerable force, and aided by the perfidious Edric, Canute ravaged the southern part of England and threatened London. A new division of the kingdom resulted; but, Edmund having been assassinated by Edric, Canute was finally recognized as king of all England. Afterward he sailed to conquer Norway, from which country he returned to attack Scotland. When he died, he divided the kingdom between his three children, according to the usage of the times.

Five years after Canute's death, the English assigned the crown to their Anglo-Saxon princes; but Edward, to whom it fell, was better fitted to be a monk than to save a kingdom a prey to such commotions. He died in 1066, leaving to Harold a crown which the chief of the Normans settled in France contested with him, and to whom, it is said, Edward had made a cession of the kingdom. Unfortunately for Harold, this chief was a great and ambitious man.

The year 1066 was marked by two extraordinary expeditions. While William the Conqueror was preparing in Normandy a formidable armament against Harold, the brother of the latter, having been driven from Northumberland for his crimes, sought support in Norway, and, with the King of Norway, set out with thirty thousand men on five hundred vessels, and landed at the mouth of the Humber. Harold almost entirely destroyed this force in a bloody battle fought near York; but a more formidable storm was about to burst upon his head. William took advantage of the time when the Anglo-Saxon king was fighting the Norwegians, to sail from St. Valery with a very large armament. Hume asserts that he had three thousand transports; while other authorities reduce the number to twelve hundred, carrying from sixty to seventy thousand men. Harold hastened from York, and fought a decisive battle near Hastings, in which he met an honorable death, and his fortunate rival soon reduced the country to submission.

At the same time, another William, surnamed Bras-de-fer, Robert Guiscard, and his brother Roger, conquered Calabria and Sicily with a handful of troops, (1058 to 1070.)

Scarcely thirty years after these memorable events, an enthusiastic priest animated Europe with a fanatical frenzy and precipitated large forces upon Asia to conquer the Holy Land.

At first followed by one hundred thousand men, afterward by two hundred thousand badly-armed vagabonds who perished in great part under the attacks of the Hungarians, Bulgarians, and Greeks, Peter the Hermit succeeded in crossing the Bosporus, and arrived before Nice with from fifty to sixty thousand men, who were either killed or captured by the Saracens.

An expedition more military in its character succeeded this campaign of religious pilgrims. One hundred thousand men, composed of French, Burgundians, Germans, and inhabitants of Lorraine, under Godfrey of Bouillon, marched through Austria on Constantinople; an equal number, under the Count of Toulouse, marched by Lyons, Italy, Dalmatia, and Macedonia; and Bohemond, Prince of Tarentum, embarked with a force of Normans, Sicilians, and Italians, and took the route by Greece on Gallipolis.

This extensive migration reminds us of the fabulous expeditions of Xerxes. The Genoese, Venetian, and Greek fleets were chartered to transport these swarms of Crusaders by the Bosporus or Dardanelles to Asia. More than four hundred thousand men were concentrated on the plains of Nice, where they avenged the defeat of their predecessors. Godfrey afterward led them across Asia and Syria as far as Jerusalem, where he founded a kingdom.

All the maritime resources of Greece and the flourishing republics of Italy were required to transport these masses across the Bosporus and in provisioning them during the siege of Nice; and the great impulse thus given to the coast states of Italy was perhaps the most advantageous result of the Crusades.

This temporary success of the Crusaders became the source of great disasters. The Mussulmans, heretofore divided among themselves, united to resist the infidel, and divisions began to appear in the Christian camps. A new expedition was necessary to aid the kingdom which the brave Noureddin was threatening. Louis VII. and the Emperor Conrad, each at the head of one hundred thousand Crusaders, marched, as their predecessors had done, by the route of Constantinople, (1142.) But the Greeks, frightened by the recurring visits of these menacing guests, plotted their destruction.

Conrad, who was desirous of being first, fell into the traps laid for him by the Turks, and was defeated in detachments in several battles by the Sultan of Iconium. Louis, more fortunate, defeated the Turks on the banks of the Mender; but, being deprived of the support of Conrad, and his army being annoyed and partially beaten by the enemy in the passage of defiles, and being in want of supplies, he was confined to Attalia, on the coast of Pamphylia, where he endeavored to embark his army. The means furnished by the Greeks were insufficient, and not more than fifteen or twenty thousand men arrived at Antioch with the king: the remainder either perished or fell into the hands of the Saracens.

This feeble reinforcement soon melted away under the attacks of the climate and the daily contests with the enemy, although they were continually aided by small bodies brought over from Europe by the Italian ships; and they were again about to yield under the attacks of Saladin, when the court of Rome succeeded in effecting an alliance between the Emperor Frederick Barbarossa and the Kings of France and England to save the Holy Land.

The emperor was the first to set out. At the head of one hundred thousand Germans, he opened a passage through Thrace in spite of the formal resistance of the Greeks, now governed by Isaac Angelus. He marched to Gallipolis, crossed the Dardanelles, and seized Iconium. He died in consequence of an imprudent bath in a river, which, it has been pretended, was the Cydnus. His son, the Duke of Swabia, annoyed by the Mussulmans and attacked by diseases, brought to Ptolemais scarcely six thousand men.

At the same time, Richard Coeur-de-Lion [Richard sailed from England with twenty thousand foot and five thousand horsemen, and landed in Normandy, whence he proceeded by land to Marseilles. We do not know what fleet he employed to transport his troops to Asia. Philip embarked at Genoa on Italian ships, and with a force at least as large as that of Richard.] and Philip Augustus more judiciously took the route over the sea, and sailed from Marseilles and Genoa with two immense fleets, (1190.) The first seized Cyprus, and both landed in Syria,—where they would probably have triumphed but for the rivalry which sprang up between them, in consequence of which Philip returned to France.

Twelve years later, a new Crusade was determined upon, (1203.) Part of the Crusaders embarked from Provence or Italy; others, led by the Count of Flanders and the Marquis of Montferrat, proceeded to Venice, with the intention of embarking there. The party last mentioned were persuaded by the skillful Dandolo to aid him in an attack upon Constantinople, upon the pretext of upholding the rights of Alexis Angelus, the son of Isaac Angelus, who had fought the Emperor Frederick and was the successor of those Comnenuses who had connived at the destruction of the armies of Conrad and Louis VII.

Twenty thousand men had the boldness to attack the ancient capital of the world, which had at least two hundred thousand defenders. They assailed it by sea and land, and captured it. The usurper fled, and Alexis was replaced upon the throne, but was unable to retain his seat: the Greeks made an insurrection in favor of Murzupha, but the Latins took possession of Constantinople after a more bloody assault than the first, and placed upon the throne their chief, Count Baldwin of Flanders. This empire lasted a half-century. The remnant of the Greeks took refuge at Nice and Trebizond.

A sixth expedition was directed against Egypt by John of Brienne, who, notwithstanding the successful issue of the horrible siege of Damietta, was obliged to give way before the constantly-increasing efforts of the Mussulman population. The remains of his splendid army, after a narrow escape from drowning in the Nile, deemed themselves very fortunate in being able to purchase permission to re-embark for Europe.

The court of Rome, whose interest it was to keep up the zeal of Christendom in these expeditions, of which it gathered all the fruits, encouraged the German princes to uphold the tottering realm at Jerusalem.

The Emperor Frederick and the Landgrave of Hesse embarked at Brundusium in 1227, at the head of forty thousand chosen soldiers. The landgrave, and afterward Frederick himself, fell sick, and the fleet put in at Tarentum, from which port the emperor, irritated by the presumption of Gregory IX., who excommunicated him because he was too slow in the gratification of his wishes, at a later date proceeded with ten thousand men, thus giving way to the fear inspired by the pontifical thunders.

Louis IX., animated by the same feeling of fear, or impelled, if we may credit Ancelot, by motives of a higher character, set out from Aigues-Mortes, in 1248, with one hundred and twenty large vessels, and fifteen hundred smaller boats, hired from the Genoese, the Venetians and the Catalans; for France was at that time without a navy, although washed by two seas. This king proceeded to Cyprus, and, having there collected a still larger force, set out, according to Joinville's statement, with more than eighteen hundred vessels, to make a descent into Egypt. His army must have numbered about eighty thousand men; for, although half of the fleet was scattered and cast away upon the coast of Syria, he marched upon Cairo a few months later with sixty thousand fighting-men, twenty thousand being mounted. It should be stated that the Count of Poictiers had arrived also with troops from France.

The sad fortune experienced by this splendid army did not prevent the same king from engaging in a new Crusade, twenty years later, (1270.) He disembarked upon that occasion at the ruins of Carthage, and besieged Tunis. The plague swept off half his army in a few months, and himself was one of its victims. The King of Sicily, having arrived with powerful reinforcements at the time of Louis's death, and desiring to carry back the remains of the army to his island of Sicily, encountered a tempest which caused a loss of four thousand men and twenty large ships. This prince was not deterred by this misfortune from desiring the conquest of the Greek empire and of Constantinople, which seemed a prize of greater value and more readily obtained. Philip, the son and successor of Saint Louis, being anxious to return to France, would have nothing to do with that project. This was the last effort. The Christians who were abandoned in Syria were destroyed in the noted attacks of Tripoli and Ptolemais: some of the remnants of the religious orders took refuge at Cyprus and established themselves at Rhodes.

The Mussulmans, in their turn, crossed the Dardanelles at Gallipolis in 1355, and took possession, one after the other, of the European provinces of the Eastern Empire, to which the Latins had themselves given the fatal blow.

Mohammed II., while besieging Constantinople in 1453, is said to have had his fleet transported by land with a view to placing it in the canal and closing the port: it is stated to have been large enough to be manned by twenty thousand select foot-soldiers. After the capture of this capital, Mohammed found his means increased by all those of the Greek navy, and in a short time his empire attained the first rank of maritime powers. He ordered an attack to be made upon Rhodes and upon Otranto on the Italian main, whilst he proceeded to Hungary in search of a more worthy opponent (Hunniades.) Repulsed and wounded at Belgrade, the sultan fell upon Trebizond with a numerous fleet, brought that city to sue for terms, and then proceeded with a fleet of four hundred sail to make a landing upon the island of Negropont, which he carried by assault. A second attempt upon Rhodes, executed, it is stated, at the head of a hundred thousand men, by one of his ablest lieutenants, was a failure, with loss to the assailants. Mohammed was preparing to go to that point himself with an immense army assembled on the shores of Ionia, which Vertot estimates at three hundred thousand men; but death closed his career, and the project was not carried into effect.

About the same period England began to be formidable to her neighbors on land as well as on the sea; the Dutch also, reclaiming their country from the inroads of the sea, were laying the foundations of a power more extraordinary even than that of Venice.

Edward III. landed in France and besieged Calais with eight hundred ships and forty thousand men.

Henry V. made two descents in 1414 and 1417: he had, it is stated, fifteen hundred vessels and only thirty thousand men, of whom six thousand were cavalry.

All the events we have described as taking place, up to this period, and including the capture of Constantinople, were before the invention of gunpowder; for if Henry V. had cannon at Agincourt, as is claimed by some writers, they were certainly not used in naval warfare. From that time all the combinations of naval armaments were entirely changed; and this revolution took place—if I may use that expression—at the time when the invention of the mariner's compass and the discovery of America and of the Cape of Good Hope were about to turn the maritime commerce of the world into new channels and to establish an entirely new system of colonial dependencies.

I shall not mention in detail the expeditions of the Spaniards to America, or those of the Portuguese, Dutch, and English to India by doubling the Cape of Good Hope. Notwithstanding their great influence upon the commerce of the world,—notwithstanding the genius of Gama, Albuquerque, and Cortez,—these expeditions, undertaken by small bodies of two or three thousand men against tribes who knew nothing of fire-arms, are of no interest in a military point of view.

The Spanish navy, whose fame had been greatly increased by this discovery of a new world, was at the height of its splendor in the reign of Charles V. However, the glory of the expedition to Tunis, which was conquered by this prince at the head of thirty thousand fine soldiers transported in five hundred Genoese or Spanish vessels, was balanced by the disaster which befell a similar expedition against Algiers, (1541,) undertaken when the season was too far advanced and in opposition to the wise counsels of Admiral Doria. The expedition was scarcely under way when the emperor saw one hundred and sixty of his ships and eight thousand men swallowed up by the waves: the remainder was saved by the skill of Doria, and assembled at Cape Metafuz, where Charles V. himself arrived, after encountering great difficulties and peril.

While these events were transpiring, the successors of Mohammed were not neglecting the advantages given them by the possession of so many fine maritime provinces, which taught them at once the importance of the control of the sea and furnished means for obtaining it. At this period the Turks were quite as well informed with reference to artillery and the military art in general as the Europeans. They reached the apex of their greatness under Solyman I., who besieged and captured Rhodes (1552) with an army stated to have reached the number of one hundred and forty thousand men,—which was still formidable even upon the supposition of its strength being exaggerated by one-half.

In 1565, Mustapha and the celebrated Dragut made a descent upon Malta, where the Knights of Rhodes had made a new establishment; they carried over thirty-two thousand Janissaries, with one hundred and forty ships. John of Valetta, as is well known, gained an enduring fame by repulsing them.

A more formidable expedition, consisting of two hundred vessels and fifty-five thousand men, was sent in 1527 to the isle of Cyprus, where Nicosia was taken and Famagosta besieged. The horrible cruelties practiced by Mustapha increased the alarm occasioned by his progress. Spain, Venice, Naples, and Malta united their naval forces to succor Cyprus; but Famagosta had already surrendered, notwithstanding the heroic defense of Bragadino, who was perfidiously flayed alive by Mustapha's order, to avenge the death of forty thousand Turks that had perished in the space of two years spent on the island.

The allied fleet, under the orders of two heroes, Don John of Austria, brother of Philip II., and Andrea Doria, attacked the Turkish fleet at the entrance of the Gulf of Lepanto, near the promontory of Actium, where Antony and Augustus once fought for the empire of the world. The Turkish fleet was almost entirely destroyed: more than two hundred vessels and thirty thousand Turks were captured or perished, (1571.) This victory did not put an end to the supremacy of the Turks, but was a great check in their career of greatness. However, they made such vigorous efforts that as large a fleet as the former one was sent to sea during the next year. Peace terminated this contest, in which such enormous losses were sustained.

The bad fortune of Charles V. in his expedition against Algiers did not deter Sebastian of Portugal from wishing to attempt the conquest of Morocco, where he was invited by a Moorish prince who had been deprived of his estates. Having disembarked upon the shores of Morocco at the head of twenty thousand men, this young prince was killed and his army cut to pieces at the battle of Alcazar by Muley Abdulmalek, in 1578.

Philip II., whose pride had increased since the naval battle of Lepanto on account of the success he had gained in France by his diplomacy and by the folly of the adherents of the League, deemed his arms irresistible. He thought to bring England to his feet. The invincible Armada intended to produce this effect, which has been so famous, was composed of an expeditionary force proceeding from Cadiz, including, according to Hume's narrative, one hundred and thirty-seven vessels, armed with two thousand six hundred and thirty bronze cannon, and carrying twenty thousand soldiers, in addition to eleven thousand sailors. To these forces was to be added an army of twenty-five thousand men which the Duke of Parma was to bring up from the Netherlands by way of Ostend. A tempest and the efforts of the English caused the failure of this expedition, which, although of considerable magnitude for the period when it appeared, was by no means entitled to the high-sounding name it received: it lost thirteen thousand men and half the vessels before it even came near the English coast.

After this expedition comes in chronological order that of Gustavus Adolphus to Germany,(1630.) The army contained only from fifteen to eighteen thousand men: the fleet was quite large, and was manned by nine thousand sailors; M. Ancillon must, however, be mistaken in stating that it carried eight thousand cannon. The debarkation in Pomerania received little opposition from the Imperial troops, and the King of Sweden had a strong party among the German people. His successor was the leader of a very extraordinary expedition, which is resembled by only one other example mentioned in history: I refer to the march of Charles X. of Sweden across the Belt upon the ice, with a view of moving from Sleswick upon Copenhagen by way of the island of Funen,(1658.) He had twenty-five thousand men, of whom nine thousand were cavalry, and artillery in proportion. This undertaking was so much the more rash because the ice was unsafe, several pieces of artillery and even the king's own carriage having broken through and been lost.

After seventy-five years of peace, the war between Venice and the Turks recommenced in 1645. The latter transported an army of fifty-five thousand men, in three hundred and fifty vessels, to Candia, and gained possession of the important post of Canea before the republic thought of sending succor. Although the people of Venice began to lose the spirit which made her great, she still numbered among her citizens some noble souls: Morosini, Grimani, and Mocenigo struggled several years against the Turks, who derived great advantages from their numerical superiority and the possession of Canea. The Venetian fleet had, nevertheless, gained a marked ascendency under the orders of Grimani, when a third of it was destroyed by a frightful tempest, in which the admiral himself perished.

In 1648, the siege of Candia began. Jussuf attacked the city furiously at the head of thirty thousand men: after being repulsed in two assaults, he was encouraged to attempt a third by a large breach being made. The Turks entered the place: Mocenigo rushed to meet them, expecting to die in their midst. A brilliant victory was the reward of his heroic conduct: the enemy were repulsed and the ditches filled with their dead bodies.

Venice might have driven off the Turks by sending twenty thousand men to Candia; but Europe rendered her but feeble support, and she had already called into active service all the men fit for war she could produce.

The siege, resumed some time after, lasted longer than that of Troy, and each campaign was marked by fresh attempts on the part of the Turks to carry succor to their army and by naval victories gained by the Venetians. The latter people had kept up with the advance of naval tactics in Europe, and thus were plainly superior to the Mussulmans, who adhered to the old customs, and were made to pay dearly for every attempt to issue from the Dardanelles. Three persons of the name of Morosini, and several Mocenigos, made themselves famous in this protracted struggle.

Finally, the celebrated Coprougli, placed by his merits at the head of the Ottoman ministry, resolved to take the personal direction of this war which had lasted so long: he accordingly proceeded to the island, where transports had landed fifty thousand men, at whose head he conducted the attack in a vigorous manner. (1667.)

In this memorable siege the Turks exhibited more skill than previously: their artillery, of very heavy caliber, was well served, and, for the first time, they made use of trenches, which were the invention of an Italian engineer.

The Venetians, on their side, greatly improved the methods of defense by mines. Never had there been seen such furious zeal exhibited in mutual destruction by combats, mines, and assaults. Their heroic resistance enabled the garrison to hold out during winter: in the spring, Venice sent reinforcements and the Duke of Feuillade brought a few hundreds of French volunteers.

The Turks had also received strong reinforcements, and redoubled their efforts. The siege was drawing to a close, when six thousand Frenchmen came to the assistance of the garrison under the leadership of the Duke of Beaufort and Navailles,(1669.) A badly-conducted sortie discouraged these presumptuous young men, and Navailles, disgusted with the sufferings endured in the siege, assumed the responsibility, at the end of two months, of carrying the remnant of his troops back to France. Morosini, having then but three thousand exhausted men to defend a place which was open on all sides, finally consented to evacuate it, and a truce was agreed upon, which led to a formal treaty of peace. Candia had cost the Turks twenty-five years of efforts and more than one hundred thousand men killed in eighteen assaults and several hundred sorties. It is estimated that thirty-five thousand Christians of different nations perished in the glorious defense of the place.

The struggle between Louis XIV., Holland, and England gives examples of great maritime operations, but no remarkable descents. That of James II. in Ireland (1690) was composed of only six thousand Frenchmen, although De Tourville's fleet contained seventy-three ships of the line, carrying five thousand eight hundred cannon and twenty-nine thousand sailors. A grave fault was committed in not throwing at least twenty thousand men into Ireland with such means as were disposable. Two years later, De Tourville had been conquered in the famous day of La Hogue, and the remains of the troops which had landed were enabled to return through the instrumentality of a treaty which required their evacuation of the island.

At the beginning of the eighteenth century, the Swedes and Russians undertook two expeditions very different in character.

Charles XII., wishing to aid the Duke of Holstein, made a descent upon Denmark at the head of twenty thousand men, transported by two hundred vessels and protected by a strong squadron. He was really assisted by the English and Dutch navies, but the expedition was not for that reason the less remarkable in the details of the disembarkation. The same prince effected a descent into Livonia to aid Narva, but he landed his troops at a Swedish port.

Peter the Great, having some cause of complaint against the Persians, and wishing to take advantage of their dissensions, embarked (in 1722) upon the Volga: he entered the Caspian Sea with two hundred and seventy vessels, carrying twenty thousand foot-soldiers, and descended to Agrakhan, at the mouths of the Koisou, where he expected to meet his cavalry. This force, numbering nine thousand dragoons and five thousand Cossacks, joined him after a land-march by way of the Caucasus. The czar then seized Derbent, besieged Bakou, and finally made a treaty with one of the parties whose dissensions at that time filled with discord the empire of the Soofees: he procured the cession of Astrabad, the key of the Caspian Sea and, in some measure, of the whole Persian empire.

The time of Louis XV. furnished examples of none but secondary expeditions, unless we except that of Richelieu against Minorca, which was very glorious as an escalade, but less extraordinary as a descent.

[In 1762, an English fleet sailed from Portsmouth: this was joined by a portion of the squadron from Martinico. The whole amounted to nineteen ships of the line, eighteen smaller vessels of war, and one hundred and fifty transports, carrying ten thousand men. The expedition besieged and captured Havana.—TRS.]

The Spaniards, however, in 1775, made a descent with fifteen or sixteen thousand men upon Algiers, with a view of punishing those rovers of the sea for their bold piracies; but the expedition, for want of harmonious action between the squadron and the land-forces, was unsuccessful, on account of the murderous fire which the troops received from the Turkish and Arab musketeers dispersed among the undergrowth surrounding the city. The troops returned to their vessels after having two thousand men placed *hors de combat*.

The American war (1779) was the epoch of the greatest maritime efforts upon the part of the French. Europe was astonished to see this power send Count d'Estaing to America with twenty-five ships of the line, while at the same time M. Orvilliers, with a Franco-Spanish fleet of sixty-five ships of the line, was to cover a descent to be effected with three hundred transports and forty thousand men, assembled at Havre and St. Malo.

This new armada moved back and forth for several months, but accomplished nothing: the winds finally drove it back to port.

D'Estaing was more fortunate, as he succeeded in getting the superiority in the Antilles and in landing in the United States six thousand Frenchmen under Rochambeau, who were followed, at a later date, by another division, and assisted in investing the English army under Cornwallis at Yorktown, (1781:) the independence of America was thus secured. France would perhaps have gained a triumph over her implacable rival more lasting in its effects, had she, in addition to the display made in the English Channel, sent ten ships and seven or eight thousand men more to India with Admiral Suffren.

During the French Revolution, there were few examples of descents: the fire at Toulon, emigration, and the battle of Ushant had greatly injured the French navy.

Hoche's expedition against Ireland with twenty-five thousand men was scattered by the winds, and no further attempts in that quarter were made. (1796.)

At a later date, Bonaparte's expedition to Egypt, consisting of twenty-three thousand men, thirteen ships, seventeen frigates, and four hundred transports, obtained great successes at first, which were followed by sad reverses. The Turks, in hopes of expelling him, landed fifteen thousand men at Aboukir, but were all captured or driven into the sea, notwithstanding the advantages this peninsula gave them of intrenching themselves and waiting for reinforcements. This is an excellent example for imitation by the party on the defensive under similar circumstances.

The expedition of considerable magnitude which was sent out in 1802 to St. Domingo was remarkable as a descent, but failed on account of the ravages of yellow fever.

Since their success against Louis XIV., the English have given their attention more to the destruction of rival fleets and the subjugation of colonies than to great descents. The attempts made in the eighteenth century against Brest and Cherbourg with bodies of ten or twelve thousand men amounted to nothing in the heart of a powerful state like France. The remarkable conquests which procured them their Indian empire occurred in succession. Having obtained possession of Calcutta, and then of Bengal, they strengthened themselves gradually by the arrival of troops in small bodies and by using the Sepoys, whom they disciplined to the number of one hundred and fifty thousand.

The Anglo-Russian expedition to Holland in 1799 was composed of forty thousand men, but they were not all landed at once: the study of the details of the operations is, however, quite interesting.

In 1801, Abercrombie, after threatening Ferrol and Cadiz, effected a descent into Egypt with twenty thousand Englishmen. The results of this expedition are well known.

General Stuart's expedition to Calabria, (1806,) after some successes at Maida, was for the purpose of regaining possession of Sicily. That against Buenos Ayres was more unfortunate in its results, and was terminated by a capitulation.

In 1807, Lord Cathcart attacked Copenhagen with twenty-five thousand men, besieged and bombarded the city, and gained possession of the Danish fleet, which was his object.

In 1808, Wellington appeared in Portugal with fifteen thousand men. After gaining the victory of Vimeira, and assisted by the general rising of the Portuguese, he forced Junot to evacuate the kingdom. The same army, increased in numbers to twenty-five thousand and placed under Moore's command, while making an effort to penetrate into Spain with a view of relieving Madrid, was forced to retreat to Corunna and there re-embark, after suffering severe losses. Wellington, having effected another landing in Portugal with reinforcements, collected an army of thirty thousand Englishmen and as many Portuguese, with which he avenged Moore's misfortunes by surprising Soult at Oporto, (May, 1809,) and then beating Joseph at Talavera, under the very gates of his capital.

The expedition to Antwerp in the same year was one of the largest England has undertaken since the time of Henry V. It was composed of not less than seventy thousand men in all,—forty thousand land-forces and thirty thousand sailors. It did not succeed, on account of the incapacity of the leader.

A descent entirely similar in character to that of Charles X. of Sweden was effected by thirty Russian battalions passing the Gulf of Bothnia on the ice in five columns, with their artillery. Their object was to take possession of the islands of Aland and spread a feeling of apprehension to the very gates of Stockholm. Another division passed the gulf to Umeå, (March, 1809.)

General Murray succeeded in effecting a well-planned descent in the neighborhood of Tarragona in 1813, with the intention of cutting Suchet off from Valencia: however, after some successful operations, he thought best to re-embark.

The expedition set on foot by England against Napoleon after his return from Elba in 1815 was remarkable on account of the great mass of *matériel* landed at Ostend and Antwerp. The Anglo-Hanoverian army contained sixty thousand men, but some came by land and others were disembarked at a friendly port.

The English engaged in an undertaking in the same year which may be regarded as very extraordinary: I refer to the attack on the capital of the United States. The world was astonished to see a handful of seven or eight thousand Englishmen making their appearance in the midst of a state embracing ten millions of people, taking possession of its capital, and destroying all the public buildings,—results unparalleled in history. We would be tempted to despise the republican and unmilitary spirit of the inhabitants of those states if the same militia had not risen, like those of Greece, Rome, and Switzerland, to defend their homes against still more powerful attacks, and if, in the same year, an English expedition more extensive than the other had not been entirely defeated by the militia of Louisiana and other states under the orders of General Jackson.

If the somewhat fabulous numbers engaged in the irruption of Xerxes and the Crusades be excepted, no undertaking of this kind which has been actually carried out, especially since fleets have been armed with powerful artillery, can at all be compared with the gigantic project and proportionate preparations made by Napoleon for throwing one hundred and fifty thousand veterans upon the shores of England by the use of three thousand launches or large gun-boats, protected by sixty ships of the line.

From the preceding narrative the reader will perceive what a difference there is in point of difficulty and probability of success between descents attempted across a narrow arm of the sea, a few miles only in width, and those in which the troops and *matériel* are to be transported long distances over the open sea. This fact gives the reason why so many operations of this kind have been executed by way of the Bosporus.

[The following paragraphs have been compiled from authentic data:—

In 1830, the French government sent an expedition to Algiers, composed of an army of thirty-seven thousand five hundred men and one hundred and eighty pieces of artillery. More than five hundred vessels of war and transports were employed. The fleet sailed from Toulon.

In 1838, France sent a fleet of twenty-two vessels to Vera Cruz. The castle of San Juan d'Ulloa fell into their hands after a short bombardment. A small force of about one thousand men, in three columns, took the city of Vera Cruz by assault: the resistance was slight.

In 1847, the United States caused a descent to be made upon the coast of Mexico, at Vera Cruz, with an army of thirteen thousand men, under the command of General Scott. One hundred and fifty vessels were employed, including men-of-war and transports. The city of Vera Cruz and the castle of San Juan d'Ulloa speedily fell into the possession of the forces of the United States. This important post became the secondary base of operations for the brilliant campaign which terminated with the capture of the city of Mexico.

In 1854 commenced the memorable and gigantic contest between Russia on the one side and England, France, Sardinia, and Turkey on the other. Several descents were made by the allied forces at different points of the Russian coast: of these the first was in the Baltic Sea. An English fleet sailed from Spithead, under the command of Sir Charles Napier, on the 12th of March, and a French fleet from Brest, under the command of Vice-Admiral Parseval Deschênes, on the 19th of April. They effected a junction in the Bay of Barosund on the 11th of June. The allied fleet numbered thirty ships and fifty frigates, corvettes, and other vessels. The naval commanders wished to attack the defenses of Bomarsund, on one of the Aland Isles, but, after a reconnoissance, they came to the conclusion that it was necessary to have land-forces. A French corps of ten thousand men was at once dispatched to Bomarsund under General Baraguay-d'Hilliers, and the place was speedily reduced.

Later in the same year, the great expedition to the Crimea was executed; and with reference to it the following facts are mentioned, in order to give an idea of its magnitude:—

September 14, 1854, an army of fifty-eight thousand five hundred men and two hundred pieces of artillery was landed near Eupatoria, composed of thirty thousand French, twenty-one thousand five hundred English, and seven thousand Turks. They were transported from Varna to the place of landing by three hundred and eighty-nine ships, steamers, and transports. This force fought and gained the battle of the Alma, (September 20,) and thence proceeded to Sebastopol. The English took possession of the harbor of Balaklava and the French of Kamiesch: these were the points to which subsequent reinforcements and supplies for the army in the Crimea were sent.

November 5, at the battle of Inkermann, the allied army numbered seventy-one thousand men.

At the end of January, 1855, the French force was seventy-five thousand men and ten thousand horses. Up to the same time, the English had sent fifty-four thousand men to the Crimea, but only fifteen thousand were alive, present, and fit for duty.

February 4, the French numbered eighty-five thousand; the English, twenty-five thousand fit for duty; the Turks, twenty-five thousand.

May 8, 1855, General La Marmora arrived at Balaklava with fifteen thousand Sardinians.

In the latter part of May, an expedition of sixteen thousand men was sent to Kertch.

In August, the French force at Sebastopol had risen to one hundred and twenty thousand men.

September 8, the final assault took place, which resulted in the evacuation of the place by the Russians. The allies had then in battery more than eight hundred pieces of artillery.

The fleet which co-operated with the land-forces in the artillery attack of October 17, 1854, consisted of twenty-five ships. There were present and prepared to attack in September, 1855, thirty-four ships.

October, 1855, an expeditionary force of nine thousand men was sent to Kinburn, which place was captured.

Marshal Vaillant, in his report, as Minister of War, to the French emperor, says there were sent from France and Algeria three hundred and ten thousand men and forty thousand horses, of which two hundred and twenty-seven thousand men returned to France and Algeria.

The marshal's report gives the following striking facts, (he refers only to French operations:-)

The artillery *matériel* at the disposal of the Army of the East comprised one thousand seven hundred guns, two thousand gun-carriages, two thousand seven hundred wagons, two millions of projectiles, and nine million pounds of powder. There were sent to the army three thousand tons of powder, seventy millions of infantry-cartridges, two hundred and seventy thousand rounds of fixed ammunition, and eight thousand war-rockets.

On the day of the final assault there were one hundred and eighteen batteries, which during the siege had consumed seven million pounds of powder. They required one million sand-bags and fifty thousand gabions.

Of engineer materials, fourteen thousand tons were sent. The engineers executed fifty miles of trenches, using eighty thousand gabions, sixty thousand fascines, and one million sand-bags.

Of subsistence, fuel, and forage, five hundred thousand tons were sent.

Of clothing, camp-equipage, and harness, twelve thousand tons.

Hospital stores, six thousand five hundred tons.

Provision-wagons, ambulances, carts, forges, etc., eight thousand tons.

In all, about six hundred thousand tons.

It is not thought necessary to add similar facts for the English, Sardinian, and Turkish armies.

In 1859, the Spaniards made a descent upon Morocco with a force of forty thousand infantry, eleven squadrons of cavalry, and eighty pieces of artillery, using twenty-one vessels of war with three hundred and twenty-seven guns, besides twenty-four gun-boats and numerous transports.

In 1860, a force of English and French was landed on the coast of China, whence they marched to Pekin and dictated terms of peace. This expedition is remarkable for the smallness of the numbers which ventured, at such a great distance from their sources of supply and succor, to land upon a hostile shore and penetrate into the midst of the most populous empire in the world.

The French expedition to Syria in 1860 was small in numbers, and presented no remarkable features.

Toward the close of the year 1861, the government of the United States sent an expedition of thirteen thousand men to Port Royal, on the coast of South Carolina, one of the seceding States. The fleet of war-vessels and transports sailed from Hampton Roads, under command of Captain Dupont, and was dispersed by a violent gale: the losses of men and *matériel* were small, however, and the fleet finally reached the rendezvous. The defenses of the harbor having been silenced by the naval forces, the disembarkation of the land-troops took place, General Sherman being in command.

England, France, and Spain are now (January 16, 1862) engaged in an expedition directed against Mexico. The first operations were the capture, by the Spanish forces, of Vera Cruz and its defenses: the Mexicans offered no resistance at that point. The future will develop the plans of the allies; but the ultimate result of a struggle (if, indeed, one be attempted by the Mexicans) cannot be doubted, when three of the most powerful states of Europe are arrayed against the feeble and tottering republic of Mexico.]